AUTOMATION

Edited by **Florian Kongoli**

Automation
Edited by Florian Kongoli

Published by InTech

Janeza Trdine 9, 51000 Rijeka, Croatia

Publishing Process Manager Romina Skomersic

Technical Editor Teodora Smiljanic

Cover Designer InTech Design Team

Additional hard copies can be obtained from orders@intechopen.com

Automation, Edited by Florian Kongoli
p. cm.
ISBN 978-953-51-0685-2

For EU product safety concerns:
IN TECH d.o.o., Prolaz Marije Krucifikse Kozulić 3, 51000 Rijeka, Croatia,
info@intechopen.com or visit our website at intechopen.com.

Meet the editor

Dr Florian Kongoli (BSc, MSc, PhD) is CEO of FLOGEN Technologies Inc., USA/Canada and Chairman of FLOGEN STAR OUTREACH. He has about 20 years of rich technological and scientific experience gained through invited engagements with more than 47 well-known chemical and metallurgical companies and well-known universities around the world.

Dr Kongoli has published 21 books and about 90 scientific articles in the last 5 years and has delivered in numerous countries about 130 plenary, keynote and invited presentations as well as articles and technical reports. He has served in many leadership positions in national and international organizations. He is Editorial Board member of several International Journals. He has organized several major International Symposiums dedicated to famous scientists.

Contents

Preface

Automation is closely related to the modern need for sustainable development in the 21st century. One of the principles of sustainability is "Doing More with Less" which in other words, is also one of the goals of automation. By replacing the routine part of human labor with the use of machines, automation not only increases productivity and the quality of products beyond what can be achieved by humans but also frees space, time and energy for humans to deal with the new, non-routine challenge of developing innovative and more advanced technologies. This magnificent cycle in which established developments are automated and the free resources achieved by this automation are used to develop newer technologies that are subsequently automated is one of the most successful recipes for the human race towards the goal of sustainable development.

Automation as a new discipline emerged in the second half of the 20th century and quickly attracted some of the brightest scientists in the world. Although in the beginning its theory and industrial applications were mostly in the electrical, mechanical, hydraulic and pneumatic fields, the computer revolution has made possible the "invasion" of automation in all fields of life without exception. This phenomenon has perpetuated itself by such a degree that today computerization and automation can hardly be clearly distinguished from one another. This is also reflected in the contents of this book where interesting contributions cover various fields of technical as well as social sciences. At first sight it seems that the chapters of the book have nothing in common since they cover different fields and are directed only to specialists in their respective domains. In fact, they are bound together by the need for automation in the various routine aspects of our lives and it is here where the overlapping interest of the reader is expected. In order to facilitate and encourage broad interest from scientists of various fields, each chapter starts similarly with a plain and simple description of the related scientific field where automation is applied, a short review of previous achievements, and a capsule detailing what the new developments are.

Dr. Florian Kongoli, BSc (Honors), MASc, PhD,
MTMS, MGDMB, MCIM, MSME, MAIST,MISIJ,
MSigmaXi, MIFAC, MACS, MASM, MMRS, MACerS, MECS
Executive President (CEO), FLOGEN Technologies Inc.
Chairman, FLOGEN Star OUTREACH

1

Optimization of IPV$_6$ over 802.16e WiMAX Network Using Policy Based Routing Protocol

David Oluwashola Adeniji
University of Ibadan
Nigeria

1. Introduction

Internet application needs to know the IP address and port number of the remote entity with which it is communicating during mobility. Route optimization requires traffic to be tunneled between the correspondent node(CN) and the mobile node (MN).). Mobile IPv6 avoids so-called triangular routing of packets from a correspondent node to the mobile node via the Home Agent. Correspondent nodes now can communicate with the mobile node without using tunnel at the Home Agent..The fundamental focal point for Optimization of IPv6 over WiMAX Network using Policy Based Routing Protocol centered on the special features that describe the goal of optimization mechanism during mobility management. To reiterate these features there is need to address the basic concept of optimization as related to mobility in mobile IPv6 Network. WiMAX which stands for Worldwide Interoperability for Microwave Access, is an open, worldwide broadband telecommunications standard for both fixed and mobile deployments. The primary purpose of WiMAX is to ensure the delivery of wireless data at multi-megabit rates over long distances in multiple ways. Although WiMAX allows connecting to internet without using physical elements such as router, hub, or switch. It operates at higher speeds, over greater distances, and for a greater number of people compared to the services of 802.11(WiFi).A WiMAX system has two units. They are WiMAX Transmitter Tower and WiMAX Receiver. A Base Station with WiMAX transmitter responsible for communicating on a point to multi-point basis with subscriber stations is mounted on a building. Its tower can cover up to 3,000 Sq. miles and connect to internet. A second Tower or Backhaul can also be connected using a line of sight, microwave link. The Receiver and antenna can be built into Laptop for wireless access.This statement brings to the fact that if receiver and antenna are built into the laptop, optimization can take place using a routing protocol that can interface mobile IPv6 network over WiMAX 802.16e.However the generic overview of optimization possibilities most especially for a managed system can be considered.

What then is Optimization? Basically optimization is the route update signaling of information in the IP headers of data packets which enable packets to follow the optimal path and reach their destination intact. The generic consideration in designing route optimization scheme is to use minimumsignaling information in the packet header. Furthermore the delivery of managed system for optimization describes the route optimization operation and the mechanism used for the optimization. In order for optimization to take place, a protocol called route optimization protocol must be introduced. Route optimization protocol is used basically to improve performance. Also route optimization is a technique that enables mobile node and a correspondent node to communicate directly, bypassing the home agent completely; this is based on IPv6 concept.

The use of domains enables a consistent state of deployment to be maintained. The main benefits of using policy are to improve scalability and flexibility for the management system. Scalability is improved by uniformly applying the same policy to large sets of devices , while flexibility is achieved by separating the policy from the implementation of the managed system. Policy can be changed dynamically, thus changing the behavior and strategy of a system, without modifying its implementation or interrupting its operation. Policy-based management is largely supported by standards organizations such as the Internet Engineering Task Force (IETF) and the Distributed Management Task Force (DMTF) and most network equipment vendors.However the Architectures for enforcing policies are moving towards strongly distributed paradigms, using technologies such as mobile code, distributed objects, intelligent agents or programmable networks.

Mobile IP is the standard for mobility management in IP networks. New applications and protocols will be created and Mobile IP is important for this development. Mobile IP support is needed to allow mobile hosts to move between networks with maintained connectivity. However internet service driven network is a new approach to the provision of network computing that concentrates on the services you want to provide. These services range from the low-level services that manage relationships between networked devices to the value-added services provided to the end-users. The complexity of the managed systems results in high administrative costs and long deployment cycles for business initiatives, and imposes basic requirements on their management systems. Although these requirements have long been recognized, their importance is now becoming increasingly critical. The requirements for management systems have been identified and can be facilitated with policy-based management approach where the support for distribution, automation and dynamic adaptation of the behaviour of the managed system is achieved by using policies. IPV6 is one of the useful delivery protocols for future fixed and wireless/mobile network environment while multihoming is the tools for delivering such protocol to the end users. Optimization of Network must be able to address specific market requirements, deployment geographical, end-user demands, and planned service offerings both for today as well as for tomorrow.

2. IP mobility

The common mechanism that can manage the mobility of all mobile nodes in all types of wireless networks is the main essential requirement for realizing the future ubiquitous computing systems. Mobile IP protocol V_4 or V_6 considered to be universal solutions for mobility management because they can hide the heterogeneity in the link-specific technology used in different network. Internet application needs to know the IP address and port number of the remote entity with which it is communicating during mobility. From a network layer perspective, a user is not mobile if the same link is used, regardless of location. If a mobile node can maintain its IP address while moving, it makes the movement transparent to the application, and then mobility becomes invisible. From this problem the basic requirement for a mobile host is the

Mobile IP works in the global internet when the mobile node (MN) which belong to the home agent (HA) moves to a new segment, which is called a foreign network (FN). The MN registers with the foreign agent (FA) in FN to obtain a temporary address i.e a care of address(COA).The MN updates the COA with the HA in its home network by sending BU update message. Any packets from the corresponding node (CN) to MN home address are intercepted by HA. HA then use the BU directly to the CN by looking at the source address of the packet header.

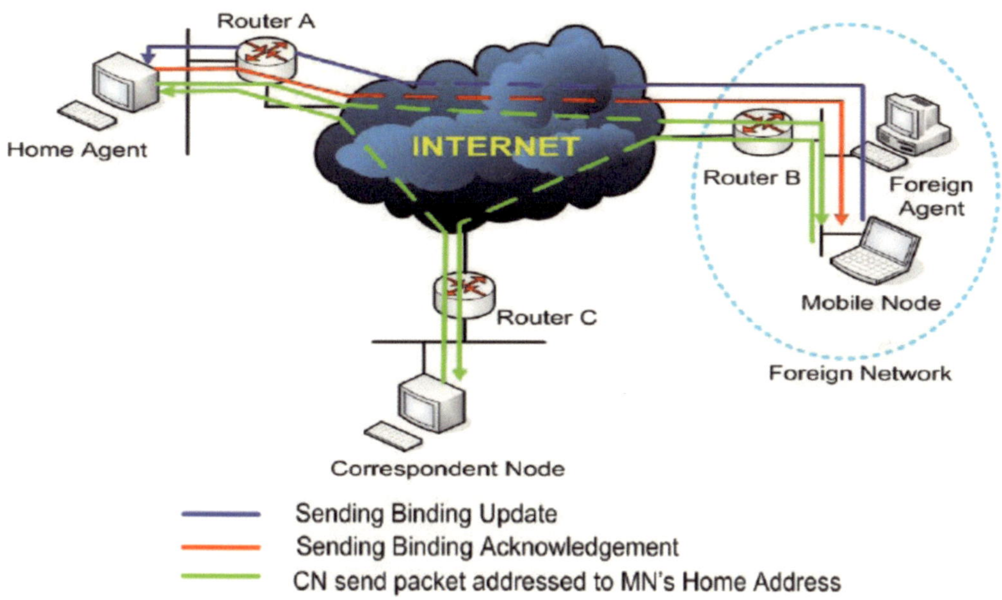

Fig. 1. Mobile IP Network

The most critical challenges of providing mobility at the IP layer is to route packets efficiently and securely. In the mobile IP protocol all packets are sent to a mobile node while away from home are intercepted by its home agent and tunneled to the mobile node using IP encapsulation within IP.

2.1 Limitation of mobile IP

Mobile IP can only provide continuous Internet connectivity and optimal routing to a mobile host, and are not suitable to support a mobile network. The reasons is that, not all devices in a mobile network is sophisticated enough to run these complicated protocols.Secondly, once a device has joined a mobile network, it may not see any link-level handoffs even as the network moves.

2.2 Detailed description of NEMO

Network Mobility describes the situation of a router connecting an entire network to the Internet dynamically changes its point of attachment. The connections of the nodes inside the network to the Internet are also influenced by this movement. A mobile network can be connected to the Internet through one or more MR, (the gateway of the mobile network), there are a number of nodes (Mobile Network Nodes, MNN) attached behind the MR(s). A mobile network can be local fixed node, visiting or nested. In the case of local fixed node: nodes which belong to the mobile network and cannot move with respect to MR. This node are not able to achieve global connectivity without the support of MR. Visiting node belong to the mobile network and can move with respect to the MR(s).Nested mobile network allow another MR to attached to its mobile network. However the operation of each MR remains the same whether the MR attaches to MR or fixed to an access router on the internet. Furthermore in the case of nested mobile network the level mobility is unlimited, management might become very complicated. In NEMO basic support it is important to note that there are some mechanism that allow to allow mobile network nodes to remain connected to the Internet and continuously reachable at all times while the mobile network they are attached to changes its point of attachment.Meanwhile, it would also be meaningful to investigate the effects of Network Mobility on various aspects of internet communication such as routing protocol changes, implications of real-time traffic, fast handover and optimization. When a MR and its mobile network move to a foreign domain, the MR would register its care-of-address (CoA) with it's HA for both MNNs and itself. An IP-in-IP tunnel is then set up between the MR and it's HA. All the nodes behind the MR would not see the movement, thus they would not have any CoA, removing the need for them to register anything at the HA. All the traffic would pass the tunnel connecting the MR and the HA. Figure 2.3 describes how NEMO works.

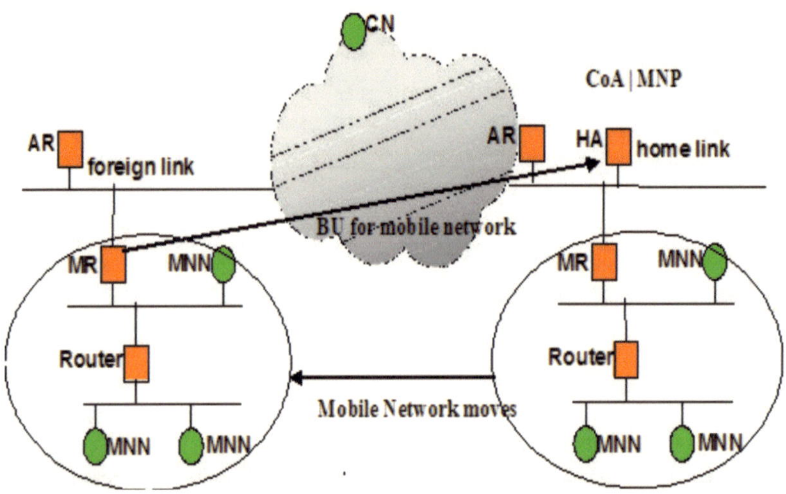

Fig. 2. Network Mobility

2.3 Micro mobility and Macro mobility

This discussion on Micromobility and Macromobility is centered on wireless communication architecture that focuses on the designed of IP Micromobility protocol that compliment an IETF standard for Macromobility management which is usually called Mobile IP. From this point of view, Macro-mobility concerns with the management of users movements at a large scale, between different wide wirelesses accesses networks connected to the Internet. Macro-mobility is often assumed to be managed through Mobile IP. On the other hand, Micro-mobility covers the management of users movements at a local level, inside a particular wireless network.The standard Internet Protocol assumes that an IP address always identifies the node's location in the Internet. This means that if a node moves to another location in the Internet, it has to change its IP address or otherwise the IP packets cannot be routed to its new location anymore. Because of this the upper layer protocol connections have to be reopened in the mobile node's new location.

The Technologies which can be Hierarchical Mobile IP,Cellular IP,HAWAII at different micro mobility solutions could coexist simultaneously in different parts of the Internet. Even at that the Message exchanges are asymmetric on Mobile IP. In cellular networks, message exchanges are symmetric in that the routes to send and receive messages are the same. Considering the mobile Equipment the appropriate location update and registration separate the global mobility from the local mobility. Hence Location information is maintained by routing cache. During Routing most especially in macro mobility scenario a node uses a gateway discovery protocol to find neighboring gateways Based on this information a node decides which gateway to use for relaying packets to the Internet. Then, packets are sent to the chosen gateway by means of unicast. With anycast routing, a node leaves the choice of gateway to the routing protocol.

The routing protocol then routes the packets in an anycast manner to one of the gateways.In the first case, a node knows which gateway it relays its packets to and thus is aware of its macro mobility.

The comparative investigation of different requirement between Micro mobility and Macro mobility are based on following below:

- ***Handoff Mobility Management Parameter***:The interactions with the radio layer, initiator of the handover management mechanism, use of traffic bicasting were necessary. The handoff latency is the parameter time needed to complete the handoff inside the network. Also potential packet losses were the amount of lost packets during the handoff must be deduced. Furthermore the involved stations: the number of MAs that must update the respective routing data or process messages during the handover are required.
- ***Passive Connectivity*** with respect to paging required an architecture that can support via paging order to control traffic against network burden.This architecture is used to support only incoming data packet.Therefore the ratio of incoming and outcoming communication or number of handover experienced by the mobiles are considered for efficiency support purposes.To evaluate this architecture an algorithm is used to perform the paging with respect to efficiency and network load.
- ***Intra Network Traffic*** basic in micro mobility scenario are the exchanged of packet between MNs connected to the same wireless network. This kind of communication is a large part of today's GSM communications and we can expect that it will remain an important class of traffic in future wireless networks.
- **Scalability and Robustness**: The expectation is that future large wireless access networks will have the same constrains in terms of users load. These facts are to be related to the increasing load of today's Internet routers: routing tables containing a few hundreds of thousandsentries have become a performance and optimization problem.

The review of micromobility via macromobility of key management in 5G technology must addressed the following features:

- 5G technology offer transporter class gateway with unparalleled consistency.
- The 5G technology also support virtual private network
- Remote management offered by 5G technology a user can get better and fast solution.
- The high quality services of 5G technology based on Policy to avoid error.

5G technology offer high resolution for crazy cell phone user and bi-directional large bandwidth shaping.

2.4 Concept of Multihoming

Mobile networks can have multiple points of attachment to the internet, in this case they are said to be multihomed. Multihoming arises when the MR has multiple addresses, multiple egress interfaces on the same link, or multiple egress interfaces on different links. Basically the classification of configuration can be divided into :***Configuration-Oriented Approach, Ownership-oriented Approach*** and ***Problem-Oriented Approach.*** The multihoming analysis classifies all these configuration of multihomed mobile networks using (x, y, z) notation. Variables x, y, and z respectively mean the number of MRs connected to the Internet (so called root MRs), the number of HAs, and the number of Mobile Network Prefix (MNP) s. In case of 1, each variable implies that there exists a single node or prefix. If the variable is N, then it means that one or more agents or prefixes exist in a single mobile network. From different combinations of the 3-tuple (x, y, z), various types of multihoming scenarios are possible. For example the (N, 1, 1) scenario means there is multiple MRs at the mobile network, but all of MRs are managed by single HA and use same MNP.

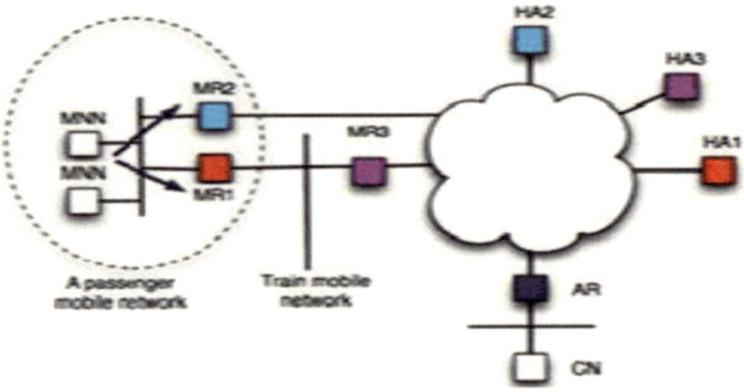

Fig. 3. A Multihoming of Nested Mobile Network

The Figure above shows how a train provide a Wifi network to the passengers with MR3, the passenger could connect to MR3 with MR1 (for example his Laptop). The passenger could also connect directly to Internet with MR2 (his Phone with its GPRS connectivity).The train is connected to Internet with Wimax connectivity. The MNNs can be a PDA and some sensors.

Multihoming from the above nested mobile network provides the advantages of session preservation and load sharing. During optimization the key data communication that must be taking into consideration are:

- Session preservation by redundancy.-the session must be preserved based on the available stable mobile environment either via wireless or wired.

- Load balancing by selecting the best available interface or enabling multiple interfaces simultanousely.Traffic load balancing at the MR is critical since in mobile networks, all traffic goes through the MR.

The apprehension from the above can be justified by specifying the mobile message notification mobile node as well as the procedure for node joining. A mobility notification message contains two important information:(i) the notification interval for multihoming; and (ii)the prefix of the access network that the sending gateway belongs. The optimal choice of the notification interval depends on the mobility of the nodes as well on the amount of traffic sent.

Processing of packets from multihomed nodes is more complex and requires the gateway to perform two tasks. First,the gateway has to verify if a node has recently been informed that its packets are relayed through this access network. If this does not take place, the gateway sends a mobility notification message to the mobile node to inform it about the actual access network. For reducing the amount of mobility notification messages, the gateway records the node address combined with a time stamp in a lookup table. After a notification interval, the gateway deletes the entry and if it is still relaying packets for this node, notifies the mobile node again.Secondly the gateway substitutes the link-local address prefix of the IP source address of the packet with the prefix of the access network it belongs to and forwards the packet to the Internet. When a multihoming node receives a mobility notification message, it adjusts its address prefix to topologically fit the new access network. Subsequently,it informs about its address change using its IP mobility management protocols. In the case where packets of a node are continuously forwarded over different access networks,multihoming support is an advantage to prevent continuous address changes. When a multihoming node receives a mobility notification message, it checks if it already is aware of X or Y access network.

2.5 Requirement of Multihoming configuration

The requirement for Multihomed configurations can be classified depending on how many MRs are present, how many egress interfaces, Care-of Address (CoA), and Home Addresses (HoA) the MRs have, how many prefixes (MNPs) are available to the mobile network nodes, etc. The reader of this chapter should note that there are eight cases of configuration of multihomed mobile network. The 3 key parameter associated to differentiate the configuration are referred to 3-tuple X, Y, Z. To describe any of this requirement configurationin respect to macro mobility, a detection mechanism and notification protocol is required. The below table present the most significant features of the eight classification approach for NEMO. Although there are several configuration but NEMO does not specify any particular mechanism to manage multihoming.

	Configuration	Class	Requirement	Prefix Advertisement
1	Configuration 1, 1, 1	Class 1,1,1	MR, HA, MNP	1 MNP
2	Configuration 1, 1, 1	Class 1,1,n	1 MR, 1 HA, More MNP	2 MNP
3	Configuration 1, 1, 1	Class 1,n,1	1 MR, More HA, 1 MNP	1 MNP
4	Configuration 1, 1, 1	Class 1,n,n	1 MR, More HA, More MNP	Multiple MNP
5	Configuration 1, 1, 1	Class n,1,1	More MR, 1 HA, 1 MNP	MNP
6	Configuration 1, 1, 1	Class n,1,n	More MR, 1 HA, More MNP	Multiple MNP
7	Configuration 1, 1, 1	Class n,n,1	More MR, More HA, MNP	1 MNP
8	Configuration 1, 1, 1	Class n,n,n	More MR, More HA, More MNP	Multiple MNP

Table 1. Analysis of Eight cases of multihoming configuration

What then about the reliability of these configurations during NEMO? Internet connection through another interface must be reliable. The levels of ***redundancy*** cases can be divided into two: if the mobile node's IP address is not valid any more, and the solution is to use another available IP address; the order is that the connection through one interface is broken, and the solution is to use another. If one of the interfaces is broken then the solution is to use another interface using a ***Path Exploration***. However in the case of multiple HAs, the redundancy of the HA is provided, if one HA is broken, another one could be used. The important note here is the broken of one interface can also lead to failure. ***Failure Detection*** **in** all the cases in which the number of MNP is larger than 1, because the MNN could choose its own source address, if the tunnel to one MNP is broken, related MNNs have to use another source address which is created from another MNP. In order to keep sessions alive, both failure detection andredirection of communication mechanisms are needed. If those mechanisms could not perform very well, the transparent redundancy can not be provided as well as in the cases where only one MNP is advertised.

The only difference between using one MR with multiple egress interfaces and using multiple MRs each of which only one egress interface is ***Load sharing.*** Multiple MRs could share the processing task comparing with only one MR, and of course it provides the redundancy of the disrupting of the MR. Therefore the mechanisms for managing and cooperating between each MRs are needed.Also the common problem related to all the configuration is where the number of MNP are larger than 1 and at the same time the number of MR or the number of HA or both is larger than one. So a mechanism for solving the ***ingress filtering*** problem should be used. In most cases the solution is to use second binding on the ingress interface by sending a Prefix-BU through the other MRs and then the HA(s) get(s) all other CoAs.

How do we then distinguish between CoAs.? We use ***Preference Settings*** One solution is to use an extra identifier for different CoAs and include the identifier information in the update message. This kind of situation exists a lot, except for the cases in which one MNP is only allowed to be controlled by one CoA.

2.6 Policy Based Routing Protocol

Policy is changing the behavior and strategy of a system, without modifying its implementation or interrupting its operation. Policy-based management is largely supported by Standards organizations such as the Internet Engineering Task Force ***(IETF)*** and the Distributed Management Task Force ***(DMTF)*** and most network equipment vendors. The focal point in the area of policy-based management is the notion of policy as a means of driving management procedures. Although the technologies for building management building management systems are available, work on the specification and deployment of policies is still scarce.Routing decisions and interface selection are based entirely on IP/network layer information.In order to provide adequate information the level of hierarchy must be considered.The specific information during optimization and deployment is based on the following approach:

Link Layer Information: Interface selection algorithm should take into account all available information and at the same time minimize resource consumption and make decisions with as light computation as possible .However , link quality must be constantlymonitored and the information must be made available for the network layer and user applications in a form that suits them best.

IP layer Information: Several attributes can be retrieved from the IPv6 header without looking into the data, e.g., source address and destination address etc.Some attributes can also be retrieved from IPv6 extension headers (e.g. HOA) only transport protocols like TCP and UDP can be identified directly from the IP header.

Network Originated Information: A service provider may disseminate information about cost, bandwidth and availability of the Internet access in an area using WiMax,WLAN, GPRS and Bluetooth. To advertise such information the default gateway or the access router can send information on cost and bandwidth. The mobile users could then have preferences for connections, like maximize bandwidth or minimize price and the host would select the appropriate interface satisfying these preferences.

This section of the chapter considered the below Algorithm for message notification of mobile node.

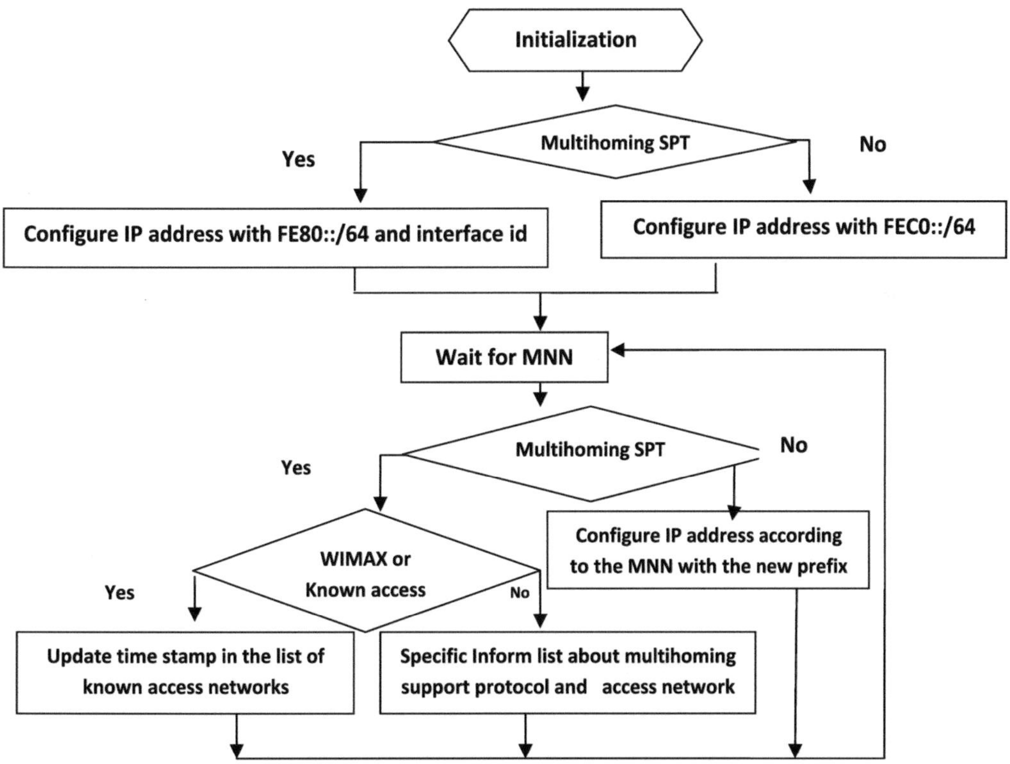

Fig. 4. Algorithm for mobility notification messages at a mobile node during optimization.

2.7 Mechanism for interface selection

The separation of policy and mechanism makes it possible to implement a dynamic interface selection system. The mechanism evaluates connection association and transport information against the actions in policies, using principles.The interface selection system is based on four basic components, ***entities, action, policy and mechanism***. ***Entities*** define actions. ***An entity*** may be a user, peer node or 3rd party, e.g., operator. Action is an operation that is defined by an entity and is controlled by the system. ***Actions*** specify the interfaces to be used for connections on account of entity's requirements. Actions can be presented as conditional statements. Policy governs the actions of an entity. Only one action can take place at a time in a policy. ***A policy*** set contains several policies possible defined by different entities. ***Mechanism*** evaluates actions against connection related information and decides which interface is to be used with a specific connection.

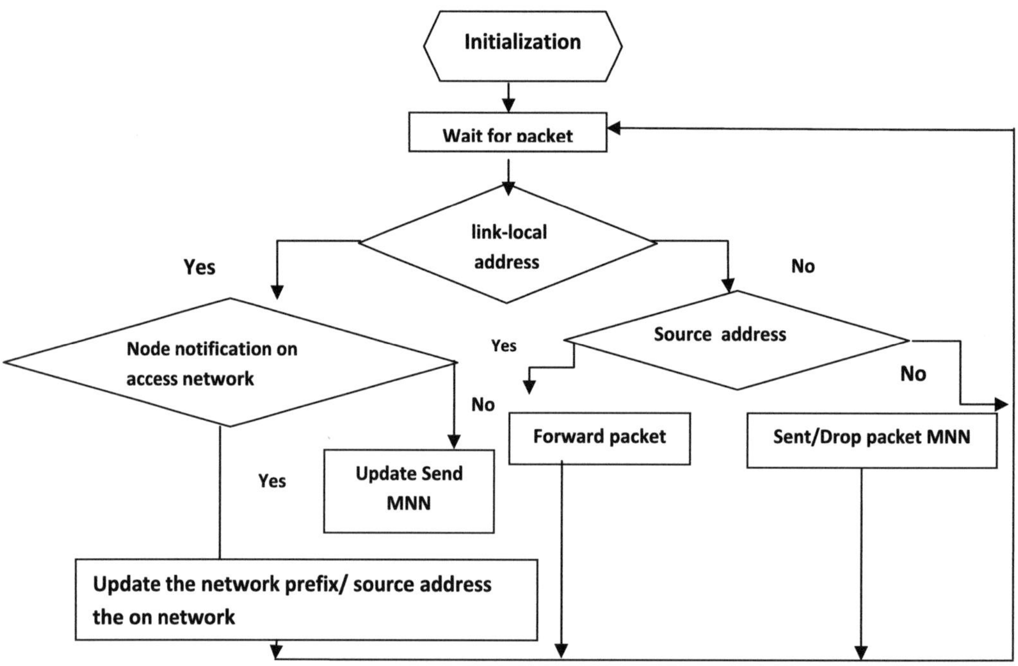

Fig. 5. Algorithm to optimizing the interface selection mechanism.

2.7.1 Network model and optimization

Network model topology to be optimized must contain futures that addressed parameter such as assurance of service delivery and security. Since Wimax is a Flexible Access Point System that delivers on the promise of personal broadband and rich service delivery. Paired with a converged IP core and communicating with feature-rich, multimodal devices combining one network, one service delivery platform and seamless experience that is transparent to the end users. The below fig 6 consider the network model topology for optimization of IPv6 over wimax.

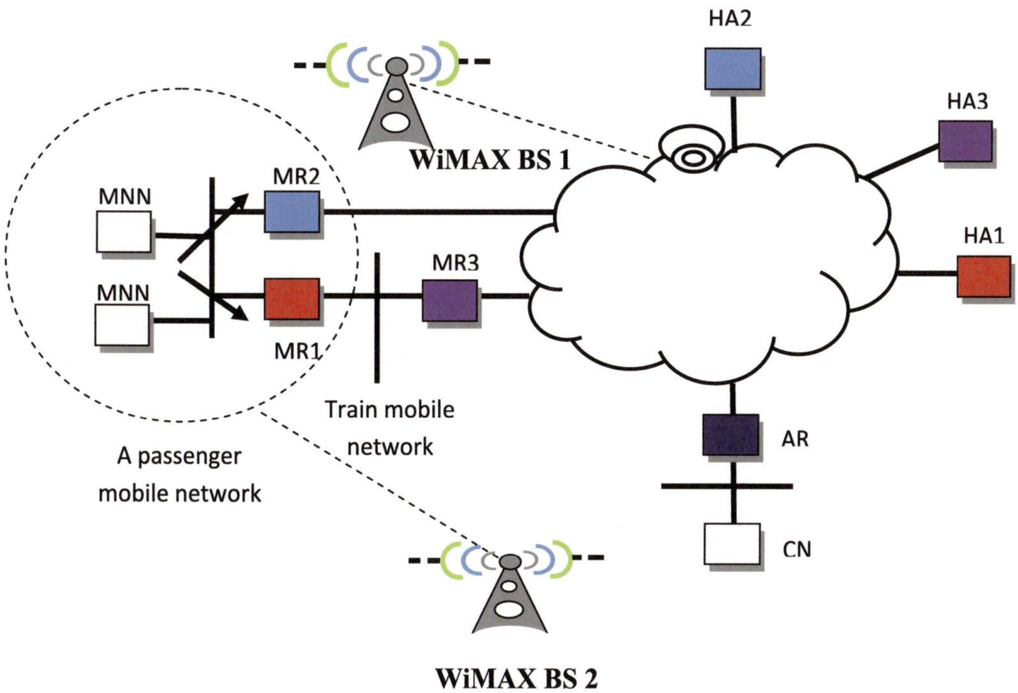

Fig. 6. Network Model Topology For Optimization of IPv6 over WIMAX

From the Network model the wimax BS1, wimax BS2,AR was coined from the architectural specification that depict the concept of Wimax deployment. The access Service Network (ASN) mainly was used for regrouping of BS and AR.The connectivity service Network (CSN) offers connectivity to the internet. To optimize using policy based routing protocol. The link layer information,IP layer information,Network originated information are initialized.

2.8 Standard for WiMAX architecture

WiMAX is a term coined to describe standard, interoperable implementations of IEEE 802.16 wireless networks, similar to the way the term Wi-Fi is used for interoperable implementations of the IEEE 802.11 Wireless LAN standard. However, WiMAX is very different from Wi-Fi in the way it works. The architecture defines how a WiMAX network connects with other networks, and a variety of other aspects of operating such a network, including address allocation, authentication. An overview of this specification for different architectures in order to deploy IPv6 over WiMAX is depicted below in Fig 7 by WiMAX forum .

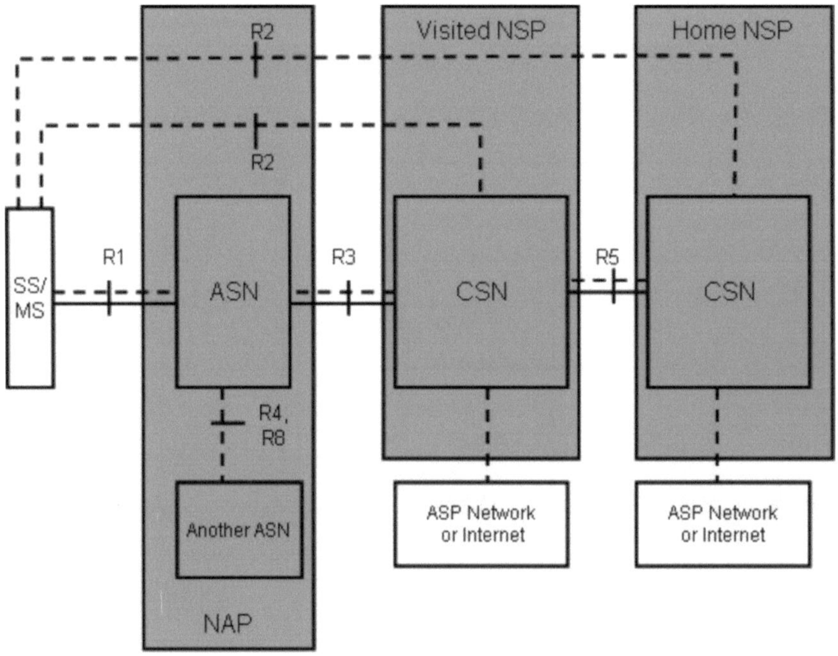

Fig. 7. Architectural Specification for Deployment of IPv6 over WiMAX.

In the proposed network model and optimization the reminder should note that : Regrouping of BS and AR into one entity is named the Access Service Network (ASN) for WiMAX. It has a complete set of functions such as AAA (Authentication, Authorization, Accounting), Mobile IP Foreign agent, Paging controller, and Location Register to provide radio access to a WiMAX Subscriber. The Connectivity Service Network (CSN) offers connectivity to the internet. In the ASN, the BS and AR (or ASN-Gateway) are connected by using either a Switch or Router. The ASN has to support Bridging between all its R1 interfaces and the interfaces towards the network side; forward all packets received from any R1 to a network side port and flood any packet received from a network side port destined for a MAC broadcast or multicast address to all its R1 interfaces. The SS are now considered as mobile (MS), the support for dormant mode is now critical and a necessary feature. Paging capability and optimizations are possible for paging an MS are neither enhanced nor handicapped by the link model itself. However, the multicast capability within a link may cause for an MS to wake up for an unwanted packet.

The solution can consist of filtering the multicast packets and delivering the packets to MS that are listening for particular multicast packets. To deploy IPv6 over IEEE 802.16, SS enters the networks and auto-configure its IPv6 address. In IEEE 802.16, when a SS enters the networks it gets three connection identifier (CID) connections to set-up its global configuration. The first CID is usually used for transferring short, sensitive MAC and radio link control messages, like those relating to the choice of the physical modulations. The second CID is more tolerant connection, it is considered as the primary management connection. With this connection, authentication and connection set-up messages are exchanged between SS and BS. Finally, the third CID is dedicated to the secondary management connection.

2.8.1 WiMAX security

WiMAX Security is a broad and complex subject most especially in wireless communication networks. The subject mechanism of Wimax Technology must meet the requirement design for security architecture in Wimax. Each layer handles different aspects of security, though in some cases, there may be redundant mechanisms. As a general principle of security, it is considered good to have more than one mechanism providing protection so that security is not compromised in case one of the mechanisms is broken. Security goals for wireless networks can be summarized as follows. Privacy or confidentiality is fundamental for secure communication, which provides resistance to interception and eavesdropping.

Message authentication provides integrity of the message and sender authentication, corresponding to the security attacks of message modification and impersonation. Anti-replay detects and disregards any message that is a replay of a previous message. Non-repudiation is against denial and fabrication. Access control prevents unauthorized access. Availability ensures that the resources or communications are not prevented from access by DoS attack. The 802.16 standard specifies a security sub layer at the bottom of the MAC layer. This security sub layer provides SS with privacy and protects BS from service hijacking. There are two component protocols in the security sub layer: an encapsulation protocol for encrypting packet data across the fixed BWA, and a Privacy and Key Management Protocol (PKM) providing the secure distribution of keying data from BS to SS as well as enabling BS to enforce conditional access to network services. The model below was adapted based on security in wimax. This chapter is still investigating the protocol in the sub layer that can mitigate encapsulation of packet data.

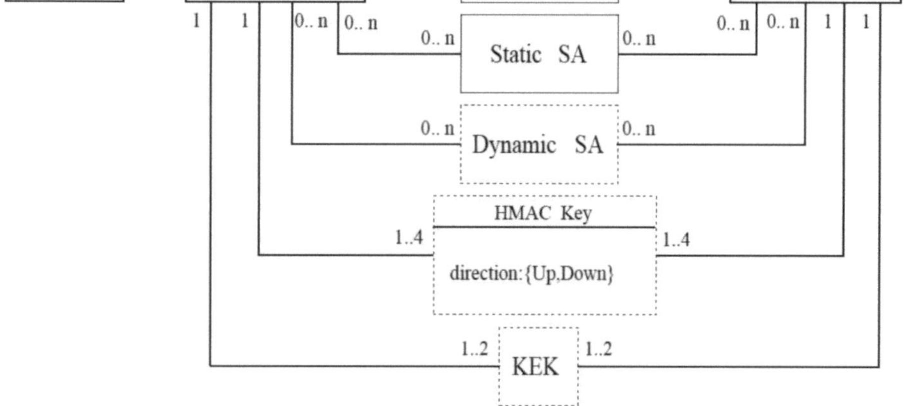

Fig. 8. Security model

2.8.2 Authentication in WiMAX

Basically in WIMAX/802.16 there are three options for authentication: device list based, X.509 based or EAP-based. If device list-based authentication is used only, then the likelihood of a BS or MA masquerading attack is likely. Impact can be high. The risk is therefore high and there is a need for countermeasures. If X.509-based authentication is used, the likelihood for a user (a MS) to be the victim of BS masquerading is possible because of the asymmetry of the mechanism.

There are specific techniques that identify theft and BS attack. Identity theft consists of reprogramming a device with the hardware address of another device. This is a well know problem in unlicensed services such as WiFi/802.11, but in cellular networks because it had been made illegal and more difficult to execute with subscriber ID module (SIM) cards. The exact method of attack depends on the type of networks.

The proposed policy based routing protocol for optimization evaluates connection association and transport information against the actions in policies, using the following principles:

- The mechanism must allow dynamic management of policies such as add, update and remove operations.
- The evaluation of policies should always result in exactly one interface for any traffic flow or connection. This is reached by having a priority order for actions.
- All attribute-value pairs in an action must match for a traffic flow or connection for the action to take place.
- The mechanism selects an interface based on the priority order of interfaces in an action.
- The mechanism uses default actions which match to all flows and connections if no other matching action is found. The mechanism should support distributed policy management and allow explicit definition of priorities. The below table consider our model for optimization during Authentication of WIMAX/802.16.

Threat	DoS on BS or MS
Kind	Mechanism
Device	Device List : RSA / X.509 Certificate
User Level	EAP + EAP – TLS (X.509) or EAP – SIM (subscriber ID module)
Data Traffic	AES-CCM CBC-MAC
Physical Layer Header	None
MAC Layer Header	None
Management messages	SHA - 1 Based MAC AES Based MAC

Table 2. Authentication in WIMAX 802.16

The intended proposed concept is to mitigate and prevent Dos on the BS or MS by introducing **Policy Repository.** In a WiMax/802.16 network, it is more difficult to do these because of the time division multiple access model. The attacker must transmit while the legitimate BS is transmitting. The signal of the attacker, however, must arrive at the targeted receiver MS(s) with more strength and must put the signal of the legitimate BS in the background, relatively speaking. Again, the attacker has to capture the identity of a legitimate BS and to build a message using that identity. The attacker has to wait until a time slot allocated to the legitimate BS starts. The attacker must transmit while achieving a receive signal strength. The receiver MSs reduce their gain and decode the signal of the attacker instead of the one from the legitimate BS.

2.8.3 Key management in Wimax for 5G technology

5G technology has a bright future because it can handle best technologies. The primary concern that should be focused on in 5G is the automated and optimization capability to support software. Although the issue of handover is being address since the Router and switches in this Network has high connectivity capability. Security is under studied in this regard. The knowledge base for the key management for 5G technology centered on physical layer, privacy sub layer threat, mutual Authentication, Threat of identity theft, water Torture and Black hat threat in wimax technology. The protocol used is not rolled out because some flexible framework created by the IETF (RFC 3748), allows arbitrary and complicated **authentication protocols** to be exchanged between the supplicant and the authentication server. **Extensible Authentication Protocol (EAP)** is a simple encapsulation that can run over not only PPP but also any link, including the WiMAX link. A number of **Extensible Authentication Protocol (EAP)** methods have already been defined to support authentication, using a variety of credentials, such as passwords, certificates, tokens, and smart cards. For example, **Protected Extensible Authentication Protocol (PEAP)** defines a password- based EAP method, EAP-transport-layer security (EAP-TLS) defines a certificate-based **Extensible Authentication Protocol (EAP)** method, and EAP-SIM (subscriber identity module) defines a SIM card-based EAP method. EAP-TLS provides strong mutual authentication, since it relies on certificates on both the network and the subscriber terminal.(Chong li).

3. Conclusion

Considering the complex issues and areas that have been addressed in this book chapter. The main focus of the chapter is how to provide techniques on automation and optimization using Algorithm based on policy based routing protocol. However, the various issues on this subject matter have been addressed. Analysis of micro mobility via Macro mobility based on comparative investigation and requirement was advanced. Furthermore the key optimization and data communication of IPv6 over wimax deployment must consider: *session preservation and interface selection mechanism.* The account of policy based routing protocol must provide: *link layer information, IP layer information and network originated information.* Our network model topology for optimization evaluates connection association and transport information against the actions in the policies using the aforementioned Algorithm. The remainder of this report should note that there are limitations in wimax deployment such as: low bit rate, speed of connectivity and sharing of bandwidth.

Finally, the chapter provides the basic Algorithm for optimization of IPv6 over wimax deployment using policy based routing protocol.

4. References

[1] Mobile IPv6 Fast Handovers over IEEE 802.16e Networks H. Jang, J. Jee, Y. Han, S. Park, J. Cha, June 2008
[2] IPv6 Deployment Scenarios in 802.16 Networks M-K. Shin, Ed., Y-H. Han, S-E. Kim, D. Premec, May 2008.
[3] Transmission of IPv6 via the IPv6 Convergence Sublayer over IEEE 802.16 Networks B. Patil, F. Xia, B. Sarikaya, JH. Choi, S. Madanapalli, February 2008.
[4] Mobility Support in IPv6 D. Johnson, C. Perkins, J. Arkko,RFC 3775. June 2004.
[5] Analysis of IPv6 Link Models for 802.16 Based Networks S. Madanapalli, Ed. ,August 2007.
[6] Threats Relating to IPv6 Multihoming Solutions E. Nordmark, T. Li, October 2005.
[7] Koodli, "Fast Handovers for Mobile IPv6", IETF RFC- 4068, July 2005.

2

Towards Semantic Interoperability in Information Technology: On the Advances in Automation

Gleison Baiôco, Anilton Salles Garcia and Giancarlo Guizzardi
Federal University of Espírito Santo
Brazil

1. Introduction

Automation has, over the years, assumed a key role in various segments of business in particular and, consequently, society in general. Derived from the use of technology, automation can reduce the effort spent on manual work and the realization of activities that are beyond human capabilities, such as speed, strength and precision. From traditional computing systems to modern advances in information technology (IT), automation has evolved significantly. At every moment a new technology creates different perspectives, enabling organizations to offer innovative, low cost or custom-made services. For example, advents such as artificial intelligence have enabled the design of intelligent systems capable of performing not only predetermined activities, but also ones involving knowledge acquisition. On the other hand, customer demand has also evolved, requiring higher quality, lower cost or ease of use. In this scenario, advances in automation can provide innovative automated services as well as supporting market competition in an effective and efficient way. Considering the growing dependence of automation on information technology, it is observed that advances in automation require advances in IT.

As an attempt to allow that IT delivers value to business and operates aligned with the achievement of organizational goals, IT management has evolved to include IT service management and governance, as can be observed by the widespread adoption of innovative best practices libraries such as ITIL (ITIL, 2007) and standards such as ISO/IEC 20000 (ISO/IEC, 2005). Nonetheless, as pointed out by Pavlou & Pras (2008), the challenges arising from the efforts of integration between business and IT remain topic of various studies. IT management, discipline responsible for establishing the methods and practices in order to support the IT operation, encompasses a set of interrelated processes to achieve this goal. Among them, configuration management plays a key role by providing accurate IT information to all those involved in management. As a consequence, semantic interoperability in the domain of configuration management has been considered to be one of the main research challenges in IT service and network management (Pras et al., 2007). Besides this, Moura et al. (2007) highlight the contributions that computer systems can play in terms of process automation, especially when they come to providing intelligent solutions, fomenting self-management. However, as they emphasize, as an emerging paradigm, this initiative is still a research challenge.

According to Pras et al. (2007), the use of ontologies has been indicated as state of the art for addressing semantic interoperability, since they express the meaning of domain concepts and relations in a clear and explicit way. Moreover, they can be implemented, thereby enabling process automation. In particular, ontologies allow the development of intelligent systems (Guizzardi, 2005). As a result, they foment such initiatives as self-management. Besides this, it is important to note that ontologies can promote the alignment between business and IT, since they maximize the comprehension regarding the domain conceptualization for humans and computer systems. However, although there are many works advocating their use, there is not one on IT service configuration management that can be considered as a de facto standard by the international community (Pras et al., 2007).

As discussed in Falbo (1998), the development of ontologies is a complex activity and, as a result, to build high quality ontologies it is necessary to adopt an engineering approach which implies the use of appropriate methods and tools. According to Guizzardi (2005, 2007), ontology engineering should include phases of conceptual modeling, design and implementation. In a conceptual modeling phase, an ontology should strive for expressiveness, clarity and truthfulness in representing the domain conceptualization. These characteristics are fundamental quality attributes of a conceptual model responsible for its effectiveness as a reference framework for semantic interoperability. The same conceptual model can give rise to different ontology implementations in different languages, such as OWL and RDF, in order to satisfy different computational requirements. Thus, each phase shall produce different artifacts with different objectives and, as a consequence, requires the use of languages which are appropriate to the development of artifacts that adequately meet their goals. As demonstrated by Guizzardi (2006), languages like OWL and RDF are focused on computer-oriented concerns and, for this reason, improper for the conceptual modeling phase. Philosophically well-founded languages are, conversely, committed to expressivity, conceptual clarity as well as domain appropriateness and so suitable for this phase.

Considering these factors, Baiôco et al. (2009) present a conceptual model of the IT service configuration management domain based on foundational ontology. Subsequently, Baiôco & Garcia (2010) present an implementation of this ontology, describing how a conceptual model can give rise to various implementation models in order to satisfy different computational requirements. The objective of this chapter is to provide further details about this IT service configuration management ontology, describing the main ontological distinctions provided by the use of a foundational ontology and how these distinctions are important to the design of models aligned with the universe of discourse, maximizing the expressiveness, clarity and truthfulness of the model and consequently the semantic interoperability between the involved entities. Moreover, this chapter demonstrates how to apply the entire adopted approach, including how to generate different implementations when compared with previous ones. This attests the employed approach, makes it more tangible and enables to validate the developed models as well as demonstrating their contributions in terms of activity automation.

It is important to note that the approach used in this work is not limited to the domain of IT service configuration management. In contrast, it has been successfully employed in many fields, such as oil and gas (Guizzardi et al., 2009) as well as medicine (Gonçalves et al., 2011). In fact, the development of a computer system involves the use of languages able to adequately represent the universe of discourse. According to Guizzardi (2005), an imprecise representation of state of affairs can lead to a false impression of interoperability, i.e.

although two or more systems seem to have a shared view of reality, the portions of reality that each of them aims to represent are not compatible. As an alternative, ontologies have been suggested as the best way to address semantic interoperability. Therefore, in particular, the ontological evaluation realized in this work contributes to the IT service configuration management domain, subsidizing solutions in order to address key research challenges in IT management. In general, this chapter contributes to promote the benefits of the employed approach towards semantic interoperability in IT in various areas of interest, maximizing the advances in automation. Such a contribution is motivated in considering that although recent research initiatives such as that of Guizzardi (2006) have elaborated on why domain ontologies must be represented with the support of a foundational theory and, even though there are many initiatives in which this approach has been successfully applied, it has not yet been broadly adopted. As reported by Jones et al. (1998), most existing methodologies do not emphasize this aspect or simply ignore it completely, mainly because it is a novel approach.

In this sense, this chapter is structured as follows: Section 2 briefly introduces the IT service configuration management domain. Section 3 discusses the approach to ontology development used in this work. Section 4 presents the conceptual model of the IT service configuration management domain. Section 5 shows an implementation model of the conceptual model presented in Section 4 and finally Section 6 relates some conclusions and future works.

2. IT service configuration management

The business of an organization requires quality IT services economically provided. According to ITIL, to be efficient and effective, organizations need to manage their IT infrastructure and services. Configuration management provides a logical model of an infrastructure or service by identifying, controlling, maintaining and verifying the versions of configuration items in existence. The logical model of IT service configuration management is a single common representation used by all parts of IT service management and also by other parties, such as human resources, finance, suppliers and customers. A configuration item, in turn, is an infrastructure component or an item that is or will be under the control of configuration management (ITIL, 2007; ISO/IEC, 2005). For innovative IT management approaches such as ITIL and ISO/IEC 20000, configuration items are viewed not only as individual resources but as a chain of related and interconnected resources compounding services. Thus, just as important as controlling each item is managing how they relate to each other. These relationships form the basis for activities such as impact assessment.

According to ITIL and ISO/IEC 20000, a configuration item and the related configuration information may contain different levels of detail. Examples include an overview of all services or a detailed view of each component of a service. Thus, a configuration item may differ in complexity, size and type, ranging from a service, including all hardware, software and associated documentation, to a single software module or hardware component. Configuration items may be grouped and managed together, e.g. a set of components may be grouped into a release. Furthermore, configuration items should be selected using established selection criteria, grouped, classified and identified in such a way that they are manageable throughout the service lifecycle.

As with any process, IT service configuration management is associated with goals that in its case include: (i) supporting, effectively and efficiently, all other IT service management processes by providing configuration information in a clear, precise and unambiguous way; (ii) supporting the business goals and control requirements; (iii) optimizing IT infrastructure settings, capabilities and resources; (iv) subsidizing the dynamism imposed on IT by promoting rapid responses to necessary changes and by minimizing the impact of changes in the operational environment. To achieve these objectives, configuration management should, in summary, define and control the IT components and maintain the configuration information accurately. Based on best practices libraries such as ITIL and standards such as ISO/IEC 20000 for IT service management, the activities of an IT service configuration management process may be summarized as: (i) planning, in order to plan and define the purpose, scope, objectives, policies and procedures as well as the organizational and technical context for configuration management; (ii) identification, aiming to select and identify the configuration structures for all the items (including their owner, interrelationships and configuration documentation), allocate identifiers and version numbers for them and finally label each item and enter it on the configuration management database (CMDB); (iii) control, in order to ensure that only authorized and identified items are accepted and recorded, from receipt to disposal, ensuring that no item is added, modified, replaced or removed without appropriate controlling documentation; (iv) status accounting and reporting, which reports all current and historical data concerned with each item throughout its life cycle; (v) verification and audit, which comprises a series of reviews and audits that verify the physical existence of items and check that they are correctly recorded in the CMDB.

As an attempt to promote efficiency and effectiveness, IT management has evolved to include IT service management and governance, which aims to ensure that IT delivers value to business and is aligned with the achievement of organizational goals. As emphasized by Sallé (2004), in this context, IT processes are fully integrated into business processes. Thus, one of the main aspects to be considered is the impact of IT on business processes and vice versa (Moura et al., 2008). As a consequence, IT management processes should be able to manage the entire chain, i.e. from IT to business. For this reason, the search for the effectiveness of such paradigms towards business-driven IT management has been the topic of several studies in network and service management (Pavlou & Pras, 2008). According to Moura et al. (2007), one of the main challenges is to achieve the integration between these two domains. Configuration management, in this case, should be able to respond in a clear, precise and unambiguous manner to the following question: what are the business processes and how are they related to IT services and components (ITIL, 2007)? Furthermore, as cited by ITIL, due to the scope and complexity of configuration management, keeping its information is a strenuous activity. In this sense, research initiatives consider automation to be a good potential alternative. In fact, the automation of management processes has been recognized as one of the success factors to achieve a business-driven IT management, especially when considering intelligent solutions promoting self-management (Moura et al., 2007). Besides its scope and complexity, configuration management is also closely related to all other management processes. In IT service management and governance, this close relationship includes the interaction among the main entities involved in this context, such as: (i) business, (ii) people, (iii) processes, (iv) tools and (v) technologies (ITIL, 2007). Thus, semantic interoperability among such entities has been characterized as one of the main research challenges, not only in terms of the configuration management process, but in the whole chain of processes that comprise the discipline of network and service management (Pras et al., 2007).

According to Pras et al. (2007), the use of semantic models, in particular, the use of ontologies, has been regarded as the best way with respect to initiatives for addressing issues related to semantic interoperability problems in network and service management. According to these authors, ontologies make the meaning of the domain concepts such as IT management, as well as the relationships between them, explicit. Additionally, this meaning can be defined in a machine-readable format, making the knowledge shared between humans and computer systems, enabling process automation, as outlined by these authors. From this point of view, it is worth mentioning that ontologies are considered as potential tools for the construction of knowledge in intelligent systems (Guizzardi, 2005). Thus, they allow the design of intelligent and above all interoperable solutions, fomenting initiatives as self-management. Finally, it is important to note that ontologies can promote the alignment between business and IT when applied in the context of IT service management and governance since they maximize the expressiveness, clarity and truthfulness of the domain conceptualization for humans and computer systems. However, Pras et al. (2007) point out that despite the efforts of research initiatives, there are still many gaps to be addressed.

Several studies claim that the use of ontologies is a promising means of achieving interoperability among different management domains. However, an ontology-based model and formalization of IT service configuration management remains a research challenge. Regarding limitations, it should be mentioned that ontologies are still under development in the management domain. In fact, the technology is not yet mature and there is not an ontology that can be considered as a *de facto* standard by the international community (Pras et al., 2007). In general, the research initiatives have not employed a systematic approach in the development of ontologies. According to Falbo (1998), the absence of a systematic approach, with a lack of attention to appropriate methods, techniques and tools, makes the development of ontologies more of an art rather than an engineering activity. According to Guizzardi (2005, 2007), to meet the different uses and purposes intended for the ontologies, ontology engineering should include phases of conceptual modeling, design and implementation. Each phase should have its specific objectives and thus would require the use of appropriate languages in order to achieve these goals. However, in most cases, such research initiatives are engaged with the use of technologies and tools such as Protégé and OWL. Sometimes these technologies and tools are used in the conceptual modeling phase, which can result in various problems relating to semantic interoperability, as shown in Guizzardi (2006). At other times, however, they are employed in the implementation phase, ignoring previous phases such as conceptual modeling and design. As a result, such initiatives are obliged to rely on models of low expressivity. Moreover, in most cases, such initiatives propose the use of these technologies and tools for the formalization of network management data models, such as MIB, PIB and the CIM schema. It is noteworthy that data models are closely related to the underlying protocols used to transport the management information and the particular implementation in use. In contrast, information models work at a conceptual level and they are intended to be independent of any particular implementation or management protocol. Working at a higher level, information models usually provide more expressiveness (Pras et al., 2007). Following this approach, Lopez de Vergara et al. (2004) propose an integration of the concepts that currently belong to different network management data models (e.g. MIB, PIB and the CIM schema) in a single model, formalized by ontology languages such as OWL. In an even more specific scenario, i.e. with no intention to unify the various models but only to formalize a particular model, Majewski et al. (2007) suggest the formalization of the CIM schema through ontology languages such as OWL. Similarly, while differentiating the type of data model, Santos (2007) presents an ontology-based network configuration management system. In his work, the proposed ontology was developed according to the MIB data model concepts. As MIB is limited in describing a single system, a view of the entire infrastructure, including the relationships between its components, is not supported by the model. In practice, this gap is often filled by functionalities provided by SNMP-based network management tools which, for example, support the visualization of network topologies (Brenner et al., 2006). Aside from the fact that, in general, the research initiatives are committed to the use of technologies and tools, it is also observed that they are characterized by specific purposes in relation to peculiar applications in information systems that restrict their conceptualizations. In Xu and Xiao (2006), an ontology-based configuration management model for IP network devices is presented, aiming at the use of ontology for the automation of this process. In Calvi (2007), the author presents a modeling of the IT service configuration management described by the ITIL library based on a foundational ontology. The concepts presented and modeled in his work cover a specific need regarding the demonstration of the use of ITIL processes for a context-aware service platform. Finally, there are approaches that seek to establish semantic interoperability among existing ontologies by means of ontological mapping techniques, as evidenced in Wong et al. (2005). However, it is not within the scope of such approaches to develop an ontology but rather to integrate existing ones.

Therefore, in considering the main challenges as well as the solutions which are considered to be state of the art and in analyzing the surveyed works, it is observed that there are gaps to be filled, as highlighted by Pras et al. (2007). In summary, factors such as the adoption of *ad hoc* approaches, the use of inappropriate references about the domain, the intention of specific purposes and, naturally, the integration of existing ontologies, all result in gaps. As a consequence, such factors do not promote the conception of an ontology able to serve as a reference framework for semantic interoperability concerning the configuration management domain in the context of IT service management and governance. This scenario demonstrates the necessity of a modeling which considers the gaps and, therefore, promotes solutions in line with those suggestions regarded as state of the art for the research challenges discussed earlier in this chapter. In particular, it demonstrates the necessity of an appropriate approach for the construction of ontologies as a subsidy for such modeling. In this sense, the next section of this chapter presents an approach for ontology development.

3. Ontology engineering

In philosophy, ontology is a mature discipline that has been systematically developed at least since Aristotle. As a function of the important role played by them as a conceptual tool, their application to computing has become increasingly well-known (Guizzardi et al., 2008). According to Smith and Welty (2001), historically there are three main areas responsible for creating the demand for the use of ontologies in computer science, namely: (i) database and information systems; (ii) software engineering (in particular, domain engineering); (iii) artificial intelligence. Additionally, Guizzardi (2005) includes the semantic web, due to the important role played by this area in the current popularization of the term.

According Guizzardi et al. (2008), an important point to be emphasized is the difference between the senses of the term ontology when used in computer science. In conceptual modeling, the term has been used as its definition in philosophy, i.e. as a philosophically well-founded domain-independent system of formal categories that can be used to articulate domain-specific models of reality. On the other hand, in most other areas of computer science, such as artificial intelligence and semantic web, the term ontology is generally used as: (i) an engineering artifact designed for a specific purpose without giving much importance to foundational issues; (ii) a representation of a particular domain (e.g. law, medicine) expressed in some language for knowledge representation (e.g. RDF, OWL).

From this point of view, the development of ontologies should consider the various uses and, consequently, the different purposes attributed to ontologies as well as any existing interrelationship in order to enable the construction of models that satisfactorily meet their respective goals. However, despite the growing use of ontologies and their importance in computing, the employed development approaches have generally not considered these factors, resulting in inadequate models for the intended purpose. In considering such distinctions Guizzardi (2005, 2007) elaborates and discusses a number of questions in order to elucidate such divergences and thus provide a structured way with respect to the use of ontologies. In addition, Guizzardi and Halpin (2008) describe that the interest in proposals for foundations in the construction of ontologies has been the topic of several studies and they report some innovative and high quality research contributions. It is based on such questions that are elaborated the further discussions contained in this section and thus the approach used for the construction of the ontological models proposed in this work.

As discussed in Falbo (1998), the development of ontologies is a complex activity and, hence, in order to build high quality ontologies, able to adequately meet their various uses and purposes, it is necessary to adopt an engineering approach. Thus, unlike the various *ad hoc* approaches, the construction of ontologies must use appropriate methods and tools. Falbo (2004) proposes a method for building ontologies called SABiO (Systematic Approach for Building Ontologies). This method proposes an life cycle by prescribing an iterative process that comprises the following activities: (i) purpose identification and requirements specification, which aims to clearly identify the ontology's purpose and its intended use by means of competence questions; (ii) ontology capture, viewing to capture relevant concepts existing within the universe of discourse as well as their relationships, properties and constraints, based on the competence questions; (iii) ontology formalization, which is responsible for explicitly representing the captured conceptualization by means of a formal language, such as the definition of formal axioms using first-order logic; (iv) integration with existing ontologies, in order to search for other ones with the purpose of reuse and integration; (v) ontology evaluation, which aims to identify inconsistencies as well as verifying truthfulness in line with the ontology's purpose and requirements; (vi) ontology documentation. Noticeably, the competence questions form an important concept within SABiO, i.e. the questions the ontology should be able to answer. They provide a mechanism for defining the scope and purpose of the ontology, guiding its capture, formalization and evaluation - regarding this last aspect, especially with respect to the completeness of the ontology.

The elements that constitute the relevant concepts of a given domain, understood as domain conceptualization, are used to articulate abstractions of certain states of affairs in reality, denominated as domain abstraction. As an example, consider the domain of product sales.

A conceptualization of this domain can be constructed by considering concepts such as, *inter alia*: (i) customer, (ii) provider, (iii) product, (iv) is produced by, (v) is sold to. By means of these concepts, it is possible to articulate domain abstractions of certain facts extant in reality, such as: (i) a product is produced by the provider and sold to the customer. It is important to highlight that conceptualizations and abstractions are abstract entities which only exist within the mind of a user or a community of users of a language. Therefore, in order to be documented, communicated and analyzed, they must be captured, i.e. represented in terms of some concrete artifact. This implies that a language is necessary for representing them in a concise, complete and unambiguous way (Guizzardi, 2005). Figure 1-a presents "Ullmann's triangle" (Ullmann, 1972), which illustrates the relation between a language, a conceptualization and the part of reality that this conceptualization abstracts. The relation "represents" concerns the definition of language semantics. In other words, this relation implies that the concepts are represented by the symbols of language. The relation "abstracts", in turn, denotes the abstraction of certain states of affairs within the reality that a given conceptualization articulates. The dotted line between language and reality highlights the fact that the relation between language and reality is always intermediated by a certain conceptualization. This relation is elaborated in Figure 1-b, which depicts the distinction between an abstraction and its representation, as well as their relationships with the conceptualization and representation language. The representation of a domain abstraction in terms of a representation language is called model specification (or simply model, specification or representation) and the language used for its creation is called modeling language (or specification language).

Fig. 1. Ullmann's triangle and relations between conceptualization, abstraction, modelling language and model, according to Guizzardi (2005).

Thus, in addition to the adoption of appropriate methods, able to systematically lead the development process, ontology engineering as an engineering process aims at the use of tools, which should be employed in accordance with the purpose of the product that is being designed. In terms of an ontology development process such tools include modeling languages or even ontology representation languages. According to Guizzardi (2005), one of the main success factors regarding the use of a modeling language is its ability to provide its users with a set of modeling primitives that can directly express the domain conceptualization. According to the author, a modeling language is used to represent a conceptualization by compounding a model that represents an abstraction, which is an instance of this conceptualization. Therefore, in order for the model to faithfully represent an abstraction, the modeling primitives of the language used to produce the model must accurately represent the domain conceptualization used to articulate the abstraction represented by the model. According to Guizzardi (2005), if a conceptual modeling language is imprecise and coarse in the description of a given domain, then there can be specifications of the language which, although grammatically valid, do not represent admissible state of affairs. Figure 2-a illustrates this situation. The author also points out that a precise representation of a given conceptualization becomes even more critical when it is necessary to integrate different independently developed models (or systems based on these models). As an example, he mentions a situation in which it is necessary to have the interaction between two independently developed systems which commit to two different conceptualizations. Accordingly, in order for these two systems to function properly together, it is necessary to ensure that they ascribe compatible meanings to the real world entities of their shared subject domain. In particular, it is desirable to reinforce that they have compatible sets of admissible situations whose union (in the ideal case) equals the admissible states of affairs delimited by the conceptualization of their shared subject domain. The ability of entities (in this case, systems) to interoperate (operate together) while having compatible real-world semantics is known as semantic interoperability (Vermeer, 1997). Figure 2-b illustrates this scenario.

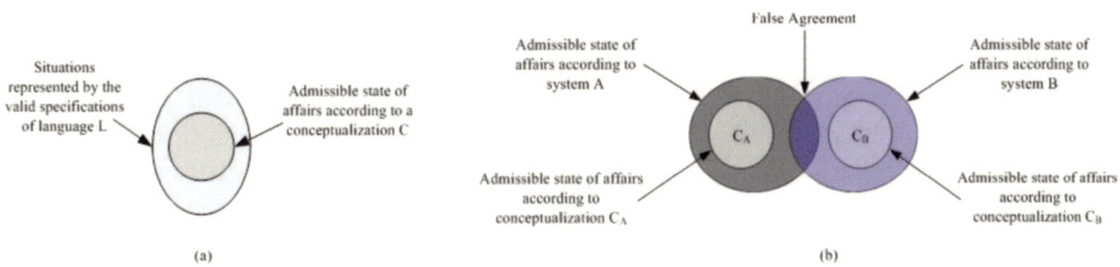

Fig. 2. Consequences of an imprecise and coarse modelling language (Guizzardi, 2005).

In Figure 2-b, C_A and C_B represent the conceptualizations of the domains of systems A and B, respectively. As illustrated in this figure, these conceptualizations are not compatible. However, because these systems are based on poor representations of these conceptualizations, their sets of considered possible situations overlap. As a result, systems A and B agree exactly on situations that are neither admitted by C_A nor by C_B. In summary, although these systems appear to have a shared view of reality, the portions of reality that each of them aims to represent are not compatible. Therefore, the more it is known about a given domain and the more precisely it is represented, the bigger the chance of obtaining interpretations that are consistent with the reality of that domain and, therefore, of achieving semantic interoperability between the entities involved in these interpretations. Thus, Guizzardi (2005) concludes that, on the one hand, a modeling language should be sufficiently expressive to adequately characterize the conceptualization of the domain and, on the other hand, the semantics of the produced specifications should be clear, allowing users to recognize what language constructs mean in terms of domain concepts. Moreover, the specification produced by means of the language should facilitate the user in understanding and reasoning about the represented state of affairs.

In view of the different purposes, Guizzardi (2005, 2007) highlights that ontology engineering, analogous to software engineering and information systems, must include phases of conceptual modeling, design and implementation. Each phase has its specific objectives and thus requires different types of methods and tools to meet its particular characteristics. As mentioned, during a conceptual modeling phase, an ontology must strive for expressivity, clarity and truthfulness in representing the domain conceptualization. Therefore, the conceptual modeling phase requires specialized languages so as to create ontologies that approximate as closely as possible to the ideal representation of the domain. The same conceptual model can give rise to different implementation models in different languages, such as OWL and RDF, in order to satisfy different non-functional requirements, such as decidability and completeness. The section delimited as *Level* in Figure 3 illustrates this approach based on relations between conceptualization, abstraction, modeling language and model, shown in Figure 1-b. According to Guizzardi (2006), semantic web languages such as OWL and RDF are focused on computation-oriented concerns and are therefore inadequate for the conceptual modeling phase. Philosophically well-founded languages, on the other hand, are engaged in expressivity, conceptual clarity and domain appropriateness and are therefore suitable for this phase. To support his assertion, Guizzardi (2006) presents several problems of semantic interoperability from the use of semantic web languages in the representation of the domain and demonstrates how philosophically well-founded languages are able to address these problems.

As shown in Guizzardi (2005), while domain conceptualizations and, consequently, domain ontologies are established by the consensus of a community of users with respect to a material domain, a conceptual modeling language (which can be used to express these domain ontologies) must be rooted in a domain independent system of real-world categories, philosophically and cognitively well-founded, i.e. a foundational ontology. Foundational ontologies aggregate contributions from areas such as descriptive metaphysics, philosophical logic, cognitive science and linguistics. The theories inherent to these areas are called (meta-) conceptualizations and describe knowledge about reality in a way which is independent of language and particular states of affairs. A foundational ontology, in turn, is the representation of these theories in a concrete artifact. Thus, foundational ontologies, in the philosophical sense, can be used to provide real-world semantics for modeling languages as well as to constrain the possible interpretations of their modeling primitives, increasing the clarity of interpretation and, consequently, reducing ambiguities (which are key success factors in achieving semantic interoperability). Accordingly, it is possible to build domain ontologies by means of conceptual modeling languages based on foundational ontologies. In this sense, the *Meta-level* section in Figure 3, in addition to the *Level* section, represents the approach proposed by Guizzardi (2005, 2007).

An example of a foundational ontology is UFO (Unified Foundational Ontology). UFO was initially proposed in Guizzardi and Wagner (2004) and its most recent version is presented by Guizzardi et al. (2008). It is organized in three incrementally layered compliance sets: (i) UFO-A, which is essentially the UFO's core, defining terms related to endurants (objects, their properties etc); (ii) UFO-B, which defines as an increment to UFO-A terms related to perdurants (events etc); (iii) UFO-C, which defines as an increment to UFO-B terms explicitly related to the spheres of social entities.

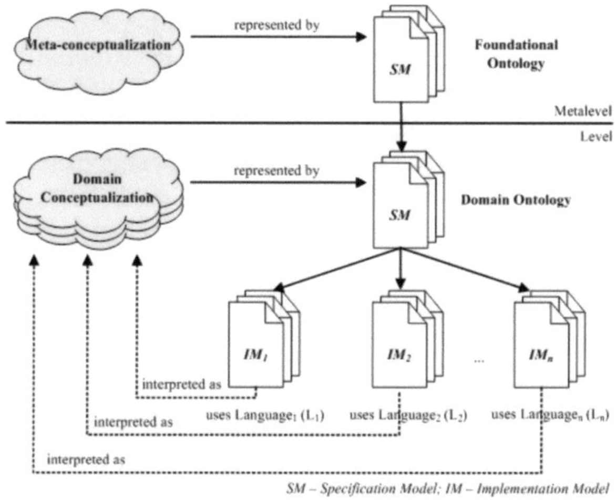

Fig. 3. Ontology engineering approach proposed by Guizzardi (2005, 2007).

As a conclusion, the ontology engineering approach described in this section considers the distinctions of the term ontology as well as their interrelationships, thereby establishing the phases, their respective objectives, as well as the methods and tools appropriate for the characteristic of each phase, allowing thus the construction of models capable of meeting the various purposes intended for them. Therefore, based on the study about IT service configuration management, as well as the study about ontology development, it is possible to construct an ontology of this domain as the objective of this chapter.

4. Conceptual model of the IT service configuration management domain

In considering the main research challenges, as well as the initiatives for solutions which are regarded state of the art, which properly lead the identification of gaps, as much as the appropriated approach to fulfil them, this section presents the conceptual models proposed in this work. On the basis of Figure 3, which shows the ontology development approach adopted in this work, the conceptual model proposed in this section concerns the domain ontology, whose conceptualization in discussion is the IT service configuration management. Regarding the foundational ontology, this work uses UFO, which represents the meta-conceptualization responsible for promoting the philosophical base of the work.

As discussed in Section 2, configuration management is responsible for maintaining information about configuration items and providing them to all the other management processes. In the context of IT service management and governance, configuration management must be able to answer questions such as: what are the business processes and how do they relate to the IT services and components? Based on this question and given that the main goal of this ontology is to describe a theory of the domain of IT service configuration management independent of specific applications, the defined competency questions reflect this intention. In this case, they lead to a mapping between IT and business concepts, as follows:

CQ1: How do the IT services and the business processes of an organization relate?

CQ2: How do the IT services and the IT components such as hardware and software relate?

To answer these questions, it is necessary to address others, such as: (i) what is an IT service? (ii) what is a business process? For this reason, Baiôco et al. (2009) propose an IT service configuration management ontology, which addresses such questions in order to provide a basis for the ontology presented in this chapter. As described through the SABiO method, if the domain of interest is too complex, a decomposing mechanism should be used in order to better distribute this complexity. In this case, a potentially interesting approach is to consider sub-ontologies. Therefore, to answer the competency questions, the following sub-ontologies were developed: (i) business process; (ii) IT service; (iii) IT component and lastly (iv) configuration item. These sub-ontologies complement each other in constituting the IT service configuration management ontology discussed in this work.

In terms of reusing existing ontologies, it is important to mention that besides the adopted literature, the conceptual modeling of this section also takes into consideration the discussions inherent to processes in general done in Falbo (1998) and Guizzardi et al. (2008), as well as the discussions inherent to IT services done in Calvi (2007) and Costa (2008). Still in line with the SABiO method, during the capture of the ontology the use of a graphic representation is essential to facilitate the communication between ontology engineers and domain experts. However, a graphical model is not enough to completely capture an ontology. This way, axioms should be provided to reinforce the semantics of the terms and establish the domain restrictions. Thus, the sub-ontologies developed in this work are connected by relations between their concepts and by formal axioms. Due to limitations of space, will be shown only those axioms also used in the next section. To distinguish the subject domain concepts and the UFO concepts, these last ones are presented in blank in the conceptual model that follows. Figure 4 presents part of the proposed ontology.

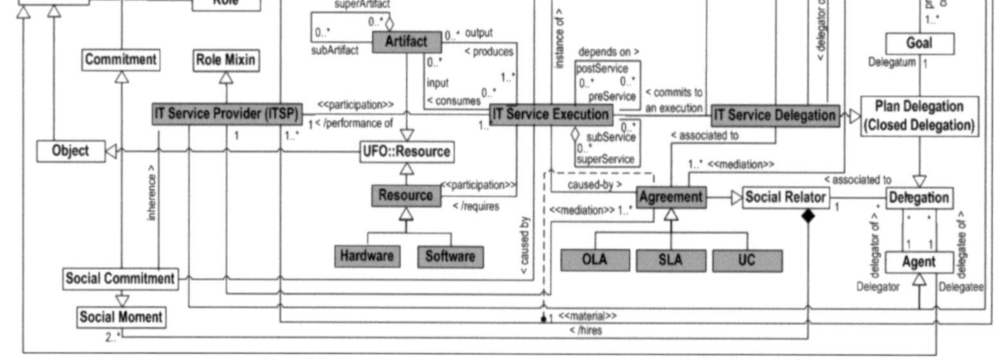

Fig. 4. Part of the IT service configuration management ontology.

According to the ITIL library, an IT service is a service which is provided by an IT organization to one or more clients. Therefore, in accordance with the UFO terminology, an IT service is characterized as a type of plan, i.e. an intentional event. As a result, the properties inherent to events are applied to IT services. As established in UFO, events are possible changes from one portion of reality to another, which means they can transform reality by altering the state of affairs from a pre-state to a post-state. Consequently they can produce, direct or indirectly, situations that satisfy the necessary conditions for other events to happen. On that account, events can cause other events, including then the service executions as represented in the model by the relation causes. An IT service execution (ITSE), in turn, denotes one or more particular actions that occur at specific time intervals, aiming to satisfy the propositional content of a commitment. On this note, an IT service execution is an action that instantiates a type of plan, in this case an IT service. This distinction, derived from the foundational ontology, provides greater adequacy to the domain, making it possible to distinguish services from their executions and allowing the comparison, for example, between achieved and planned results.

Because it denotes one or more actions, an IT service execution can be atomic or complex. As a complex execution, it is decomposed into other smaller service executions, termed subservices. In this way, a subservice is a service execution that is part of a bigger service execution, its super-service. As the properties inherent to events are applied to the IT services, the decomposition of service executions, as any other event decomposition, is characterized as a transitive, asymmetric and irreflexive relation. This inheritance of properties from the foundational ontology facilitates modeling decisions and minimizes the possibilities of incoherent descriptions of the domain.[1]

According to the ITIL library, an IT service execution aims to produce resources in order to satisfy the needs of its customers. On the conceptual model proposed in this section, the produced resources are said to be outputs. On the other hand, an IT service execution can consume resources, seen as raw material, to produce results. On the model, these resources are said to be inputs. From the point of view of UFO, an artifact of a service execution is a type of resource (UFO::Resource) which in turn is mapped to the notion of an object. As such, the subartifact and superartifact relations are then governed by the axioms defined for the (different types of) parthood relations between substantials, as described in Guizzardi (2005).

According to UFO, a resource (UFO::Resource) is a role that an object plays in an event. Thus, the artifacts of a service execution are roles played by objects in the scope of this service instance. This being said, it is important to highlight a contribution from UFO attributed to the model. As the notions of objects and roles are defined, it becomes possible to represent real situations of the domain, as with those where the same object plays the role of an output to a service execution and input to another, distinguishing only the type of participation performed by the object and keeping its identity throughout its existence. This is because, according to UFO, an object is a type of endurant which, in contrast to a

[1] UFO makes explicit distinctions often ignored by many languages. For example, while it is possible to consider that an event x is a part of an event z because x is a part of an event y that is a part of z, it is not the case that the musician's hand (and so a part thereof) is a part of the band within which the musician is a part. In the first case, there is transitivity, but in the latter this does not exist. In this sense, parthood relations denote distinctions that should be considered.

perdurant (e.g. an event), is an individual that keeps its identity throughout its existence. On the model, the participation of the output is represented through the relation "produces" while the participation of the input is represented through the relation "consumes".

As objects, the participation of artifacts in a service execution should correspond to the types of participation of objects in an action, defined in UFO. In fact, being produced or consumed does not express the exact notion on the participation of an object in a service execution. Namely, outputs can be created or modified. Inputs, on the other hand, can be used, modified into new products or else terminated. As such, the relation "produces", as well as the relation "consumes", designates distinct notions which should be considered. Therefore, in light of the UFO terminology and the subject domain, the relation "produces" denotes the participation of creation or change, while the relation "consumes" indicates the participation of usage, change or termination as defined in UFO. This demonstrates the foundation of the domain concepts and relations in terms of philosophically well-founded concepts and relations, making the representation of the universe of discourse even more clear, expressive and coherent with reality.

As described by the ITIL library, an IT service is based on the use of the information technology, which includes, among other things, IT components such as hardware and software. Therefore, an IT service execution, as any other activity, presumes the use of resources, in particular IT components, in order to achieve results. Essentially, under the UFO's perspective, the resources of a service execution are types of resources (UFO::Resource), i.e. objects participating in an action. Therefore, not only artifacts but also resources are roles played by objects within the scope of a service execution. Again, the distinction between objects and roles contributes to the representation of real situations in the universe of discourse, including those where the same object is produced by a service execution but is required by another, though remaining as the same individual. In the model, the participation of a resource is represented by the relation "requires". Once these resources are used as support tools in a service execution, the foundation associated with this relation leads to only one of the types of resource participation defined in UFO, i.e. the usage participation.

In considering the definitions of artifacts and resources, especially regarding the foundation provided by UFO, it is noted that the inputs and the resources are objects which can play the same type of participation in a service execution, i.e. the usage participation. However, taking a service execution as a transformation primitive, inputs indicate raw materials which are incorporated into the product. Resources, on the other hand, refer to components that support a service execution, but are not intended as products of this execution. Thus, the resources employed in a service execution cannot be considered as products within the scope of this execution. Therefore, by clearly representing the participation of inputs and resources in a service execution, this foundation promoted the identification with regard to the similar type of participation of these roles. As a consequence, it required the use of domain definitions able to characterize such roles, since fundamentally they are similar. Hence, as pointed out, this observation becomes evident, in this case, especially due to the foundation of the domain concepts in terms of UFO. This demonstrates the contribution of a foundational ontology, in general, and of UFO, in particular, in supporting the construction of appropriate models regarding domain and comprehension.

As described in Costa (2008), an IT service execution is an intended and orderly execution of one or more actions in order to satisfy the propositional content (goal) of a commitment agreed with an agent. This ordering, in some cases, is governed according to certain situations that culminate into the execution of the service. In this sense, the causality relation between events leads to a dependence relation between them. This happens because, by changing the state of affairs of the reality from one state (pre-state) to another (post-state), events can generate (directly or indirectly) a situation which satisfies a necessary condition for other events to occur. In summary, an event depends on another if the first is caused, directly or indirectly, by the second. Hence, service executions can depend on events (or on other service executions) in order to occur. It is relevant to mention that artifacts and resources can imply a dependency relation between service executions. In fact, as described in UFO, situations are complex entities that agglutinate other entities (including objects) and denote the pre- and post-state of an event. So, as a type of object, artifacts and resources are present in situations that symbolize the pre- and post-states of service executions. As an example, a service may require resources or consume inputs that are products of other services, characterizing thus a dependence relation between them because of their respective objects. Besides this, it is important to highlight that if an event depends on the other then there is a temporal relation between them that should be taken into account. Concerning dependence between service executions, this temporal relation is represented by the pre-service and post-service relations. As the causality relation between events, the dependence relation between them (including the IT service executions) is transitive, asymmetric and irreflexive. These properties are also valid for the relations pre- and post-service. Temporal relations between events make possible to define the temporal ordering in which the events (including the service executions) are submitted and, therefore, establish the order in which they occur, even when there is no dependence relation defined between them. This is because, according to UFO, events are framed into temporal intervals, from which originate the temporal relations. Thus, the model proposed in this section is based on a framework of concepts and relations defined in UFO which makes possible to specify the flow in which the events are associated and, consequently, the ordering associated with the IT service executions. In this way, this structure supports the modeling decisions at the same time as it contributes to the creation of a more expressive, clear and truthful model with regard to the universe of discourse.

An IT service is described by a normative description, termed IT service description. The description of an IT service, for instance, describes the roles played by each agent in a service execution. Agents, as well as objects, are substantial from the UFO point of view. However, agents differ from objects because of the fact that they can possess beliefs, desires and intentions. Intentions are characterized as desired states of affairs for which the agent commits itself to pursuing, i.e. an internal commitment. For this reason, intentions cause the agent to perform actions. In this sense, the participation of an agent in an action is characterized as an action contribution. Consequently, service executions are performed by agents. Indeed, the action contribution of an agent in a service execution is caused by a social commitment of the agent in performing this service execution (or part thereof, the subservice) with its consequent permissions and obligations. Therefore, the role modeling pattern described by UFO applies to the IT service domain. In fact, as advanced in Guizzardi (2006), as a domain independent knowledge representation language, foundational ontologies in general and UFO in particular aim to support the construction of domain-dependent models by acting as a reference framework and thus guiding modeling decisions and allowing the creation of models that clearly and accurately represent, as much as possible, the real situations in a universe of discourse. This pattern has resolved various role modeling problems in the literature. Thus, the execution of a service instance is characterized by an agent playing a role in an IT service, in this case, acting as a service provider. It is worth noting that not all types of agents are responsible for the service executions. According to ITIL, an IT service is one provided to one or more customers by an IT organization. On the other hand, not all the instances of service providers are IT organizations, but possibly another type of agent. Therefore, the execution of an IT service occurs by means of an IT organization acting as a service provider. Nonetheless, given that IT organizations can play other roles, such as the role of a customer, it is not the case that all instances of IT organizations act as service providers in every situation, but only when they perform IT services. So, the relation between service provider and IT organization cannot be direct in any sense. On the contrary, it should be intermediated by a role (in this case, IT Service Provider - ITSP) that aggregates the criteria of identity of the species (in this case, IT Organization) and performs the mixed role (in this case, Service Provider). In this way, it is defined that not all service provider is characterized as an IT organization, which may be another type of agent. Furthermore, it is defined that there are IT organizations which in certain circumstances are service providers but not in others, possibly playing other roles. The responsibility of each played role is described by the IT service description.

The execution of an IT service occurs by means of a request, which is motivated according to the requestor's needs. Therefore, a customer is a type of requestor, i.e. an agent who requests an IT service. The process of requesting an IT service, as well as other concepts inherent to the domain of service-level management, is discussed and modeled in Costa (2008). In summary, the author describes that an IT service is appropriate for a certain need when the product from its execution satisfies the requestor's requirements, i.e. its needs. As such, an IT service can achieve a need if, and only if, the post-state of an occurrence of this service is a situation that satisfies the propositional content of the referred need, as formalized by the axiom A1.

$$\text{(A1)}\ \forall x,y\ (\text{IT-Service}(x) \wedge \text{Need}(y) \wedge \text{can-achieve}(x,y) \leftrightarrow \exists a,b,c\ (\text{ITSE}(a) \wedge \text{Situation}(b) \wedge \text{Proposition}(c) \wedge \text{instance-of}(a,x) \wedge \text{postState}(b,a) \wedge \text{satisfies}(b,c) \wedge \text{propositional-content-of}(c,y)))$$

An IT service is requested by means of a document that describes the need of the requestor, termed service-level requirement (SLR). Thus, given a SLR that describes a certain need which can be achieved by an IT service, then this SLR is used to request this service (COSTA, 2008). This definition is formalized by the axiom A2.

$$\text{(A2)}\ \forall x,y,z\ (\text{IT-Service}(x) \wedge \text{Need}(y) \wedge \text{SLR}(z) \wedge \text{can-achieve}(x,y) \wedge \text{describes}(z,y) \rightarrow \text{used-to-ask-for}(z,x))$$

So, through a SLR, a requestor can search for an IT service that satisfies its needs. This allows the requestor to find out whether or not there are services that can achieve its needs and request, in case there is, the most adequate, according to its demand. This definition is formalized by the axiom A3.

(A3) $\forall$x,y,z,w (IT-Service(x) $\wedge$ Need(y) $\wedge$ SLR(z) $\wedge$ Requestor(w) $\wedge$ can-achieve(x,y) $\wedge$ describes(z,y) $\wedge$ used-to-ask-for(z,x) $\wedge$ characterizes(y,w) $\rightarrow$ asks-for(w,x))

Once the requester asks for services that can achieve its necessities and finds such services offered by the providers, this requestor is able to request such services. According to Costa (2008), when an IT service is requested, the requestor hires the provider responsible for this service, as formalized by the axiom A4.

(A4) $\forall$x,y,z (Requestor(x) $\wedge$ ITSP(y) $\wedge$ IT-Service(z) $\wedge$ asks-for(x,z) $\wedge$ requests(x,z) $\wedge$ provides(y,z) $\rightarrow$ hires(x,y))

According to the ITIL library, the provision of an IT service to a requestor is mediated by an agreement. Thus, once the requesting process has been established, there will be an agreement that will mediate all the IT service provision, establishing the claims and obligations related to both parts, the requester and the provider. This relation is formalized by the axiom A5.

(A5) $\forall$x,y (Requestor(x) $\wedge$ ITSP(y) $\wedge$ hires(x,y) $\rightarrow$ $\exists$z (Agreement(z) $\wedge$ mediates(z,x) $\wedge$ mediates(z,y)))

With regard to UFO, an agreement is a type of social relation. A social relation is composed of social moments, called claims and commitments. A social commitment is a type of commitment and thus the motivating cause of an action performed by an agent. Thus, the agreement established between the requestor and the provider will cause the execution of the service requested by the requestor, for it is composed of social commitments inherent to the provider.

The requesting of a service is motivated by a need of a requestor and it is fulfilled through a service execution performed by a provider. In this context, there is a relation of dependence between the requestor and the provider. From the UFO point of view, a dependence relation between agents leads to a delegation relation. According to UFO, an agent *a* depends on an agent *b* regarding a goal *g* if *g* is a goal of agent *a*, but *a* cannot achieve *g* and agent *b* can achieve *g*. This matter may be the reason why agent *a* decides to delegate such goal achievement to agent *b*. A delegation is thus associated with a dependency, but it is more than that. As a material relation, it is founded on more that its connected elements. In this case, the connected elements are two agents, namely, the requestor (delegator) and the provider (delegatee), as well as a goal (delegatum), which is the delegation object that represents the needs of the requestor. The foundation of this material relation is the social relator (a pair of commitments and claims), i.e. the agreement established between the two agents involved in this delegation, entitled requestor and provider. In other words, when a requestor delegates a goal to a provider, besides the fact that the requestor depends on the provider with regard to the goal, the provider commits itself to achieving the goal on behalf of the requester. This commitment is established by means of an agreement. As described in Calvi (2007), an IT service delegation (ITSD) is a delegation committed to achieve a goal according to a specific plan, the IT service. Therefore, according to UFO, an IT service delegation is a closed delegation, since it describes a specific plan which the provider should adopt to achieve the goal delegated by the requestor. In summary, if there is an agreement between the requestor and the provider which causes the execution of a service, then there will be a service delegation associated with the agreement. In addition, this delegation will be committed with the service and consequently with its execution. In this context, the requestor assumes the role of delegator while the provider assumes the role of delegatee. This relation is formalized by the axiom A6.

(A6) $\forall$x,y,z (Requestor(x) $\wedge$ ITSP(y) $\wedge$ Agreement(z) $\wedge$ mediates(z,x) $\wedge$ mediates(z,y) $\rightarrow$ $\exists$w (ITSD(w) $\wedge$ associated-to(w,z) $\wedge$ delegator-of(x,w) $\wedge$ delegatee-of(y,w)))

As defined by UFO, when an agent is said to be able to achieve a certain goal it means that such agent can achieve this goal by itself or else delegate it to another agent that would be able to achieve it on its behalf. Thus, when a provider receives a delegation by the means of a service-level agreement, this provider analyses the delegated service and, if needed, delegates that service to other service providers, termed internal providers (e.g. an IT infrastructure department) and external providers (e.g. suppliers). In this manner, each subservice execution contributes to the service delegated by the customer, in this case, the super-service.

In view of this discussion, it is defined the relationship between business process and IT service inherent to the competency question CQ1. In summary, information technology is frequently used to support the business process activities through IT services. Given that a business process activity is an activity that is owned and performed by the business (commonly by a business unit) and an IT service is a service provided by an IT organization, the relation between business process and IT service occurs as a result of the dependence relation between the respective agents, which means business unit and IT organization. In other words, the fact of a certain business process activity occurrence being performed by the business denotes a social commitment of this agent in performing this occurrence. This social commitment contains a propositional content, i.e. a goal. Thereby this agent has a commitment to perform a certain action which satisfies this goal. In case this agent, i.e. the business, needs an IT service to achieve such a goal, this agent can delegate the goal (or part thereof) to the agent responsible for providing the IT service. In this context, the agent responsible for the business process activity assumes the role of requestor (more specifically, the role of customer) and the agent responsible for providing the service assumes the role of IT service provider. Therefore, the relation between business process and IT service is derived from the dependency relation between its respective agents, leading to a delegation relation. In this sense, it is worth noting the UFO contribution as it fundaments the entire service delegation process. Regarding the relationship between IT services and IT components, which is inherent to the competence question CQ2, the discussions during this section have defined that such components are seen as resources under the UFO point of view, assuming the usage participation. Moreover, in considering the definition of hardware and software presented by IEEE 610.10 (IEEE, 1994) as well as by ISO/IEC 2382-1 (ISO/IEC, 1993) it is possible to conclude that a hardware is a physical component that processes the instructions described by a software. In terms of UFO, a software is a type of normative description which describes a computer process, i.e. a type of event (an event universal). An instance of this process denotes an occurrence of such a process, termed computer process occurrence, i.e. a predetermined course of events whose execution includes the participation of a hardware. In this sense, the concept of hardware, as well as of software, is mapped to the notion of substantial. As such, if a service execution requires a software resource which is processed by a hardware resource, then this service execution also requires this hardware resource. In other words, if a service execution x requires a software resource y and this last resource describes a computer process z whose occurrence w includes the participation of a hardware q, then the service execution x also requires the hardware resource q, as formalized by the axiom A7.

(A7) ∀x,y,z,w,q (ITSE(x) ∧ Software(y) ∧ Computer-Process(z) ∧ Computer-Process-Occurrence(w) ∧ Hardware(q) ∧ requires(x,y) ∧ describes(y,z) ∧ instance-of(w,z) ∧ participation-of(w,q) → requires(x,q))

In this sense, it is useful to note the UFO contribution, as it allows the distinction between the notions pertinent to software, its processes and its occurrences, as well as the proper participations of hardware components, granting more expressiveness, clarity and veracity to the model. Indeed, the use of appropriate tools such as UFO and methods such as SABiO promoted o development of important ontological distinctions, as discussed in this section. Considering the ontology engineering approach adopted in this chapter, the next section explores the benefits from the implementation of the conceptual model.

5. An implementation and application of the proposed ontology

In view of the contributions that ontologies provide towards automated solutions, including important features such as artificial intelligence and, above all, interoperability, this section aims to present a case study in order to: (i) perform a proof of concept of the models developed in the previous section and also (ii) demonstrate how the concepts modeled by using ontologies can be implemented and applied in a computational system, in order to enable automation, including important features such as cited in this paragraph. Taking into accounting Figure 3, which illustrates the ontology development approach adopted in this work, the ontology proposed in this section concerns to the implementation model, which represents an implementation of the conceptual model proposed in the previous section.

As discussed in Guizzardi (2005, 2007), each ontology engineering phase requires the use of appropriate languages in relation to the context within which the model is being designed. In the implementation phase, the choice of a language must be conducted by the end-application requirements. This refers to languages focused on computational requirements, such as decidability and efficient automated reasoning. In terms of the configuration management process, factors such as decidability, completeness and expressiveness are considered to be key requirements because of its role in relation to all other processes in service management. Thus, for the implementation of the conceptual models, this work used the OWL DL sublanguage, since it allows a greater degree of expressiveness, as compared with OWL Lite, while maintaining computational guarantees such as completeness and decidability, features not guaranteed by OWL Full (Bechhofer et al., 2004). In addition to OWL DL, the implementation of the models also used SWRL (Horrocks et al., 2003). The SWRL language allows the representation of the axioms defined in the conceptual models presented in Section 4 in an integrated way with the concepts and relations implemented by means of OWL. Finally, for the implementation of the models in OWL and the definition of the axioms in SWRL, this work used the Protégé tool (Protégé, 2011), an ontology editor that enables the integration of different languages such as OWL and SWRL inside the same implementation environment.

Once defined the development environment, the implementation models were developed according to the modular structure of the conceptual models, as follows: (i) UFO.owl; (ii) BusinessProcess.owl; (iii) ITService.owl; (iv) ITComponents.owl and finally (v) ConfigurationItem.owl. Due to the expressivity restrictions inherent in the implementation languages, the main issue concerning the mapping from conceptual models into implementation models is related to the treatment of the reduction in semantic precision. In order to maintain this reduction at an acceptable level, the most relevant losses that were found were related to the transformation of all ontologically well-founded concepts and relations into OWL classes and properties, respectively. Regardless of the application scenario, this mapping must consider the information contained in the notation used for the development of the conceptual models, such as cardinality, transitivity, domain and range. With respect to cardinality and transitivity, in OWL it is not possible to represent them simultaneously (Bechhofer et al., 2004). As a result, this work considers that the representation of cardinality restrictions is more relevant to the implementation models developed in this section. In addition, to represent the cardinality restrictions in both directions inverse relations were used. For instance, the relation "requests" is represented by the pair of relations "requests" and "is requested by". However, according to Rector and Welty (2001), the use of inverse relations significantly increases the complexity of automated reasoning. Thus, they should be used only when necessary. With respect to domain and range, an issue that should be considered is how to organize and represent many generic relations. For example, if a generic relation "describes" is created, it is not possible to restrict the domain and the range. In this case, the design choice was to use specific relations like "describes_Software_ComputerProcess", which is represented as a sub-relation of a generic relation "describes". Finally, with respect to SWRL restrictions, this language has neither negation operators nor existential quantifiers (Horrocks et al., 2003). In addition, the SWRL language might lead to undecidable implementation models. Nevertheless, this issue may be worked around by restricting the use of rules and manipulating only those that are DL-safe (Motik et al., 2005). As an attempt to make this tangible, consider an implementation of the axiom A7, which concerns the competence question QC2, discussed in Section 4. This implementation is represented by the rule R7a.

(R7a) IT_Service_Execution(?IT-SERVICE-EXECUTION) ∧ Software(?SOFTWARE) ∧ ComputerProcess(?COMPUTER-PROCESS) ∧ ComputerProcessOccurrence(?COMPUTER-PROCESS-OCCURRENCE) ∧ Hardware(?HARDWARE) ∧ requires_ITServiceExecution_Resource(?IT-SERVICE-EXECUTION,?SOFTWARE) ∧ describes_Software_ComputerProcess(?SOFTWARE,?COMPUTER-PROCESS) ∧ isInstanceOf_ComputerProcessOccurrence_ComputerProcess(?COMPUTER-PROCESS-OCCURRENCE,?COMPUTER-PROCESS) ∧ hasParticipationOf_ComputerProcessOccurrence_Hardware(?COMPUTER-PROCESS-OCCURRENCE,?HARDWARE) → requires_ITServiceExecution_Resource(?IT-SERVICE-EXECUTION,?HARDWARE)

The axiom A7 constitutes the set of axioms that establishes the relationship between the computational resources that are required by an IT service execution as a response to the competence question QC2. Thus, this axiom involves concepts such as IT service execution, hardware and software, as well as the interrelationship between these concepts, such as the relations "requires" and "describes". As discussed earlier, concepts are implemented as classes, while relations are implemented as properties, according to OWL. According to described throughout this work, the same conceptual model can give rise to a variety of implementation models in order to meet different requirements, in accordance with the purpose of the application scenario. The rule R7a is intended to meet configuration management activities, especially the activities of identification and inference of managerial information. In Baiôco and Garcia (2010), the axiom A7 is implemented mainly in order to meet the latter activity. In this case, the axiom is implemented as formalized by the rule R7b.

(R7b) IT_Service_Execution(?IT-SERVICE-EXECUTION) ∧ Software(?SOFTWARE) ∧ ComputerProcess(?COMPUTER-PROCESS) ∧ ComputerProcessOccurrence(?COMPUTER-PROCESS-OCCURRENCE) ∧ Hardware(?HARDWARE) ∧ requires_ITServiceExecution_Resource(?IT-SERVICE-EXECUTION,?SOFTWARE) ∧ describes_Software_ComputerProcess(?SOFTWARE,?COMPUTER-PROCESS) ∧ isInstanceOf_ComputerProcessOccurrence_ComputerProcess(?COMPUTER-PROCESS-OCCURRENCE,?COMPUTER-PROCESS) ∧ hasParticipationOf_ComputerProcessOccurrence_Hardware(?COMPUTER-PROCESS-OCCURRENCE,?HARDWARE) → query:select(?IT-SERVICE-EXECUTION,?SOFTWARE, ?HARDWARE) ∧ query:orderByDescending(?IT-SERVICE-EXECUTION)

There are numerous contributions offered by ontology engineering for the construction of autonomous, intelligent and above all interoperable computational applications. Although done in a different area, Gonçalves et al. (2008) presents an application for the interpretation of electrocardiogram results where the use of an ontology model provides a graphical simulation of the heart behavior of an individual and the correlation of the heart behavior with the known pathologies. Regarding configuration management, this process identifies, controls, maintains and checks the versions of the existing configuration items and reports the information of the IT infrastructure to all those involved in the management. Thus, this section aims to demonstrate how implementation models can be applied in a computational environment in order to support management activities in an automated manner. In addition, the results will provide a proof of concept of the developed ontology.

The first part comprises the mapping between business processes activities and business units responsible for these activities. In addition, it includes the needs that characterize these business units. Figure 5-a shows, for example, that the business process activity BPAO_Sales is composed of the activity BPAO_Ordering, which is owned by the business unit BU_Sales, as shown in Figure 5-b. The business unit BU_Sales, in turn, has need inherent to this activity, as presented in Figure 5-c. It is worth mentioning that such information, as well as any assertion that appears highlighted in blue, concerns information previously inserted into the implementation models. On the other hand, assertions that appear highlighted in yellow are information automatically inferred by the implementation models, which denote knowledge acquisition. Such inferences are performed by the Pellet reasoner (Sirin et al., 2007).

As discussed in Section 4, IT services can achieve business needs by supporting its activities. Thus, Figure 5-d illustrates the IT services that can achieve the needs of the business. This information is inferred by executing the axiom A1, implemented in this section. Figure 5-d illustrates, for example, that the service IT_Service_Ordering can achieve the need

Need_Ordering. Figure 5-e, in turn, presents the requirements used for service requests. This inference is performed by executing the implemented axiom A2. This scenario motivates the requestor, in this case the business unit, to ask for services which can achieve its needs, as shown in Figure 5-f. This inference is performed by means of the implemented axiom A3. It should be mentioned that information of this nature is fundamental to processes such as service level management, which interacts with configuration management in requesting information in order to find services able to meet the needs of the requestors.

According to the axiom A4, once a service capable meeting the need is found, the requestor can then initiate the delegation process by contracting the service. Thus, Figure 5-g shows the provider hired by the business unit. As shown in Figure 5-g, the provider hired by the business unit BU_Sales is the IT_Department, because it is the provider responsible for providing the requested service. As discussed in Section 4, the hiring process is mediated by an agreement, as illustrated in Figure 5-h. This inference is performed by means of the implemented axiom A5. Figure 5-h illustrates, for example, that the agreement SLA_Ordering mediates the requestor BU_Sales and the provider IT_Department. This agreement, in turn, characterizes the delegation process, which has a delegator (in this case the requestor) and a delegatee (in this case the provider), as shown in Figure 5-i. This inference is performed by means of the implemented axiom A6. Figure 5-i illustrates, for example, that the delegation ITSD_Ordering is associated to the agreement SLA_Ordering and has as the delegator the BU_Sales and as the delegatee the IT_Department.

As described in Section 4, the hired provider receives the service delegation from the requestor and provisions the necessary resources for the service execution. In this sense, the received service execution is characterized as a complex action which is delegated to the support groups by means of subservices. In this context, the hired provider, in a manner similar to that of the business unit, plays the role of requestor and the support groups, in turn, play the role of service provider. Thus, Figure 5-j shows, for instance, that the execution ITSE_Ordering, which represents an instance of the service IT_Service_Ordering, is composed of the sub-executions ITSE_Ordering_Processing and ITSE_Ordering_Printing. The delegation performed by the provider to the support groups is mediated by agreements. Figure 5-k shows, for example, that the agreement OLA_Ordering_Processing mediates the delegation process of the sub-execution ITSE_Ordering_Processing between the IT_Department and the IT_Department_System. In this case, IT_Department plays the role of delegator while the support groups play the role of delegatee, as shown in Figure 5-l.

IT services are based on the use of information technology. Thus, Figure 5-m relates the software required by each service execution. Figure 5-m shows, for example, that the execution ITSE_Ordering_Processing requires the software Software_Ordering_Processing. As discussed in Section 4, software is processed by hardware. Thus, Figure 5-n presents the hardware associated with the processing of the concerned software. In addition, Section 4 states that if a service execution requires a software and this software is processed by a hardware, then this service execution also requires this hardware, as illustrated in Figure 5-o. This inference is performed by means of the implemented axiom A7. Figure 5-o illustrates, for example, that the execution ITSE_Ordering_Processing requires the hardware Hardware_Sales, since such hardware processes the software Software_Ordering_Processing (as shown in Figure 5-n) required by such an execution.

Thus, this case study concludes the mapping between business and IT. This mapping is fundamental to other management processes. As an illustration, the configuration management process of this case study is able to correlate and determine that a particular event of unavailability on the hardware Hardware_Sales affects the software Software_Ordering_Processing, which is used by the service execution ITSE_Ordering_Processing. This execution is part of the execution ITSE_Ordering which is instance of the service IT_Service_Ordering and, in turn, supports related activity of the business process activity BPAO_Sales. This correlation, provided by the implementation model, is the basis for activities such as: (i) event correlation in event management; (ii) workaround identification in incident management; (iii) root cause analysis in problem management and (iv) impact analysis in change management.

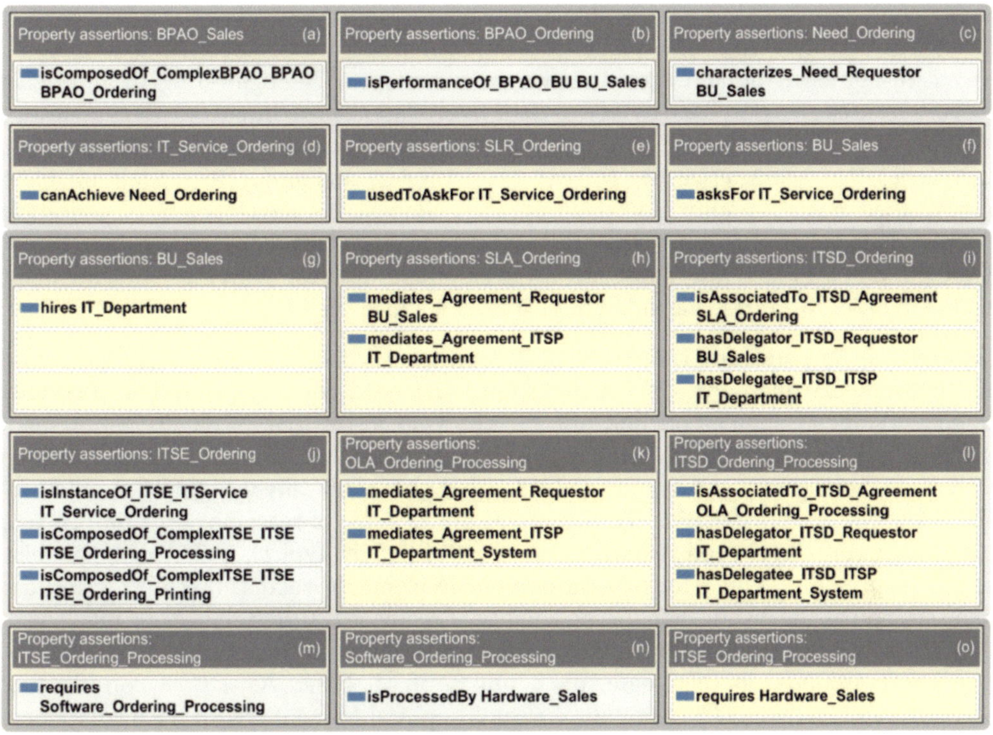

Fig. 5. Application of the implementation model.

To complete this case study, consider intelligent software agents playing the roles of requestor and provider and, consequently, negotiating the provision of services that meet the needs of the environment. Regarding the role of provider, such activities refer to the various management disciplines, as presented throughout this section. In particular, this scenario denotes implementation models subsidizing paradigms known as autonomous networks. In general, it denotes implementation models promoting automation in various areas of interest.

6. Conclusion

As discussed in this chapter, automation enables organizations to explore opportunities as well as supporting challenges in an effective and efficient way. The important role played by

IT as an instrument for automation has made automation increasingly dependent on IT, consequently rising the demands for advances, as observed by the growing challenges arising from the conception of systems continuously more complex, intelligent and, above all, interoperable. Moreover, the need for an efficient and effective IT management has grown substantially, as evidenced by widespread adoption of innovative best practices libraries and standards. For this reason, this work presented an ontology of IT service configuration management. The objective was not only to adopt the state of the art in order to address key research challenges in IT management, but also to foment novel approaches which can be applied in IT in various areas of interest.

The diverse uses attributed to the ontologies in computer science and the interrelation between their purposes promote the search for approaches capable of providing the construction of ontological models able to achieve the various objectives assigned to them. Based on innovative and high quality research initiatives, this chapter discussed about a systematic approach for building ontologies known as Ontology Engineering. In considering the various uses and purposes, as well as their interrelationships, these initiatives attempt to establish a structured means of development as an alternative to the various *ad hoc* approaches that characterize the current developments and imply in models unable to achieve their goals. In summary, this approach allowed the development of conceptual models which are application-independent artifacts and, as a result, it enabled their use as a reference ontology for the subsequent development phases, deriving implementation models in order to address the different purposes of end applications.

According to Guizzardi and Halpin (2008), the practice of conceptual modeling is permeated by philosophical questions. This demonstrates the need for an appropriate theoretical foundation for conceptual modeling languages so as to ensure that the quality requirements of domain and comprehensibility appropriateness can be fulfilled by the produced conceptual models. In this sense, they advance that philosophically well-founded ontologies play a key role in this initiative. They complement this line of reasoning by citing Guarino and Guizzardi (2006) and emphasizing that although typical conceptual modeling languages provide facilities for structuring domain elements, such as taxonomies and data value structures, the justification for the validity of many structuring choices, as much as the justification for the grammar of many natural language sentences, can only be made on ontological grounds, in this sense, on a philosophical basis. As a final consideration, Guizzardi and Halpin (2008) point out that philosophical foundations are vital components with respect to conceptual modeling, in general, and domain ontology engineering, in particular, as mature disciplines with sound principles and practices. Thus, in quoting the physicist and philosopher of science Mario Bunge, "every science presupposes some metaphysics", they conclude that a scientific field can either choose to develop and make explicit its philosophical foundations or to remain oblivious to its inevitable and often *ad hoc* ontological and epistemological commitments.

Accordingly, in addition to the appropriate methods and techniques, such as the SABiO method, this approach used a philosophically well-founded ontology, termed UFO. The SABiO method provided a systematic approach that led the development of ontology proposed in this chapter, describing an iterative process, closely related to evaluation. Thus, with emphasis on the concept of competence questions, the SABiO method provided a means for defining the scope and purpose of the ontology, leading its capture and formalization, the reuse of existing ontologies, as well as its evaluation and documentation, enabling the ontology proposed in this chapter to adequately meet the requirements for which it was designed, as confirmed by proof of concept. The UFO ontology, in turn, was useful in building a conceptual model committed to maximizing the expressivity, clarity and truthfulness of the modeled domain. These characteristics are key quality attributes of a conceptual model, responsible for its effectiveness as a reference framework for the tasks of semantic interoperability and reuse. In fact, as a knowledge representation language philosophically well-founded, the UFO ontology aims to provide a sound basis for the representation of a conceptualization and therefore inhibit arbitrary descriptions of concepts and relationships of a universe of discourse. As discussed during the development of the conceptual models proposed in this chapter, the UFO ontology guided diverse modeling decisions, contributing to the derivation of new knowledge or the identification and elucidation of ambiguous and inconsistent representations of the domain, often represented in various literatures.

As discussed in this chapter, the more it is known about a universe of discourse and the more precisely it is represented, the bigger the chance of producing models that reflect, as much as possible, the appropriate conceptualization of the domain. In this sense, besides the use of appropriate languages, such as ontologies philosophically well-founded, as a common and shared specification, it is important that the specification of the domain considers appropriate literatures and, especially, that it considers its main concepts and relationships as well as application independence. Considering the universe of study of this chapter, these aspects associated with the use of appropriate methods and tools enabled the development of models able to maximize the alignment between IT and business for humans and computers. Moreover, these aspects allude to a point of view which should be mentioned. By maximizing the capacity of a model in acting as a common and shared source about a universe of discourse, conceptual models, representing norms and standards, can be potentially used as an addendum to such literatures. This is because, in general, these libraries are described in natural languages, which are susceptible to ambiguities and inconsistencies, as opposed to conceptual models, which are formally described.

Considering the importance of automation, as well as the contributions that an implementation provides in terms of ontology evaluation, it was developed an implementation model, derived from the conceptual model proposed in this chapter. In addition, by applying the entire approach discussed in this chapter, it is possible to attest it, as well as making it more tangible, promoting its benefits. It should be mentioned that the development of conceptual models followed by the development of implementation models became evident the distinction between ontology representation languages, as discussed throughout this chapter. This demonstrates that the approach adopted in this chapter shows itself appropriate by considering the various uses and purposes assigned to ontologies. In this way, despite the expressivity restrictions inherent in implementation languages, it was possible to perform a proof of concept of the ontology developed in this chapter as well as demonstrating how such models can be derived and implemented in computing environments with a view to the different computational requirements. In particular, it was possible to show how implementation models can support automation, including special characteristics such as knowledge acquisition and interoperability. From this point of view it is important to note that the concepts inherited from UFO are important in promoting paradigms such as artificial intelligence by making explicit, for computational agents, concepts that express the daily lives of human agents, such as intentions, goals and actions. These factors are especially important in face of the increasing need for integration of complex as well as heterogeneous systems and also when considering autonomous systems.

Therefore, the contributions of this chapter are not restricted to the domain of configuration management. Instead, they promote semantic interoperability in IT in diverse areas of interest, maximizing the advances in automation. Additionally, this chapter makes possible other researches. In this sense, future works include: (i) the extension of the ontology for covering other business-driven IT management concepts, such as IT services metrics and business measures as well as their relationships, improving the alignment between these two domains; (ii) the extension of the ontology in order to cover other configuration management concepts, such as baseline, version and variant; (iii) the extension of the ontology to cover other management process, such as change management and release management; (iv) the application of conceptual models as an addendum to norms and standards; (v) the application of the implementation models, especially with techniques like artificial intelligence, promoting paradigms such as autonomous networks.

7. References

Baiôco, G., Costa, A.C.M., Calvi, C.Z. & Garcia, A.S. (2009). IT Service Management and Governance - Modeling an ITSM Configuration Process: a Foundational Ontology Approach, In *4th IFIP/IEEE International Workshop on Business-driven IT Management, 11th IFIP/IEEE International Symposium on Integrated Network Management*, 2009

Baiôco, G. & Garcia, A.S. (2010). Implementation and Application of a Well-Founded Configuration Management Ontology, In *5th IFIP/IEEE International Workshop on Business-driven IT Management (BDIM), 12th IFIP/IEEE Network Operations and Management Symposium (NOMS)*, Osaka, 2010

Bechhofer, S. et al. (2004). OWL Web Ontology Language Reference, In *W3C Recommendation*, Oct 2011, Available from http://www.w3.org/TR/owl-ref/

Brenner, M., Sailer, M., Schaaf, T. & Garschhammer, M. (2006). CMDB - Yet Another MIB? On Reusing Management Model Concepts in ITIL Configuration Management, In *17th IFIP/IEEE Distributed Systems Operations and Management (DSOM)*, 2006

Calvi, C. Z. (2007). IT Service Management and ITIL Configuration Process Modeling in a Context-Aware Service Platform (in Portuguese), Master Dissertation, UFES, 2007

Costa, A.C.M. (2008) ITIL Service Level Management Process Modeling: An Approach Using Foundational Ontologies and its Application in Infraware Platform (in Portuguese), Master Dissertation, UFES, 2008

Falbo, R.A. (1998). Knowledge Integration in a Software Development Environment (in Portuguese), Doctoral Thesis, COPPE/UFRJ, 1998

Falbo, R.A. (2004). Experiences in Using a Method for Building Domain Ontologies, In *16th Conference on Software Engineering and Knowledge Engineering (SEKE)*, Canada, 2004

Gonçalves, B. et al. (2009). An Ontology-based Application in Heart Electrophysiology: Representation, Reasoning and Visualization on the Web, In *ACM Symposium on Applied Computing*, 2009

Gonçalves, B.N., Guizzardi, G. & Pereira Filho, J.G. (2011). Using an ECG reference ontology for semantic interoperability of ECG data, In *Journal of Biomedical Informatics*, Special Issue on Ontologies for Clinical and Translational Research, Elsevier, 2011

Guarino, N., Guizzardi, G. (2006). In the Defense of Ontological Foundations for Conceptual Modeling, In *Scandinavian Journal of Information Systems*, ISSN 0905-0167, 2006

Guizzardi, G. & Wagner, G. (2004). A Unified Foundational Ontology and some Applications of it in Business Modeling, In *Open INTEROP Workshop on Enterprise Modelling and Ontologies for Interoperability, 16th CAiSE*, Latvia, 2004

Guizzardi, G. (2005). *Ontological Foundations for Structural Conceptual Models*, Ph.D. Thesis, University of Twente, ISBN 90-75176-81-3, The Netherlands, 2005

Guizzardi, G. (2006). The Role of Foundational Ontology for Conceptual Modeling and Domain Ontology Representation, In *Companion Paper for the Invited Keynote Speech, 7th International Baltic Conference on Databases and Information Systems*, 2006

Guizzardi, G. (2007). On Ontology, ontologies, Conceptualizations, Modeling Languages, and (Meta)Models, In *Frontiers in Artificial Intelligence and Applications, Databases and Information Systems IV*, IOS Press, ISBN 978-1-58603-640-8, Amsterdam, 2007

Guizzardi, G., Falbo, R.A. & Guizzardi, R.S.S. (2008). The Importance of Foundational Ontologies for Domain Engineering: The case of the Software Process Domain (in Portuguese), *IEEE Latin America Transactions*, Vol. 6, No. 3, July 2008

Guizzardi, G. & Halpin, T. (2008). Ontological foundations for conceptual modeling, In *Journal of Applied Ontology*, v.3, p.1-12, 2008

Guizzardi, G., Lopes, M., Baião, F. & Falbo, R. (2009). On the importance of Truly Ontological Distinctions for Ontology Representation Languages: An Industrial Case Study in the Domain of Oil and Gas, In *Lecture Notes in Business Information Processing*, 2009

Horrocks, I. et al. (2003). SWRL: A Semantic Web Rule Language Combining OWL and RuleML, In *DAML*, Oct 2011, From http://www.daml.org/2003/11/swrl/

IEEE (1994). *IEEE Standard Glossary of Computer Hardware Terminology*, IEEE Std 610.10-1994

ISO/IEC (1993). *IT – Vocabulary – Part 1: Fundamental terms*, ISO/IEC 2382-1:1993

ISO/IEC (2005). *Information technology – Service management*, ISO/IEC 20000, 2005

ITIL (2007). *ITIL Core Books*, Office of Government Commerce (OGC), TSO, UK, 2007

Jones, D.M., Bench-Capon, T.J.M. & Visser, P.R.S. (1998). Methodologies For Ontology Development, In *15th IFIP World Computer Congress*, Chapman-Hall, 1998

Lopez de Vergara, J.E., Villagra, V.A. & Berrocal, J. (2004). Applying the Web Ontology Language to management information definitions, *IEEE Communications Magazine*

Majewska, M., Kryza, B. & Kitowski, J. (2007). Translation of Common Information Model to Web Ontology Language, In *International Conference on Computational Science*, 2007

Moura, A., Sauve, J. & Bartolini, C. (2007). Research Challenges of Business Driven IT Management, In *2nd IEEE/IFIP Business-driven IT Management, 10th IFIP/IEEE IM*

Moura, A., Sauve, J. & Bartolini, C. (2008). Business-Driven IT Management - Upping the Ante of IT: Exploring the Linkage between IT and Business to Improve Both IT and Business Results, In *IEEE Communications Magazine*, vol. 46, issue 10, October 2008

Motik, B., Sattler, U. & Studer, R. (2005). Query Answering for OWL-DL with Rules, In *Journal of Web Semantics*, Vol. 3, No. 1, pp. 41-60, 2005

Pavlou, G. & Pras, A. (2008). Topics in Network and Service Management, In *IEEE Communications Magazine*, 2008

Pras, A., Schönwälder, J., Burgess, M., Festor, O., Pérez, G.M., Stadler, R. & Stiller, B. (2007). Key Research Challenges in Network Management, *IEEE Commun Magazine*, 2007

Protégé (2011). The Protégé Ontology Editor and Knowledge Aquisition System, October 2011, Available from http://protege.stanford.edu/

Rector, A. & Welty C. (2001). Simple part-whole relations in OWL Ontologies, In *W3C Recommendation*, Oct 2011, Available from http://www.w3.org/2001/sw/BestPractices/OEP/SimplePartWhole

Sallé, M. (2004). IT Service Management and IT Governance: Review, Comparative Analysis and their Impact on Utility Computing, In *HPL-2004-98*, 2004

Santos, B.C.L. (2007). O-bCNMS: An Ontology-based Network Configuration Management System (in Portuguese), Master Dissertation, UFES, 2007

Sirin E. et al. (2007). Pellet: a Practical OWL-DL Reasoner, In *Journal of Web Semantics: Science, Services and Agents on the World Wide Web*, Vol. 5, No. 2, pp 51-53, 2007

Smith, B. & Welty, C. (Eds.). (2001). Ontology: Towards a new synthesis, Chris Welty and Barry Smith, Formal Ontology in Information Systems, ACM Press, 2001

Ullmann, S. (1972). Semantics: An Introduction to the Science of Meaning, Basil Blackwell, Oxford, 1972

Vermeer, M.W.W. (1997). Semantic interoperability for legacy databases, PhD Thesis, University of Twente, The Netherlands, 1997

Wong, A.K.Y., Ray, P., Parameswaran, N. & Strassner, J. (2005). Ontology Mapping for the Interoperability Problem in Network Management, *IEEE Journal on Selected Areas in Communications (JSAC)*, Vol. 23, No. 10, October 2005

Xu, H. & Xiao, D. (2006). A Common Ontology-Based Intelligent Configuration Management Model for IP Network Devices, *In Proceedings of the First International Conference on Innovative Computing, Information and Control (ICICIC)*, 2006

3

Automatic Restoration of Power Supply in Distribution Systems by Computer-Aided Technologies

Daniel Bernardon[1], Mauricio Sperandio[2],
Vinícius Garcia[1], Luciano Pfitscher[1] and Wagner Reck[2]
[1]*Federal University of Santa Maria*
[2]*Federal University of Pampa*
Brazil

1. Introduction

The need to improve quality and reliability of Power Systems has contributed to the advancement of research on Smart Grids and related topics. Some challenges that motivate the deployment of such researches include the increasing in energy consumption, environmental aspects, the integration of distributed and renewable generation, and new advances in computer aided technologies and automation. Smart Grids are characterized by a series of integrated technologies, methodologies and procedures for planning and operation of electrical networks. A survey of the main projects and researches related to Smart Grid is presented in (Brown, 2008).

Some of the main desired features in a Smart Grid are the low operating and maintenance costs and the ability to self-healing. In this context, utilities have concentrated significant efforts in order to improve the continuity of the supplied electrical energy, especially because of regulatory policies, besides the customer satisfaction and the improvement of the amount of energy available to commercial and industrial activities. However, supply interruptions are inevitable due to implementation of the expansion of the system, preventive maintenance on network equipment, or even by the action of protective devices due to defects.

A distribution network can have its topology changed by opening or closing switches, allowing to isolate faults and to restore the supply in contingency situations, and also in case of scheduled shutdowns. In addition, the change of topology allows a better load balancing between feeders, transferring loads of heavily loaded feeders to other feeders, thus improving the voltage levels and reducing losses and increasing levels of reliability.

The reconfiguration of a distribution network is considered an optimization problem in which the objective is to seek for one configuration, among several possible solutions, that leads to better performance, considering the ultimate goal of the reconfiguration and observing the network constraints. One factor that increases the complexity of the problem is the large number of existing switches in a real distribution network, which leads to a lot of different possible configurations to be analyzed.

The stages of planning the reconfiguration of distribution networks usually involve the collection of information of the network and load, the application of methodologies for estimating loads and calculation of power flows, and the application of the methodology for network optimization.

The behavior of the load is essential in the study of reconfiguration of distribution networks, because load variations cause demand peaks at different periods, and thus the reconfiguration at a particular period may not meet the objectives and constraints at later periods.

Several researches are related to the reconfiguration of distribution networks, based on mathematical methods, heuristics and artificial intelligence techniques, as shown in (Takur & Jaswanti, 2006). Generally, the reconfiguration aims to meet the maximal number of consumers as much as possible, or the maximum energy demand, with loss minimization and load balancing. The reduction of losses is often taken as the primary objective of the reconfiguration. When more than one objective function is defined, it is necessary to apply multicriteria methods for decision making. Among the restrictions, there is the need to maintain a radial network, and the operation within the limits of voltage and current capacity of equipment and conductors of the line. The coordination of the protection system of the new network configuration should be taken into consideration.

In recent years, new methodologies of reconfiguration of distribution networks have been presented, exploring the greater capacity and speed of computer systems, the increased availability of information and the advancement of automation, in particular, the SCADA (Supervisory Control and Data Acquisition). With the increased use of SCADA and distribution automation - through the use of switches and remote controlled equipment - the reconfiguration of distribution networks become more viable as a tool for planning and control in real time.

Act as soon as possible in a contingency situation may result in a minimum cost to the utility and consumers. When a fault is identified at any point of the network, the following procedures must be performed: identify the location of the problem; isolate the minimum portion of the distribution system by opening normally closed switches; restore the power supply to consumers outside the isolated block reclosing the feeder breaker and/or normally open switches; correct the problem; re-operate the switches to get back to the normal network status. To do that, generally a maintenance crew has to travel long distances, and may be affected by traffic jams and unfavorable accesses.

The automation of distribution systems, with the installation of remote controlled switches, plays an important role on reducing the time to implement a service restoration plan (Sperandio et al., 2007). These devices have shown to be economically viable due to the growth of a large number of automation equipment suppliers and new communication technologies.

The commitment of an efficient system to operate these devices is quite important for the utilities, aiming to guarantee the technical feasibility of the network reconfiguration with minimal time necessary to restore the energy supply of the affected consumers.

In this chapter, a methodology for the automatic restoration of power supply in distribution systems by means of remote controlled switches is presented. It includes a validation of the technical feasibility for the reconfiguration of the network in real time using computer simulations. Since there may be many configuration options with different gains, an algorithm based on a multiple criteria decision making method, the Analytic Hierarchical Process - AHP (Saaty, 1980), is employed to choose the best option for load transfers after contingencies. The AHP method has proven to be effective in solving multi-criteria problems, involving many kinds of concerns including planning, setting priorities, selecting the best choice among a number of alternatives and allocating resources. It was developed to assist in making decisions where competing or conflicting evaluation criteria difficult to make a judgment (Saaty, 1994). Additionally, the algorithms for load modeling and load flow are also presented, since they are essential for analysis of the maneuvers.

As a result, load transfers are carried out automatically, being preceded by computer simulations that indicate the switches to be operated and that ensure the technical feasibility of the maneuver, with the characteristics of agility and safety for the power restoration and in agreement with the Smart Grids concepts. The developed tool has been applied in a pilot area of a power utility in Brazil. The reduction in the displacement of maintenance teams and improving indices of continuity characterize the greatest benefits for the company, making a difference in the market and, consequently, generating economic and productivity gains.

2. Modeling of power load profiles

The most common information on power loads profile comes from utilities charges, based on measurements of monthly energy consumption. Unfortunately, these data are insufficient for the analysis of the distribution systems, since they do not reflect the daily power behavior. Therefore, a methodology for power load modeling is need, which usually employs typical load curves for their representation. A technique for building the typical load curves that has advantages in relation to the traditional statistical methods was presented in (Bernardon et al., 2008). It reduces the influence of random values and also the amount of measurements required to form a representative sample of the load types. Instead of using the simple average to determine the active and reactive power values for a typical load curve ordinate, the following equation is used:

$$X_t = \frac{1}{5} \cdot \left[2M\{X_t\} + 2Me\{X_t\} + Mo\{X_t\} \right] \tag{1}$$

where:
X_t - active (P_t) or reactive (Q_t) power value for the time t of the typical load curve;
$M\{X_t\}$ - sample average;
$Me\{X_t\}$ - sample median (central number of a sample);
$Mo\{X_t\}$ - sample mode (most frequently occurring value of a sample).

According to the monthly power consumption data and economic activity developed, each customer is associated to a typical load curve. Based on the load factor (LF) and monthly energy consumption values (W), the maximum power demand (P_{max}) for a k group of consumers is calculated:

$$P_{max_k} = \frac{W_k}{T \cdot LF_k} \tag{2}$$

The typical curves utilized are normalized in relation to the maximum active demand, and the load curve for a *k* group of consumers is built by multiplying each ordinate by this value:

$$P_{kUt} = P_{max_k} \cdot P^*_{kUt}$$

$$P_{kSt} = P_{max_k} \cdot P^*_{kSt} \tag{3}$$

$$P_{kDt} = P_{max_k} \cdot P^*_{kDt}$$

$$Q_{kUt} = P_{max_k} \cdot Q^*_{kUt}$$

$$Q_{kSt} = P_{max_k} \cdot Q^*_{kSt} \tag{4}$$

$$Q_{kDt} = P_{max_k} \cdot Q^*_{kDt}$$

where:

P^*_{kt} and Q^*_{kt} - active and reactive power values, normalized for ordinate *t* of typical curve *k*;

U, *S* and *D* – Working days, Saturdays, Sundays/holidays, respectively.

The integral load curves for Working days, Saturdays and Sundays for the distribution transformer *j* is done through the sum of the load curves of the different groups of N_k consumers connected to it:

$$P_{jUt} = \sum_{i=1}^{N_k} P_{iUt} \; ; \; P_{jSt} = \sum_{i=1}^{N_k} P_{iSt} \text{ and } P_{jDt} = \sum_{i=1}^{N_k} P_{iDt} \tag{5}$$

$$Q_{jUt} = \sum_{i=1}^{N_k} Q_{iUt} \; ; \; Q_{jSt} = \sum_{i=1}^{N_k} Q_{iSt} \text{ and } Q_{jDt} = \sum_{i=1}^{N_k} Q_{iDt} \tag{6}$$

Thus, it is possible to consider the load levels corresponding to the period in which the failure occurred in the distribution network. Usually, the load transfers are analyzed considering the fault's time of the occurrence and the next four consecutive hours, thus ensuring the technical feasibility of the load transfers until the network returns to its original configuration.

3. Load flow method

A version of the classical backward/forward sweep method by (Kersting & Mendive, 1976) was performed to calculate the load flow in radial distribution networks. Since the electrical loads are defined by a constant power according to the applied voltage, the circulating current varies with the voltage drops. So, the solution is found only iteratively. The resulting procedure is described as follows:

Step 1. It is considered that the voltage in all points of the feeder is the same as the voltage measured in the substation bar. This information can be automatically received by the remote measurement systems installed at the substations. Voltage drops in the branches are not taken into account in this step.

Step 2. Active and reactive components of the primary currents absorbed and/or injected in the system by the electrical elements are calculated.

Step 3. The procedure to obtain the current in all network branches consists of two stages: (a) a search in the node set is performed adding the current values in the set of branches and (b) currents from the final sections up to the substation are accumulated.

Step 4. Voltage drops in primary conductors are determined.

Step 5. From the substation bus it is possible to obtain the voltage drops accumulated at any other part of the primary network, and, consequently, the voltage values at any point.

Step 6. The difference between the new voltage values for all nodes and the previous values is checked. If this difference is small enough comparing to a previously defined threshold, the solution for the load flow calculation was found and the system is said to be convergent. Otherwise, steps 2 to 6 are repeated, using the calculated voltages to obtain the current values. The threshold of 1% was chosen, because it leads to accurate values for the status variables without requiring too much processing time.

At the end of the process, the active and reactive powers and the technical losses in the primary conductors are defined for all branches of the feeder.

This load flow method was implemented in the proposed methodology for the automatic power supply restoration. It was used for analyzing the technical feasibility of the load transfers and the results were considered as constraints in the optimization procedure. That is, such transfers may neither cause an overload on the electrical elements (conductors and transformers), nor reach the pickup threshold of the protective devices, nor exceed the limits of voltage range of the primary network. The checking of the constraints is performed by considering the load profile compatible with the period of the failure.

4. Methodology for automatic operation of remote controlled switches to restore power supply

The logic for power restoration is presented considering the hypothetical example of the simplified distribution network illustrated in Figure 1. Normally close (NC) switch and normally open (NO) switches in Figure 1 are remotely controlled. Assuming that an outage has occurred in feeder FD-1, the procedure for electric power restoration is:

- Fault downstream of NC-1 switch: in the event of this fault, the current values of short-circuit will be flagged online in the SCADA (Supervisory Control and Data Acquisition) system. So, it is assumed that the failure occurred downstream of NC-1 switch; then, this switch is operated automatically to isolate the defect.

- Fault upstream of NC-1 switch: in the event of this fault, the current values of short-circuit will not be flagged in the SCADA system. So, it is assumed that the failure occurred upstream of NC-1 switch, automatically operating the remote controlled switches to open NC-1 and to close NO-1 or NO-2 in order to transfer consumers downstream of NC-1 to another feeder.

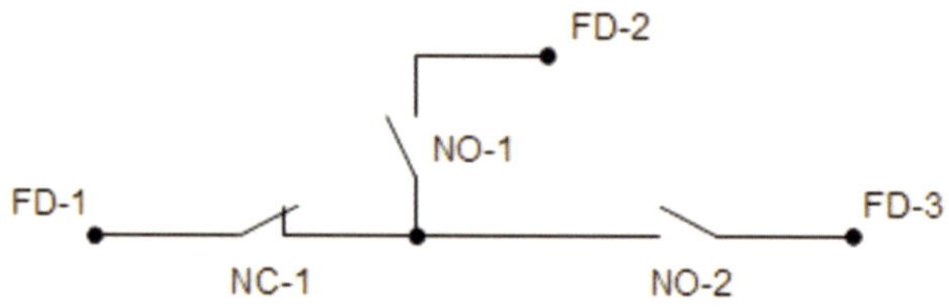

Fig. 1. Example of switches in a distribution network (FD - feeder, NC - normally closed switch, NO - normally open switch).

The technical and operational feasibility of load transfers using the remote controlled switches is verified by computer simulations. If there is more than one option of load transfer (e.g. to FD-2 or FD-3), the best option will be chosen considering the defined objective functions and constraints by a multiple criteria decision making algorithm. After this analysis, the developed tool automatically sends the necessary commands to maneuver the equipment.

Moreover, the automatic operation of remote controlled switches are carried out only after all attempts to restart the protection devices have been tried, i.e., they are done only in case of permanent fault, after a maximum of 3 minutes needed to complete the computer simulations and maneuvers of the switches since the instant of the fault identification. Figure 2 illustrates the architecture of the proposed system.

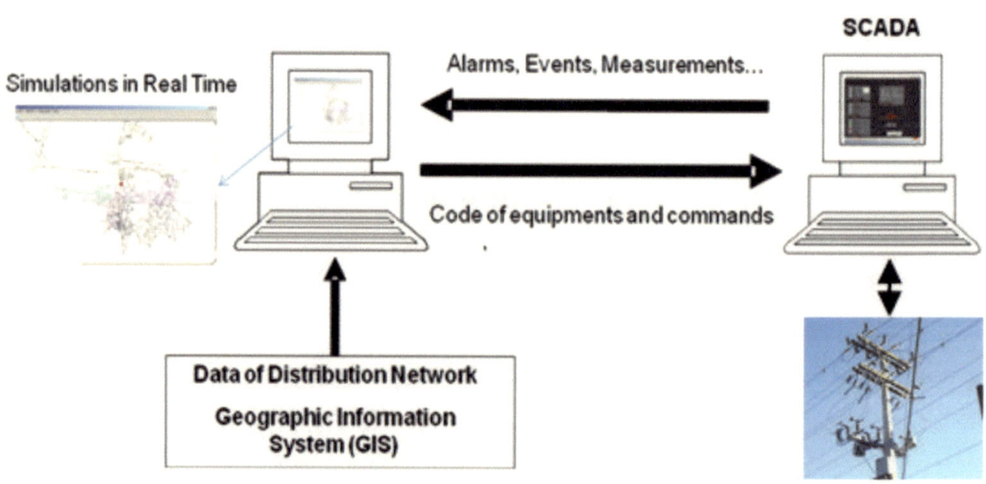

Fig. 2. Architecture of the developed system.

The flowchart of the proposed methodology is shown in Figure 3.

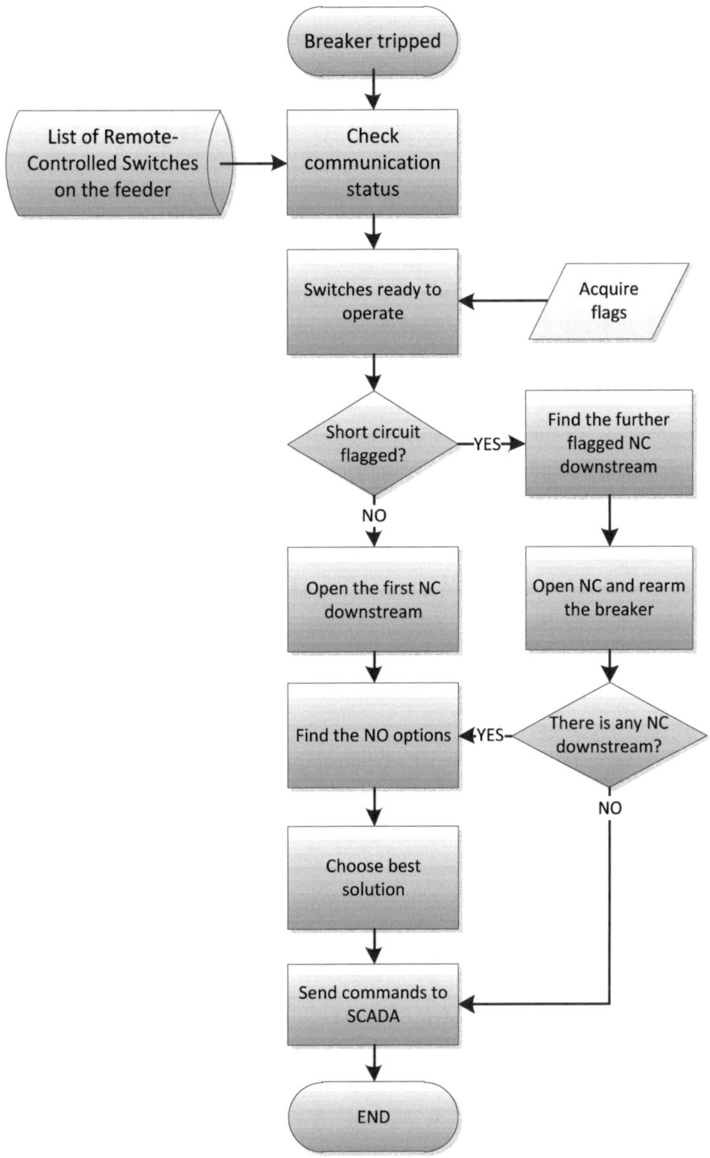

Fig. 3. Flowchart of the proposed methodology.

5. Methodology to determine the optimized reconfiguration based on the AHP method

After a contingency, the challenge is to decide which is the best load transfer scheme using the remote controlled switches among all possibilities, depending on previously defined objective functions and constraints. This is a multiple criteria decision making problem, since various types of objective functions can be considered.

The most common objectives are the maximization of restored consumers and of restored energy; however, generally it is not possible to optimize the grid for both objectives simultaneously. Furthermore, it is also important to ensure the reliability of distribution systems, through continuity indicators. The basic parameters are the SAIDI (System Average Interruption Duration Index) and the SAIFI (System Average Interruption Frequency Index), according to (Brown, 2009). In this approach, we adopt expected values based on the system's failure probability.

The constraints considered are the maximum loading of electrical elements, the protection settings and the allowable voltage drop in the primary network. Typically, the last two restrictions are the hard ones. On the other hand, a percentage of overloading of the network elements is acceptable in a temporary situation, assuming that the fault can be fixed in a couple of hours.

In our approach the following objective functions and constraints were defined to be used in the analysis of load transfers in case of contingencies:

Objective functions:

- Maximization of the number of restored consumers;
- Maximization of the amount of restored energy;
- Minimization of the expected SAIFI:

$$\text{ESAIFI} = \frac{\text{Expected Total Number of Customer Interruptions}}{\text{Total Number of Customers Served}} \tag{7}$$

Constraints:

- Current magnitude of each element must lie within its permissible limits:

$$|I_i| \le I_{i\,max} \tag{8}$$

- Current magnitude of each protection equipment must lie within its permissible limits:

$$|I_i| \le I_{jprot} \tag{9}$$

- Voltage magnitude of each node must lie within its permissible ranges:

$$V_{jmin} \le V_j \le V_{jmax} \tag{10}$$

where:

ESAIFI - expected value of system average interruption frequency (failures/year);

I_i - current at branch i;

$I_{i\,max}$ - maximum current accepted through branch i;

I_{jprot} - pickup current threshold of the protection device j;

V_j - voltage magnitude at node j;

V_{jmin} - minimum voltage magnitude accepted at node j;

V_{jmax} - maximum voltage magnitude accepted at node j.

The verification of the objective functions and constraints is made by calculation of the load flow for the various alternatives in real time. The ESAIFI is obtained by applying the classical equations of reliability during the process of calculating the load flow (Tsai, 1993).

Identifying the best option for load transfers is not simple since three objective functions are employed. For example, one particular option may have the largest number of consumers to be transferred, the other the largest amount of energy to be transferred, and the other the least expected value of consumers interrupted per year.

To solve this, the Analytic Hierarchical Process - AHP method was chosen, because of its efficiency in handling quantitative and qualitative criteria for the problem resolution. The first step of the AHP is to clearly state the goal and recognize the alternatives that could lead to it. Since there are often many criteria considered important in making a decision, the next step in AHP is to develop a hierarchy of the criteria with the more general criteria at the top of it. Each top level criteria is then examined to check if it can be decomposed into subcriteria.

The next step is to determine the relative importance of each criterion against all the other criteria it is associated with, i.e., establish weights for each criterion. The final step is to compare each alternative against all others on each criterion on the bottom of the hierarchy. The result will be a ranking of the alternatives complying with the staged goal according to the defined hierarchy of the criteria and their weights (Baricevic et al., 2009).

In the proposed approach, the main criterion is to choose the best option for load transfers and the subcriteria are the proposed objective functions. The alternatives are the options for load transfers.

An example of the AHP algorithm is defined in (Saaty, 1980):

1. The setup of the hierarchy model.
2. Construction of a judgment matrix. The value of elements in the judgment matrix reflects the user's knowledge about the relative importance between every pair of factors. As shown in Table 1, the AHP creates an intensity scale of importance to transform these linguistic terms into numerical intensity values.

Intensity of Importance	Definition
1	Equal importance
3	Weak importance of one over another
5	Essential or strong importance
7	Very strong or demonstrated importance
9	Absolute importance
2, 4, 6, 8	Intermediate values between adjacent scale values

Table 1. Intensity scale of importance (Yang & Chen, 1989).

Assuming $C_1, C_2, \ldots, C_n$ to be the set of objective functions, the quantified judgments on pairs of objectives are then represented by an n-by-n matrix:

$$\mathbf{M} = \begin{matrix} C_1 \\ C_2 \\ \vdots \\ C_n \end{matrix} \begin{bmatrix} 1 & a_{12} & \cdots & a_{1n} \\ 1/a_{12} & 1 & \cdots & a_{2n} \\ \vdots & \vdots & \ddots & \vdots \\ 1/a_{1n} & 1/a_{2n} & \cdots & 1 \end{bmatrix} \quad (11)$$
$$\qquad\qquad C_1 \quad C_2 \quad \cdots \quad C_n$$

Where n is the number of objective functions and the entries $a_{i,j}$ (i, j = 1, 2, …, n) are defined by the following rules:

- if $a_{i,j} = \alpha$, then $a_{j,i} = 1/\alpha$, where α is an intensity value determined by the operators, as shown in Table 1;
- if C_i is judged to be of equal relative importance as C_j, then $a_{i,j} = 1$, and $a_{j,i} = 1$; in particular, $a_{i,i} = 1$ for all i.

3. Calculate the maximal eigenvalue and the corresponding eigenvector of the judgment matrix **M**. The weighting vector containing weight values for all objectives is then determined by normalizing this eigenvector. The form of the weighting vector is as follows:

$$W = \begin{bmatrix} w_1 \\ w_2 \\ \vdots \\ w_n \end{bmatrix} \quad (12)$$

4. Perform a hierarchy ranking and consistency checking of the results. To check the effectiveness of the corresponding judgment matrix an index of consistency ratio (CR) is calculated as follow (Saaty & Tran, 2007):

$$CR = \frac{\left(\dfrac{\lambda_{max} - n}{n-1} \right)}{RI} \quad (13)$$

where:
λ_{max} = the largest eigenvalue of matrix **M**;
RI = random index.

A table with the order of the matrix and the RI value can be found in Saaty, 1980. In general, a consistency ratio of 0.10 or less is considered acceptable.

The AHP method was implemented in the proposed methodology. Its application is presented considering the example of distribution network illustrated in Figure 4. Normally close (NC) switch and normally open (NO) switches are remotely controlled.

The main goal is to define the best option for load transfer, considering that an outage has occurred in the feeder F1 (fault upstream of NC-1). In this case, there are two transfer options:

a. open the NC-1 switch and close the NO-1 switch; or
b. open the NC-1 switch and close the NO-2 switch.

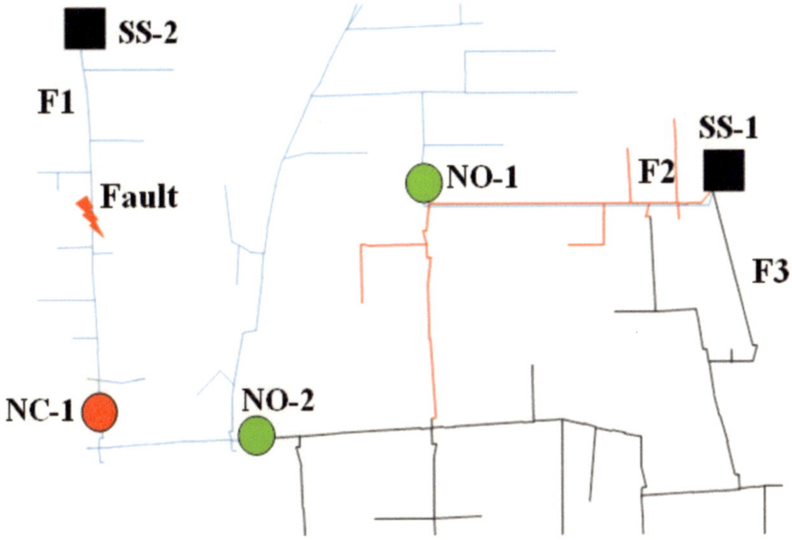

Fig. 4. Distribution network with three feeders.

First, the judgment matrix was obtained:

$$\mathbf{M} = \begin{array}{c} \\ C_1 \\ C_2 \\ C_3 \end{array}\begin{array}{c} \begin{array}{ccc} C_1 & C_2 & C_3 \end{array} \\ \begin{bmatrix} 1 & 3 & 5 \\ 1/3 & 1 & 3 \\ 1/5 & 1/3 & 1 \end{bmatrix} \end{array} \tag{14}$$

where:
M = judgment matrix;
C_1 = number of restored consumers;
C_2 = amount of restored energy;
C_3 = expected number of interrupted consumers per year.

Thus, the weight values for the three objective functions were determined:

$$W = \begin{bmatrix} 0.64 \\ 0.26 \\ 0.10 \end{bmatrix} \tag{15}$$

where:
$\lambda_{max} = 3.07$.

The consistency ratio (*CR*) was calculated by Equation 13:

$$CR = \frac{\left(\dfrac{3.07 - 3}{3 - 1}\right)}{0.52} = 0.0673 \tag{16}$$

The consistency ratio is lower than 0.10 and it is considered acceptable.

Tables 2 and 3 show the results of the analysis for each load transfer. In this example, there is no violation on the constraints.

Options	Number of Restored Consumers	Amount of Restored Energy (kW)	Expected Number of Interrupted Consumers per Year
1 (open NC-1 and close NO-1)	14,000	1,930.00	1,800
2 (open NC-1 and close NO-2)	14,000	1,930.00	2,300
Base Selected	14,000	1,930.00	1,800

Table 2. Results of the analysis for each load transfer.

Options	Number of Restored Consumers	Amount of Restored Energy	Expected Number of Interrupted Consumers per Year
1	1.00	1.00	1.00
2	1.00	1.00	0.78

Table 3. Normalized values of Table 2.

The results using AHP method were obtained by Equation 17:

$$\begin{bmatrix} Op.1 \\ Op.2 \end{bmatrix} = \begin{bmatrix} 1.00 & 1.00 & 1.00 \\ 1.00 & 1.00 & 0.78 \end{bmatrix} \cdot \begin{bmatrix} 0.64 \\ 0.26 \\ 0.10 \end{bmatrix} = \begin{bmatrix} 1.00 \\ 0.98 \end{bmatrix} \tag{17}$$

According to the proposed method the option "1" is considered the best solution. Thus, the system performs the commands to make the load transfer, open NC-1 and close NO-1, without violating the set of constraints.

6. Experimental analysis

To verify the performance of the proposed methodology several case studies were carried out in the concession area of a power utility in Brazil. The developed methodology was applied on the distribution system shown in Figure 5, which has 20 distribution substations, 125 feeders, 214 remote controlled switches, and 523,619 consumers.

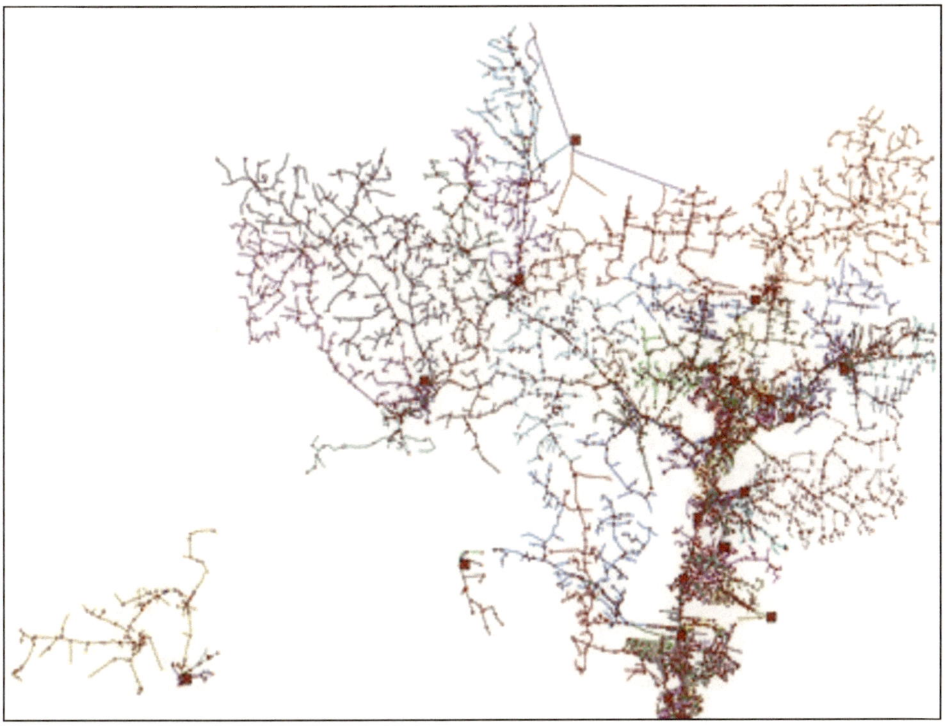

Fig. 5. Distribution network of a power utility in Brazil.

Figures 6 and 7 show the results for the calculation of load flow and the typical load curves used, respectively:

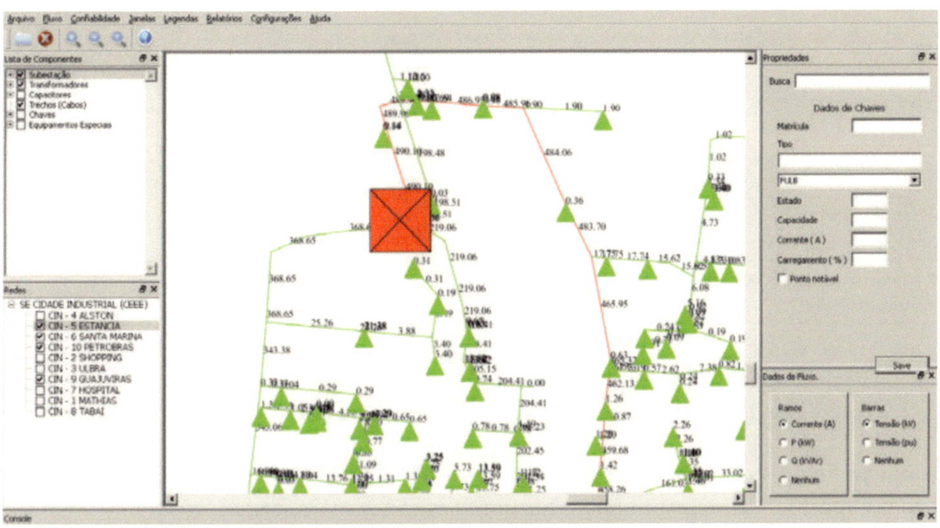

Fig. 6. Results for the calculation of load flow with the indication of values of current in branches and voltage in nodes.

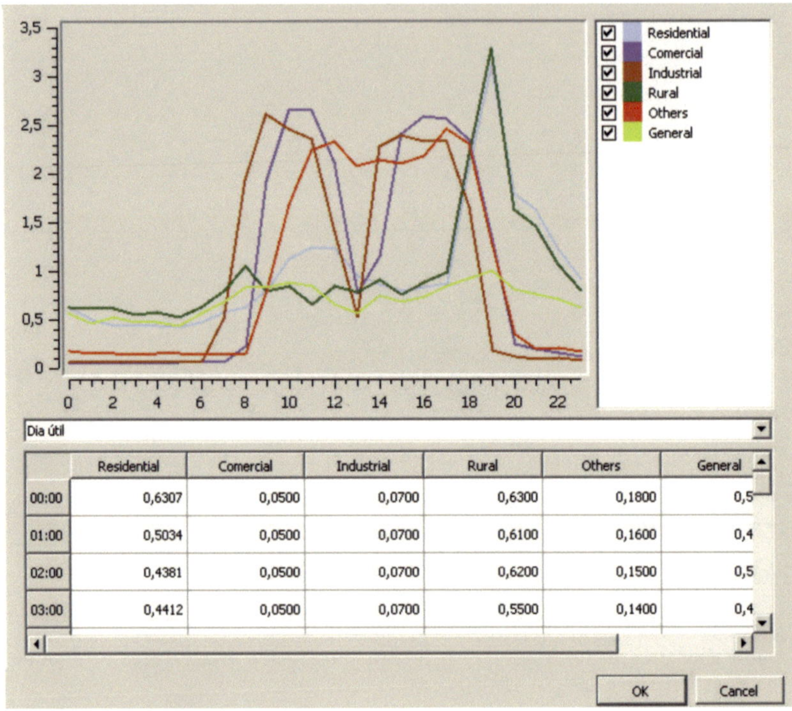

Fig. 7. Typical load curves used.

Table 4 shows the results obtained by the application of this methodology in case of outage of a feeder when considering the power restoration time.

Description	Mean Time to Restore Energy			
	Faults upstream the NC switch		Faults downstream the NC switch	
	Clients upstream the switch	Clients downstream the switch	Clients upstream the switch	Clients downstream the switch
Before installing the remote controlled switches	1h54min	58min	43min	1h34min
After installing the remote controlled switches	1h54min	0min	0min	1h34min
Reduction	-	58min	43min	-

Table 4. Results obtained with the use of remote controlled switches.

It should be noted that a reduction of approximately 30 % on the annual SAIDI index of this system is expected, assuming the number of faults in the main feeder.

Figures 8 and 9 show the picture of a pole top remote controlled switch been installed and the screen of the SCADA system, respectively:

Fig. 8. Installation of a remote controlled switch.

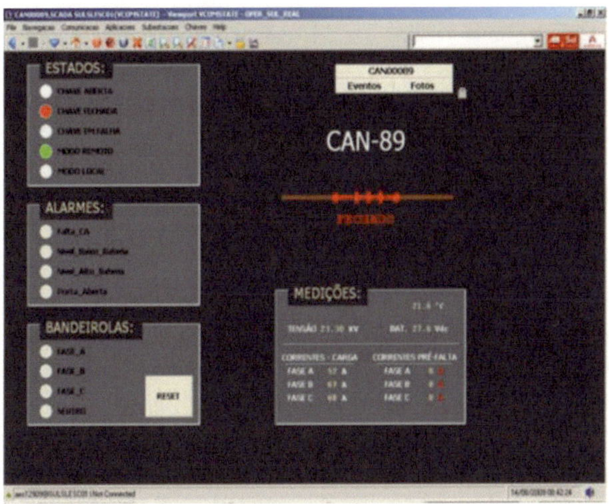

Fig. 9. Screen of the SCADA system.

7. Conclusion

This work presented a methodology developed for automatic power restoration system, which operates remote controlled switches in the distribution network. It was show how to assess the technical feasibility of the load transfers in real time by means of computer simulations, and how the best maneuver option to execute after a contingency is defined based on the AHP multicriterial method. The system will automatically handle the load transfers in accordance with the defined objective functions, without violating the established constraints. Case studies with real data from utilities were conducted to evaluate the performance of software developed presenting satisfactory results.

8. Acknowledgment

The authors would like to thank the technical and financial support of AES Sul Distribuidora Gaúcha de Energia SA, Conselho Nacional de Desenvolvimento Científico e Tecnológico (CNPq), and Coordenação de Aperfeiçoamento de Pessoal de Nível Superior (CAPES).

9. References

Baricevic, T., Mihalek, E., Tunjic, A. & Ugarkovic, K. (2009). AHP Method in Prioritizing Investments in Transition of MV Network to 20kV, *CIRED 2009 - 20th International Conference on Electricity Distribution*, pp. 1-4, Jun. 2009.

Bernardon, D.P., Comassetto, L. & Canha, L.N. (2008). Studies of parallelism in distribution networks served by different-source substations, *Electric Power Systems Research*, Elsevier, v. 78, p. 450-457, 2008.

Brown, R.E. (2008). Impact of Smart Grid on distribution system design, *IEEE Power and Energy Society General Meeting - Conversion and Delivery of Electrical Energy in the 21st Century*, pp.1-4, 2008.

Brown, R.E. (2009). *Electric Power Distribution Reliability*, CRC Press, Second Edition, ISBN 978-0-8493-7567-5, New York, 2009.

Kersting, W.H. & Mendive, D.L. (1976). An application of ladder network theory to the solution of three-phase radial load-flow problems, *IEEE Power Engineering Society Winter Meeting*, vol. A76 044-8, pp. 1-6, 1976.

Saaty, T.L. & Tran, L.T. (2007). On the invalidity of fuzzifying numerical judgments in the Analytic Hierarchy Process, *Mathematical and Computer Modelling*, vol. 46, pp. 962–975, 2007.

Saaty, T.L. (1980). *The Analytic Hierarchy Process: Planning, Priority Setting, Resource Allocation*, McGraw-Hill, ISBN 0-07-054371-2, New York, 1980.

Saaty, T.L. (1994). Highlights and Critical Points in the Theory and Application of the Analytic Hierarchy Process, *European Journal of Operational Research*, vol. 52, pp. 426-447, 1994.

Sperandio, M., Coelho, J., Carmargo, C.C.B., et al. (2007). Automation Planning of Loop Controlled Distribution Feeders, *2nd International Conference on Electrical Engineering (CEE'07)*, Coimbra, 2007.

Thakur, T. & Jaswanti (2006). Study and Characterization of Power Distribution Network Reconfiguration, *Proc. 2006 IEEE Transmission & Distribution Conference and Exposition: Latin America*, pp. 1-6.

Tsai, L. (1993). Network reconfiguration to enhance reliability of electric distribution systems, *Electric Power Systems Research*, Elsevier, no. 27, pp. 135-140, 1993.

Yang, H.T. & Chen, S.L. (1989). Incorporating a multi-criteria decision procedure into the combined dynamic programming/production simulation algorithm for generation expansion planning, *IEEE Transaction Power System*, vol. 4, pp. 165–175, Feb. 1989.

4

Automation of Subjective Measurements of Logatom Intelligibility in Classrooms

Stefan Brachmanski
Wroclaw University of Technology
Poland

1. Introduction

A great number of rooms are dedicated to a voice communication between a singular speaker and a group of listeners. Those rooms could be as small as meeting room and classrooms or larger like auditoriums and theatres. There is a major demand for them to assure the highest possible speech intelligibility for all listeners in the room. Classrooms are an example of rooms where a very good speech intelligibility is required (a teacher talks to a group of students who want to understand the teacher's utterance). To determine the intelligibility degree (its maps) of rooms, it is necessary to take measurements in many points of those rooms. The number of measurement points depends on room's size and precision of created intelligibility room's maps. Despite of the crucial progress in the instrumental measurement techniques, the only reliable method subjective speech intelligibility measurement is still very time consuming, expensive, demanding high skills and specially trained group of listeners. The first part of this chapter presents the idea of speech intelligibility subjective measurements. The measurements with properly trained team are taken in described standards, conditions have to be controlled and repeated. The subsequent sections of the chapter are focusing on one of the classic subjective speech intelligibility measurement method in rooms (classrooms) and its automated version which is named as the modified intelligibility test with forced choice (MIT-FC). Finally, in the last section of the chapter, there compared results of subjective speech intelligibility measurements in rooms taken with classic and automated methods are and the relation between intelligibility taken with the forced choice method is given as well. The presented relation let us compare results taken with both methods and use relations known from earlier research carried in domain of speech intelligibility. However, the biggest advantage of the speech intelligibility measurement automation is the shortening of measurement time and the possibility of taking simultaneous measurements in several points of the room. The result of speech intelligibility is obtained just after the end of measurements, it is then possible to obtain the intelligibility map in few minutes. Of course the precision is growing with number of listeners in particular points.

Speech quality is a multi-dimensional term and a complex psycho-acoustic phenomenon within the process of human perception. Every person interprets speech quality in a different way. The pioneering work on speech intelligibility was carried out by Fletcher at Bell Labs in the early 1940s. Fletcher and his team not only established the effects of bandwidth on intelligibility but also the degree to which each octave and ⅓-octave band contributed.

One of the fundamental parameters for quality assessment of speech signal transmission in analogue and digital telecommunication chains as well as in rooms, sound reinforcement systems and at selection of aural devices is speech intelligibility (ANSI, 2009; Basciuk & Brachmanski, 1997; Brachmanski, 2002, 2004; Davies, 1989; International Organization for Standardization - [ISO], 1991; Majewski, 1988, 1998, 2000; Polish Standard, 1991, 1999; Sotschek 1976). Satisfactory speech intelligibility should be provided by telecommunication channels, rooms and hearing aids. It is obvious, that speech intelligibility concerns only the linguistic information (i.e. what was said) and does not take into account such features of speech, like its naturalness or the speaker voice individuality. Nevertheless, intelligibility should be and - up till now - is viewed as a basic and most important aspect of the quality of all systems which transmit, code, enhance and process the speech signals. Satisfactory speech intelligibility requires adequate audibility and clarity.

In general the evaluation of the speech quality may be done by subjective (intelligibility, quality rating) (Farina, 2001; Howard & Angus, 2009; Möller, 2010) and objective methods e.g. Articulation Index - AI (American National Standards Institute - [ANSI], 1997; French & Steinberg, 1947; Kryter, 1962), Speech Transmission Index - STI (Brachmanski, 1982, 2004, 2006, Houtgast & Steeneken, 1973; International Electrotechnical Commission [IEC], 1991; Lam & Hongisto, 2006; Steeneken & Houtgast, 1980, 2002), Perceptual Evaluation of Speech Quality - PESQ (International Telecommunication Union - Telecommunication Standardization Sector [ITU-T], 2003), Perceptual Objective Listening Quality Assessment - POLQA, (ITU-T, 2011) . Designers of devices and systems intended for speech transmission incline to usage of objective measurement methods, not always taking into consideration the application limitations and exactitude which depends on type of examined object and measure conditions. However the final verifier of quality of speech transmission devices is their user, that is to say man.

The aim of aural evaluation is quantitative evaluating and qualitative differentiation of acoustical signals reaching a listener. The process of aural evaluation can be presented as follows

$$B \rightarrow S \rightarrow R \tag{1}$$

where: B - stimulus reaching the listener, S - listener, R - listener's reaction

The reaction R is dependent on the signal S reaching the listener's receptors and on conditions in which the listener is based. Generally it can be assumed that the reaction R depends on the sum of external stimuli having the effect on the listener and internal factors having the effect on his organism. That assumption, however, doesn't take into consideration the listener's characteristic features such as cognitive abilities, rate of information processing, memory etc. The reaction can be then presented as a function

$$R = f(B,S) \tag{2}$$

Physiological and psychological process connected with a reaction to sound signal (audio) consists of sensational reaction and emotional reaction. The total listener's reaction is a sum of both types of reaction.

The sensational reaction is the effect of a physiological process which occurs during the listening. It arises when a certain stimulus overdraws sensitivity levels or aural sensation category levels. The emotional reaction is more complex and difficult in analysis because it isn't a direct result of received signal features but the listener's habits and individuality. As a conclusion it can be stated that the sensational reaction is the reflection of an acoustical picture created in a person's (listener's) mind, whereas the emotional reaction is the reflection of a person's attitude to that picture. Psychological and psychoacoustical research have proven that when provided stable in time conditions of evaluation, the differences in sensational reactions of particular listeners are substantially smaller than the differences in their emotional reactions. Therefore, one aim of objectivisation of aural evaluation is limiting the influence of emotional reaction on the final assessment result. That aim is achieved by introducing appropriately numerous assessments statistics, a proper choice and training of a listeners team and proper choice of testing material and rules of carrying listeners tests. The results analysis also has the big role in minimization the influence of the emotional reaction on the assessment result.

Among the different subjective methods that have been proposed for assessment of speech quality in rooms, the preferred are methods based on intelligibility tests or listening-only tests (ITU-T, 1996a, 1996b; ITU-R, 1997). The subjective measure results should be mostly dependent on physical parameters of the tested room and not on the structure of the tested language material. The elimination of semantic information is done by means of logatom[1] (i.e. pseudo-words) lists on the basis of which the logatom intelligibilities are obtained. The problem ist hat speech (logatom) intelligibility is not a simply parameter to measure.

2. The traditional method of logatom intelligibility

The measurement of logatom intelligibility consists in the transmission of logatom lists, read out by a speaker, through the tested channel (room), which are then written down by listeners and the correctness of the record is checked by a group of experts who calculate the average logatom intelligibility. It is recommended to use lists of 50 or 100 logatoms (Fig.1).

Logatom lists are based on short nonsense word of the CVC type (consonant-vowel-consonant). Sometimes only CV, but also CVVC, CVCVC-words are used. The logatoms are presented in isolation or in carrier phrase e.g. "Now, please write down the logatom you hear". Each list should be phonetically and structurally balanced (Fig.2).

---

[1] Logatom – (*logos* (gr) - spoken phrase, *atom* (gr) – indivisible) vocal sound, generally insignificant, usually made by the sound of a consonant or the first consonant, then by an intermediate vowel, finally by a consonant or a final consonant sound.

LOGATOM LIST 3.1

1 - 10	11 - 20	21 - 30	31 - 40	41 - 50
dufcze	cze	teń	jo	czalmy
ajtes	wać	foli	es	ny
wesk	niacko	delak	gapysz	nij
pi	wespa	szo	tyr	dnaf
żewan	szkio	penia	geja	uzy
pracło	łare	si	lo	nio
nujec	le	weru	ju	towir
wajt	wnieśma	krwolta	dziorka	ła
be	myśro	ponta	jeskol	woszki
szonie	ri	tyn	ca	krynom

51 - 60	61 - 70	71 - 80	81 - 90	91 - 100
chfypa	dy	pa	gruto	słynej
ca	zjaciech	piestma	ke	jentuś
czniesa	wyzo	fa	todzi	jos
szaźmo	żam	żnota	ply	bep
neni	fjesztech	gzał	abe	du
se	ro	dojsto	tnązy	wo
błunki	kiet	ga	kamu	cioch
dnozger	desi	vrum	ksi	zał
bomuć	mo	iżo	wnici	ną
dzany	bejce	du	nu	sze

Fig. 1. Illustrative 100-logatom list from set 3 for the Polish language.

The measurement should be carried out in rooms in which level of internal noise together with external noise (not introduced on purpose) does not exceed 40 dBA. If no requirements as to background noise are specified for a tested chain, articulation should be measured at a noise level of 60 dBA in the receiving room and for the Hoth spectrum (Fig. 3).

The listeners should be selected from persons who have normal, good hearing and normal experience in pronunciation in the language used in the test. A person is considered to have normal hearing if her/his threshold does not exceed 10 dB for any frequency in a band of 125 Hz –4000 Hz and 15 dB in a band of 4000 Hz– 6000 Hz. Hearing threshold should be tested by means of a diagnostic audiometer.

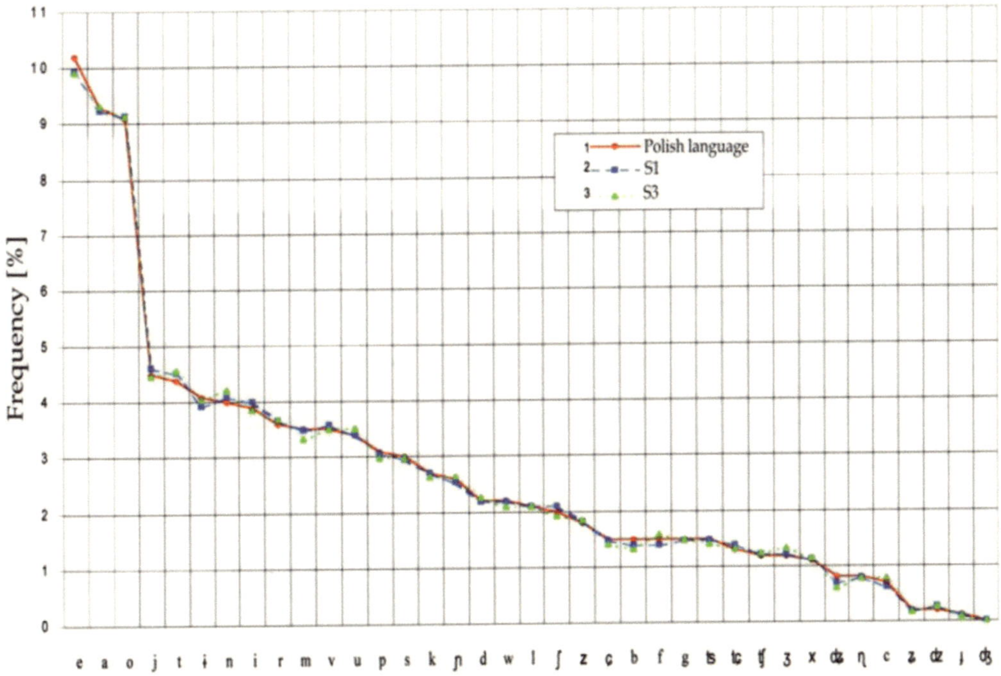

Fig. 2. Example of phonetic balance of two sets, three lists of 100 logatom each for the Polish language.

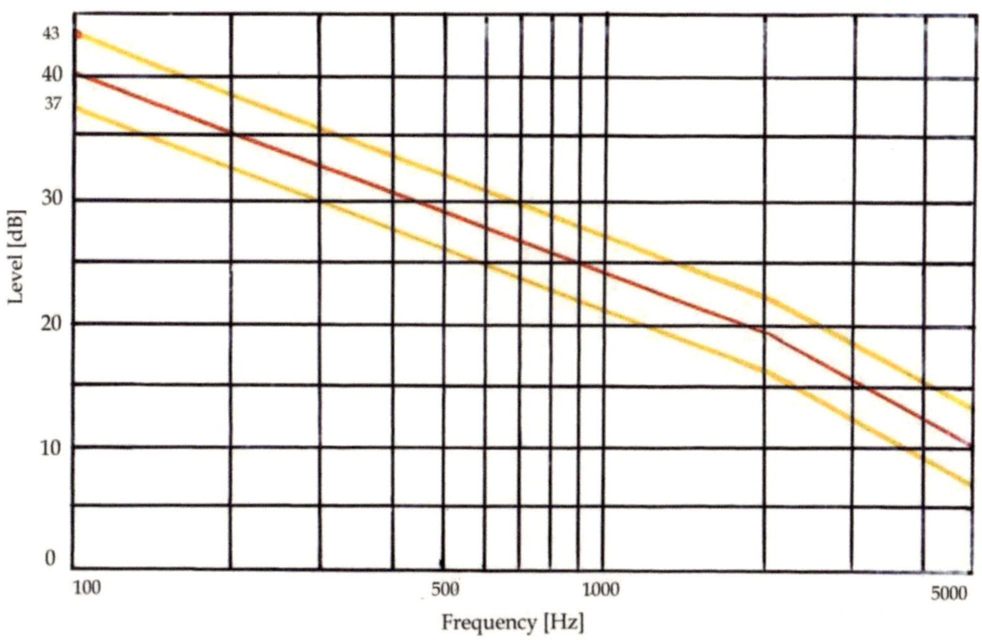

Fig. 3. The room noise power density spectrum (Hoth noise) (Polish Standard 1991, 1999).

The size of the listening group should be such that the obtained averaged test results do not change as the group size is further increased (minimum 5 persons). The group of listeners who are to take part in logatom intelligibility measurements should be trained (2-3 training sessions are recommended). Logatoms should be spoken clearly and equally loudly without accenting their beginnings or ends. The time interval between individual logatoms should allow the listener to record the received logatom at leisure. It is recommended that logatoms should be spoken with 3-5 sec. pauses in between. The time interval between sessions should not be shorter than 24 h and not longer than 3 days. The total duration of a session should not exceed 3 hours (including 10 minute breaks after each 20 minute listening period).

Listeners write the received logatoms on a special form on which also the date of the test, the test list number, the speaker's name or symbol (no.), the listener's name and additional information which the measurement manager may need from the listener is noted. The recording should be legible to prevent a wrong interpretation of the logatom. The received logatoms may be written in phonetic transcription (a group of specially trained listeners is needed for this) or in an orthographic form specific for a given language. In the next step, the group of experts checks the correctness of received logatoms and the average logatom intelligibility is calculated in accordance to the equation (3) and (4)

$$W_L = \frac{1}{N \cdot K} \sum_{n=1}^{N} \sum_{k=1}^{K} W_{n,k} \; [\%] \tag{3}$$

N - number of listeners, K - number of test lists, $W_{n,k}$ - logatom intelligibility for n-th listener and k-th logatom list,

$$W_{n,k} = \frac{P_{n,k}}{T_k} \cdot 100 \, [\%] \tag{4}$$

$P_{n,k}$ - number of correctly received logatoms from k-th logatom list by n-th listener, T_k - number of logatoms in k-th logatom list.

Standard deviation s, calculated in accordance to Eq.(3), expresses the distribution of logatom intelligibility values W_L over listeners.

$$s = \left[\frac{1}{N \cdot K - 1} \sum_{n=1}^{N} \sum_{k=1}^{K} (W_{n,k} - W_L)^2 \right]^{1/2} \tag{5}$$

If $|W_{n,k} - W_L| > 3s$, the result of measurement is not taken into account, when an average intelligibility is calculated and calculation of W_L and s must be done in accordance to Eq. (1) and (2) for reduced number of measurements.

The obtained average logatom intelligibility value can be used to determine quality classes according to Table 1 (Polish Standard 1991).

Quality class	Description of quality class	Logatom intelligibility [%]
I	Understanding transmitted speech without slightest concentration of attention and without subjectively detectable distortions of speech signal	above 75
II	Understanding transmitted speech without difficulty but with subjectively detectable distortions of speech	60 ÷ 75
III	Understanding transmitted speech with concentration of attention but without repetitions and return queries	48 ÷ 60
IV	Understanding transmitted speech with great concentration of attention and with repetitions and return queries	25 ÷ 48
V	It is impossible to fully understand transmitted speech (breakdown of communication)	to 25
For each quality class lowest logatom articulation values are lowest admissible values		

Table 1. Speech intelligibility quality classes for analog channels in the traditional logatom intelligibility method.

3. Modified intelligibility test with forced choice (MIT-FC)

In the traditional intelligibility tests the listeners write down (in ortographic form) received utterences on a sheet of paper and next a professional team revises the results and calculates the average intelligibility. This is the most time-consuming and difficult operation. To eliminate "hand-made " revision of the tests a method, called *"modified intelligibility test with forced choice"* (MIT-FC) has been designed and investigated in the Institute of Telecommunications, Teleinformatics and Acoustics at the Wroclaw University of Technology .

In the MIT-FC method all experiments are controlled by a computer. The automation of the subjective measurement is connected with the basic change in generation of logatoms and in making decision by a listener. The computer generates logatoms and presents the utterances (for logatom test the list consists of 100 phonetically balanced nonsense words), via a D/A converter and loudspeaker to the listeners subsequently and for each spoken utterance several logatoms that have been previously selected as perceptually similar are visually presented.

It has been found (Brachmanski, 1995) that the optimal number of logatoms presented visually to the listeners is seven (six alternative logatoms and one transmitted logatom to be recognized). The listener chooses one logatom from the list visually presented on the computer monitor. The computer counts the correct answers and calculates the average logatom intelligibility and standard deviation. The measurement time for one logatom set (3 lists) consisted of 300 logatoms is 20 minutes.

All measurement procedures are fully automatized and an operator has a flexible possibilities to set the measurement parameters and options. It is also possible to upgrade the application which realizes the MIT-FC method with more sophisticated scores processing. Block diagram system for the subjective measurements of logatom intelligibility by MIT-FC method in the rooms is presented in Fig.4.

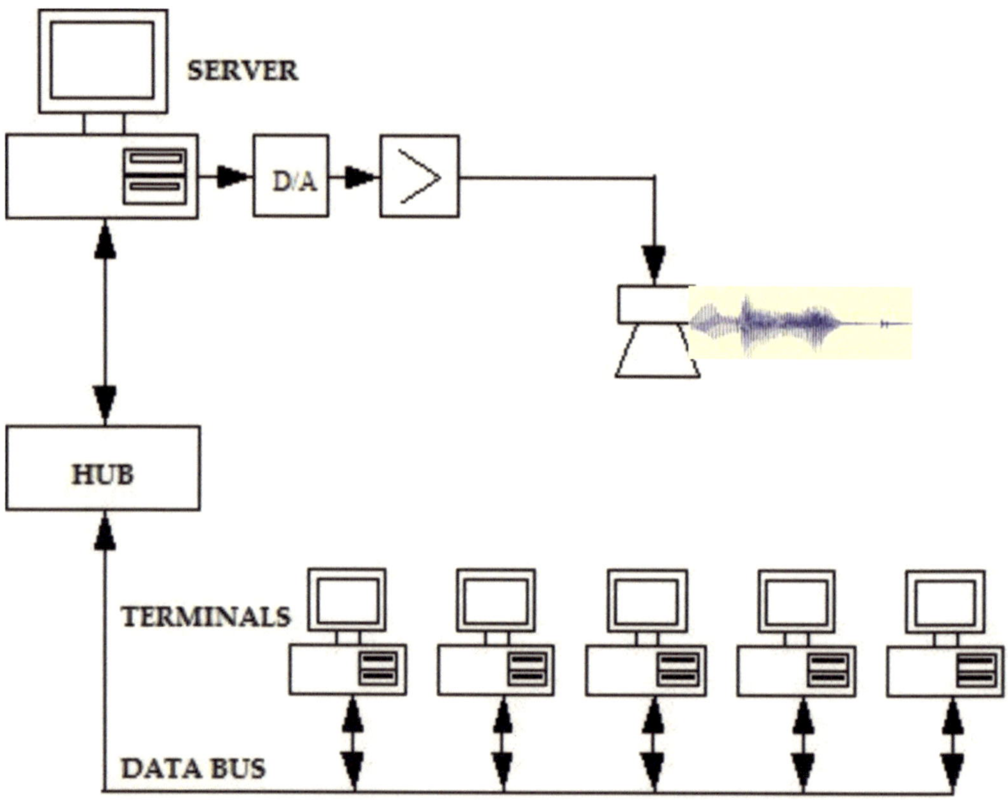

Fig. 4. The measuring system for the assessment of logatom intelligibility in room.

The obtained with MIT-FC method average logatom intelligibility value can be used to determine quality classes according to Table 2.

Quality class	Description of quality class	Logatom intelligibility [%]
I	Understanding transmitted speech without slightest concentration of attention and without subjectively detectable distortions of speech signal	above 70
II	Understanding transmitted speech without difficulty but with subjectively detectable distortions of speech	60 ÷ 70
III	Understanding transmitted speech with concentration of attention but without repetitions and return queries	54 ÷ 60
IV	Understanding transmitted speech with great concentration of attention and with repetitions and return queries	40 ÷ 54
V	It is impossible to fully understand transmitted speech (breakdown of communication)	to 40
For each quality class lowest logatom articulation values are lowest admissible values		

Table 2. Classes of speech intelligibility quality for analog channels for the MIT-FC method.

4. MIT-FC measurement system

The program for the subjective assessment of speech transmission in rooms with logatom intelligibility method with forced choice (MIT-FC) is based on TCP Client/Server technology i.e. the communication is done by local network. Requirements of the program are following: PC computer with Windows 9x, a network card and hub for communication between the Server and Clients (members of the team of listeners).

The work with the program starts with the installation on the Server computer a Server program. The Server is supervised by the person leading the subjective assessment. The next step is the installation of the Client program on the Client (listener) computers. The Client computers are used by members of the team of listeners. Before starting the assessment the Server and Client programs should be configured. During the configuration of the Server program it is necessary to give the path to the directory with signal files (testing signals - logatoms), number of logatoms per a session, intervals between reproduced logatoms, number of sessions and the port for the communication with the Client (usually 3000) (Fig.5).

The configuration of the Client program is done by giving the name of the measure point (e.g. a room, location of listener in the room etc.), listeners login, IP address and the port number (usually: 3000) (Fig.6).

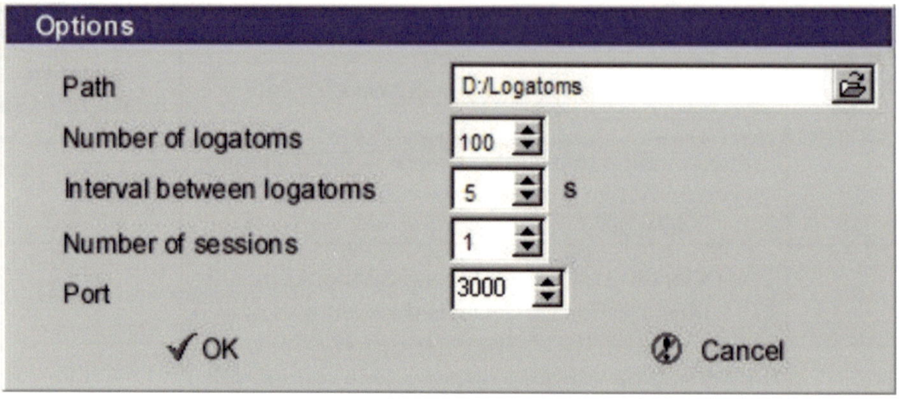

Fig. 5. The example of the window of the Server program

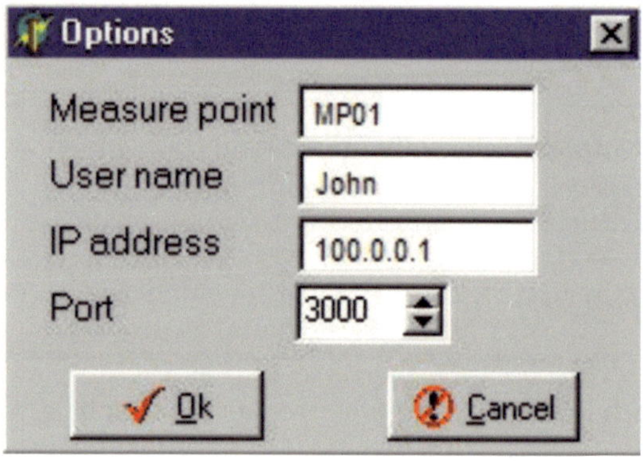

Fig. 6. The example of the window of the Client program

The next step after the configuration is the connection process. The number of people being logged on can be seen on the right-hand side of the main window of the program. During a multi-session assessment (the number of session should be chosen during the configuration of the Server), after finishing of the first session the program waits for reconfiguration of measurement positions. For example, changing the listener on certain measurement position, we should first disconnect the Client and change the user name. After this and choosing the Continue option, the next session can be started. The computer generates logatoms and for each spoken utterance visually shows six alternative logatoms and one transmitted logatom to be recognized (Fig.7a). The listener chooses one logatom from the list visually presented on the computer monitor (Fig. 7b). During the tests, the listener is confirming his response by using key from '1' to '7'. Other keys are non-active during all testing session.

After finishing all measurement sessions, the dialog window presenting results in two options shows up:

1. Result of session nr - in this option the number of session for which the results will be shown should be given.
2. Summary - in this option the summary of all sessions with detailed list of listeners and measurement points shows up.

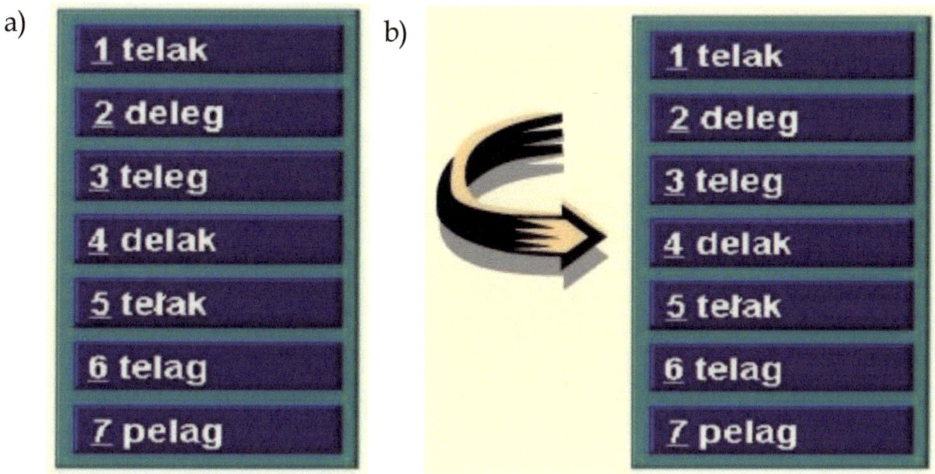

Fig. 7. The example of the listener's window.

5. Experiments

The goals of experiment

- decision if the results of traditional and modified with forced choice methods let finding the relation which would allow to convert results from one method to the other and the classification of rooms tested with both methods,
- measurement of experimental relations between traditional and modified logatom intelligibility methods.

Taking into account the comparative character of the experiment it was planned to be done only in a function of used logatom intelligibility method. With this end in view each measurement was done with both traditional and modified methods, not changing:

- listeners,
- surroundings.
- measurement system (only the logatom lists),

The subjective tests were done according to Polish Standard PN-90/T-05100 with the team of listeners made up of 12 listeners in age from 18 to 25 years. The listening team was selected from students at Wroclaw University of Technology with normal hearing. The qualification was based on audiometric tests of hearing threshold. The measurements of logatom intelligibility were done using the traditional method and the MIT-FC method. The measurements were taken in two unoccupied rooms (Fig. 8). In each room, four measure point (Mp) were selected. These positions were chosen in the expectation of yielding a wide range of logatom intelligibility. Sound sources (voice and white or rose noise) were positioned in the part of the room normally used for speaking. One loudspeaker was the voice source and the second - the noise source. The various conditions were obtained by combination 14 level of white and rose noise and four measure point. The 14 signal-to-noise ratios (SNR) were used: 39 (without noise, only background noise), 36, 33, 30, 27, 24, 21, 18, 15, 12, 9, 6, 3, 0dB. As a result

112 different transmission conditions were obtained (14 SNR*4Mp*2rooms). The speech and noise signal levels were controlled by means of a 2606 Bruel & Kjaer instrument by measuring them on a logarithmic scale according to correction curve A.

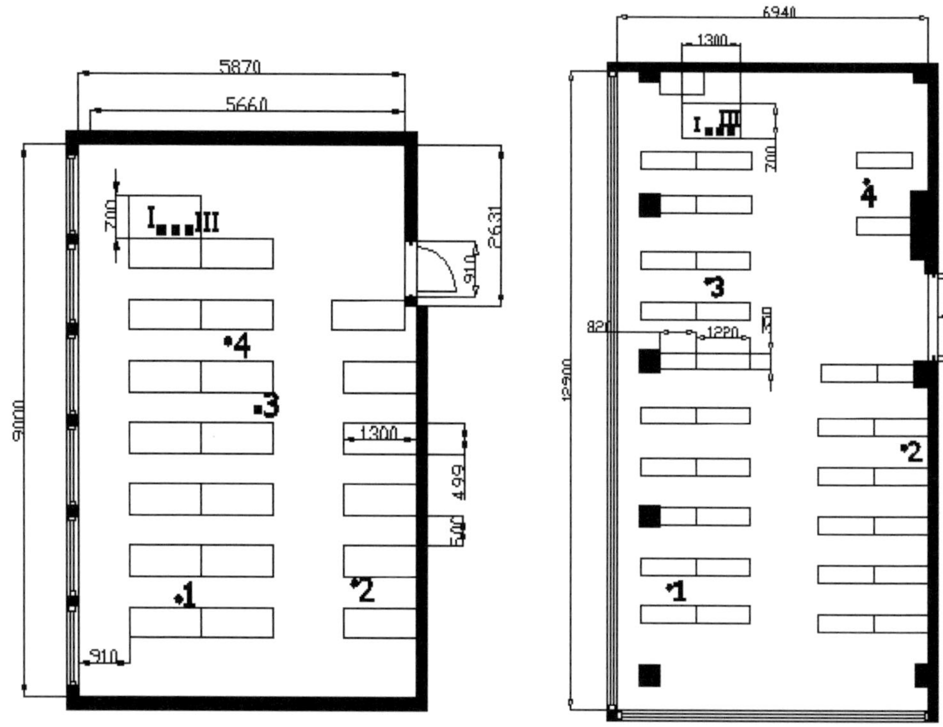

Fig. 8. Plan view of the rooms 1 and 2 showing source position and four receiver positions (**I** - loudspeaker – source of the logatoms, **III** - loudspeaker – source of the noise, **1, 2, 3, 4** – measurement points).

The testing material consisted of phonetically and structurally balanced logatoms and sentences lists uttered by professional male speaker, whose native language was Polish. Logatom lists reproduced through the loudspeaker were recorded on the digital tape recorder in an anechoic chamber using a linear omnidirectional microphone. The microphone was positioned 200mm from the speaker's lips. The active speech level was controlled during recording with a meter conforming to Recommendation P.56 (ITU-T, 1993). At the beginning of each logatom set recording, 20 seconds 1000Hz calibration tone and 30 seconds rose noise are inserted at a level equal to the mean active speech level.

For each measure point (Mp) (the place where the measure position was situated) a list of 300 logatoms has been prepared. The same logatom lists were never played for any condition day after day. The logatom lists at the four listener locations were recorded on the digital tape recorder. These recordings were played back over headphones to the subject afterward. This way of subjective measurements realization provides the same listening conditions for both traditional and with choice methods. In each room for each position of listener (Mp) and for each signal-to-noise ratio (SNR) the logatom intelligibility was obtained by averaging out the group of listeners results.

The listening tests were done in the studio of Institute of Telecommunications, Teleinformatics and Acoustics of Wroclaw University of Technology (Fig.9). Background noise level didn't exceed 40 dBA. Subjective measurements of logatom intelligibility were taken in conditions of binaural listening using headphones for the optimum speech signal level of 80 dBA.

Prior to the proper measurements the listening team was subjected to a 6 - hour training (two 3 - h sessions). The measurement sessions duration did not exceed 3h (together with 10 min breaks after every 20 min of listening).

Fig. 9. Speech quality subjective measurement stand.

The subjective logatom intelligibility measurement results, obtained for different speech to noise ratio, are presented graphically in fig. 10. The curves, representing the relationship between logatom intelligibility and the signal-to-noise ratio in a room were approximated by a four-degree polynomial calculated on the basis of the least-squarees method. The obtained relations are presented in fig. 10. which also includes values of correlation coefficient R^2 - a measure of the conformity between the polynomial and the results obtained from the subjective tests. As one can see there is very good agreement between the theoretical curve and the empirical results; the value of correlation coefficient R^2 exceeds 0.99 in each case.

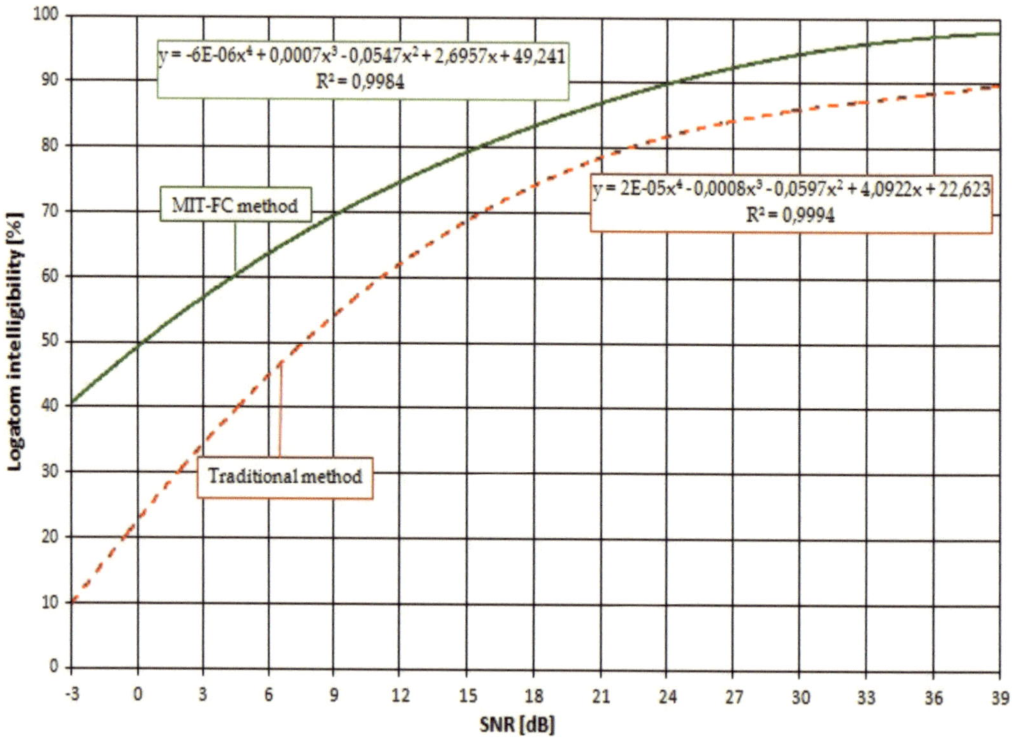

Fig. 10. Relationship between logatom intelligibility and signal-to-noise ratio (SNR) for rooms (Logatom intelligibility measured with traditional and MIT-FC method).

For a few randomly selected measuring points the distribution of $W_{n,k}$ values was compared with the normal distribution. The agreement between the $W_{n,k}$ distribution and the normal distribution were tested by applying the Kolmogorov-Smirnov test. It has been found that at significance level $\alpha = 0.05$ there are no grounds to reject the hypothesis about the goodness of fit of the distributions. Thus, it is reasonable to use average logatom hypothesis W_L as an estimator of the logatom intelligibility for a given measuring point.

The main goal of presented research was assessment if there exists a relation between measurement results of traditional method and modified intelligibility test with forced choice (MIT-FC) in rooms, and if such a relation exists, its finding out. The obtained results revealed that there exists monotonic relation between results of traditional method and MIT-FC. The results were presented on the surface: logatom intelligibility MIT-FC - traditional, and approximated by a fourth order polynomial. The obtained relations are presented in fig. 11. which also includes values of correlation coefficient R^2 - a measure of the conformity between the polynomial and the results obtained from the subjective tests.

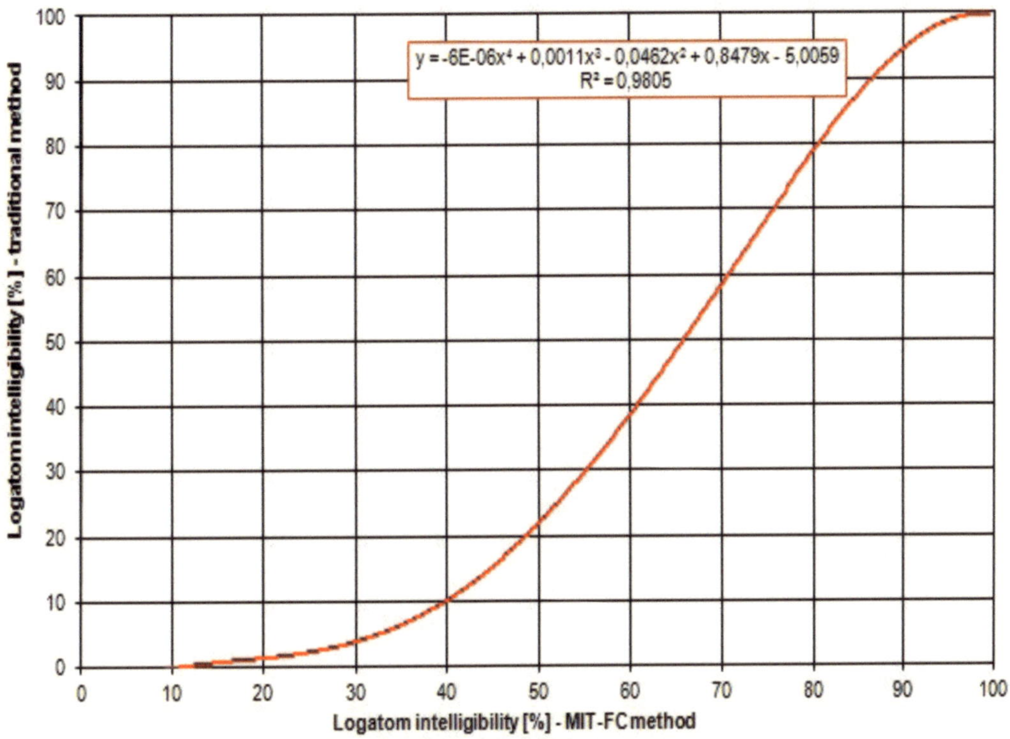

Fig. 11. Relationship between logatom intelligibility measured with traditional and MIT-FC method for rooms.

6. Conclusion

The presented MIT-FC method offers a simple, easy to use, stable, and fully automatized speech system to assessment of speech quality in rooms. The results of the experiments have shown that the MIT-FC method is very useful in the evaluation of speech quality in rooms. The time needed to carry out the measurement with MIT-FC method is the same as in traditional one but we obtain the results right after finishing the measurement process.

The experiments carried out in finding the relations between logatom intelligibility measured with traditional and semi-automatic with forced choice methods for the rooms have shown that there exist the multi-value and repetitive relation between them. It allows using both methods interchangeably and converting results between them.

The obtained relations are applied in the Institute of Telecommunications, Teleinformatics and Acoustics of Wrocław University of Technology to the design of subjective tests for the verification of results yielded by a new objective method based on automatic speech recognition techniques.

7. References

ANSI S 3.2, (2009), *Methods for Measuring the Intelligibility of Speech over Communication Systems*, American National Standards Institute.

ANSI S 3.5, (1997), *Methods for the Calculation of the Speech Intelligibility index (SII)*. American National Standards Institute

Basciuk K., Brachmanski S., (1997), The Automation of the Subjective Measurements of Logatom Intelligibility, Prep. 4407, *102-nd Convention AES*, Munich, Germany, March 22-25, 1997.

Brachmanski S., (1982), *Modulation Transfer Function (MTF) as a Measure of Polish Speech Transmission Quality* (in Polish), Ph. D. Thesis, Wrocław.

Brachmanski S., (1995), Choosing Optimum Number of Test Items in Subjective Logatom Measurements (in Polish). *Proc. XLII Open Seminar on Acoustics*, pp. 423-428, ISBN 83-902146-3-6, Warszawa-Bialowieza, September 12-15, 1995

Brachmanski S., (2002) The Automation of Subjective Measurements of Speech Intelligibility in Rooms, *112th Conv. AES*, Munich, Germany, May 10-13, 2002.

Brachmanski S., (2004) Estimation of Logatom Intelligibility with the STI Method for Polish Speech Transmitted via Communication Channels, *Archives of Acoustics*, Vol. 29, No. 4, (2004), pp. 555-562, , ISSN 0137-5075.

Brachmanski S., (2004), The Subjective Measurements of Speech Quality in Rooms, *Proc. of Subjective and Objective Assessment of Sound*, Poznan, September 1-3, 2004.

Brachmanski S., (2006), Experimental Comparison between Speech Transmission Index (STI) and Mean Opinion Scores (MOS) in Rooms,. *Archives of Acoustics*, Vol .31, No.4, (2006), pp. 171-176, ISSN 0137-5075.

Davies D.D., Davies C., (1989), Application of Speech Intelligibility to Sound Reinforcement. J. *Audio Eng. Soc.*, Vol. 37, No.12, (December 1989), pp. 1002-1018.

Farina A., (2001), Acoustic Quality of Theatres: Correlations between Experimental Measures and Subjective Evaluations, *Applied Acoustic*, Vol. 62, No 8, (August 2001), pp. 889-916, ISSN: 0003682X.

French, N.R., Steinberg, J.C., (1977), Factors Governing the Intelligibility of Speech Sounds, , In: *Speech Intelligibility and Speaker Recognition*, Hawley M. E., pp. 128 - 152, Dowden, Hutchinson & Ross Inc., ISBN 0-470-99303-0, Stroudsburg, Pennsylvania.

Houtgast T., Steeneken H.J.M., (1973), The Modulation Transfer Function in Room Acoustics as a Predictor of Speech Intelligibility, *J. Acoust. Soc. Am.* Vol. 54, No. 2, (August 1973), pp. 557-557, pp. 66-73, ISSN 0001-4966

Howard D.M, Angus J.A.S, *Acoustics and Psychoacoustics*, Elsevier-Focal Press, 2009, ISBN 978-0-240-52175-6, Oxford, UK.

IEC 60268-16, (2003), *Sound system equipment – Part16: Objective Rating of Speech Intelligibility by Speech Transmission Index*, International Electrotechnical Commission, Geneva, Switzerland.

ISO/TR 4870, (1991), *Acoustics – The Construction and Calibration of Speech Intelligibility Tests*. International Organization for Standardization.

ITU-T Recommendation P.56, (1993), *Objective Measurement of Active Speech Level*, International Telecommunication Union, Geneva, Switzerland.

ITU-T Recommendation P.800, (1996), *Method for Subjective Determination of Transmission Quality*, International Telecommunication Union, Geneva, Switzerland.

ITU-T Recommendation P.830, (1996), *Method for Objective and Subjective Assessment of Quality*, International Telecommunication Union, Geneva, Switzerland.

ITU-T Recommendation P.862, (2003), *Perceptual Evaluation of Speech Quality (PESQ), an Objective Method for End-To-End Speech Quality Assessment of Narrowband Telephone Networks and Speech Codecs*, International Telecommunication Union, Geneva, Switzerland.

ITU-T Recommendation P.863, (2011), *Methods for Objective and Subjective Assessment of Speech Quality. Perceptual Objective Listening Quality Assessment*, International Telecommunication Union, Geneva, Switzerland.

ITU-R Recom. BS.1116-1, (1997), *Method for the Subjective Assessment of Small Impairments in Audio Systems Including Multichannel Sound Systems*, International Telecommunication Union, Geneva, Switzerland.

Kryter, K.D., (1962), Methods for the Calculation and Use of the Articulation Index, *J. Acoust. Soc. Am.*, Vol. 34, No. 11, (November 1962), pp. 1689-1697, ISSN 0001-4966.

Lam P., Hongisto V., (2006), Experimental Comparison between Speech Transmission Index, Rapid Speech Transmission Index, and Speech Intelligibility Index, *J. Acoust. Soc. Am.*, Vol. 119, No. 2, (February 2006), pp.1106-1117, ISSN 0001-4966.

Majewski W., Basztura Cz., Myślecki W., (1988), Relation between Speech Intelligibility and Subjective Scale ff Speech Transmission Quality., *Proc. 7th FASE Symposium, Proceedings SPEECH'88*, 719-726, Edinburgh, August 22-26, 1988.

Majewski W., Myślecki W., Baściuk K., Brachmański S., (1998), Application of Modified Logatom Intelligibility Test in Telecommunications, Audiometry and Room Acoustics, *Proceedings 9th Mediterranean Electrotechnical Conference Melecon'98*, pp. 25-28, Tel-Aviv, Israel, May 18-20, 1998.

Majewski W., Myślecki W., Brachmański S., (2000), Methods of Assessing the Quality of Speech Transmission (in Polish), *Proceedings of 47th Open Seminar on Acoustics*, pp. 66-75, ISBN 83-914391-0-0, Rzeszow-Jawor, September 19-22, 2000.

Möller S. *Assessment and Prediction of Speech Quality in Telecommunications*, Kluwer Academic Publisher, ISBN 978-1-4419-4989-9, Dordrecht, Netherlands.

Polish Standard PN-90/T-05100, (1991), *Analog Telephone Chains. Requirements and Methods of Measuring Logatom Articulation* (in Polish), Polski Komitet Normalizacyjny (Polish Committee for Standardization), 1990, Warszawa, Poland

Polish Standard PN-V-90001, (1999), *Digital Communication Systems. Requirements and Methods for Measurement of Logatom Articulation* (in Polish), Polski Komitet Normalizacyjny (Polish Committee for Standardization), 1999, Warszawa, Poland

Sotschek J., (1976), Methoden zur Messung der Sprachgüte I: Verfahren zur Bestimmung der Satz- und der Wortverständlichkeit, *Der Fernmelde Ingenieur*, Vol. 30, No. 10, pp. 1-31.

Srinivasan S.H., (2004), Speech Quality Measure Based on Auditory Scene Analysis, *Proc. IEEE 6th Workshop on Multimedia Signal Processing*, pp. 371 - 374, ISBN: 0-7803-8578-0 September 29- October 1, 2004, Siena, Italy.

Steeneken H.J.M., Houtgast T., (1980), A Physical Method for Measuring Speech-Transmission Quality, *J. Acoust. Soc. Am.*, Vol. 67, No. 1, (January 1980), pp. 318 - 326, ISSN 0001-4966.

Steeneken H.J.M., Houtgast T., (2002), Validation of Revised STI Method, *Speech Communication*, Vol. 38, No. 3-4, (November 2002), pp.413-425, ISSN:0167-6393.

5

Automation in Aviation

Antonio Chialastri
Medicair, Rome,
Italy

1. Introduction

An aircraft landed safely is the result of a huge organizational effort required to cope with a complex system made up of humans, technology and the environment. The aviation safety record has improved dramatically over the years to reach an unprecedented low in terms of accidents per million take-offs, without ever achieving the "zero accident" target. The introduction of automation on board airplanes must be acknowledged as one of the driving forces behind the decline in the accident rate down to the current level.

Nevertheless, automation has solved old problems but ultimately caused new and different types of accidents. This stems from the way in which we view safety, systems, human contribution to accidents and, consequently, corrective actions. When it comes to aviation, technology is not an aim in itself, but should adapt to a pre-existing environment shared by a professional community.

The aim of this paper is to show why, when and how automation has been introduced, what problems arise from different ways of operating, and the possible countermeasures to limit faulty interaction between humans and machines.

This chapter is divided into four main parts:

1. Definition of automation, its advantages in ensuring safety in complex systems such as aviation;
2. Reasons for the introduction of onboard automation, with a quick glance at the history of accidents in aviation and the related safety paradigms;
3. Ergonomics: displays, tools, human-machine interaction emphasizing the cognitive demands in fast-paced and complex flight situations;
4. Illustration of some case studies linked to faulty human-machine interaction.

2. What is automation

According to a shared definition of automation, the latter may be defined in the following way: "Automation is the use of control systems and information technologies to reduce the need for human work in the production of goods and services". Another plausible definition, well-suited the aviation domain, could be: "The technique of controlling an apparatus, a process or a system by means of electronic and/or mechanical devices that replaces the human organism in the sensing, decision-making and deliberate output" (Webster, 1981).

The Oxford English Dictionary (1989) defines automation as:

1. Automatic control of the manufacture of a product through a number of successive stages;
2. The application of automatic control to any branch of industry or science;
3. By extension, the use of electronic or mechanical devices to replace human labour.

According to Parasumaran and Sheridan, "automation can be applied to four classes of functions:

1. Information acquisition;
2. Information analysis;
3. Decision and action selection;
4. Action implementation."

Information acquisition is related to the sensing and registration of input data. These operations are equivalent to the first human information processing stage, supporting human sensory processes. If we adopt a decision-making model based on perception, identification, mental process, decision, action, follow-up and feedback, information acquisition could be likened to the first step: perception. Let's imagine a video camera and the aid it offers in monitoring activity. It helps to replace continuous, boring, monotonous human observation with reliable, objective and detailed data on the environment.

Automation may handle these functions, as it is more efficient in detecting compared to humans, while - at the same time - it offers the possibility of positioning and orienting the sensory receptors, sensory processing, initial data pre-processing prior to full perception, and selective attention (e.g.: the focus function in a camera).

Information analysis is related to cognitive functions such as working memory and inferential processes. It involves conscious perception and manipulation of processed items. It allows for quick retrieval of information in the working memory. In aviation, this kind of system is broadly used to provide pilots with predictive information, such as how much fuel will be available at destination, where the top of climb or top of descent will be in order to optimize the flight path, and so forth.

With regard to decision and action selection, automation is useful because it involves varying levels of augmentation or replacement of human decision-making with machine decision-making. It is generally acknowledged that human decision-making processes are subject to several flaws, among them a tendency to avoid algorithmic thought, a biased development of pros and cons based on the laws of logic, a partial view of the overall system and, often, the heavy influence of emotions.

The fourth stage involves the implementation of a response or action consistent with the decision taken. Generally, it this stage automation replaces the human hands or voice. Certain features in the cockpit allow automation to act as a substitute for pilots. For instance, this occurs when - following an alert and warning for windshear conditions - the automation system detects an imminent danger from a power setting beyond a pre-set threshold. In this case, the autopilot automatically performs a go-around procedure, which avoids a further decline in the aircraft's performance.

Besides being applicable to these functions, automation has different levels corresponding to different uses and interactions with technology, enabling the operator to choose the optimum level to be implemented based on the operational context (Parasumaran, Sheridan, 2000). These levels are:

1. The computer offers no assistance; the human operator must perform all the tasks;
2. The computer suggests alternative ways of performing the task;
3. The computer selects one way to perform the task and
4. Executes that suggestion if the human operator approves, or
5. Allows the human operator a limited time to veto before automatic execution, or
6. Executes the suggestion automatically then necessarily informs the human operator, or
7. Executes the suggestion automatically then informs the human operator only if asked.
8. The computer selects the method, executes the task and ignores the human operator.

3. History of accidents

Automation in the aviation world plays a pivotal role nowadays. Its presence on board airplanes is pervasive and highly useful in improving the pilots' performance and enhancing safety. Nevertheless, certain issues have emerged in the recent past that evidence automation misuse by pilots. This could depend on a series of factors, among them human performance, capabilities and limitations on one side, and poor ergonomics on the other.

We should first investigate the reasons leading to the introduction of onboard automation.

During the Fifties and Sixties, the main causes of aviation accidents were believed to be related to the human factor. The immediate cause of an accident was often to be found in "active failures", e.g. loss of control of the aircraft in which pilots failed to keep the aircraft under control, reaching over-speed limits, stalling, excessive bank angles, etc.

In these cases the root cause was a flawed performance that eventually caused the loss of control (the effect). Factors related to human performance, e.g. the impact of fatigue, attention, high workload sustainability, stress mismanagement, etc. were consequently addressed. Technological solutions were sought to help pilots manage these factors. Innovation at that time eventually led to the introduction of the auto-pilot, auto-throttle, flight director, etc. After the mid-Fifties, as a result of these innovations, the accident curve dropped sharply.

Looking at the graph below, we can clearly notice the impact of such innovations on flight safety. The vertical axis corresponds to the number of accidents per million take-offs, while the horizontal axis corresponds to the relative decades.

As we can observe, after a dramatic improvement the accident curve rose again during the mid-Seventies. Aviation safety experts were faced with accidents involving a perfectly functioning aircraft, with no evidence whatsoever of malfunctions. In these cases (known as "Controlled Flight Into Terrain" - CFIT), the aircrafts were hitting obstacles with the pilots in full control. The accidents were caused by loss of situational awareness, either on the horizontal or on the vertical path. The evidence showed that an improper interaction between pilots was the main cause behind the accident, so this time the solution came from psychology. Human factor experts developed techniques and procedures to enhance cooperation between pilots, and specific non-technical training aspects became mandatory for pilots. For instance, a Cockpit Resource Management (CRM) course is nowadays mandatory in a pilot's curriculum: its topics may include leadership, cross-checking and criticizing fellow colleagues, assertiveness, resolution of conflicts, communication skills, etc.

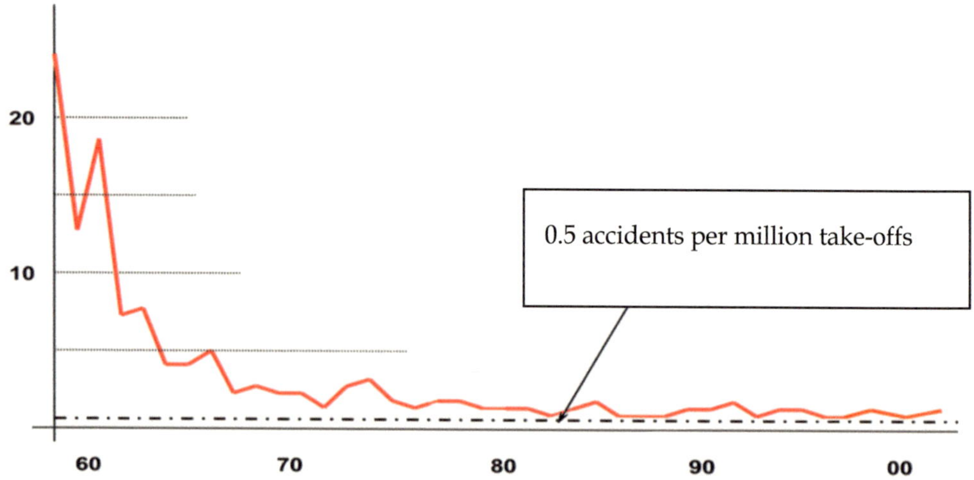

Source: ICAO doc. 9683/950. Accident rate over the years

In the last decade, the pendulum has swung back to loss of control as a major cause of aviation accidents, however, compared to the accidents occurring during the Fifties, the factors leading to loss of control appear to be different. Whereas in the beginnings of aviation, human performance was impaired by "under-redundancy", that is, insufficient aids available to pilots for avoiding the effects of factors like fatigue, distraction, workload and stress which reduced the pilots' performance, nowadays many domain experts are pointing at possible cases of "over-redundancy". This means that increasing automation might be putting the pilot out-of-the-loop, thus causing reduced situational awareness, automation complacency or over-confidence and loss of skills, due to lack of practice in manually flying the aircraft. As a result, pilots may not be able to regain control once automation has failed, or may be incapable of effectively monitoring the performance of automated systems (and questioning it when required).

The safety philosophy behind the adoption of increasing onboard automation is based on the assumption that human error is the main cause of accidents. Therefore, since the human (*liveware*) component of the system is the flawed link in the accident chain, we ought to look for a substitute capable of handling the tasks once performed by pilots. This is partially true, as we'll see later on. It is first necessary to understand what are the pros and cons of human contribution to safety, at what levels of operation does automation offer undoubted advantages and where the latter should end to leave room for pilots' decisions. Pilots and machines are not alternatives, but complementary factors in ensuring flight safety. Achieving the correct balance between these components of the aviation domain benefits safety, since the role of technology appears to have reached a standstill. Until the mid-1990s, it appears that pilots tried to adjust their behaviour to a given challenge (new automation), which was conceived regardless of their actual need and deeply rooted habits. In fact, one of the main drivers of the recent cockpit design philosophy was the reduction of costs related to better performances, lower fuel consumption, cheaper maintenance and flexible pilot training. Concern for the adaptability of pilots to these new solutions only came at a later stage and following some severe mishaps. To achieve this balance, we'll briefly analyse what levels of operation are involved in flight and where automation - primarily conceived to replace certain human operator tasks - should give way to the pilot's intervention.

4. Skill, rules and knowledge

According to a paradigm proposed by Rasmussen and also developed by James Reason, human activity can be grouped into three main fields: skill-based action, rule-based action and knowledge-based action.

The first field is based on the human capability to accomplish physical tasks, such as providing correct input to flight control (so-called "stick and rudder training"), responding to external stimuli in a quick and consistent way, and coordinating the body in order to obtain a desired result. It is mainly an area in which the psycho-physiological aspect is paramount. Moreover, we could also include monitoring tasks in this field, such as detection, identification and response to external signals stemming from habits (body automatism or conditional reaction). Automation has played an important role at this level by replacing human performance rather well. As a result, autopilots, auto-throttle (and later on, auto-thrust computers) have come to gradually replace pilots in "hand-flying". Generally speaking, an autopilot can tolerate workloads that are hardly sustainable by a pilot. Let's imagine an oceanic flight during the night; an autopilot is able to maintain (with no effort at all) altitude, speed, track and so forth, whereas pilots are subject to tiredness, attention lapses, distraction, etc. On the other hand, the systematic replacement of basic flying skills has led to the erosion of competence, because, as Germans put it: *"Die übung macht den meister"* ("Use makes master"). In the U.S. the FAA (Federal Aviation Authority) has suggested adopting "back to basics training", in which pilots are taught how to fly without the help of automatisms and how to retrieve the elementary notions of aerodynamics, in order to avoid grossly misreading altitude, speed and power.

The second level - the rule base action - is the conceptual layer. It indicates compliance with the rules, norms, laws, and everything laid down in the official documentation. It is unproblematic to apply a given rule whenever conditions warrant it. This is the case of a limit set for a device, e.g. the maximum temperature for operating an engine (EGT). When the upper limit is exceeded, something happens: red indications on instruments, alarms, flashing light on to attract the pilot's attention, automatic exclusion of the failed system and so forth. A machine can easily detect whether the operating conditions are normal or abnormal, by matching the real values with an operating envelope. Since the pilot may forget some rules, apply them incorrectly, or fail to apply valid norms, certain functions (especially those relating to the monitoring activity that induces boredom and complacency) are assigned to automation. It is a consequence of automation, therefore, that the flight engineer is no longer required in the cockpit. Some problems were initially detected in the normal flying activity of newly designed cockpits, since two pilots were required to manage a three-pilot cockpit, with automation playing the role of a "silent crew-member".

The third level - the knowledge-base - includes the sound judgement of pilots in deciding if and when the given rules are applicable. This implies the notion of a complex system. Complexity is evident at every level of reality, from physics to biology, from thermodynamics to meteorology (Morin, 2001). Different conceptions of complexity emerge in the current scientific debate, but generally speaking, we may highlight some commonalities between the different theories: refusal of reductionism, different level of system description (be it physical or biological or even man-made) according to the level of observation, emergent proprieties, etc.

Since aviation is a complex system made up of complex subsystems such as humans, technology and the environment, it is almost impossible to govern everything in advance through rules and norms. There will always be a mismatch between the required task and the final outcome (Hollnagel, 2006). The resilience engineering approach to safety is aware of this complexity and focuses on the ambivalent role - in such a system - of man, who simultaneously constitutes a threat and resource in coping with unexpected events, unforeseeable situations and flawed procedures. Much of this activity, which continuously and strategically adapts the means to the goal, is undetected either by the top management, or by the front-line operators themselves (pilots). These micro-corrections are so pervasive that the person involved in accomplishing a task fails to even realize how much he/she deviates from a given rule. The front-end operator should always seek to compromise between efficiency and thoroughness. Hollnagel calls this compromise the ETTO (Efficiency, Thoroughness Trade Off) principle. The paradox emerging from the blind application of rules - the so-called "white strike" or "working by the rule" - is that it leads to a paralysis of the entire production activity (Hollnagel, 2009). The effects of such shortcuts during normal operations are another area of concern affecting flight safety, due to the systems' opacity, the operator's superficial knowledge, uncertainties and ambiguities of the operational scenario. If we turn to the initial introduction of onboard automation, we may detect some commonalities regarding the way pilots have coped with the innovations.

5. Evolution of automation

Automation seems to follow an evolutionary path rather than a revolutionary approach. Its adoption on board aircrafts does not respond to the planned purpose of enhancing safety "from scratch" in a consistent way, but rather resembles a biological organism trying to continuously adapt to the challenges posed by its environment (fly-fix-fly).

This trial-and-response approach can be observed regardless of the fact that innovation introduced on board generally lags a step behind the overall level achieved by the industry. In fact, one of the requirements for a certain technology to be implementable in the aviation domain is its reliability; it is preferable to have a slower yet reliable system rather than a high-speed one that is not completely tested or tried in an operational environment.

We may identify three main generations of onboard automation systems: mechanical, electrical and electronic.

In the beginnings of commercial flight there were no instrumental aids to help pilots to fly. A piece of string was attached to the wing to indicate whether the airflow over it was sufficient to sustain flight. Later on, the first anemometers and altimeters were introduced to indicate pilots the airspeed and altitude, respectively. These were the first steps toward the "virtualization of the environment" (Ralli, 1993). The invention of the pneumatic gyroscope (replaced, shortly after, by the electric gyroscope), used to stabilize an artificial horizon, helped pilots to understand their situation even in meteorological conditions characterised by extremely poor visibility, while at the same time preventing dangerous vestibular illusion (a false sense of equilibrium stemming from the inner ear). These simple instruments were merely capable of providing basic indications. Early signs of automation were introduced on board aircrafts during the decade from 1920 to 1930, in the form of an autopilot based on a mechanical engineering concept that was designed to keep the aircraft flying straight: a very basic input to control the flight at a "skill" level. Moreover, as airplanes became bigger and bigger, it became necessary to apply some form of amplification of the pilots' physical force, because of the airflow over large aerodynamic surfaces. Servo-mechanisms were introduced on board, alongside certain devices aimed at facilitating perception of the force acting on such surfaces (artificial feel load, mach trim compensator) and absorbing the effects of the so-called Dutch-roll, an abnormal behaviour whereby the airplane yaws and oscillates in an uncoordinated manner (yaw damper). This (mechanical) innovation was the first of multiple steps that began widening the gap between the pilot's input (action on the yoke) and the final outcome (aerodynamic movement). Instead of direct control, with the yoke mechanically attached to the ailerons, airplanes began to be constructed with a series of mechanisms intervening between the pilot's input and the expected output. In this case, the virtualization of flight controls accompanied the parallel virtualization of flight instruments introduced by the artificial horizon (Attitude Display Indicator). At this stage, automation aided pilots mainly in their skill-based activity.

The second generation of automation included electric devices replacing the old mechanisms. Electric gyroscopes instead of pneumatic ones, new instruments such as the VOR (Very High Frequency Omni-directional Range) to follow a track based on ground aids, the ILS (Instrumental Landing System) to follow a horizontal and a vertical path till the runway threshold, and so forth. The 1960s saw plenty of innovations introduced on board aircrafts that enhanced safety: electric autopilots, auto-throttle (to manage the power setting in order to maintain a selected speed, or a vertical speed), flight directors (used to show pilots how to manoeuvre to achieve a pre-selected target such as speed, path-tracking and so forth), airborne weather radars, navigation instruments, inertial platforms, but also improved alarming and warning systems capable of detecting several parameters of engines and other equipment. Whereas the first generation of automation (mechanical) managed the pilot's skill-based level, the second generation managed the skill-based and rule-based levels previously assigned to pilots. The airplane systems were monitored through a growing number of parameters and this gave rise to a new concern: the inflation of information with hundreds of additional gauges and indicators inside the cockpit, reaching almost 600 pieces (Boy, 2011). At this stage, pilots used the technology in a tactical manner. In other words, their inputs to automation were immediately accessible, controllable and monitored in the space of a few seconds. For example, if the pilot wanted to follow a new heading, he would use a function provided by the autopilot. The desired heading value was selected in the glare-shield placed in front of the pilot's eye and the intended outcome would be visible within a few seconds: the airplane banked to the left or right to follow the new heading. End of the task. At this stage, automation helped pilots also at a rule-based level, since monitoring of thousands of parameters required efficient alarming and warning systems, as well as recovery tools.

The third generation of innovation involved electronics, and was mainly driven by the availability of cheap, accessible, reliable and usable technology that invaded the market, bringing the personal computer into almost every home. The electronic revolution occurring from the mid-80s also helped to shape the new generation of pilots, who were accustomed to dealing with the pervasive presence of technology since the early years of their life. Electronics significantly helped to diminish the clutter of instruments on board and allowed for replacing old indicators - gauges in the form of round-dial, black and white mechanical indicators for every monitored parameter - with integrated coloured displays (e.g: CRT: Cathode Ray Tube, LCD: Liquid Crystal Display) capable of providing a synthetic and analytic view of multiple parameters in a limited area of the cockpit.

It is worthy mentioning that the type of operations implied by the Flight Management System shifted from tactical to strategic. In fact, whilst in the previous electrical automation stage, pilots were accustomed to receiving immediate feedback visible shortly after the entered input, in the new version, a series of data entered by the operator would show their effects at a distance of hours. The data was no longer immediately accessible and visible, therefore this new way of operating placed greater emphasis on crew co-ordination, mutual cross-checking, operational discipline, not only in the flying tasks but also in the monitoring activity. The Flight Management System database contains an impressive amount of data, from navigational routes, to performance capabilities and plenty of useful information that can be retrieved from pilots. Further on we analyze the traps hidden behind this kind of automation.

However, it is important to point out that the actual discontinuity introduced by this generation of automation: was the notion of electronic echo-system. Compared to the past, when pilots were acquainted with the inner logic of the systems they used, their basic components, and normal or abnormal procedures for coping with operational events, in the new cockpit pilots are sometimes "out of the loop". This occurrence forces them to change their attitude towards the job. On an airplane such as the A-320 there are almost 190 computers located in almost every area of the fuselage. They interact with each other without the pilot being aware of this interaction. Every time the pilot enters an input to obtain a desired goal (e.g. activating the hydraulic system), he/she starts a sequence in which not only the selected system is activated, but also a number (unknown to the pilot) of interactions between systems depending on the flight phase, operational demands, the airplane's conditions, etc.

The unmanageable complexity of the "electronic echo-system" is a genuine epistemological barrier for the pilot. Whereas before, the pilot had thorough knowledge of the entire airplane and could strategically operate in a new and creative manner whenever circumstances required, the evolution of the cockpit design and architecture has brought about a new approach to flight management that is procedural and sequential. Only actions performed in accordance with computer logic and with the given sequence are accepted by the system. In acting as a programmer who cannot perform tasks that are not pre-planned by the computer configuration, the pilot has lost most of his expertise concerning the hardware part of the system (the aircraft). He too is constrained by the inner logic which dictates the timing of operations, even in high-tempo situations. Consequently, in the international debate on automation and the role of pilots, the latter are often referred to as "system operators". This holds true up to a certain point but, generally speaking, it is an incorrect assessment. Recalling what has been said about the levels of operation, we may say that at the skill-based level, pilot have become system operators because flying skills are now oriented to flight management system programming. As a flight instructor once said: "we now fly with our fingers, rather than with a hands-on method". What he meant was that the pilot now appears to "push-the-button" rather than govern the yoke and throttle.

At the skill-based level, automatisms may be in charge for the entire flight, relegating the pilot to a monitoring role. However, at a rule-based level, computers manage several tasks once accomplished by pilots, including monitoring the pressurization system, air conditioning system, pneumatic system and so forth. This statement is no longer valid when referred to the knowledge-based level. Here, the pilot cannot be replaced by any computer, no matter how sophisticated the latter is. Sound judgment of an expert pilot is the result of a series of experiences in which he/she fills the gap between procedures and reality. Since this paper discusses automation rather than the human factor in aviation, anyone wanting to investigate concepts such as flexibility, dealing with the unexpected, robustness, etc. may refer to authors adopting a Resilience Engineering approach (Hollnagel, 2008) (Woods, Dekker, 2010).

The fact that pilots are no longer so acquainted with the airplane as in the past, has led them to only adopt the procedural way to interact with the airplane. This is time-consuming, cognitively demanding, and above all, in some cases it may lead to miss the "big picture", or situation awareness. This new situation introduces two major consequences: automation intimidation (ICAO, 1998) and a restriction of the available tools for coping with unexpected events. In this sense, we can compare the natural echo-system - made up of a complex network of interactions, integrations, retro-actions that make it mostly unforeseeable and unpredictable - with the electronic system. On a modern aircraft, the electronic echo-system is mostly not known by its user. Pilots only know which button they have to press, what the probable outcome is, and ignore what lies is in between. It is a new way of operating that has pros and cons. To assess the real impact on safety, we ought to analyze the relationship between man and technology, by shifting from HMI (Human Machine Interaction) to HCI (Human Computer Interaction). A further step will lead to Human Machine Engineering and/or Human Machine Design (Boy, 2011). The core of this approach is that the focus of the research should be on human-centred design; in other words, the final user - with his/her mental patterns activated in real scenarios - should be regarded as the core of the entire project for a new form of automation.

From an engineering perspective, it is very strange that concepts routinely used to describe the role of onboard automation miss a basic focal point: the final user. In fact, when we talk about technology, we refer to bolt-on versus built-in systems. These expressions indicate the different pattern of integration of onboard technology. Bolt-on indicates the introduction of a new technology on board an airplane conceived without automation. It is a reactive mode that strives to combine the old engineering philosophy with new devices. It is a kind of "patchwork" which requires several local adjustments to combine new requirements with old capabilities. The expression "built-in" indicates the development of a new technology incorporated in the original project. Every function is integrated with the airplane's systems, making every action consistent with the original philosophy of use. The paradox is that the final user is left out of the original project. Nowadays, after several avoidable accidents, pilots are involved in the early stage of design in order to produce a user-friendly airplane.

6. Why automation?

Two main reasons led to the decision to adopt onboard automation: the elimination of human error and economic aspects. The first element stems from the general view whereby human performance is regarded as a threat to safety. As such a topic would require a paper in itself, it is more appropriate to briefly mention some references for students eager to investigate the topic thoroughly. The second element is easier to tackle since we can even quantify the real savings related to, say, lower fuel consumption. According to IATA estimates, "a one percent reduction in fuel consumption translates into annual savings amounting to 100,000,000 dollars a year for IATA carriers of a particular State". (ICAO, 1998). Aside from fuel, the evolution of onboard technology over the years has led to a dramatic improvement in safety, operational costs, workload reduction, job satisfaction, and so forth. The introduction of the glass cockpit concept allows airlines to reduce maintenance and overhauling costs, improve operational capabilities and ensure higher flexibility in pilot training.

a. Fuel consumption

A crucial item in an airline's balance sheet is the fuel cost. Saving on fuel is vital to remain competitive on the market. The introduction of the "fly-by-wire" concept helps to reduce fuel consumption in at least three areas: weight, balance and data predictions.

1. The fly-by-wire concept has brought a tangible innovation. Inputs coming from the pilots' control stick are no longer conveyed via cables and rods directly to the aerodynamic surfaces. In fact, the side-stick (or other devices designed to meet pilots' demands regarding a conventional yoke) provides input to a computer which – via optic fibres – sends a message to another computer placed near the ailerons or stabilizer. This computer provides input to a servo-mechanism to move the surfaces. Therefore, there is no longer any need for steel cables running through the fuselage and other weighty devices such as rods, wheels, etc. This also significantly reduces the aircraft's weight and improves fuel consumption, since less power is required to generate the required lift.
2. The second area that contributes to saving fuel is aircraft balance. The aircraft must be balanced to maintain longitudinal stability (pitch axis in equilibrium). This equilibrium may be stable, unstable or neutral. In a stable aircraft, the weight is concentrated in front of the mean aerodynamic chord. Basically, this means that the stabilizer (the tail) should "push-down" (or, technically, induce de-lifting) to compensate for the wing movements. In an unstable equilibrium, the balance of weight is shifted sensibly backwards compared to a stable aircraft. In other words, the stabilizer should generate lift to compensate for the wing movements. Where does the problem lie with unstable aircrafts? A stable aircraft tends to return to its original state of equilibrium after it deviates from the latter, but is less

manoeuvrable since the excursion of the stabilizer is narrower. On the contrary, unstable equilibrium causes an increasingly greater magnitude of oscillations as it deviates from the initial point. It makes the aircraft more manoeuvrable but unstable. In practical terms, in these kinds of airplanes pilots are required to make continuous corrections in order to keep the aircraft steady. This is why the computer was introduced to stabilize the airplane with continuous micro-corrections. This significantly reduces the pilots' workload for flying smoothly. Moreover, due to the distribution of weight concentrated on the mean aerodynamic chord, an unstable aircraft consumes less fuel.

3. The third factor helping pilots is a database capable of computing in real time any variation to the flight plan either on the horizontal path (alternative routes, short-cut, mileage calculations, etc.) or on the vertical profile (optimum altitude, top of descent to manage a low drag approach, best consumption speed, and so forth). This enhances the crew's decision-making task in choosing the best option in order to save fuel.

b. Maintenance costs

The glass cockpit concept enables airlines to reduce maintenance and overhauling costs. In conventional airplanes, every instrument had its box and spare part in the hangar. Whenever a malfunction was reported by the crew, maintenance personnel on the ground fixed it by replacing the apparatus or swapping the devices. All these actions required a new component for every instrument. If we consider that an airliner has roughly one million spare parts, we can easily understand the economic breakthrough offered by the glass cockpit concept.

In these airplanes, a single computer gives inputs to several displays or instruments. The maintenance approach is to change a single computer rather than every component or actuator. Based to this operating method, few spare parts are required in the hangar: no more altimeters, no more speed indicators, no more navigation displays (often supplied different manufacturers). Moreover, training of maintenance personnel is simplified as it focuses on a few items only which, in turn, allows for increased personnel specialisation.

c. Selection and Training costs

The fast growth pace of the airline industry over the last decades has generated concern about the replacement of older pilots, since training centres cannot provide the necessary output for airline requirements. Hiring pilots from a limited base of skilled workers creates a bottleneck in the industrial supply of such an essential organizational factor as are pilots for an airline. Automation has facilitated the hiring of new pilots, since the basic skills are no longer a crucial item to be verified in the initial phase of a pilot's career. If, barely thirty years ago, it would not have been sufficient "to have walked on the moon to be hired by a major airline", as an expert pilot ironically put it, nowadays the number of would-be pilots has increased exponentially. In the current industrial philosophy, almost everyone would be able to fly a large airliner safely with a short amount of training. This phenomenon gives rise to new and urgent problems, as we'll now see.

Besides the selection advantages, broadening the potential pilot base enables airlines to save money for the recurrent training of pilots or to reduce transition costs. Indeed, once the manufacturer sets up a standardized cockpit display, the latter is then applied to a series of aircraft. If we look at the Airbus series comprising A-319, A-320, A-321, A-330, A-340, etc., we realize how easy is to switch from one airplane to another. In this case, the transition course costs are significantly lower since pilots require fewer lessons; indeed, aside from certain specific details, the only difference involves the performance (take-off weight, cruise speed, landing distances, etc.).

Since pilot training costs make up a considerable portion of an airline's budget, it is financially convenient to purchase a uniform fleet made up of same "family" of airplanes. Often, the regulations enable airlines to use a pilot on more than one airplane belonging to a "family". Consequently, this leads to shorter transition courses, operational flexibility as pilots get to fly several aircrafts at a time, and in the long-run, better standardization among pilots.

Some authors have pointed out how automation has redefined the need for different training processes and crew interaction (Dekker, 2000).

d. Operational flexibility

A pilot flying with no aids at all, be they mechanical, electrical or electronic, is limited in many ways.

He/she must fly at low altitude, because of his/her physiological limits (hypoxia), he/she cannot fly too fast since the effort on the yoke exceeds his/her physical power, he/she cannot even fly in bad weather (clouds or poor visibility) since he/she must maintain visual contact with the ground. Automation allows for overcoming such limitations. Higher flight levels also mean lower fuel consumption and the possibility of flying out of clouds. Faster speeds allow for reaching the destination earlier and completing multiple flights a day. Onboard instruments enable pilots to achieve better performance. Let's imagine an approach in low visibility conditions. In the beginnings, pilots would rely on a Non-Directional Beacon (NDB) as an aid to find the final track to land on the runway. Safety measures were implemented such as the operating minima. These implied that during a final landing approach, the pilot had to identify the runway before reaching a certain altitude (landing minima). Obviously, as instruments became more reliable, the landing minima were lowered. Subsequently, the introduction of the VOR (Very High Frequency Omnidirectional Range), which provided a more accurate signal, allowed for performing an approach closer to the runway and at a lower "decision altitude", as the safety margins were assured. When the ILS (Instrumental Landing System) was introduced on board, pilots could also rely on vertical profile indications. This implied higher safety margins that led the regulatory bodies to once again lower the landing minima. As the ILS became increasingly precise and accurate, the landing minima were lowered till they reached the value of zero. This means that an airplane can touch down with such low visibility that pilots must rely on autopilot to perform such an approach. This is due to human limitations such as visual and vestibular illusions (white-out, wall of fog, duck under, etc.) that may impair the pilot's performance. Let's imagine an approach during the '50s: with a visibility of 1,000 metres at the destination airport, the airplane should have diverted to another airport because the pilots would not have been able to attain the airport visual reference at the decision altitude (landing minima). Nowadays, the same airport could operate with 100 metres visibility due to improved transmitting apparatus on the ground and flight automation that enhances pilot performance.

A problem evidenced by several authors concerns the shift in responsibility of the people managing the automation system. With the performances brought about by automation that exceed human capabilities, a failure in the automatic system places pilots in a situation in which they have no resources available for coping with an unexpected event. In one case, where the pilots were performing a low visibility approach, the automatic system went out of control at very low altitude, causing the airplane to pitch down and hit hard on the runway. Experts acknowledge that the pilots' recovery in that case was beyond reasonable intervention, that is to say "impossible" (Dismukes, Berman, Loukopoulos, 2008). Indeed, the time available for detecting, understanding and intervening was so short that it was virtually impossible to solve the airplane's faulty behaviour. Who is responsible in such a case? Is it correct to refer to the pilots' mismanagement, poor skill, or untimely reaction when they are operating outside their safety boundaries? If the operations are conducted beyond human capabilities we should also review the concept of responsibility when coping with automation. As Boy and Grote put it: "*Human control over technical systems, including transparency, predictability, and sufficient means of influencing the systems, is considered to be the main prerequisite for responsibility and accountability*" (Boy, 2011).

Classic T-model

7. Different cockpit presentation

In recent years, cockpit design has undergone somewhat of a revolution. Old aircraft were designed to satisfy pilots' needs and relied upon slight modifications in the previous onboard scanning pattern. A common standardized cockpit display emerged in the early '50s: the T-model. It included the basic instruments such as artificial horizon (in the top-centre), anemometer (speed indicator) on the left, the altimeter on the right and compass (indicating the heading and track) in the bottom-centre position. Two further instruments, the side-slip indicator and vertical speed indicator were added later on. The adjacent picture shows the classic T-model.

Primary flight Display

With the introduction of the glass cockpit, the traditional flight instrument display was replaced by a different presentation encompassing more information, greater flexibility, colour coding and marking, but at the same time, it could lead to information overload, as several parameters are displayed in a compact area. In a single display, known as the PFD (Primary Flight Display), multiple information is included not only for basic parameters, but also for navigation functions, approach facilities, automatic flight feedback, flight mode awareness, and so forth.

Since the pilot needs to cross-check a great number of instruments at a glance, the design of traditional instruments was developed according to a pattern of attention, from the more important information to the more marginal information. Each instrument had its own case and its functional dynamics was perfectly known by the pilot; it was easy to detect, easy to use during normal operations and easy to handle in case of failure. The switch positions and shapes were positioned on board according to a pattern of use and were instantly recognisable at the touch. It was common for pilots to perform the checklist according to the "touch and feel" principle, by detecting whether a switch was in the required configuration by confirming its on/off position. Almost every switch had a peculiar shape: i.e., rough and big for operative ones, smooth and small for less important switches. A form of training implemented in most flying schools was the "blind panel" exercise, which consisted in covering the trainee's eye with a bandage and asking him/her to activate the switch prompted by the instructor. Being familiar with the physical ergonomics of the cockpit proved very helpful. The pilot knew how much strength was required to activate a command, how much the arm had to be stretched to reach a knob and so forth. This exercise enhanced the pilot's skill, or the skill level of operation (according to Rasmussen's Skill-

Rule-Knowledge paradigms). Regarding the input to the flight controls, it was assured via flight control wheels and sticks, cables and rods, in order to enable the pilot to intervene directly on the aerodynamic surfaces such as the ailerons, rudder and stabilizer.

With regard to displays, the traditional instrument configuration - made up of drum pointers, single instrument boxes and "touch and feel" overhead panel - was replaced by a new approach towards onboard information available to the pilot. The "touch and feel" philosophy was replaced by the "dark cockpit" approach. A dark, flat, overhead panel was adopted to show the pilot that everything was OK. Failures or anomalies were detected by an illuminated button, recalled by a master caution/warning light just in front of the pilots' eyes. In the unlikely (but not impossible) situation of smoke in the cockpit severely impairing the pilots' sight, (due to the concentration of a thick layer of dense, white fog), pilots of conventional airplanes could retrieve the intended system's configuration "by heart", by detecting the switch position with a fingertip.

Vice-versa, let's imagine a dark panel with smoke on board. It is truly challenging to spot where to place your hand and, above all, what the system's feedback is, since the system configuration is not given by the switch position but by an ON/OFF light, which is often undetectable.

8. Ergonomics

The problem with innovation, especially in a critical context for safety as is aviation, lies in the interaction between a community of practice and a new concept, conceived and implemented by engineers. This implies a relationship between automation and ergonomics. Ergonomics is a word deriving from the Greek words *ergon* (work) and *nomos* (law), and it is a field of study aimed at improving work conditions so as to guarantee optimal adaptation of the worker to his/her environment. We may identify three main types of ergonomics: physical, cognitive and social (or organizational).

Initially, with the onset of the first generation of automation (mechanical), ergonomists studied physical ergonomics, namely: how to reach a control, how much force is required to operate a lever, visibility of displayed information , seat position design, and so forth. For example, several accidents occurred due to misuse of the flap lever and landing gear lever. Indeed, these were positioned near one another and had similar shapes, leading the pilots to mistake them (Koonce, 2002). The ergonomists found a straightforward and brilliant solution: they attached a little round-shaped rubber wheel to the lever tip to indicate the landing gear, and a wing-shaped plastic cover indicating the flaps. Moreover, the two levers were separated in order to avoid any misuse.

In the next generation of automation, namely electrical automation, the ergonomists aimed to improve the cockpit design standardization. A pilot flying on an airplane had difficulty in forming a mental image of the various levers, knobs and displays, as every manufacturer arranged the cockpit according to different criteria. Dekker has clearly illustrated a case of airplane standardization: the position of the propeller, engine thrust and carburettor in a cockpit (Dekker, 2006). Pilots transitioning to other aircrafts struggled to remember the position of every lever since they were adjacent to each other, and the possibility of mistaking them was very high. Moreover, during the training process there was a risk of the so-called "negative transfer", that is, the incorrect application of a procedure no longer suited to the new context. For example, the switches on Boeing aircrafts have top-down activation, while Airbus has chosen the swept-on system (bottom-up). A pilot transitioning from a Boeing 737 to an Airbus A-320 may quickly learn how to switch on the systems, but in conditions of stress, fast-paced rhythm and high workload, he/she may return to old habits and implement them incorrectly. Social ergonomics tries to eliminate such difficulties.

Cognitive ergonomics studies the adaptability of the technology to the mental patterns of the operator. The third generation of automation introduced on board aircrafts gave rise to many concerns about whether the instrument logic was suitable for being correctly used by pilots. From a cognitive perspective, the new system should be designed according to the user's need, while bearing in mind that pilot-friendly is not equivalent to user-friendly (Chialastri, 2010). A generic user is not supposed to fly, as some engineers erroneously tend to believe. Pilots belong to a professional community that share a mental pattern when coping with critical operational situations. The *modus operandi* is the result of a life-long in-flight experience. A professional community such as the pilot community is target-oriented, and aims at the final goal rather than observing all the required procedural tasks step-by-step. This is why we talk about "shortcuts", "heuristics", and so forth.

For example purposes, let's discuss the introduction of a new airplane with variable wings: the F-111. It was conceived by the designers with a lever in the cockpit to modulate the wings from straight (useful at low speed) to swept (to fly at high speed). In the mind of the designer (who is not a pilot), it seemed perfectly sensible to associate the forward movement of the lever in the cockpit with the forward movement of the wings: lever forward - straight wings, lever back - swept wings. When the airplane was introduced in flight operations, something "strange" occurred. Pilots associated the forward movement with the concept of speed. So, when they wanted high speed, they pushed the thrust levers forward, the yoke down to dive and, consequently, the wing-lever forward (incorrectly so). For a thorough analysis of this case, see Dekker (2006).

All this occurred due to the confusion between generic users and specialized users, as are pilots. As Parasumaran puts it: "Automation does not simply supplant human activity but rather changes it, often in ways unintended and unanticipated by the designers of automation and, as a result, poses new coordination demands on the human operator" (Parasumaran, Sheridan, 2000).

9. Automation surprise

The interaction between pilots and technologies on board the aircraft raises some concerns regarding an acknowledged problem: the automation surprise. This occurs when pilots no longer know what the system is doing, why it is doing what it does and what it will do next. It is an awful sensation for a pilot to feel to be lagging behind the airplane, as - ever since the early stages of flight school - every pilot is taught that he/she should be "five minutes ahead of the airplane" in order to manage the rapid changes occurring during the final part of the flight. When approaching the runway, the airplane's configuration demands a higher workload, the communication flow with air traffic controllers increases and the proximity to the ground absorbs much of the pilot's attention.

Once all the fast-pace activities are handled with the aid of automation, the pilot may shed some of the workload. When automation fails or behaves in a "strange" manner, the workload increases exponentially. This is only a matter of the effort required to cope with automation, and could be solved with additional training. A more complicated issue - arising from investigations on some accidents - is the difference between two concepts, namely: "situational surprise" and "fundamental surprise" (Wears, 2011). To summarise our explanation, we may refer to the example mentioned by Wears regarding an everyday situation: a wife returns home unexpectedly and finds her husband in bed with the waitress. She is shocked and the only thing she can say is: "Darling, I am surprised". Her husband calmly replies, "no, darling, I am surprised; you are astonished". In this example, the husband knows he is doing something regrettable, but accepts the risk even if he doesn't expect his wife to return home. He is surprised, but has calculated even the worst case (wife back home sooner than expected). The wife comes home and finds a situation she is not prepared for. She is surprised and doesn't know what to do at first, because she lacks a plan. No doubt she will find one…

Therefore, situation surprise arises when something occurs in a pre-defined context, in which the pilots know how to manage the situation. Such a situation is unexpected, being an abnormal condition, but nonetheless remains within the realm of the pilot's knowledge. Moreover, it returns to being a manageable condition once the pilot acknowledges the difference between a normal and an abnormal condition, by applying the relevant abnormal procedure. After the initial surprise due to the unexpected event, the pilot returns to a familiar pattern of operation acquired with training.

A so-called fundamental surprise is a very different circumstance: it implies a thorough re-evaluation of the situation starting from the basic assumptions. It is an entirely new situation for which there is no quick-fix "recipe" and is difficult to assess, since it is unexampled. The pilots literally do not know what to do because they lack any cognitive map of the given situation. This concept can be well summarised by the findings following an accident involving a new generation airplane. When analysing the black box after an accident, it is not rare to hear comments such as: "I don't know what's happening" - this is a something of a pilot's nightmare.

It should be noted that some failures occurring on board highly automated aircrafts are transient and are very difficult to reproduce or even diagnose on the ground. Even the manufacturer's statements regarding electrical failures are self-explaining, as a failure may produce different effects in each case or appear in different configurations depending on the flight phase, airplane configuration, speed and so forth. Sometimes, an electrical transient (also influenced by Portable Electrical Devices activated by unaware passengers) could trigger certain airplane reactions that leave no memory of the root cause.

10. Automation issues

Although automation has improved the overall safety level in aviation, some problems tend to emerge from a new way of operating. A special commission was set up in 1985 by the Society of Automotive Engineers to determine the pros and cons arising from flight deck automation. Nine categories were identified: situation awareness, automation complacency, automation intimidation, captain's command authority, crew interface design, pilot selection, training and procedures, the role of pilots in automated aircrafts (ICAO, 1998). Sometimes, pilots loose situation awareness because they lag behind the automation logic. A pilot should always know what the automation system is going to do in the next five minutes, to either detect any anomalies or take control of the airplane. In fast-paced situations - such as crowded skies, proximity with the terrain and instrumental procedures carried out in poor visibility - a high number of parameters must be monitored. As the workload increases, the pilot tends to delegate to the automation system a series of functions in order to minimise the workload. It is important to specify that the physical workload (the number of actions performed within a given time frame) differs from the cognitive workload in that the latter implies a thorough monitoring, understanding and evaluation of the data coming from the automation system. The paradox involving airplane automation is that it works as an amplifier: with low workloads, it could lead to complacency ("let the automation system do it") that reduces alertness and awareness, while the latter increase with high workloads, due to the high number of interactions and data involved in fast-paced situations. Plainly said: "When good, better; when bad, worse".

Poor user interface design is another issue. Norman, Billings et al. have studied the way humans interact with automation and have developed cognitive ergonomics. There are several studies concerning the optimal presentation (displays, alarms, indicators, etc.) for pilot use. For example, old displays were designed in such a way that each single instrument provided data for a specific domain only: the anemometer for speed, the altimeter for altitude and flight levels, and so forth. With the integrated instruments, all the required data is available at a glance in a single instrument, which not only provides colour-coded information but also on-demand information by means of pop-up features, so that it can be concealed during normal operations and recalled when needed.

Several issues are related to displays, but only two are addressed here: cognitive mapping of instruments and human limitations in applying the colour-code to perception.

We have to consider that the replacement of round-dial black and white instruments, such as the old anemometer, with coloured multi-function CRT (Cathodic Ray Tube) instruments, has determined a new approach by pilots in mentally analysing speed data. Indeed, with round-dial indicators, pilots knew at a glance in which speed area they were flying: danger area, safe operations or limit speed. Although it is not important to know the exact speed, according to a Gestalt theory concept we need to see the forest rather than every single tree.

Furthermore, from a cognitive perspective, we tend to perceive the configuration as a multiple instrument: for example, during the approach phase, the anemometer needle will point to 6 o'clock, the vertical speed indicator to 8 o'clock and so forth.

In the new speed indicator, the values appear on a vertical speed tape rolling up and down, with the actual speed magnified for greater clarity. Moreover, additional indicators are displayed on the speed tape, such as minimum speed, flap retraction speed and maximum speed for every configuration. In many ways, this system facilitates the pilot's task. Problems arise whenever failures occur (electrical failure, unreliable indication etc.), as the markings disappear.

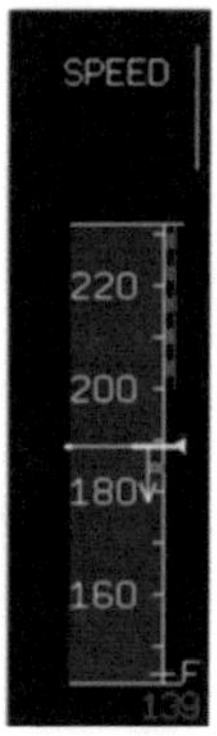

The situation worsens compared to the old indicators, as pilots needs to create a mental image of the speed field in which they are flying. For example, it is unimportant for us to know whether we are flying at 245 or 247 knots, but it becomes important to know that we are flying at 145 knots rather than 245 knots. This means that the mental mapping of the speed indication involves a higher workload, is time-consuming and also energy-consuming. Let's take a look at these two different speed indicators to grasp the concept.

The second issue is related to human performance and limitations. The human eye may detect colours, shape, light and movement with foveal vision. There are two kinds of receptors in the fovea: the cones and the rods.

The cones are responsible for detecting shapes and colours, in a roughly 1 cent large area, and account for focused vision. Instead, rods are responsible for peripheral vision and are able to detect movement and light. This means that colour-coded information should appear in an area just in front of the pilot's head, as only the cones may such information. In fact, since an input appearing to the side can be detected by the rods, which are insensitive to colour and shape, the pilot will not notice it unless he stares at it directly.

Other relevant indicators for pilots are symmetry and context. The needle in a round-dial instrument satisfies both these requirements, since we may easily detect which field of operation we are currently in (near the upper limit, in the middle, in the lower margin), in addition to the trend (fast-moving, erratic, gradual) and symmetry with nearby parameters. A digital indicator shows this information in an alternative way, by arranging numbers on a digital display, as can be clearly seen in the above picture. The configuration on the left can be grasped at first glance, while the second set of information requires a considerable effort to detect differences.

Disrupted symmetry **Digital information**

Another area worthy of attention is the greater magnitude of the errors made by pilots when flying highly automated airplanes. Pilots flying older aircrafts normally relied on their ability to obtain the required data using a heuristic and rule-of-thumb approach. This method was not extremely accurate but roughly precise. Nowadays, with the all-round presence of computers on board, flight data can be processed very precisely but with the risk of gross errors due to inaccurate entries made by pilots. An example taken from everyday life may help to explain the situation: when setting older-version alarm clocks, the user's error might have been limited to within the range of a few minutes, whereas the new digital alarm clocks are very accurate but subject to gross errors – for example 8 PM could be mistaken for 8AM.

Lower situation awareness means that pilots no longer have the big picture of the available data. The aviation truism, "trash in-trash out", is applicable to this new technology as well. Having lost the habit of looking for the "frame", namely the context in which the automation system's computed data should appear, pilots risk losing the big picture and, eventually, situation awareness. Unsurprisingly, there have been cases of accidents caused by inconsistent data provided by the computer and uncritically accepted by pilots. This is neither a strange nor uncommon occurrence. Fatigue, distractions, and heavy workload may cause pilots to lower their attention threshold. In one case, a pilot on a long haul flight entered the available parameters into the computer to obtain the take-off performance (maximum weight allowed by the runway tables, speed and flap setting). He mistook the take-off weight with other data and entered the parameters into the computer. The end result was a wrong take-off weight, wrong speed and wrong flap setting. The aircraft eventually overrun the runway and crashed a few hundred meters beyond the airport fences.

Another issue linked the introduction of onboard automation is a weakening of the hierarchy implied by the different way of using the automation system. The captain is indeed the person with the greatest responsibility for the flight and this implies a hierarchical order to establish who has the final word on board. This hierarchical order is also accompanied by functional task sharing, whereby on every flight there is a pilot flying and a pilot not flying (or monitoring pilot). Airplanes with a high level of automation require task sharing, in which the pilot flying has a great degree of autonomy in programming the Flight Management System, in deciding the intended flight path and type of approach. Above all, the pilot flying also determines the timing of the collaboration offered by the pilot not flying, even in an emergency situation. In fact, since the crew must act in a procedural way in fulfilling the demands of the automation system, both pilots must cooperate in a more horizontal way compared to the past. The hierarchical relationship between captain and co-pilot is known as the "trans-authority gradient" (Hawkins, 1987) and, from a human factor perspective, should not be too flat nor too steep.

11. Case studies

Currently, several case studies could be cited to illustrate the relationship between pilots and automation. In the recent past, accidents have occurred due to lack of mode awareness (A-320 on Mount St. Odile), automation misuse (Delhi, 1999), loss of braking leading to loss of control (S. Paulo Garulhos, 2006), loss of control during approach (B-737 in Amsterdam, 2008), and several other cases.

From these accidents, we chose a couple of peculiar cases. The first case occurred in 1991 and involved an Airbus flying at cruise altitude. While the captain was in the passenger's cabin, the co-pilot tried to "play" with the FMS in order to learn hands-on. He deselected some radio-aids (VOR: Very High Frequency Omni-directional Range) from the planned route. After the tenth VOR was deselected, the airplane de-pressurized, forcing the pilot to perform an emergency descent. From an engineering perspective there was a bug in the system but, moreover, it was inconceivable for the pilot to link the VOR deselection (a navigation function) to the pressurization system.

The second case is related to a B-727 performing a low-visibility approach to Denver. The flight was uneventful till the final phase. The sky was clear but the city was covered by a layer of fog that reduced the horizontal visibility to 350 ft. and the vertical visibility to 500 ft. The captain was the pilot flying, in accordance with the company regulations, and used the autopilot as specified in the operating manual. The chosen type of approach was an ILS Cat II. The ILS is a ground based navigation aid that emits a horizontal signal to guide the airplane to the centre of the runway together with a vertical signal to guide the aircraft along a certain slope (usually 3°) in order to cross the runway threshold at a height of 50 ft. The regulations state that once the aircraft is at 100 ft., the crew must acquire the visual references to positively identify the runway lights (or markings). If not, a go-around is mandatory. Once the captain has identified the runway lights he must proceed to land manually, by switching off the autopilot at very low altitude. There are certain risks associated with this manoeuvre, since – with impaired visibility – the pilot loses the horizon, depth perception and the vertical speed sensation, and could be subject to spatial disorientation due to the consequences of the so-called "white-out" effect. In this case, however, nothing similar occurred. The autopilot duly followed the ILS signals, but due to a random signal emitted in the last 200 ft., the aircraft pitched down abruptly. The captain tried to take over the controls, though unsuccessfully. He later reported – during the investigation – that the window was suddenly "full of lights", meaning that the aircraft assumed a very nose-down pitch attitude. The high rate of descent, coupled with the surprise factor and low visibility, did not allow the captain to recover such a degraded situation: the airplane touched down so hard that it veered off the runway, irreversibly damaging the fuselage.

This case emphasizes the relationship between pilots and liability. Operating beyond human capabilities, the pilot finds himself/herself in a no-man's land where he/she is held responsible even when he/she cannot sensibly regain control of the airplane. As stated by Dismukes and Berman, it is hardly possible to recover from such a situation.

12. Conclusion

This brief paper highlights certain aspects concerning the introduction of automation on board aircrafts. Automation has undeniably led to an improvement in flight safety. Nevertheless, to enhance its ability to assure due and consistent help to pilots, automation itself should be investigated more thoroughly to determine whether it is suitable in terms of human capabilities and limitations, ergonomics, cognitive suitability and instrument standardization, in order to gradually improve performance.

13. References

Air Pilot Manual, (2011), *Human Factors and pilot performance*, Pooleys, England

Alderson D.L., Doyle J.C., "Contrasting view of Complexity and Their Implications For Network-Centric Infrastructures", in *IEEE Transaction on Systems, Man and Cybernetics* – part A: Systems and Humans, vol 40 No. 4, July 2010

Amalberti R., "The paradox of the ultra-safe systems", in Flight Safety Australia, September-October 2000

Bagnara S., Pozzi S., (2008). Fondamenti, Storia e Tendenze dell'HCI. In A. Soro (Ed.), *Human Computer Interaction. Fondamenti e Prospettive*. Monza, Italy: Polimetrica International Scientific Publisher.

Lisanne Bainbridge, (1987), "The Ironies of Automation", in "New technology and human error", J. Rasmussen, K. Duncan and J. Leplat, Eds. London, UK: Wiley,.

Barnes C., Elliott L.R., Coovert M.D., Harville D., (2004), "Effects of Fatigue on Simulation-based Team Decision Making Performance", Ergometrika volume 4, Brooks City-Base, San Antonio TX

Boy G., a cura di, (2011), *The Handbook of human Machine Interface – A Human-centered design approach,* Ashgate, Surrey, England

Chialastri Antonio (2011), "Human-centred design in aviation", in *Proceedings of the Fourth Workshop on Human Centered Processes,* Genova, February 10-11

Chialastri Antonio (2011), "Resilience and Ergonomics in Aviation", in *Proceedings of the fourth Resilience Engineering Symposium* June 8-10, 2011, Mines ParisTech, Paris

Chialastri Antonio (2010), "Virtual Room: a case study in the training of pilots", HCI aero-conference, Cape Canaveral

Cooper, G.E., White, M.D., & Lauber, J.K. (Eds.) (1980) "Resource management on the flightdeck," Proceedings of a NASA/Industry Workshop (NASA CP-2120)

Dekker S., "Sharing the Burden of Flight Deck Automation Training", in The International Journal of Aviation Psychology, *10*(4), 317–326 Copyright © 2000, Lawrence Erlbaum Associates, Inc.

Dekker S. (2003), "Human Factor in aviation: a natural history", Technical Report 02 - Lund University School of Aviation

Dekker S., Why We need new accident models, Lund University School of Aviation,Technical Report 2005-02

Dismukes, Berman, Loukopoulos (2008), *The limits of expertise*, Ashgate, Aldershot, Hampshire

Ferlazzo F. (2005), *Metodi in ergonomia cognitiva*, Carocci, Roma

Flight Safety Foundation (2003), "The Human Factors Implications for Flight Safety of Recent Development In the Airline Industry", in *Flight Safety Digest*, March-April

Hawkins Frank (1987), *Human factor in Flight*, Ashgate, Aldershot Hampshire

Hollnagel E., Woods D., Leveson N. (2006) (a cura di), *Resilience Engineering – Concepts and Precepts*, Ashgate, Aldershot Hampshire

Hollnagel Erik, (2008) "Critical Information Infrastructures : should models represent structures or functions ?", in *Computer Safety, Reliability and Security*, Springer, Heidelberg

Hollnagel Erik, (2009), *The ETTO Principle – Efficiency-Thoroughness Trade-Off*, Ashgate, Surrey, England

Hutchins Edwin (1995), "How a cockpit remembers its speeds", Cognitive Science, n. 19, pp. 265-288.

IATA (1994), *Aircraft Automation Report*, Safety Advisory Sub-Committee and Maintenance Advisory Sub-committee.

ICAO - Human Factors Digest No. 5, Operational Implications of Automation in Advanced Technology Flight Decks (Circular 234)

ICAO (1998) – Doc. 9683-AN/950, Montreal, Canada

Köhler Wolfgang (1967), "Gestalt psychology", Psychological Research, Vol. 31, n. 1, pp. 18-30

Koonce J.M., (2002), *Human Factors in the Training of Pilots*, Taylor & Francis, London

Maurino D., Salas E. (2010), *Human Factor in aviation*, Academic Press, Elsevier, MA, USA

Morin E. (2008), *Il Metodo – Le idee: habitat, vita organizzazione, usi e costumi*, Raffaello Cortina editore, Milano

Morin E. (2004), *Il Metodo – La vita della vita*, Raffaello Cortina editore, Milano

Morin E. (2001), *Il Metodo – La natura della natura*, Raffaello Cortina editore, Milano

Morin E. (1989), *Il Metodo – La conoscenza della conoscenza*, Feltrinelli editore, Milano

David Navon (1977), "Forest before trees: the precedence of global features in visual perception", Cognitive Psychology, n. 9, pp. 353-383

Norman D. (1988), *The psychology of everyday things*, New York, NY: Basic Books, 1988.

Parasumaran, Sheridan , "A Model for Types and Levels of Human Interaction with Automation" IEEE Transactions on Systems, Man, and Cybernetics—part A: Systems and Humans, Vol. 30, No. 3, MAY 2000

Parasumaran R., Wickens C., "Humans: Still Vital After All These Years of Automation", in *Human Factors*, Vol. 50, No. 3, June 2008, pp. 511–520.

Ralli M. (1993), *Fattore umano ed operazioni di volo*, Libreria dell'orologio, Roma

Rasmussen J., *Skills, Rules, Knowledge: Signals, Sign and Symbol and Other Distinctions in Human Performance Models*, In "IEEE Transactions Systems, Man & Cybernetics", SMC-13, 1983

Reason James, (1990) *Human error*, Cambridge University Press, Cambridge

Reason James (2008), *The human contribution*, Ashgate, Farnham, England

Salas E., Maurino D., (2010), *Human factors in aviation*, Elsevier, Burlington, MA, USA

Wears Robert, Kendall Webb L. (2011), "Fundamental on Situational Surprise: a case study with Implications for Resilience" in *Proceedings of the fourth Symposium on Resilience Engineering*, Mines-Tech, Paris.

Woods D., Dekker S., Cook R., Johannesen L., Sarter N., (2010), *Behind human error*, Ashgate publishing, Aldershot, England

Woods D. D., *Modeling and predicting human error,* in J.Elkind S. Card J. Hochberg and B. Huey (Eds.), *Human performance models for computer aided engineering* (248-274), Academic Press 1990

Wright Peter, Pocock Steven and Fields Bob, (2002) "The Prescription and Practice of Work on the Flight Deck"*Department of Computer ScienceUniversity of YorkYork YO10 5DD*

6

Power System and Substation Automation

Edward Chikuni
Cape Peninsula University of Technology
South Africa

1. Introduction

Automation is "the application of machines to tasks once performed by human beings, or increasingly, to tasks that would otherwise be impossible", Encyclopaedia Britannica [1]. The term automation itself was coined in the 1940s at the Ford Motor Company. The idea of automating processes and systems started many years earlier than this as part of the agricultural and industrial revolutions of the late 18th and early 19th centuries. There is little disputing that England was a major contributor to the Industrial Revolution and indeed was the birth place of some prominent inventors, for example. in the area of textiles:

- James Hargreaves: Spinning Jenny
- Sir Richard Arkwright: Mechanical Spinning Machine
- Edmund Cartwright: Power Loom

Cartwright's power loom was powered by a steam engine. At these early stages we see the symbiotic relationships between automation, energy and power. The early forms of automation can only largely be described as mechanisation, but the emergence of electrical power systems in the late 19th century and the entry of electronic valves in the early 20th century heralded the humble beginnings of modern automation. With electronic valves came computers. One of the earliest computers was the ENIAC (Electronic Numerical Integrator and Automatic Computer) built over two years between 1943 and 1946. It occupied an area of 1000 square feet (about 93 square metres), had 18000 valves and consumed 230 kW [2].

Before the deployment of computers in industrial automation, relays and RELAY LOGIC, the wiring of circuits with relays to achieve automation tasks, was in common use. Today, however, relay logic is far less used than computer-based, PROGRAMMABLE LOGIC, which has followed the invention of the transistor, integrated circuits and microprocessors.

2. Automation in the automobile (car / truck) industry

The motor assembly line pioneered by Ransom Olds and Henry Ford, the maturity of computer technology and the structured nature of the car assembly process led early entrepreneurs in the motor industry to view automation as key to business success. Indeed there was all to be gained in automation and today automation is viewed to have, among many other attributes, the following [3]:

a. relieves humans from monotony and drudgery
b. relieves humans from arduous and dangerous tasks
c. increases productivity and speeds up work rates
d. improves product quality
e. reduces costs and prices
f. increases energy and material savings
g. improves safety
h. provides better data capture and product tracking.

The automotive industry was very competitive right from the early days and automation soon came to be seen as key to commercial success. General Motors embraced it and so did many other auto manufacturing corporations in the US, Europe and Japan. The computer system used was designed for the usually harsh industrial environments. Programmable Logic Controllers (PLC) as such computers are known typically to have many inputs and outputs; the inputs receiving sensor signals, the outputs being for displaying information or driving actuators (Figure 1).

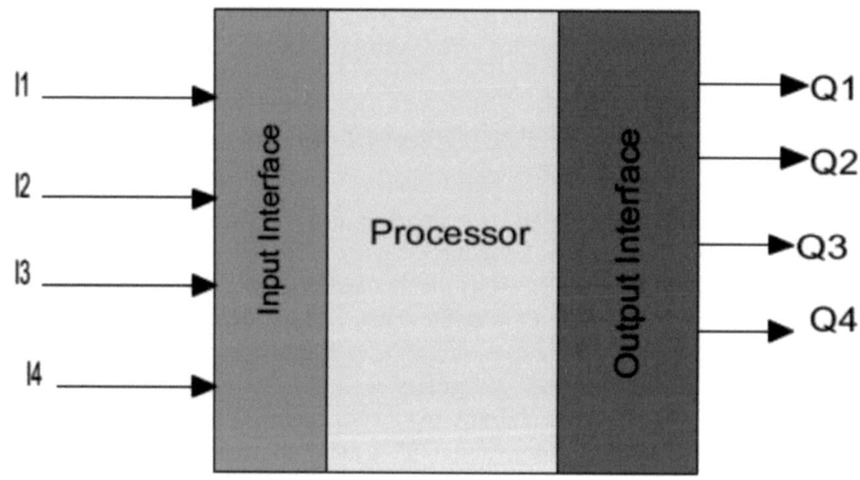

Fig. 1.

The need for standardization was realized early, especially when many PLCs are required to achieve the automation task. The standards generating body IEEE has been a driving force in computer communication standards over the years. One of its standards, the token ring 802.4 was implemented in modified form by General Motors in its Manufacturing Automation Protocol (MAP).

3. Power system automation

The early power plants had a modest number of sensor and action variables, of the order of several hundreds. Modern large power stations have in excess of hundreds of thousands, even tens of million variables [3]. It is therefore easy to see that automation took root in power and generating stations earlier than in transmission and distribution. One of the useful applications of automation is in the railway industry where remote control of power is often vital. This manifests itself in systems to control and manage power supplies to electric locomotives. The deprivation of power to any locomotives in a section could seriously disrupt schedules, inconvenience customers and in the end have serious adverse financial implications. Consider Figure 2

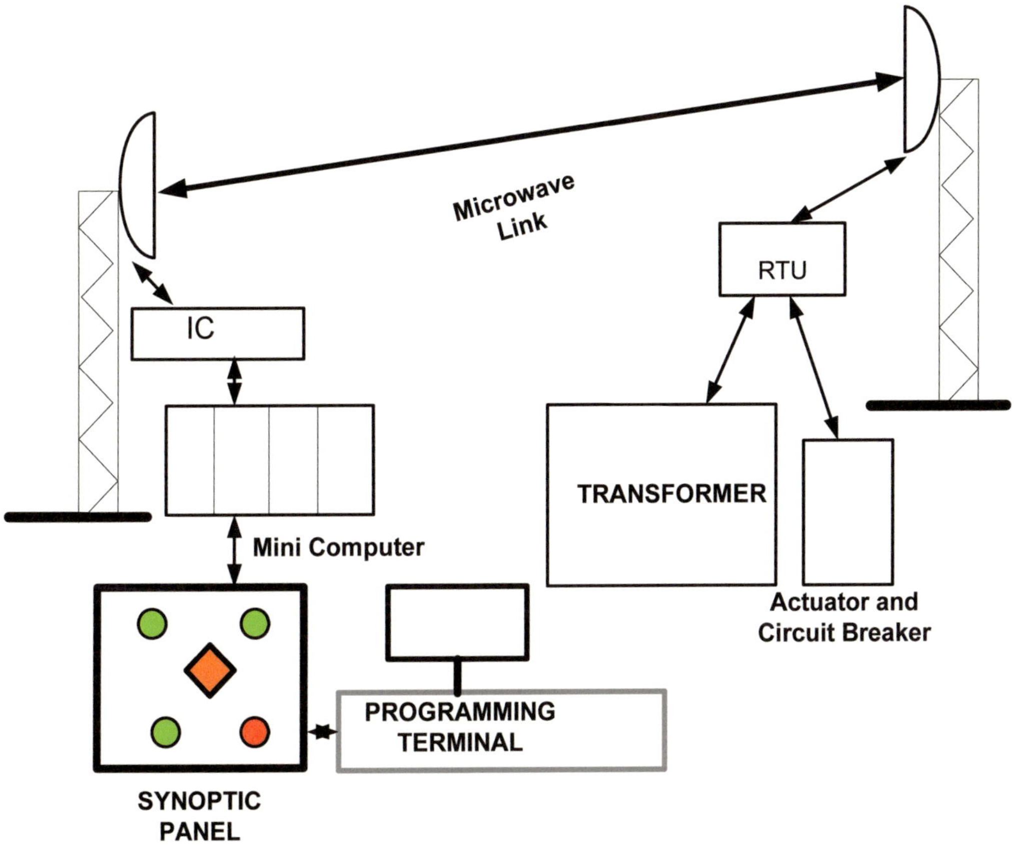

Fig. 2.

In earlier systems, the "mini computer" would have been sourced from specialised computer companies such as IBM, Perkin Elmer, Digital Equipment Corporation. The operating systems would have belonged to the same equipment provider (e.g., VMS VAX, in the case of Digital Equipment Corporation). The microwave link would have been part of the infrastructure costs for the project. The synoptic panel, allowed visual representation of the system states to the operator (and also some capabilities for remote switching). For Figure 1, the operator at the control centre is able switch a circuit breaker ON or OFF through the microwave link. Interface cards (IC) provide the necessary I/O capabilities. The remote terminal unit (RTU) is able to send a switching signal to the circuit breaker through a suitably designed actuator. The circuit breaker is also able to send its position (ON or OFF) through the RTU and the same microwave link to the operator's panel. The older type of hardwired panel is largely being replaced by electronic displays and the controller is able to receive and view system status through touch panel capabilities or through networked computers in the control room.

The proprietary nature of the old "legacy systems meant that embarking on the path to automation was not a trivial matter. The cost of ownership, (infrastructure, hardware and software costs) was very high. Also very high were the costs of maintaining and upgrading this hardware and software. The cost of software licences was often prohibitive. On the other hand cyber threats and virus attacks were unheard of. The systems themselves were broadly secure, but not necessarily reliable. Communication link (microwave failure) meant that there was at the time, as there is even in this generation, a need for "manual back-ups". This usually meant sending a technician to do manual switching operations.

4. Modern grid and substation automation

Power system automation happens in segments of the power system (Northcote-Green, Wilson) [4] which can serve different functions. One segment is bulk transmission of power which traditionally was handled by the power producer, but increasingly (in de-regulated environments), is handled by an independent transmission system operator (TSO). Bulk transmission is usually associated with outdoor switchyards and high voltage operating voltage levels (in excess of 132 kV). Bulk transmission substations play a critical role in energy trading and power exchanges. Wholesale electricity is sold through the transmission system to distributors. Figure 3 shows a portion of the 400 kV outdoor substation at AUAS near Windhoek, the capital of Namibia. Namibia is a net importer of electrical power most of it from neighbouring South Africa and this substation is of vital importance. The substation with which it connects in South Africa is over 800 km away at Aries near Kenhardt. Figure 4 shows the Namibian electrical power transmission network. The complex interconnections between equipment, such as transformers, reactors, lines and bus-bars, is such that manual operation is not a practical proposition. In the case of the AUAS substation, effective control is in the hands of Namibia Power Corporation's (Nampower) headquarter-based National Control Centre in Windhoek.

The other segment of automation is at distribution level. Large distributors are typically municipal undertakings or in countries in which electrical power is de-regulated, the so called "DISCOS". Automation has existed at the distribution level for many years, but has been restricted to situations involving either large numbers of customers or critical loads. As a result of this the quality of service given has been very good, while the rural consumers have been at a disadvantage. For power distributors, automating rural supplies was not cost-effective due to the dispersed nature of the lines and loads (Alsthom Network Protection and Application Guide) [5]. Technology changes in recent years, national power quality directives as well as increased consciousness by consumers themselves has led to radical changes in our Power System Control infrastructure and ways of operation with serious implications for those organisations that lag behind [NPAG].

Fig. 3. A group of Polytechnic of Namibia students visit the AUAS 400 kV substation

4.1 Distribution systems automation

From experience, faults at transmission levels are less frequent than at distribution levels [6]. At the same time distribution networks are not only complex, but the consequences of failure are quite severe. For this reason investment in distribution automation will increase. The elements that characterise distribution automation systems are given the definition by the IEEE. According to the IEEE, a Distribution Automation System (DAS) is "a system that enables an electric utility to remotely monitor, coordinate and operate distribution components, in a real-time mode from remote locations [7]". In this chapter we shall deal in more detail with the components involved and how they are coordinated within a DAS. In countries or situations where there are large networks, the network (primary distribution) itself is subdivided into more segments, namely, one for large consumers (no transformation provided) and for the rest at a lower HV voltage (secondary distribution) (Figure 6).

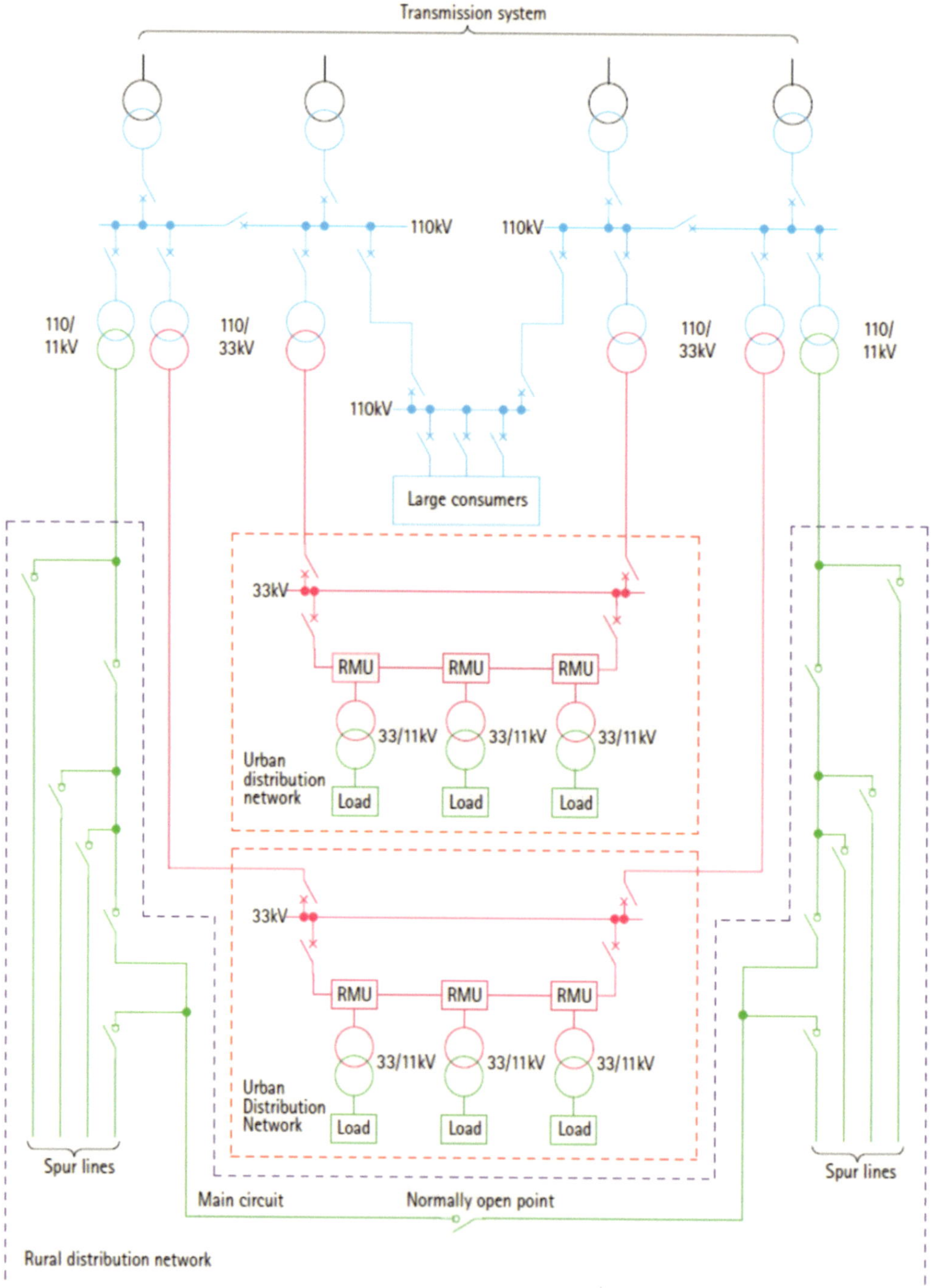

Fig. 4. (courtesy AREVA, NPAG)

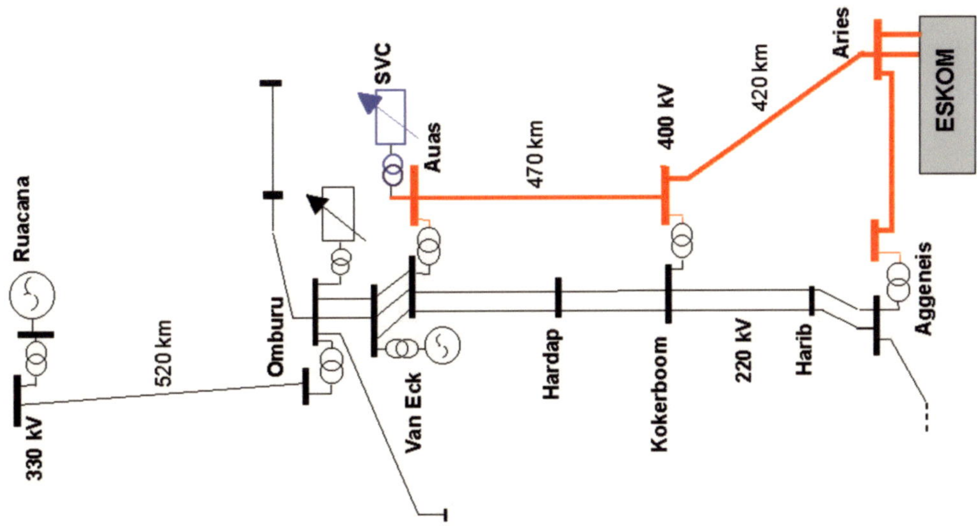

Fig. 5. (courtesy, Namibian Power Corporation)

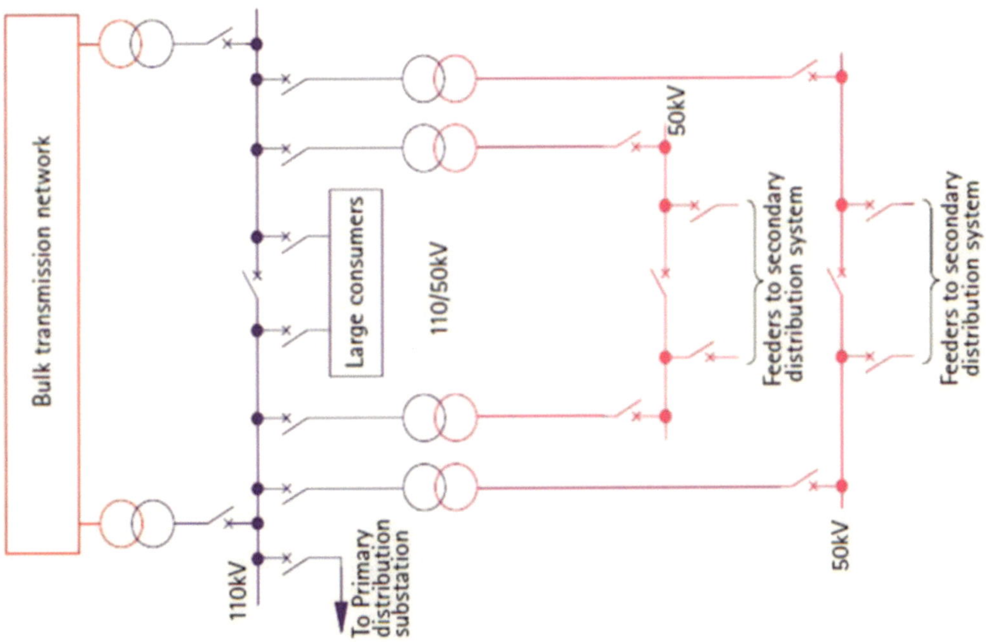

Fig. 6. (courtesy AREVA, NPAG)

5. Power system automation components

Power system automation components may be classified according to their function:

- Sensors
- Interface Equipment
- Controllers
- Actuators

Thus we see that Figure 1 is still a good representation of what is needed to effect automation, whether it is for EHV transmission, sub-transmission or distribution. Figure 7 depicts the control philosophy of a power system automation scheme.

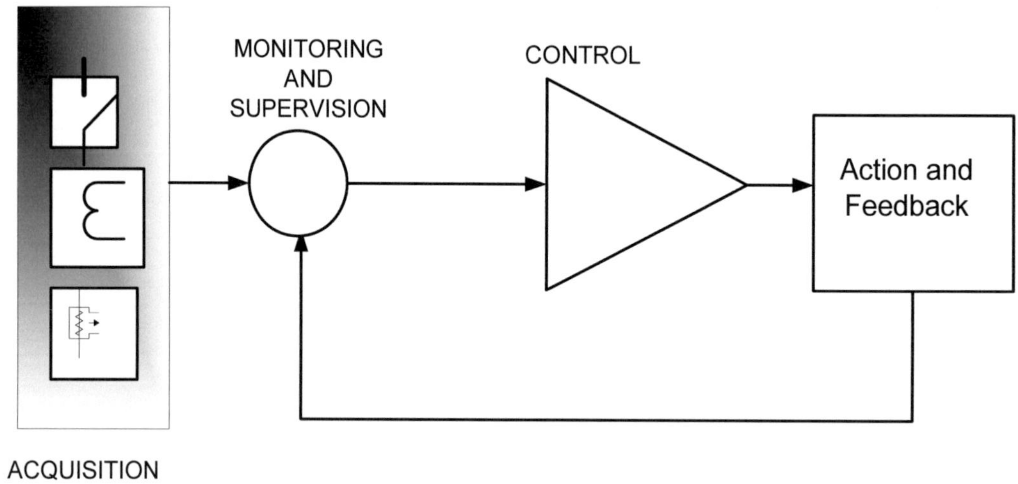

Fig. 7.

5.1 Overview of power system components

5.1.1 Sensors

5.1.1.1 Current and voltage transformers

Individually or in combination current and voltage transformers (also called instrument transformers) are used in protective schemes such as overcurrent, distance and carrier protection. Also in combination current and voltage transformers are also used for power measurements. In general custom specified voltage and current transformers are used for power metering, because of the increased accuracy requirements. Figure 8 shows instrument transformers in one of the substation areas (called bays).

Fig. 8.

5.1.1.2 Other sensors

For reliable electrical power system performance the states, stress conditions and the environmental conditions associated with the components have to be monitored. A very costly component in a substation is a transformer. For a transformer, monitoring is done, for example, for pressure inside the tank, winding temperature and oil level. For circuit breakers, sensing signals may need to be obtained from it such as gas pressure and number of operations.

5.1.2 Switches, isolators, circuit breakers

A most important function of a substation is the enabling of circuit configuration changes occasioned by, for example, planned maintenance, faults feeders or other electrical equipment. This function is of course in addition to the other important function of circuit protection which may also necessitate configuration changes. Modern switches and circuit breakers will have contacts or sensors to indicate their state or position. Figure 9 shows the ABB HH circuit breaker mechanism. The plant required to achieve the desired operation is usually quite elaborate and includes controls and protection to ensure that it operates reliably.

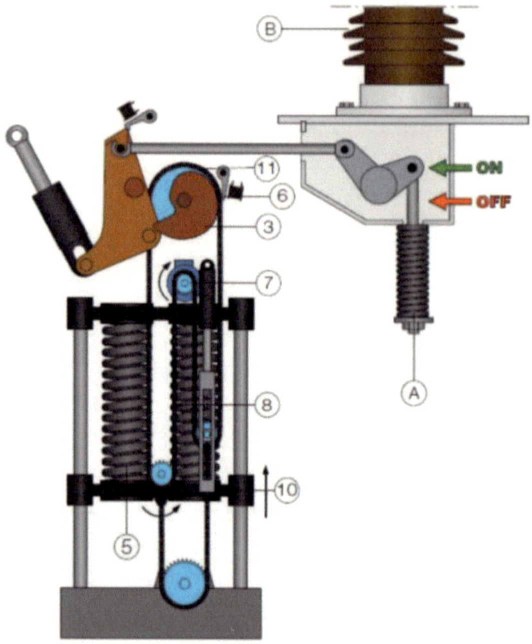

Fig. 9. Portion of HH ABB circuit breaker mechanism

6. IEC 61850 substation automation: Origin and philosophy

The International Electrotechnical Commission is one of the most recognisable standard generating bodies for the electrical power industry. Its standard the IEC 61850 "Communication Networks and Systems in Substations" is a global standard governing communications in substations. The scope of the standards is very broad and its ramifications very profound. So profound in fact that it is hard to imagine any new modern substation that would not at least incorporate parts of this standard. In addition, the standard is almost sure to be adopted albeit in customised / modified form in Generation, Distributed Energy Resources (DER) and in manufacturing. The standard has its origins in the Utility Communications Architecture (UCA), a 1988 initiative by the Electrical Power Research Institute (EPRI) and IEEE with the initial aim of achieving inter-operability between control centres and between substations and control centres. In the end it was found to be more prudent to join efforts with similar work being done by the Working Group 10 of Number 57 (TC57). The emerged document IEC 61850 used work already done by the UCA as a basis for further development.

6.1 IEC 61850 substation architecture

6.1.1 Substation bays

In an IEC 61850 compliant substation, equipment is organized into areas or zones called bays. In these areas we find switching devices (e.g., isolators and circuit breakers) that connect, for example, lines or transformers to bus-bars.

Examples of the bays would be:

- Incomer bay
- Bus-coupler bay
- Transformer bay

Figure 8 above, for example, could depict a transformer bay.

6.1.2 Merging units

Merging units are signal conditioners and processors. For example, they accept, merge and synchronise sampled current and voltage signals (all three phases' quantities of the CT/VT) from current and voltage transformers (conventional and non-conventional) and then transfer them to intelligent electronic devices (see IEDs, in the next section). So called electronic VTs and CTs are being manufactured by some companies which use new ways of sensing with the overall size being reduced. With electronic sensing, the sensing and merging are combined. Figure 10 gives an overview of the functions and associated inputs of a merging unit.

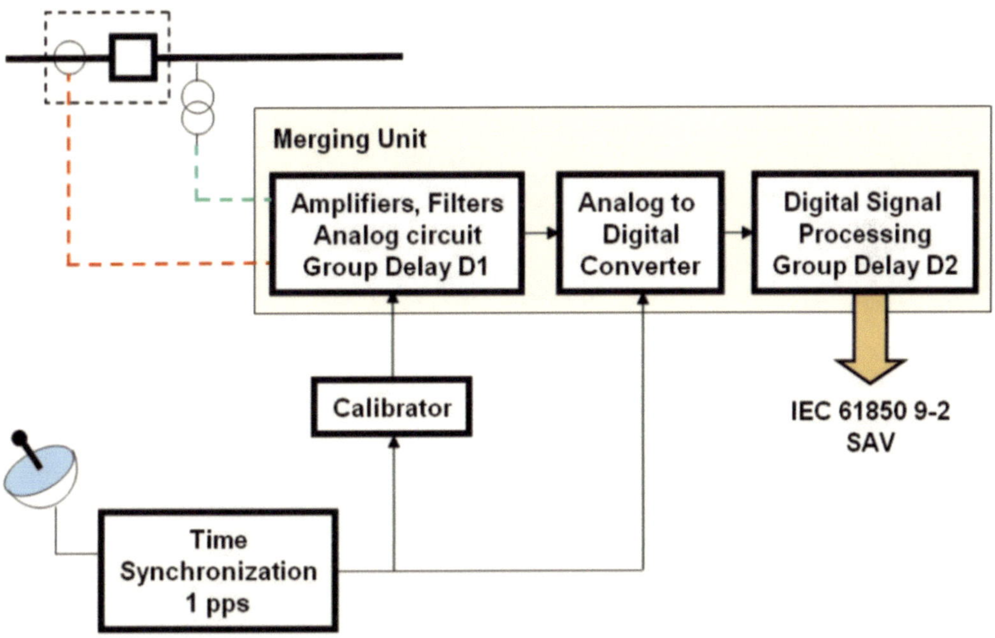

Fig. 10. (Jansen & Apostolov)

As technology progresses it is believed that there will be a move away from copper connections from field devices to the substation control room in favour of fibre. Figure 11 shows a merging unit (Brick) by vendor GE.

6.1.3 Intelligent Electronic Devices (IEDs)

An IED is any substation device which has a communications port to electronically transfer analog, status or control data via a proprietary or standard transmission format (BPL Global IEC 61850 Guide) [8]. Examples of IEDs are:

- Modern IEC 61850 protection relays (distance, over-current, etc.)
- Equipment-specific IED (e.g., for transformer bay protection and control, with tripping logic, disturbance monitoring, voltage, current, real and reactive power, energy, frequency, etc.).
- Bay controllers

Figure 11 shows some IEDs from various vendors with multiple functionality.

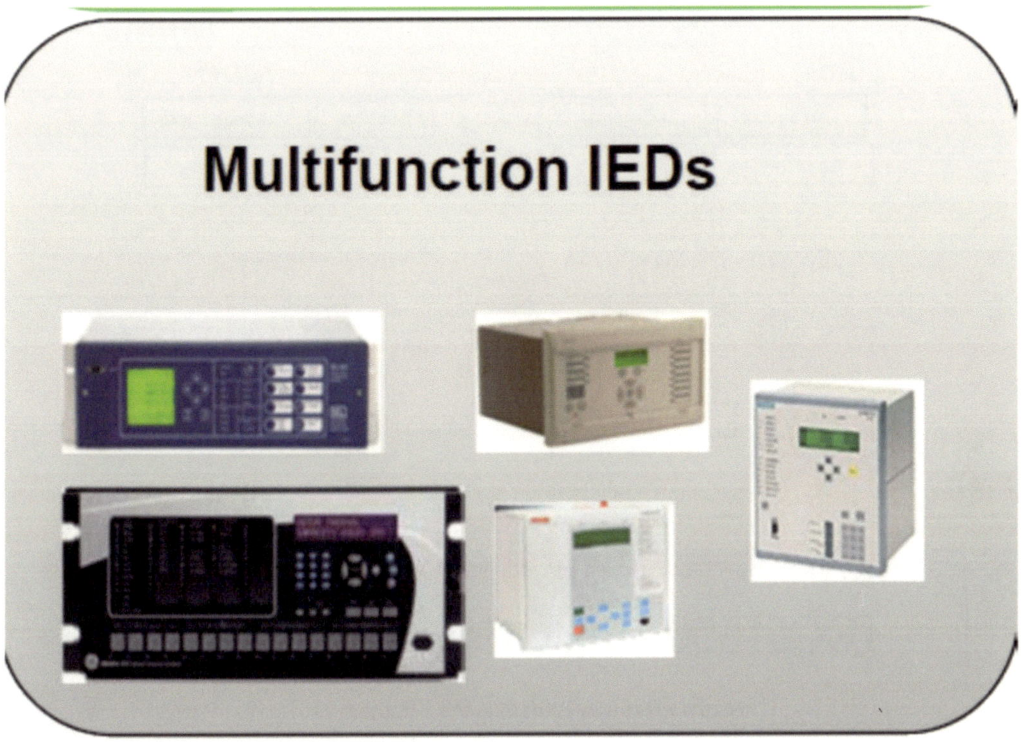

Fig. 11.

In reality today's IEDs have "mutated" to the form of programmable logical controllers (PLCs) of another kind with multiple capabilities.

6.1.4 Device/system integration: Substation functional hierarchy

An IEC 61850-designed substation has the following hierarchical zones:

- Process
- Bay
- Station

Diagrammatically this is illustrated in Figure 12 (Jansen & Apostolov) [9]. A complete representation that includes aspects, such as links to remote control centres and GIS, is given in Figure 13.

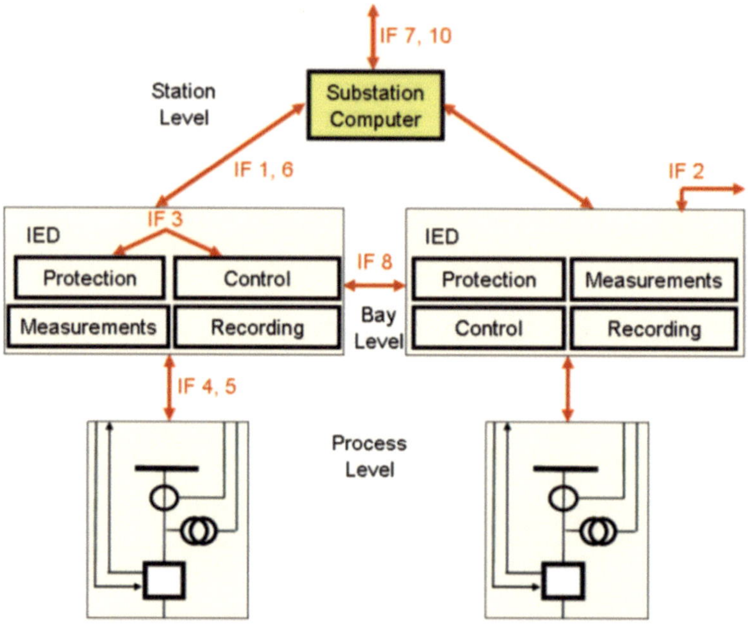

Fig. 12.

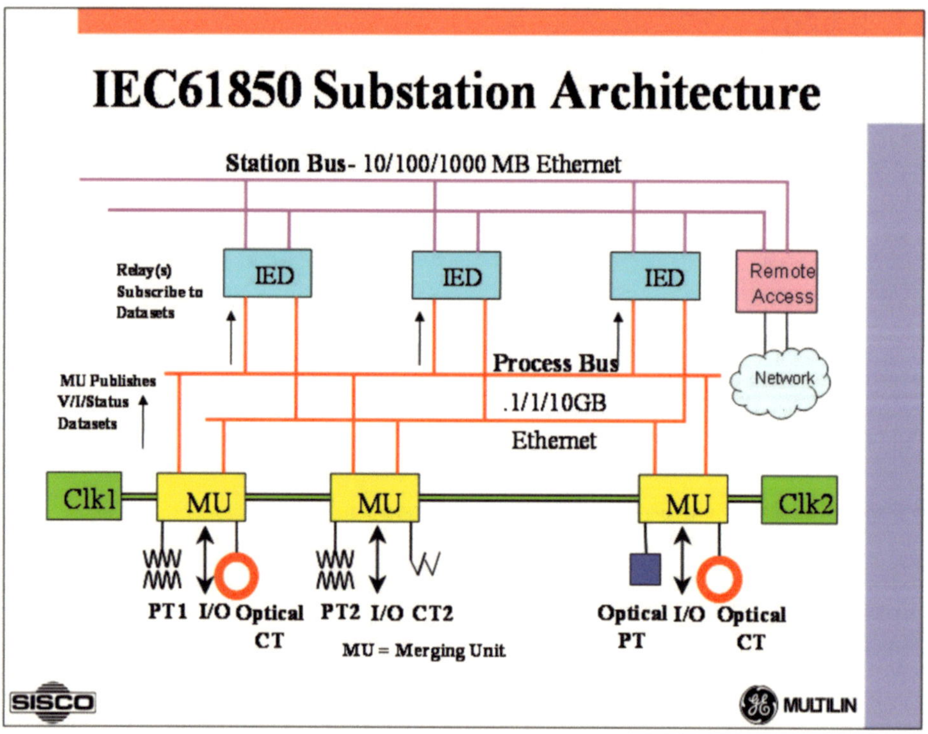

Fig. 13. (courtesy SISCO & GE)

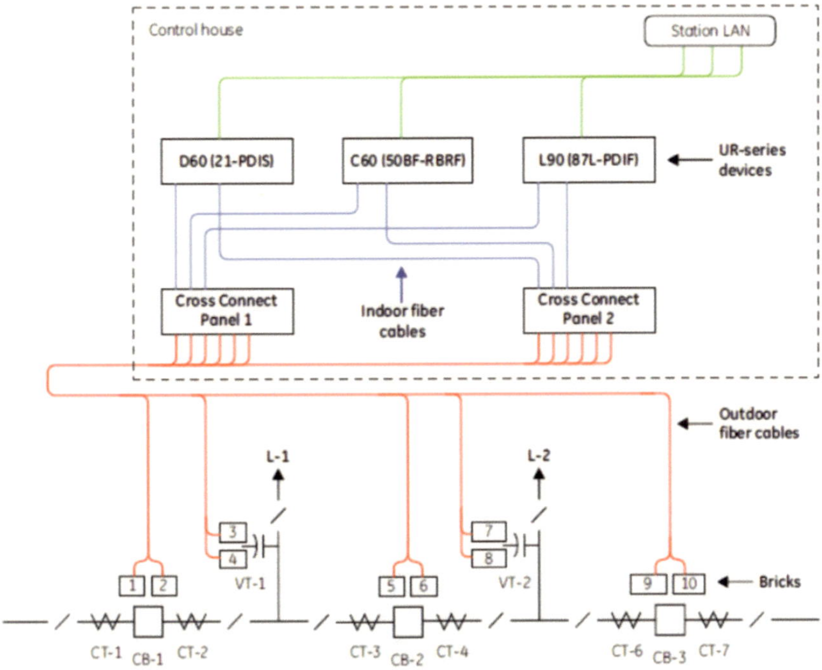

[O1]

Fig. 14. Fibre-based

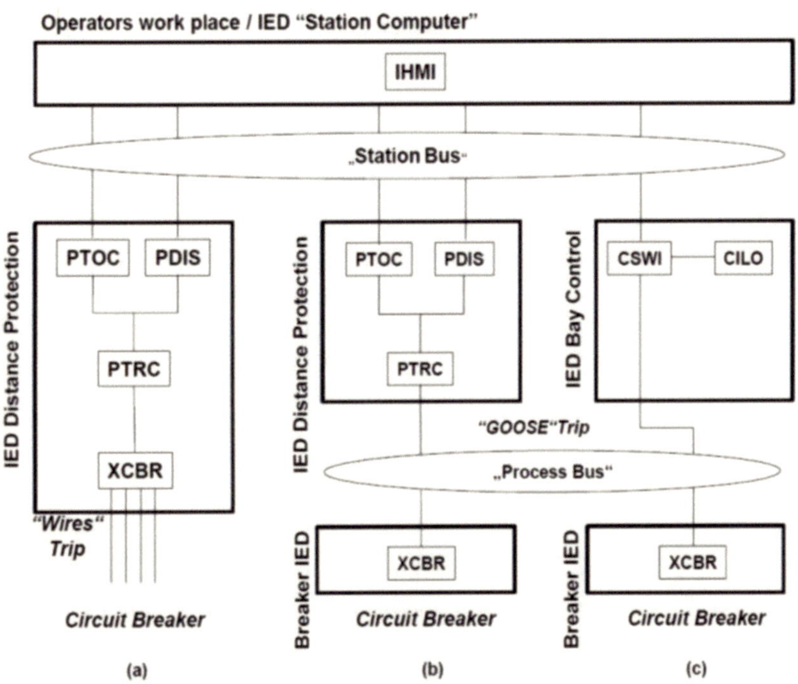

Fig. 15.

STRUCTURE OF THE IEC 61850 STANDARD

Part #	*Title*
1	Introduction and Overview
2	Glossary of terms
3	General Requirements
4	System and Project Management
5	Communication Requirements for Functions and Device Models
6	Configuration Description Language for Communication in Electrical Substations Related to IEDs
7	Basic Communication Structure for Substation and Feeder Equipment
7.1	- Principles and Models
7.2	- Abstract Communication Service Interface (ACSI)
7.3	- Common Data Classes (CDC)
7.4	- Compatible logical node classes and data classes
8	Specific Communication Service Mapping (SCSM)
8.1	- Mappings to MMS(ISO/IEC 9506 – Part 1 and Part 2) and to ISO/IEC 8802-3
9	Specific Communication Service Mapping (SCSM)
9.1	- Sampled Values over Serial Unidirectional Multidrop Point-to-Point Link
9.2	- Sampled Values over ISO/IEC 8802-3
10	Conformance Testing

7. Substation communications and protocols

With the IEC 61850 technology and with all the components and systems described in previous sections functioning normally, we have in fact a virtual substation. The remote terminal units (RTU) increasingly with IED functionality, pass on analog and digital data through either copper or fibre to IEDs in the substation control room in the form of relays or bay controllers. The process of transferring data and communicating it to various devices has been greatly simplified with the aid of the standard. The data arriving at the IEDs comes already formatted / standardized. The situation is similar to the "plug and play" philosophy applied to computer peripherals of today.

7.1 Virtualisation

With the IEC 61850 a real substation is transformed into a virtual substation, i.e., real devices transformed into objects with unique standardized codes. In Figure 16, a real device, a transformer bay is transformed into a virtual, logical device with descriptive name, e.g., Relay1. Inside the device are logical nodes (LN) named strictly in accordance with the IEC standard. For example, a circuit breaker inside this logical device is given XCBR1 [10]. In turn the breaker has other objects associated with it, e.g., status (open / closed) and health. The services associated with this data model are defined in the Abstract Communications System Interface (ACSI). The following ACSI functions are listed by Karlheinz Schwartz [11]:

- Logical Nodes are used as containers of any information (data objects) to be monitored
- Data objects are used to designate useful information to be monitored
- Retrieval (polling) of the values of data objects (GetDataObjectValues)
- Send events from a server device to a client (spontaneous reporting)
- Store historical values of data objects (logging)
- Exchange sampled values (current, voltages and vibration values)
- Exchange simple status information (GOOSE)
- Recording functions with COMTRADE files as output

7.2 Mapping

IEC 61850 is a communications standard, a main aim of which is interoperability. A good definition is:

"Interoperability is the ability of two or more IEDs (Intelligent Electronic Devices) from the same vendor, or different vendors to exchange information and uses that information for correct co-operation" [12]. Although ACSI models enable all IEDs to behave identically from a general network behaviour perspective, they still need to be made to work with practical networks in the power industry, (Baigent, Adamiak and Mackiewicz) [10]. This universal compatibility is achieved through mapping of the abstract services to universal, industry-recognised protocols. Presently the protocol most supported is the Manufacturing Message Specification (MMS). MMS was chosen because it has an established track record in industrial automation and can support the complex and service models of IEC 61850.

Table 1 gives an idea of the naming process:

IEC61850 TO MMS OBJECT MAPPING

IEC61850 Objects	MMS Object
SERVER class	Virtual Manufacturing Device (VMD)
LOGICAL DEVICE class	Domain
LOGICAL NODE class	Named Variable
DATA class	Named Variable
DATA-SET class	Named Variable List
SETTING-GROUP-CONTROL-BLOCK class	Named Variable
REPORT-CONTROL-BLOCK class	Named Variable
LOG class	Journal
LOG-CONTROL-BLOCK class	Named Variable
GOOSE-CONTROL-BLOCK class	Named Variable
GSSE-CONTROL-BLOCK class	Named Variable
CONTROL class	Named Variable
Files	Files

IEC61850 SERVICES MAPPING (PARTIAL)

IEC61850 Services	MMS Services
LogicalDeviceDirectory	GetNameList
GetAllDataValues	Read
GetDataValues	Read
SetDataValues	Write
GetDataDirectory	GetNameList
GetDataDefinition	GetVariableAccessAttributes
GetDataSetValues	Read
SetDataSetValues	Write
CreateDataSet	CreateNamedVariableList
DeleteDataSet	DeleteNamedVariableList
GetDataSetDirectory	GetNameList
Report (Buffered and Unbuffered)	InformationReport
GetBRCBValues/GetURCBValues	Read
SetBRCBValues/SetURCBValues	Write
GetLCBValues	Read
SetLCBValues	Write
QueryLogByTime	ReadJournal
QueryLogAfter	ReadJournal
GetLogStatusValues	GetJournalStatus
Select	Read/Write
SelectWithValue	Read/Write
Cancel	Write
Operate	Write
Command-Termination	Write

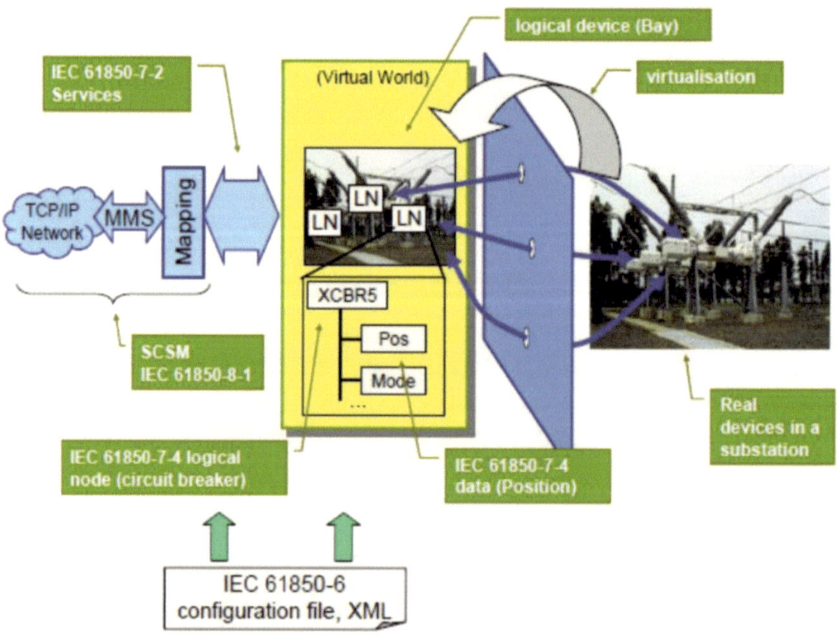

Fig. 16. Karlheinz Schwartz

8. Communication of events in an IEC 61850 substation

In his IEC 61850 Primer, Herrera states that "IEC 61850 provides a standardized framework for substation integration that specifies the communications requirements, the functional characteristics, the structure of data in devices, the naming conventions for the data, how applications interact and control the devices, and how conformity to the standard should be tested. In simpler terms, IEC 61850 it is an open standard protocol created to facilitate communications in electric substations."

8.1 The communication structure of the substation

The IEC 61850 architecture there are two busses:

- Process bus
- Station bus

IEC 61850 **station bus** interconnects all bays with the station supervisory level and carries control information such as measurement, interlocking and operations [13].

IEC 61850 **process bus** interconnects the IEDs within a bay that carries real-time measurements for protection called sampled values or sampled measured values [13].

Figure 17 shows the basic architecture.

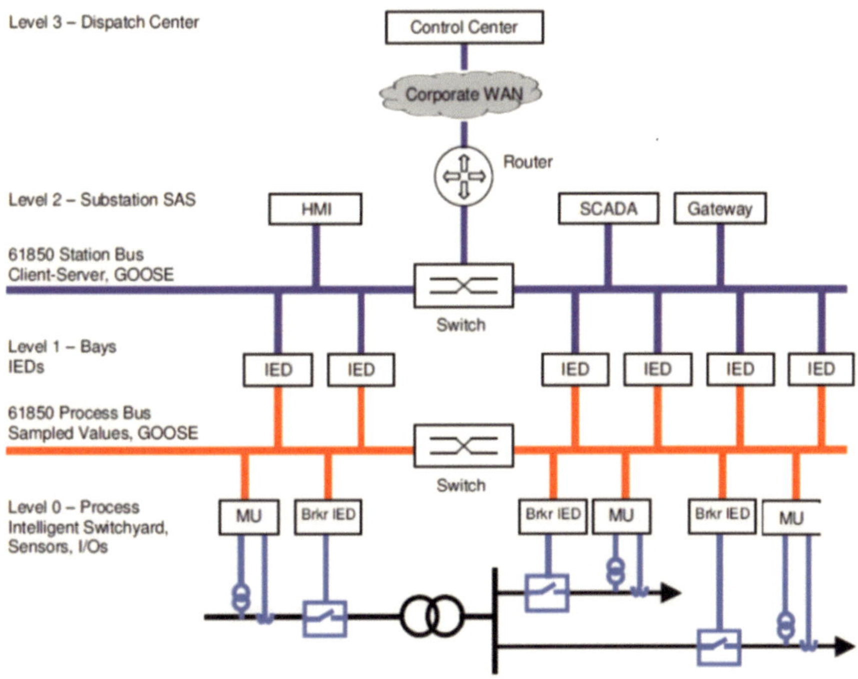

Fig. 17.

The process bus is designed to be fast since it must carry crucial I/O between IEDs and sensors/actuators.

The requirements for the process bus cited in various literature sources are as follows:

- High environmental requirements for the terminal equipment (electromagnetic compatibility, temperature, shock, where applicable) in the area of the primary system
- Adequate bandwidth for several SV data streams
- Highly prioritized trip signals for transmitting from the protection device to the CBC
- Permeability of data to the station bus/data filtering at the coupling point
- Simultaneous TCP/IP traffic for normal control and status signal traffic as well as reports on the process bus
- Download/upload channel for setting or parameterizing functions
- Highly precise time synchronization
- Redundancy
- For reasons of speed, the process bus is based on optical fibre with high data throughput of about 10Gbits/s. Because of its enhanced data capacity it is capable of carrying both GOOSE (Generic Object Oriented Substation Event) and SMV (Sampled Measured Values). The station bus is used for inter-IED communications. Only GOOSE messaging occurs in the station bus.

9. Substation control and configuration

Although the strengths of the IEC 61850 in the capturing, virtualisation, mapping and communication of substation information are undoubted, it will still be necessary to link everything together and to design a control strategy. This strategy must utilize the experience and expertise of the asset owner. The substation must also respond in accordance with the operational and safety criteria set by the organization.

9.1 Substation configuration

Automation of the substation will require in the first instance the capture of its configuration. This requires the capture of the information on all the IEDs in the substation. In some cases the IEDs could be from different vendors. The information has to be in a standardized IED Capability Description (ICD). Then, using a system configuration tool, a substation description file is created (Figure 18). The SCD (Substation Configuration Description) is then used by relay vendors to configure individual relays [14].

10. Wider implications of the IEC 61850: The Smart Grid

"Smart Grid" is a term used to describe the information driven power systems of the future. This will involve introducing new electronic, information and computer technology into the whole value chain of electrical energy systems from generation, transmission and distribution down to the consumer level. Figure 19 shows the linkages between the technology of electricity production and commerce.

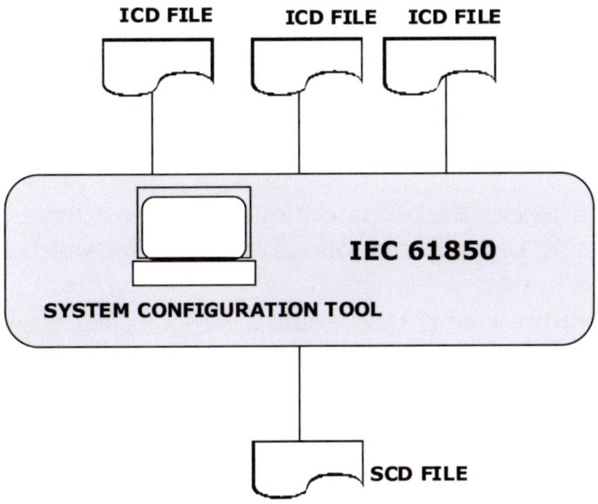

Fig. 18.

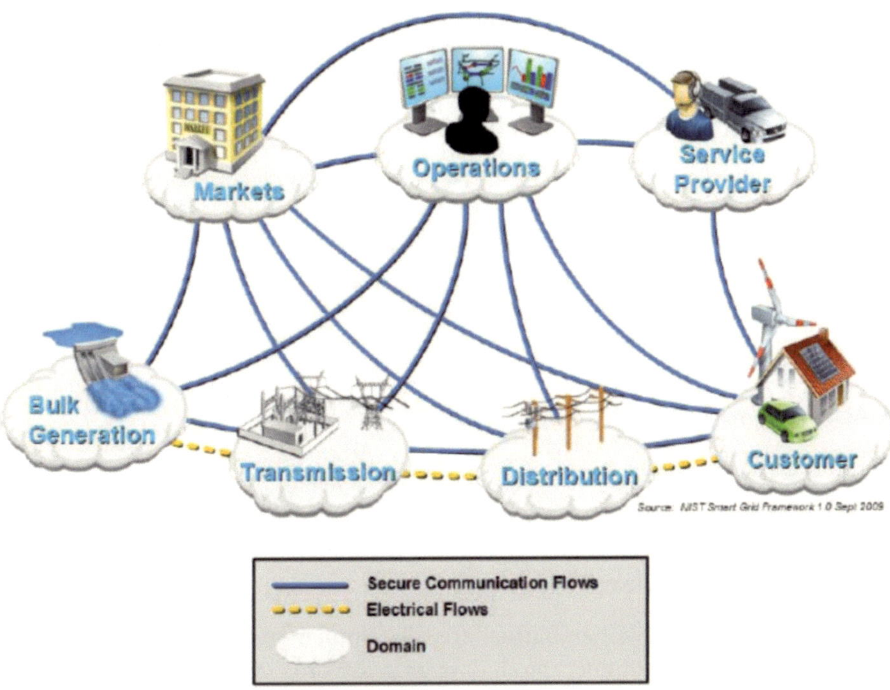

Fig. 19. Source NIST Smart Grid Framework

We have seen that automation started on the factory floor and some of the IEC 61850 functions use manufacturing protocols such as MMS. We already start to see the trend towards extending the IEC 61850 to generating stations. It is therefore not hard to imagine IEC 61850 like protocols encompassing every facet of engineering, including manufacture.

10.1 Smart Grid benefits

Among the benefits of Smart Grid are:

- Increased grid efficiency - the use of control systems to achieve optimum power flow through, for example, centrally controlled FACTS devices which can increase efficiency of the transmission system
- Better demand control - a Smart Grid would incorporate an energy management system to manage demand (e.g., managing the peaks and valleys)
- Asset optimization - the IEC 61850 information model already has the capability to store not only the status of a logical device / node, but also condition / health
- Management of renewable energy sources - renewable energy sources, such as wind and solar, tend to be unpredictable, therefore, the Smart Grid system can enable predictions on the availability of these resources at any moment and ensure proper energy scheduling decisions are taken
- Management of plug in electric vehicles - the Smart Grid can inform electric vehicle motorists of the nearest charging stations
- Smart metering - with smart metering, power usage and tariffs can be administered remotely to the advantage of both the supplier and the consumer

11. Security threats in automated power systems

In this chapter we have seen the central role computer hardware and software in the control and management of the power system bring tremendous benefits. However, investing in these high technology, information technology reliant assets also brings threats. The threats are quite serious especially when it is realized that every critical component of the substation becomes a virtual computer. The IED mentioned numerous times in this chapter is itself a computer. What are the threats?

11.1 SCADA vulnerabilities Chikuni, Dondo [15]

- Computing vulnerabilities

Hardware: RTUs, IEDs and SCADA Masters belong to the class of computer hardware and suffer from the same vulnerabilities of regular computer systems such as interruption (denial of services [DoS]) and eavesdropping

Communication links: the vulnerabilities are also similar to those in regular computer networks - if messages are not encrypted, data or passwords can be intercepted. Radiation emissions from equipment can be read by unauthorized people

- **SCADA software:** the most common attacks come in the form of interruption, interception and modification. Software bugs, if not fixed in time, can attract hobbyist hackers to attack unpatched SCADA [15]

- **Data:** SCADA data has more value to the attacker than hardware and software. Data may be stolen by competitors of saboteurs. To safe guard the data, encryption needs to be included

11.2 Other vulnerabilities

- **Equipment location**: we have seen that some IEDs and RTUs are located in usually unmanned remote locations; where this applies these must be housed or mounted securely
- **Remote access**: access to relays, controllers, IEDs and RTUs should be password protected; encryption modems are available for secure dial-up communications
- **Human element**: the employee could be the most vulnerable part of the automated power system, therefore, no unauthorized persons should have access to the control terminals. Strong authentication and smart card access are recommended
- **Integrity and confidentiality**: software and hardware should have at least the US National Computer Security Centre (NCSC) class 2 rating. In Europe criteria similar to that of NCSC is managed through the European Information Technology Security Evaluation Criteria (ITSEC)

11.3 Attack examples

Nadel et al. [16] list what they describe as generic attack categories to which all network-based threats to substation automation systems can be reduced, namely:

- Message modification
- Message injection
- Message suppression

They demonstrate the various paths that an attacker can take in a given attack. An example of circuit breaker attack scenario is shown in Figure 20. In this case an attacker may take one of the paths in the graph to prevent a circuit breaker from opening when it is supposed to.

11.4 Countermeasures

Nadel et al. [16] list some of the important assumptions / precautions necessary before any meaningful countermeasures can be instituted. These include static configuration of the SAS and the number and types of devices in the bay level are known; configuration changes only to occur during major maintenance or modification work; also that the SAS is not used for billing and no general purpose PCs are allowed at bay level.

Message modification

In this attack parts of a valid, existing message are modified in transit. Detection is facilitated through encryption and digital signatures, with the receiver having a record of all authorized senders.

Message injection and replay

The attacker sends messages which are not intended to be sent by any authorized sender. These may be entirely new messages from the attacker or original untampered with replayed messages. Digital signatures are a way of combating message injection. To mitigate against replay, a digital signature and message sequence number are required.

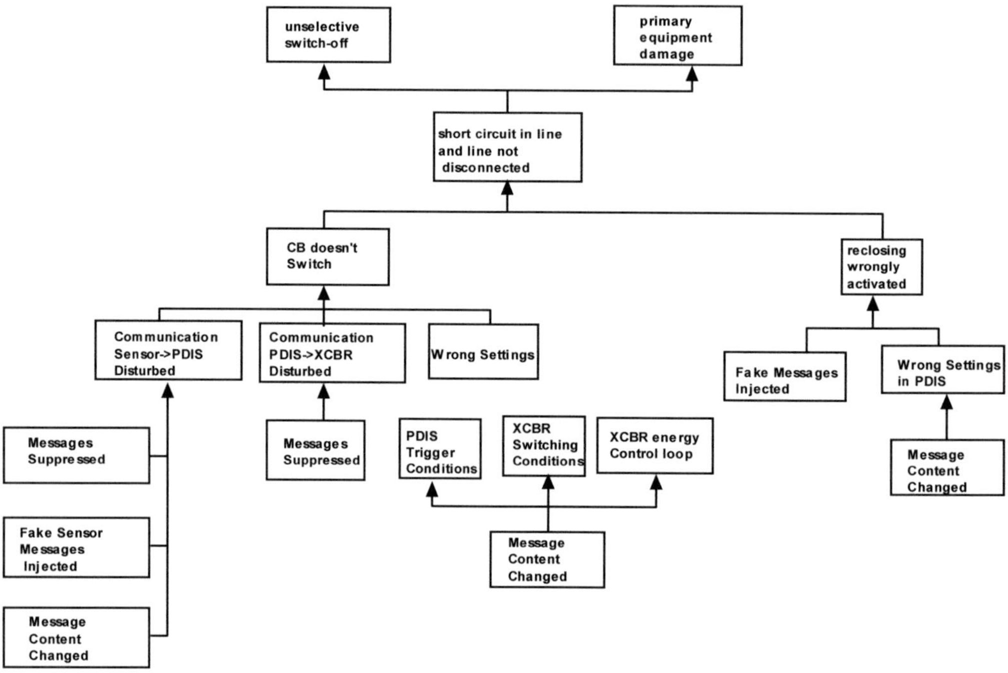

Fig. 20. Example of attack graph for a circuit breaker

Message suppression

In this attack certain messages between SAS devices are prevented from reaching the receiver, e.g., circuit breaker control devices are isolated from protection devices. Message suppression can involve several other types of attack, e.g., re-configuration of routers or switches, cutting wires or congesting the network so that genuine messages cannot get through (denial of service attack).

Security protocols

The multiplicity and varied nature of SAS attacks makes it imperative to institute robust security protocols capable of handling all eventualities. Such protocols include the use of private keys (only known to the sender), encryption and sequence numbers (initial number known between sender and receiver at the start).

Markets	
Subsystem Communication	
Intra-Markets	See Clause Markets
Intersystem Communication	
Operation	For scheduling and trading purposes, information about the availability of power (transfer power, operating reserve) or order information is transmitted to or from the operation system
Bulk Generation	For scheduling and trading purposes, information about the availability of power (transfer power, operating reserve) is transmitted from the bulk generation system
Service	Support functions for markets (e.g. forecasting for renewable generation)
Prosumer	For scheduling and trading purposes information about the availability of power or order information is transmitted to or from the markets system

System: Service

The service system offers potential for a wide range of new service developments. New business models may emerge due to the opportunities of the future Smart Grid. Therefore the service system will have and depend on various interfaces to other systems.

Service	
Subsystem Communication	
The new service application shall follow a standardized way of software development in order to seamlessly fit into an overall system. The relevant standards are not within the scope of the IEC	
Intersystem Communication	
Operation	Support functions for operation (e.g. forecasting for renewable generation)
Market	Support functions for markets (e.g. forecasting for renewable generation)
Prosumers	Customer services (Installation, Maintenance, Billing, Home & Building Management) are quite conceivable

System: Prosumer
Description

Prosumer	
Subsystem Communication	
See Clause HBES/BACS	
Process Automation	In many industries (e.g. chemical, manufacturing) process automation is applied to control and supervise not only the manufacturing process but also the energy consumption or generation
Intersystem Communication	
Service	Support functions for operation (e.g. forecasting for renewable generation)
Operation	See Clause AMI, DER
Markets	For scheduling and trading purposes information about the availability of power or order information is transmitted to or from the markets system
Distribution	Typically the distribution system infrastructure is used for the communication to DMS

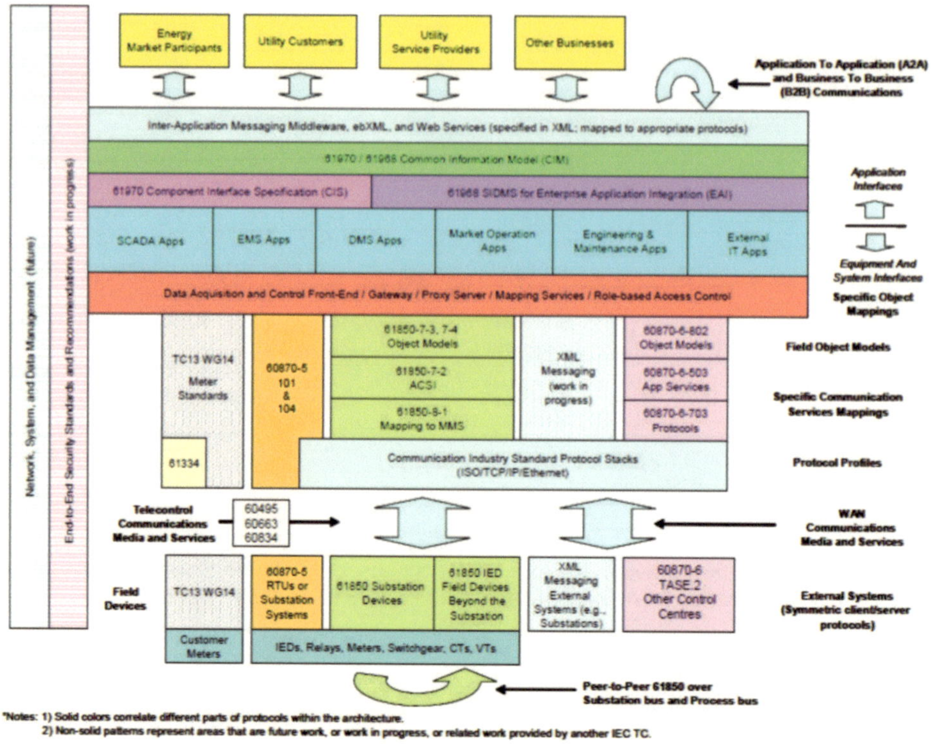

12. Effects on educational curricula

To give an idea of the profound changes the power system will have on our education systems and also to give some suggestions on how to mitigate some of the challenges, we reproduce the following excerpts from a paper by Chikuni, Engelbrecht and Dongo [17] at the PowerCon 2010 conference:

"When we analyse a modern substation incorporating the new substation technology based on IEC 61850, it would seem that the role of an electrical engineer is notable by its absence. Certainly some of the responsibilities of both engineers and technicians have shifted and there needs to be new breed of electrical engineers altogether. Some questions need to be answered:

- Should we retrain the power engineer in networks or
- Should we train network engineers so that they acquire power engineering knowledge or
- Should we work with a completely new curriculum which merges power systems, electronics and networks into one programme?

We first need to acknowledge that there are already a lot of good power engineers out there trained in the traditional manner, i.e., starting off with physics, circuits and systems, electrical machines and power systems (including electrical protection). For these engineers, one needs to identify those who can benefit both themselves and their organizations by going through this additional training. The process of training engineers has been quite formal, especially if they wish to attain professional status. One typically needs four to five years of formal training and a further two years of guided industrial training before attaining the status of chartered (CEng) or professional (PrEng) engineer. A great debate will ensue, therefore, when a computer network engineer is designated the 'responsible person' in an electrical substation, notwithstanding the obviously immense power this individual will have in making sure that the substation operates correctly, safely and efficiently.

The other route is to include networking as part of any electrical engineering curriculum. A few programmes today include industrial automation and a few even include computer networking. In the University of Zimbabwe model all electrical engineering students have a chance to complete at least the first semesters of a CISCO network academy programme. Indeed some complete the CCNA (four semesters). Whatever solution is arrived at, it is clear that electrical engineering training curricula inevitably have to include more and more electronics, sensors, automation and networking, not as peripheral subjects, but as part of the core.

13. Summary and conclusions

In this chapter we have seen the extremely rapid development of automation, starting from the years of mechanisation, production lines and the taking root of computer-based automation in the car manufacturing industry. Then we noticed rapid increases in computer power in both hardware and software forms. There has also been tremendous moves in standardization in North America and Europe. We have seen too IEC 61850 international cooperation in standards development and the benefits that are already being reaped from this. Interoperability brings some relief to customers, giving them the ability to choose hardware from an increasing variety of vendors. Quite striking is the increasing dominance of ICT in power system control and massive changes in power system operation and practice. The power systems have become more complex - more interlinked. The complexity presents new challenges. The traditionally trained power systems engineer lacks the know how to understand or tackle faults that could arise in these systems. On the other hand the network engineer may lack the underlying principles of power and energy systems. A new type of multi-discipline power systems engineer has to be trained. The Smart Grid will soon be a reality. Generation, transmission, distribution consumption and commerce will be information driven. Finally, when automation is combined with mechatronics and robotics, our lives are poised to be drastically changed.

14. References

[1] Mikel P. Groover, Britannica Online Encyclopedia

[2] Benjamin F. Shearer,. *"Home front heroes: a biographical dictionary of Americans during wartime"*, Volume 1, Greenwood Publishing Group 2007

[3] Edward Chikuni, *"Concise Higher Electrical Engineering"*, Juta Academic Publishers, 2008

[4] James Northcote-Green, Robert Wilson, *"Control and Automation of Electrical Power Distribution Systems"*, Taylor& Francis, 2006

[5] AREVA, ALSTOM, Network Protection and Application Guide, 2011 Edition

[6] Su Sheng; Duan Xianzhong; W.L. Chan, *"Probability Distribution of Fault in Distribution System* , Power Systems, IEEE Transactions on, Aug. 2008"

[7] D. Bassett, K. Clinard, J. Grainger, S. Purucker, and D.Ward, *"Tutorial course: distribution automation,"* IEEE Tutorial Publ. 88EH0280-8-PWR, 1988

[8] IEC 61850 Guide Serveron® TM8TM and TM3TM On-line Transformer Monitors 810-1885-00 Rev A August 2011

[9] M.C. Janssen,, A. Apostolov *"IEC 61850 Impact on Substation Design"*

[10] Drew Baigent, Mark Adamiak, Ralph Mackiewicz, GE and SISCO *"Communication Networks and Systems In Substations, An Overview for Users",.*

[11] Karlheinz Schwarz, SCC, *"Monitoring and Control of Power Systems and Communication Infrastructures based on IEC 61850 and IEC 61400".*

[12] Dipl.-Ing. H. Dawidczak, Dr.-Ing. H. Englert Siemens AG, Energy Automation Nuremberg, Germany *"IEC 61850 interoperability and use of flexible object modeling and naming"*

[13] R. Moore, IEEE Member, R. Midence, IEEE, M. Goraj, *"Practical Experience with IEEE 1588 High Precision Time Synchronization in Electrical Substation based on IEC 61850 Process Bus"*

[14] J. Holbach, J. Rodriguez, C. Wester, D. Baigent, L. Frisk, S. Kunsman, L. Hossenlop, *"First IEC 61850 Multivendor Project in the USA, Protection, Automation and Control World, August 2007"*

[15] Edward Chikuni, Maxwell Dondo, *"Investigating the Security of Electrical Power Systems"* SCADA, IEEE Africon 2007, Windhoek

[16] Martin Naedele , Dacfey Dzung , Michael Stanimirov, *"Network Security for Substation Automation Systems"*, Proceedings of the 20th International Conference on Computer Safety, Reliability and Security, p.25-34,September 26-28, 2001

[17] E Chikuni, F Engelbrecht, and M Dondo, "The emergence of substation automation in Southern Africa, opportunities, challenges and threats", IEEE Africon Conference, Windhoek, September, 2007

7

Virtual Commissioning of Automated Systems

Zheng Liu[1], Nico Suchold[2] and Christian Diedrich[2]
[1]*Otto-von-Guericke University, Magdeburg,*
[2]*Institut für Automation und Kommunikation (ifak), Magdeburg,*
Germany

1. Introduction

Concepts for the digitalization of products and all the production related tasks in the manufacturing and process industry have been developed for several decades. The development began with the introduction of computer-based 2D design (Figure 1).

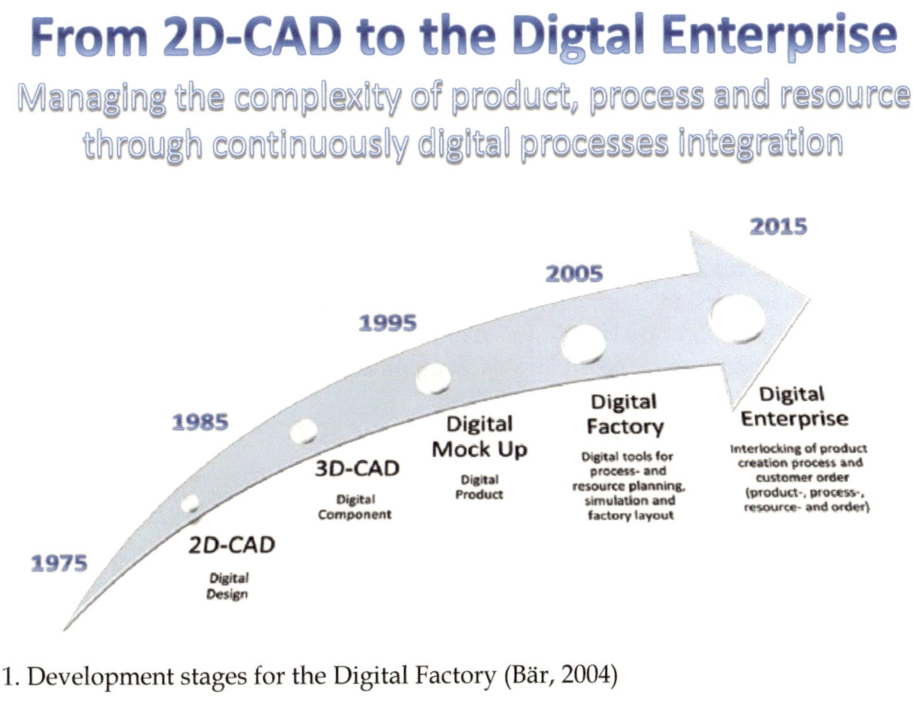

Fig. 1. Development stages for the Digital Factory (Bär, 2004)

Because of the technological developments in the information technologies field, more complex tasks in the product development process can be processed digitally (Brac, 2002). This belongs to the scope of the digital factory which includes the digitalization of models of the products and their integration into the manufacturing process chain. In (VDI 4499, 2008) the digital factory was defined as a "generic term for a comprehensive network of digital models and methods, including simulation and 3D visualization. Its purpose is the integrated planning, implementation, control and on-going improvement of all important factory processes and resources related to the product."

One of the topics in the digital factory is the Virtual Commissioning. As we know, the commissioning of the automated system is an important phase in the engineering, which makes visible whether the systems and components are planned, designed, produced and installed correctly according to the user requirements. However, this phase has been known as time-consuming and cost-consuming in practice (section 2.1) and can be improved by Virtual Commissioning.

State of the art is that the simulation of the system components is frequently used and the simulation is used without integration of automation devices and components, for example without PLCs (Programmable Logic Controller). HiL (Hardware in the loop) technologies are used only for individual components. The concept of digital factory combines both technologies.

During Virtual Commissioning the real plant is replaced by a virtual model according to the concept of digital factory. But the questions are: what digital data is necessary for the virtual plant model, how is the virtual plant implemented and configured, and how can these processes be integrated into the engineering lifecycle of manufacturing systems? This chapter gives comprehensive answers to these questions.

2. The Virtual Commissioning

The basic idea of Virtual Commissioning is to connect a digital plant model with a real plant controller (e.g. PLC- or HMI-Human Machine Interface) so that engineers from different fields such as design process and control have a common model to work with together. Thus, for instance, the PLC program can be tested virtually before the physical implementation is finished. Furthermore, the general functionalities of the plant can be validated and finally tested in an earlier phase. In this section, the general workflow of a traditional engineering process is introduced and its disadvantage discussed. Thereafter the Virtual Commissioning, as a possibility to overcome these problems, is introduced and described in detail in the subsequent subsections.

2.1 General workflow in the engineering process

The engineering of manufacturing plants begins with the construction and the rough mechanic planning. In this phase the general structure (layout) of the manufacturing system as well as the assignment of the machines in different production phases are determined, which is based on the result of product development.

The detailed planning is done in the next step. The kinematic simulation for the offline programming (OLP) of the robots is carried out, therefore the robot programs can already be created without the real robots, e.g. for the generation of a trajectory free of collision with other devices. In the phase of mechanic detail planning, a detailed CAD-based 3D cell layout (M-CAD) is created. The production workflow of a single machine is to be defined and every mechanical component in the cell can also be selected and documented in the list of materials. The results of the mechanical detail planning are used as input for the subsequent electrical detail planning. At this step the electrical energy supply as well as the control signals for the machines and their interconnection with each other is defined. This can be called E-CAD and is documented in the circuit diagram. The last phase of the detail planning is the PLC programming.

The last step in the classic engineering process is the commissioning, which can be understood as a phase of preproduction. At this phase the functionality of every single component in a function group is examined. After that it follows the interconnection check in turn from the level of function groups to the level of stations until the whole plant is complete - the bottom-up principle. The result of the commissioning is a functional plant which is ready for the production.

2.2 Today's problem in the commissioning

The above mentioned approach of the engineering process can be seen as the water fall model. A single step in this model can only be executed one after the other. There is no iteration step back to the previous steps. As a result, the debugging takes place only at the phase of commissioning. The following two main problems are identified here.

Bugs in the software

(Zäh & Wünsch, 2005) describes the time consuming for commissioning and PLC-programming in a whole project. It can be seen that the commissioning takes up 15-20% of the overall project duration, of which up to 90% is for commissioning the industrial electrics and control systems (e.g. PLC). Within the commissioning, up to 70% of the time is incurred by software errors (Figure 2).

The debugging of the PLC-Software is usually time consuming and may cause hardware damage. Since the commissioning is usually carried out under extreme time pressure, then a delay of the project must be calculated with financial penalties. Therefore it is desirable that the software debugging be carried out in an earlier phase.

Fig. 2. Contribution of control software to project delay (Wünsch, 2008)

The error is detected too late

Many errors that are detected during the commissioning arise in the earlier phases. This is a big problem since the late detection of errors and the costs of their elimination can be very high (Figure 3). When an error occurs in the early phase, an overwork of the rest phases is not avoidable. For example, if a robot is wrongly placed in the 3D cell layout and this error is not detected until the commissioning phase, then it is a great effort with extra costs to eliminate this error because all the planning data related to this robot has already been modified.

Fig. 3. Costs incurred by the correction of errors, depending on the time of error detection (Kiefer, 2007)

2.3 State of the technology

One approach to overcome the disadvantages mentioned in subsection 2.2 is the full digital planning and simulation of the production line. Concepts, concrete process sequences and the necessary resources can be defined and approved already in the early planning phase. This activity belongs to the concept of digital factory.

Fig. 4. The engineering chain in the production with Virtual Commissioning

The focus of digital factory is the digital processing of product development and production planning based on the existing CAx data with a seamless workflow in PLM (product lifecycle management). Digital simulation methods and technologies are used to secure the planning results, to optimize the process and to respond more quickly to the changes than before. In this way the product qualities are improved. Another benefit is the reduction of the planning time as well as the overall project duration. The "Time-to-Market" is thus shortened which means another advantage with respect to economy.

To test the PLC-program before the real commissioning, the Virtual Commissioning can be applied (Figure 4). It serves as a smooth transition between digital and real factory. In the case of Virtual Commissioning, the physical plant, which consists of mechatronic components, is simulated with the virtual model. This simulated system is connected to the real controller (PLC) via simple connection or real industrial communication systems. The goal is to approximate the behavior of the simulated system to that of the real physical plant by connecting the commissioned PLC to the real plant without changes (Figure 5). Therefore the development and test of automation systems can be done parallel to the electric and mechatronic development. In the case of the real commissioning, the connection can be switched to the real system again.

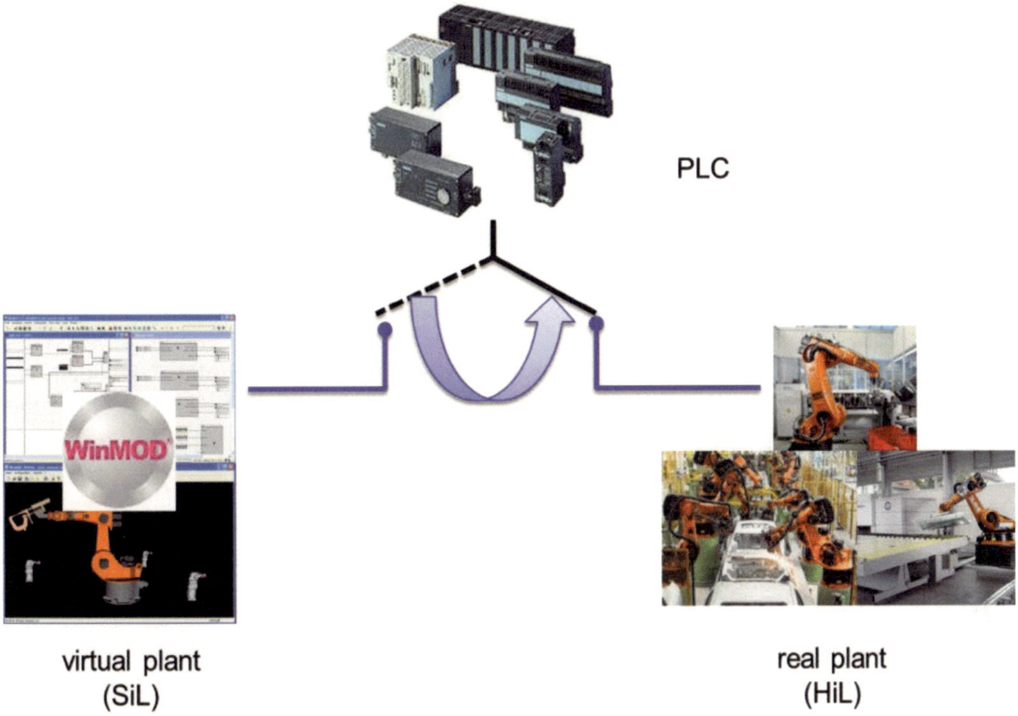

Fig. 5. From Virtual Commissioning to real commissioning

2.4 Component of the Virtual Commissioning

The virtualized plant, a mechatronic plant model, simulates the behavior of the real physical plant. It should respond correctly to the PLC control signals just as a real system. Besides, the whole process should be visualized for a better human observation. To fulfill these requests, a mechatronic model (Mühlhause et al., 2010) should be divided into a control-oriented behavior model and a kinematic 3D-model.

The behavior model simulates the uncontrolled behavior of the system. The behavioral states of the production resource are modeled and calculated by means of the logic and temporal components, which operate based on the control signals. In one PLC scan period, the behavior model should react to the control signals (output signals of PLC) according to the physical feature of the plant and give the feedback signal (input signals of PLC) back to the PLC - just as the behavior of a real system.

The kinematic 3D-model can be understood as a geometric model, which is based on the 3D-CAD model and enriched with additional information e.g. the grouping of the components which should be moved together and the definition of the DOF (degree of freedom) for every moving part. In contrast to today's mechanical CAD models, the mechatronic components do not just solely consist of the pure 3D CAD model, but also kinematic (including end positions) and electrical information like the electrical name of the respective device. For the purpose of synchronization, a signal coupling between the behavior model and kinematic 3D-model is needed.

The kinematic 3D-model is optional for the Virtual Commissioning, which helps for a better understanding by visualizing the movement behavior in 3D.

The communication between the mechatronic cell model and the real PLC can be realized via TCP/IP, field bus, Ethernet etc. There should be an interface to emulate the behavior of the communication. In the case of PROFIBUS, for example, a SIMBA-Card is needed.

2.5 Configuration of a Virtual Commissioning workstation

For the planning of a new production system, a layout plan for the necessary components (roles for the production resource to be used) and a rough workflow of the systems will be worked out based on a pattern (Figure 6). Based on this, a simulation model of the plant, including the geometric and kinematic design, will be built. This simulated layout is the starting point for the Virtual Commissioning, which is to be extended by the offline robot programming and planning process. At the step of detailed planning, the roles of resource are replaced by concrete units (function groups) under some fixed rules. For example, a certain combination of frequency inverter, the motor for a required electric moment of a torque, the type of clamp and valve cluster with the proper nominal pressure, or the robot and the weight of products to be operated, must be adhered to. These rules are to be observed by planners in the detailed planning. In a robot cell, for example, about 100 of such function groups must be matched and built. This function group can be stored in a kinematic library for the reusability. To build a simulation model, these previously created library elements can be added and connected with each other to describe the behavior of the corresponding component. Such a behavior simulation model has in average 30 inputs and outputs, which must be connected to the kinematics or to the PLC.

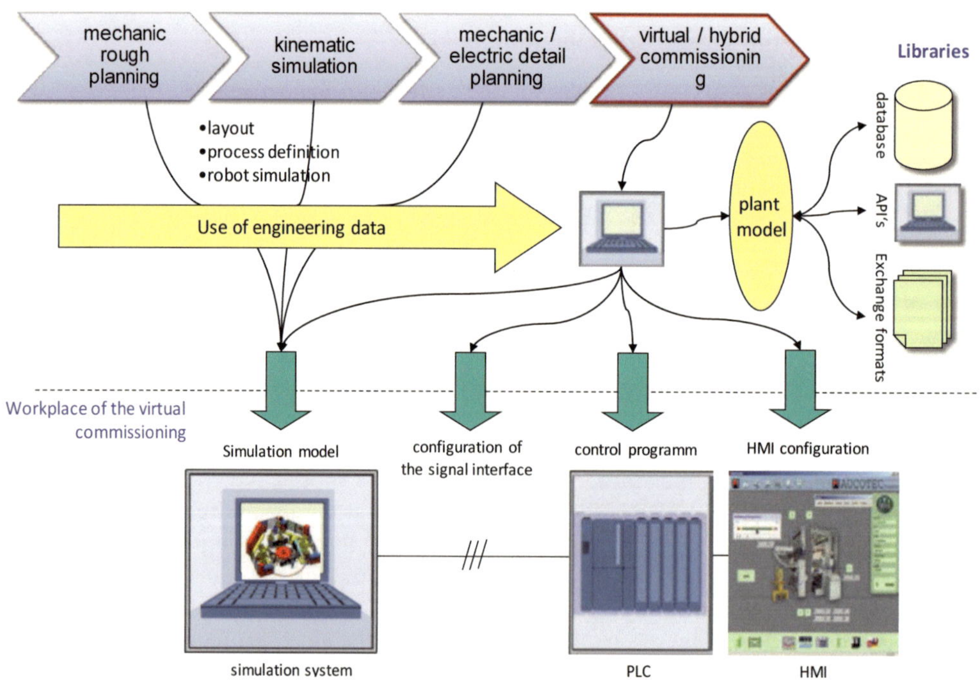

Fig. 6. Concept for integration into the Virtual Commissioning

After or in parallel to the layout planning and kinematic simulation, a suitable control program and operator visualization will also be developed, which run on a PLC or a PCS (Process Control System) system. The PLC program is developed in the following way: for each resource a corresponding control function block (Proxy FBs) is used, which is included in the control library. These Proxy FBs will be invoked in one or more SCF (Sequential Function Chart language IEC 61131-3). This control program will then be supplemented by plant-specific locks, a switch-off and parameterizations.

Afterwards, the PLC and the simulation systems are interconnected via the communication system or direct wiring. The I/O lists of the simulation system and the I/O lists of the control program must be properly connected with each other. Furthermore, each I/O has a number of features such as data type, range and unit, which need to be customized. The connection between the I/O of the Proxy FBs (or the communication configuration of the PLC) and the I/O lists of the simulation model must therefore be checked accordingly to the signal features.

A workstation for the Virtual Commissioning (Figure 6) consists of a simulation system on which the simulation model runs, a PLC which is usually connected with the simulation model via a bus system (e.g. PROFINET) and optionally a PCS.

In the phase of Virtual Commissioning, the plant which is to be tested is completely simulated in a digital model instead of a real one. This includes the automation equipment such as the clamps, robots and electric drives. When the PLC is connected to the real and digital systems at the same time, then it can be called the hybrid commissioning. For example, a robot with its controller can be integrated into a single system with other digital components, and the system should not detect any differences between the real and the digital components.

The current technology state assumes that the geometry and kinematics of the resources are provided by the manufacturer and as available libraries. In addition, the features of reused equipment may be stored in proprietary formats. These features and libraries are available in different storage systems (e.g. databases, repositories of tools and file systems).

Typical tools for Virtual Commissioning are Process Simulate/Process Simulate Commissioning and Delmia Robotics/Delmia Automation. In addition , the authors use the software WinMOD of the German company Mewes & Partner GmbH , which includes the component for 3D visualization. A circuit plan for electric planning can be edited for example with COMOS. There are also templates such as in Process Simulate, WinMOD, COMOS and Simatic S7, in which the corresponding features already exist in the library elements for converters, motors, robots etc. Today, the behavior models are not supplied by the resource producers. Instead, they are created by the users themselves for specific applications.

Besides the integration of a mechatronic plant model in the line-balancing scenario, a concept for the integration into the Virtual Commissioning was developed. The process of creating the workstation for the Virtual Commissioning has been explained above. The use of a mechatronic plant model should contribute to the construction of the workstation for Virtual Commissioning. The information which is extracted from existing data sources by using the mechatronic plant is able to represent the structure of the workstation (kinematic and behavioral simulation, PLC program, link-up between PLC and simulation model).

3. The mechatronic plant model with semantic techniques

The domain specific models in the individual life cycle phases usually have their own concepts depending on their physical phenomenon and tasks. This is essential because the models reflect only the necessary characteristics of the domains. A universal valid model is not thinkable. However, for seamless information flow it is necessary for the information consumers to have knowledge about the context of the transmitted information so that they can act in a proper way. (Rauprich et al., 2002) and (Wollschlaeger et al., 2009) point out that the historical separation of engineering disciplines, their workflows and thus the information integration is still an unsolved problem. One of the most common model structures in the manufacturing domain is the Product, Process Resource approach. It is the basis of digital planning of major tools like the Delmia tool set of Dassault Systemes (Kiefer, 2007).

Therefore, each piece of information has to be unambiguously identified and validated before it can be used in one of the engineering tasks. Additionally, its semantic including, its context information has to be specified and must be available. Mechatronic components, i.e. plant or machine parts consisting of mechanical, electrical and logical aspects, build up the modular blocks of the model, which are used in the engineering process. Thus these mechatronic components have to be represented by its semantic enriched information. Consequently, the mechatronic model must be based on semantic methods.

3.1 Mechatronic plant model

Today the information management in companies is characterized by a variety of available information, which is mostly stored distributed in a monolithic form and in different data sources. However , the size of each data source increases continuously over time, so that a connection between the information can only be realized through significant human work. As a result, much of the available information is very difficult to find by the user. Hereby a situation relevant lack of information arises within the information stream in the company. The time required for the complex information search and retrieval rises exponentially with the increasing amount of information , which results in significant costs. The mechatronic plant model is a logical layer above the design data and can be divided into the basic aspects of product, process and resource (Figure 7). Planning and other tasks in the plant life cycle can be executed by tool applications. These tools add new data to the existing data pool. These applications are modeled as so called facets (roles for the user data) in the mechatronic plant model. The resources could be further refined (see Figure 7). The basic idea of the model is to disperse all aspects (products, process and resource) in smaller features. For example, a construction group has the feature of the diameter of the pressure pipe connection or movement duration of a valve. The typical process features are process step number and number of the predecessor or successor step. If the mechatronic model is used for the acquisition of the data storage, the addressing data (e.g. data source type, URI of the data source) is also to be modeled. The mechatronic model is thus a hierarchical structure of sub-elements based on the three aspects product, process and resource. These three aspects are then to be further dissected into even smaller features, until no more division can be undertaken. In addition, these features are categorized so that semantic information about the individual elements of an aspect and its relationship with other elements of other aspects in the model can be enriched. (Figure 8)

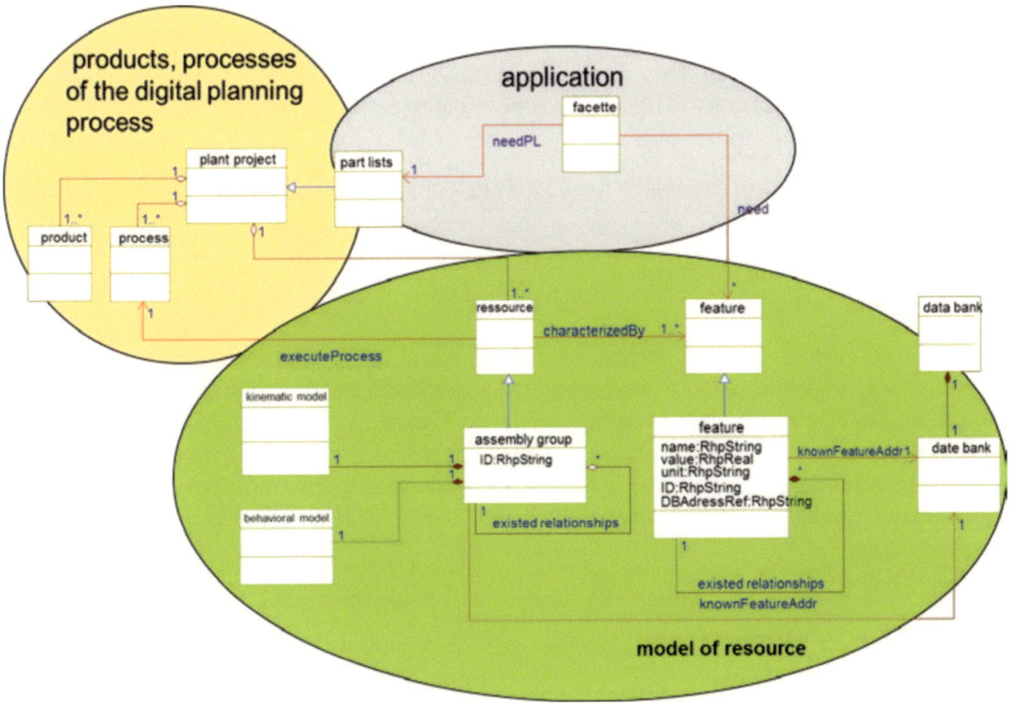

Fig. 7. Basic structure of the mechatronic plant model

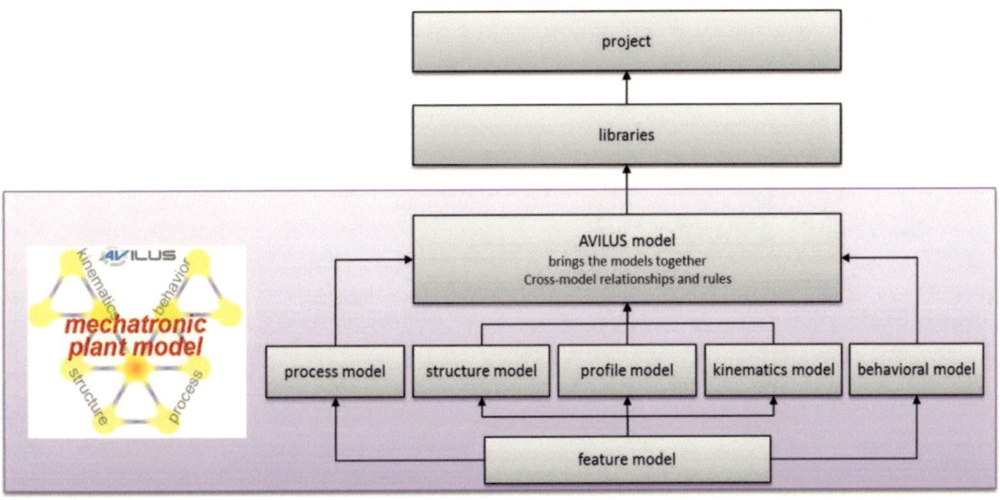

Fig. 8. The intern structure of the mechatronic plant model

3.2 Potential of the feature-based modeling approach

The basic idea of focusing on the features of engineering is to make engineering decisions and the underlying rules simpler to relationships between features. A unified set of features, for example coming from the standardization of technical product data, enables an automated execution of rules. This makes it possible to resolve the relationship with defined rules between feature carriers (e.g. valve or clamp or behavioral description and proxy FB).

This means the following assistance functions are available for the planner:

- Consistency check
- (semi-) automatic assignment
- Restrict the solution

An important element of the mechatronic plant model is the coupling between kinematics and behavior of the mechatronic components (aggregation relationship between the module as well as the kinematics and behavior model in Figure 7) and the integration of the PPR model (product, process and resource). Figure 7 shows the connection between resource and process (executeProcess). For the purpose of semantically unambiguous modeling, the description language OWL was used, which can clearly formulize the relations between the classes.

The extensive relationship between kinematics and behavior model are described in Figure 9.

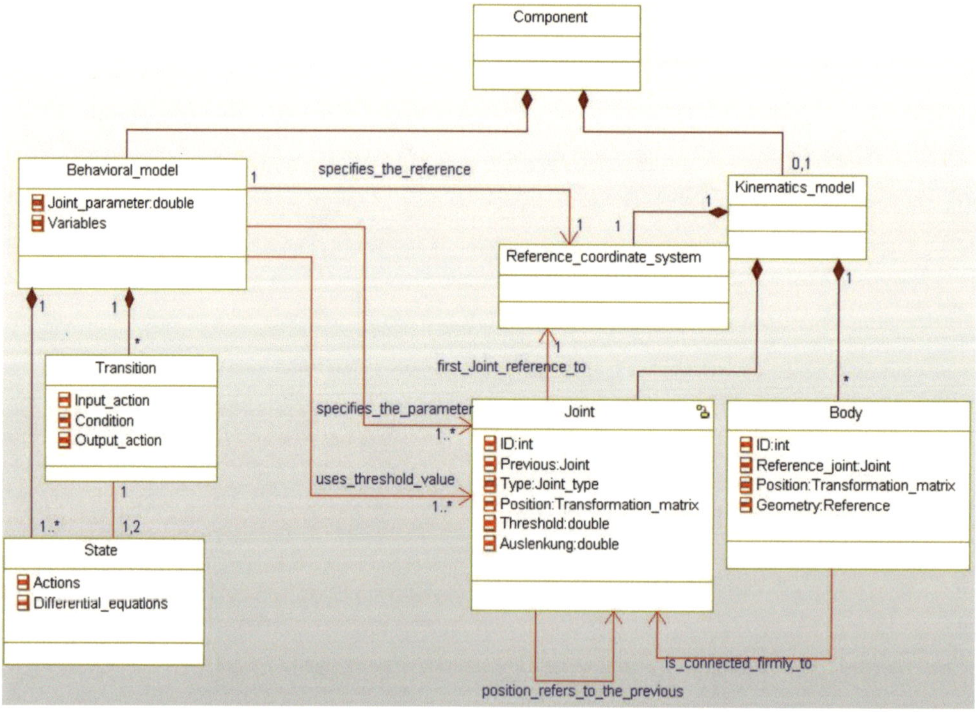

Fig. 9. Kinematic and behavioral model

In summary, the mechatronic plant model has the following essential basic elements:

- Basic structure is the grouping of product, process and resource
- Relationships are defined between the basic elements
- The PPR elements are then further specialized
- The properties of the PPR elements are described with features
- The features can be structured in a specialized hierarchy
- The features can build relationships with each other

4. Application of the model for Virtual Commissioning

4.1 Assistance by means of mechatronic cell model in the Virtual Commissioning

As described above, the starting point to establish a workplace for Virtual Commissioning is the layout and the kinematic model of the plant. In addition, it can be assumed that the information about the resources and process is located in component related files (kinematics, geometry, behavioral descriptions), in databases or tools.

The goal is to build a workplace for Virtual Commissioning. A tool should be implemented to assist the configuration of the workspace. This assistance tool is called VIB configurator or VIBCon (Figure 10).

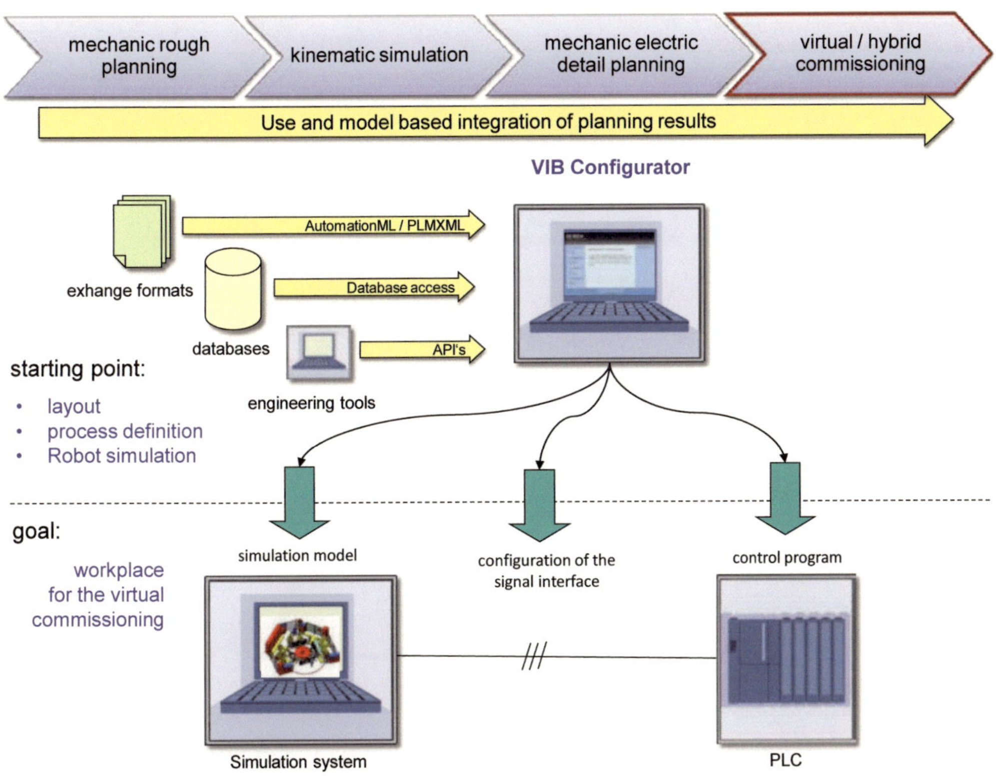

Fig. 10. The demonstrator for the Virtual Commissioning

The methodological foundations of the assistance tool allow the rule-based processing of raw data, which is related to a specific working step. The raw data can be extracted from various data sources. The result is the necessary configuration data or files for the workspace of Virtual Commissioning (Figure 11). In this way a semantic-based model is established in the mechatronic plant model. The model contains relationships such as the structure-based relationship "consistOf" or "isCharacterizedBy" (Figure 7). In addition, there are naming conventions that make, by using prefixes or suffixes, uniqueness semantic out of the mechatronic model. In this way, a check list based on the previously finished work can be generated. Based on this list, the library elements are instantiated. All these rules are defined in the VIB configurator. The following functions were implemented by the mechatronic plant model:

- Specification of the relationships between model elements and the workflow and the corresponding rules
- Generation of the needed configuration information by semantic inference or support of their derivation through a selection of possible alternatives

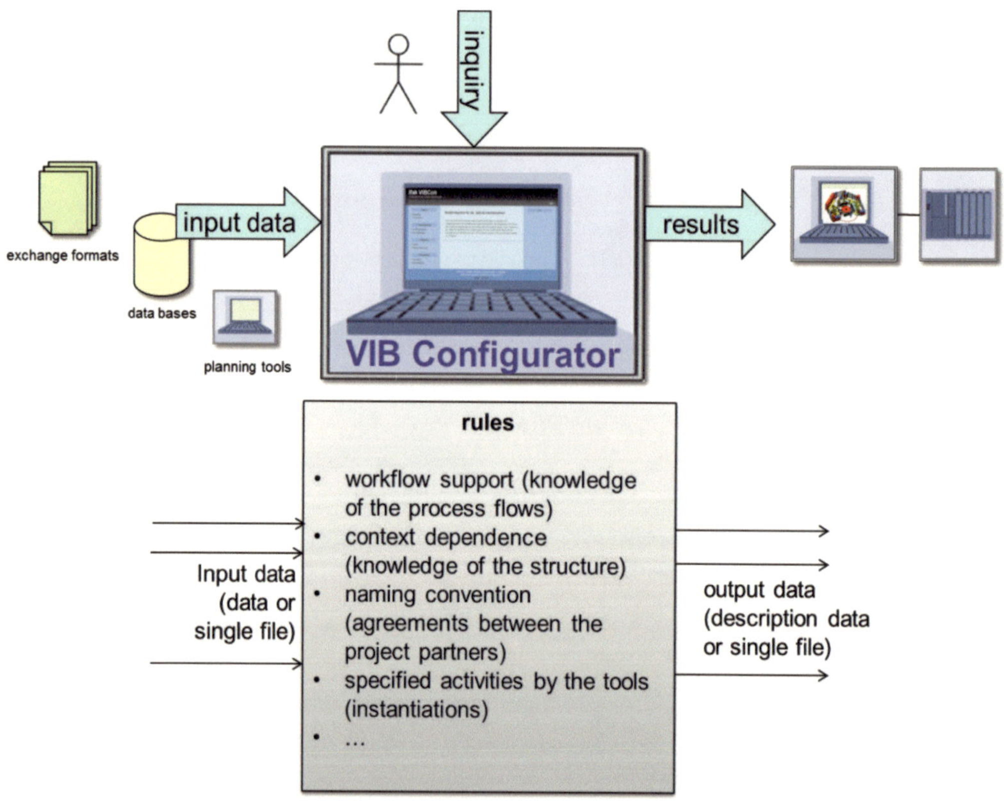

Fig. 11. Methodological approach of the rules

An example can be found at the beginning of the configuration of a Virtual Commissioning workstation (Figure 11). The needed document types (e.g. RobCAD file, VRML file, PLC program...) should be selected according to the context. These files will be earmarked and stored in the VIB configurator. When a new project is started, which uses the pattern of an old project, only the references to the types of these files and information are needed. These references are derived from the structural knowledge of the relationships between the types of resource and the types of their description.

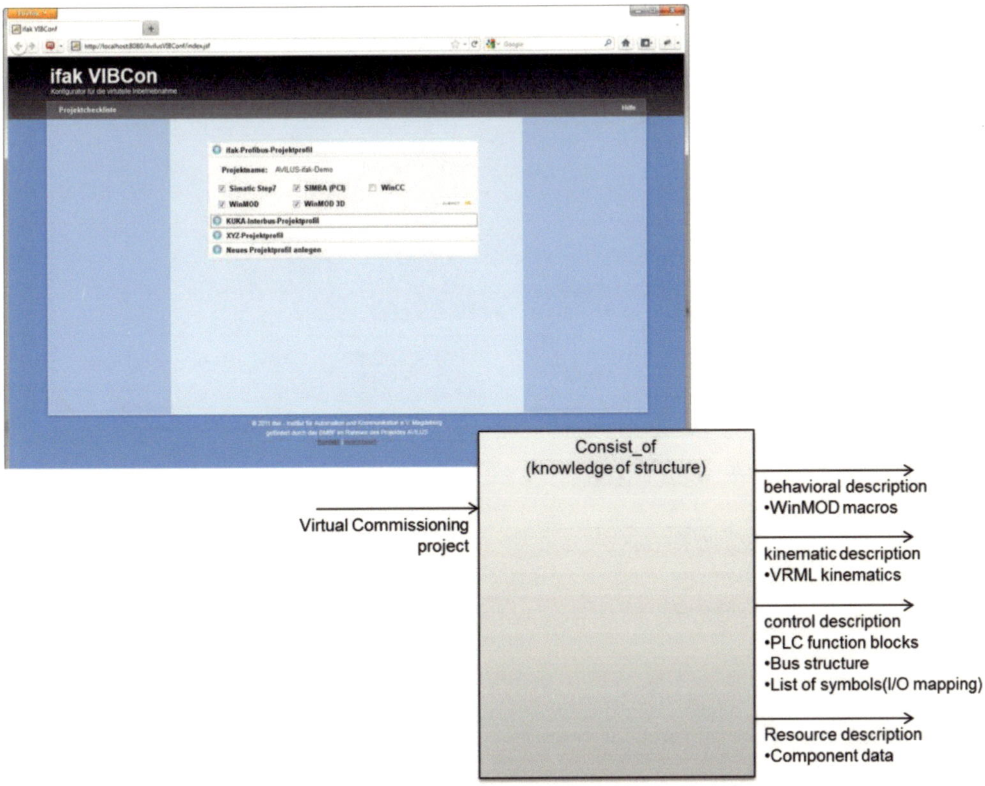

Fig. 12. Sample rule from the knowledge of structure - creating a project

A second example is the acquisition of information for all modules or components (implementation of the resources) that are included in the layout of a plant (Figure 13). The proper types of the files and information are needed for each resource in the project. The relationships between the model elements are integrated in the mechatronic plant model. Following the chain of these relationships between the elements leads to the references of the data sources. Naming conventions are applied here, such as extensions of the file name.

The mechatronic plant model serves as a knowledge model between the heterogeneous data sources and the data which is to be generated (Figure 13). The semantic model contains the know-how of the plant. The proper types of the files and information are needed for each resource in the project. The whole model is embedded in a software environment, which is described in subsection 4.2.

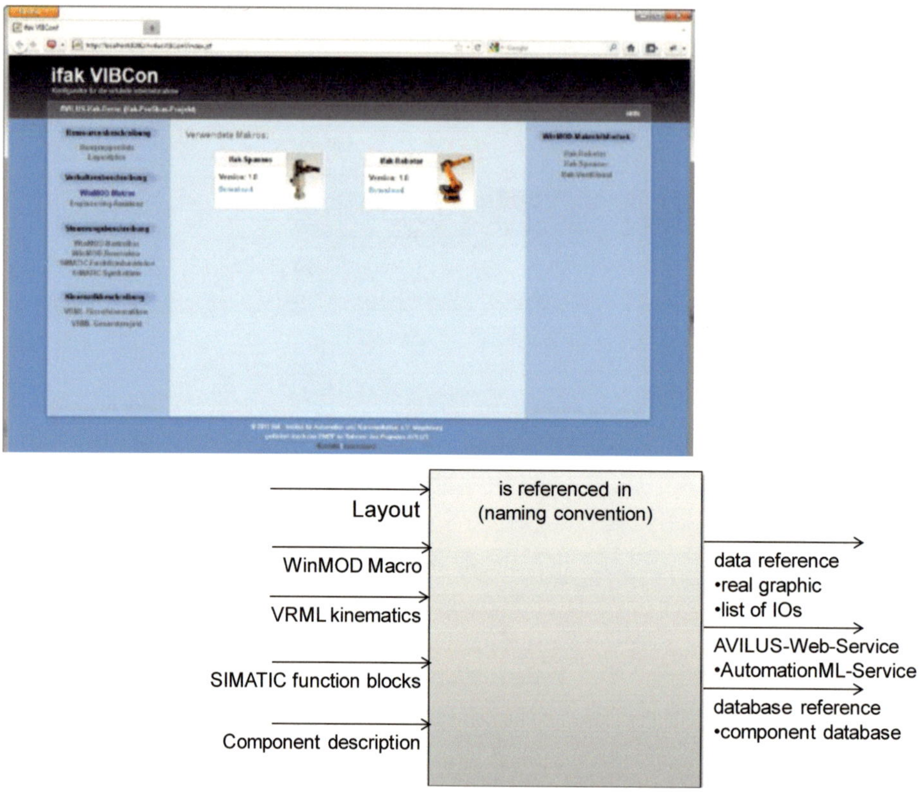

Fig. 13. Sample rule from naming convention - acquisition of information resource

4.2 Practical implementation of mechatronic cell model based on the semantic model

The main focus of the mechatronic model is the semantic modeling of the mechatronic components and the processes of the plant. This includes not only the description of the structural characteristics, e.g. the mechanical or electrical features, but also details like resource behaviors and process steps. Furthermore, a unified structure for the formal description of system properties is set (see Subsection 3.1). A framework (Figure 14) is implemented in order to access the information contained in the model like rules and relationships. Therefore, the needed information is found from different data resources (e.g. web services, exchange files of different formats). This process can be called reasoning.

The components and tasks of this framework include:

- Design and modeling (Figure 14, green)
- Standard technologies of "Semantic Web" (Figure 14, gray)
- Integration and evaluation (Figure14, orange)

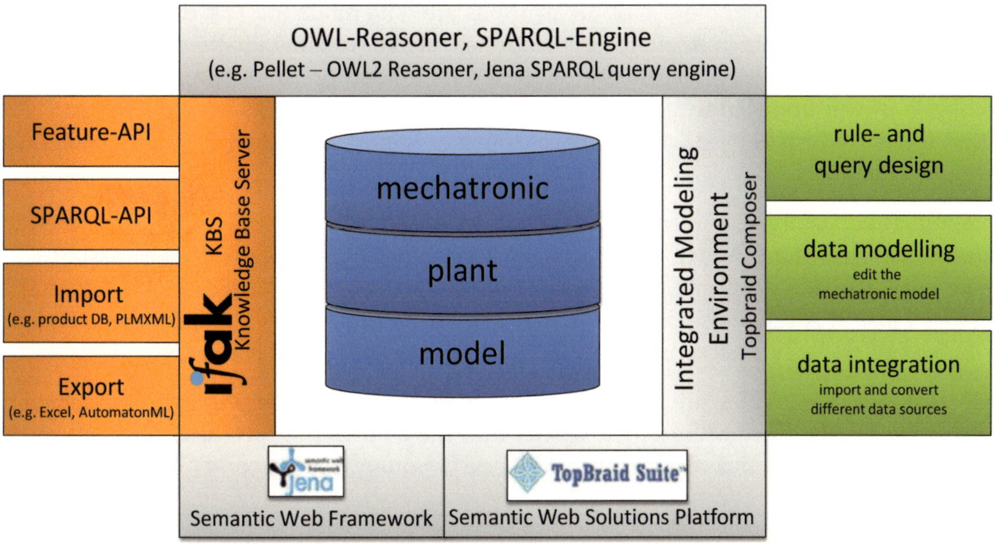

Fig. 14. Framework of the mechatronic plant model

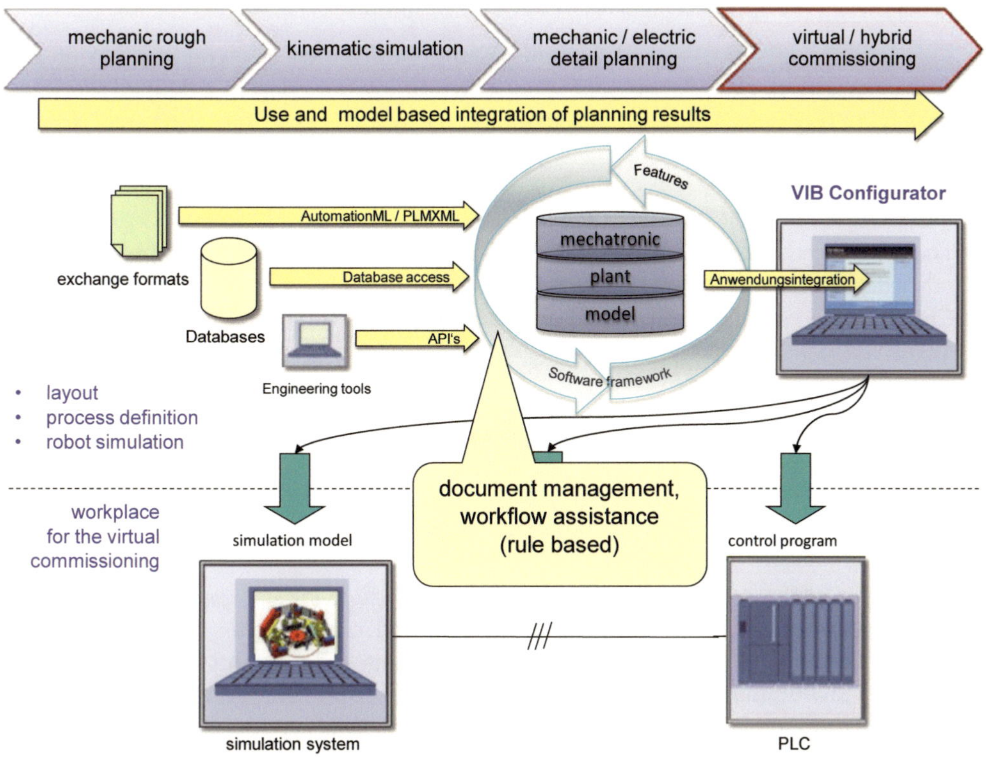

Fig. 15. Role of the mechatronic plant model in the preparation of a workstation for Virtual Commissioning

The most important technologies of the methods are data modeling and design of the rules and inquiries. For data modeling, as mentioned above, a semantic-based approach was deployed. The design of the rules and inquiries was based on this approach, which forms the basis of the knowledge processing known as automated rules-based reasoning. The instance data, which is based on the mechanisms of import and conversion, was integrated into the data model.

The integration of existing standard technologies became the focus in the development of the mechatronic plant model. The necessary technologies for knowledge processing have been implemented by applying the well-established tools and frameworks (e.g. OWL reasoned and SPARQL engines) for the Semantic Web. In this way, a semantic-based model with the functions, such as consistency check, reasoning and graph-based query, was built.

Different interfaces, which are based on a synchronous communication, are implemented as prototypes for the integration of the model into the engineering process. These interfaces enable the feature-based access (Feature API) and the query-based access (SPAQL API).

The concept, based on standardized exchange formats such as PLMXML and AutomationML (Draht et al., 2008), enables the flexible import and export of engineering data. The "Knowledge Base Server" (ifak-KBS) is a middleware which facilitates the integration of the mechatronic model using different interfaces into the engineering tools.

A conclusion of the important features of the VIB configurator is listed as follows:

- The development of a Virtual Commissioning workstation is a combination of tasks such as selection, classification and design, which are specified in a defined workflow. Assistance functions can be offered based on the knowledge of workflow and structure as well as naming conventions.
- The fundament of the assistance functions is the mechatronic plant model that was implemented in a software framework, based on semantic technologies.
- The introduced functions serve only as examples here, which can be supplemented by other detailed analysis of the process.
- Checklists and possible tests can be derived out of the available digital data.

5. Acknowledgment

This work is funded by the German Federal Ministry of Education and Research (BMBF) within the project AVILUS.

6. References

Bär, T. (2004). *Durchgängige Prozesskette vom digitalen Produkt bis zur realen Fabrik,* Congress „Digitale Fabrik" Ludwigsburg, Germany (2004)

Bergert, M.; Diedrich, C. (2008). Durchgängige Verhaltensmodellierung von Betriebsmitteln zur Erzeugung digitaler Simulationsmodelle von Fertigungssystemen, *atp – Automatisierungstechnische Praxis,* Vol.7 (2008), pp. 61-66, Oldenbourg Industrieverlag, ISSN 2190-4111

Bracht, U. (2002). Ansätze und Methoden der Digitalen Fabrik, *Tagungsband Simulation und Visualisierung,* Magdeburg, pp. 1-11

Draht, R.; Lüder, A.; Peschke, Jörn. & Hundt, L. (2008). AutomationML - The glue for seamless automation engineering, *Proceedings of the 13th international conference on Emerging Technologies and Factory Automation*, pp.616-623, ISBN 978-1-4244-1505-2, Patras, Hamburg, September 15-18, 2008

Heeg, M. (2005). *Ein Beitrag zur Modellierung von Merkmalen im Umfeld der Prozessleittechnik*, VDI-Verlag, ISBN 978-3-18-506008-3, Düsseldorf, Germany

Jain, Atul; Vera, D. A.; Harrison, R. (2010). Virtual commissioning of modular automation systems, *Proceedings of the 10th IFAC Workshop on Intelligent Manufacturing Systems*, pp. 79-84, ISBN: 978-3-902661-77-7, Lissabon, Portugal, July 1-2, 2010

Kiefer, J. (2007). *Mechatronikorientierte Planung automatisierter Fertigungszellen im Bereich Karosserierohbau*, PhD thesis, Universität des Saarlandes, Schriftenreihe Produktionstechnik, ISBN 978-3-930429-72-1, Saarbrücken, Germany

Kiefer, J.; Bergert M. & Rossdeutscher M. (2010). Mechatronic Objects in Production Engineering, *atp – Automatisierungstechnische Praxis*, Vol.12 (2010), pp. 36-45, Oldenbourg Industrieverlag, ISSN 2190-4111

McBride, B. (2010). An Introduction to RDF and the Jena RDF API, available from http://jena.sourceforge.net/tutorial/RDF_API/index.html, Retrived on 22/02/2012

McGuinness, D.; van Harmelen, F. (2011). OWL web ontology language overview, available from http://www.w3.org/TR/owl-features/

Mühlhause, M.; Suchold, N. & Diedrich, C. (2010). Application of semantic technologies in engineering processes for manufacturing systems, *Proceedings of the 10th IFAC Workshop on Intelligent Manufacturing Systems*, pp. 61-66, ISBN: 978-3-902661-77-7, Lissabon, Portugal, July 1-2, 2010

Prud'hommeaux, E.; Seaborne, A. (2008). *SPARQL Query Language for RDF*, W3C Recommendation

Rauprich G.; Haus C. & Ahrens W. (2002). PLT-CAE-Integration in gewerkeübergreifendes Engineering und Plant-Maintenance, *atp - Automatisierungstechnische Praxis*, Vol.2 (2002), pp. 50-62, Oldenbourg Industrieverlag, ISSN 2190-4111

Schreiber, S.; Schmidberger, T.; Fay, A.; May, J.; Drewes, J. & Schnieder, E. (2007). UML-based safety analysis of distributed automation systems, *Proceedings of the 12th international conference on Emerging Technologies and Factory Automation*, pp.1069-1075, ISBN 1-4244-0826-1, Patras, Greece, September 25-28, 2007

TopQuadrant (2011). *Getting Started with SPARQL Rules (SPIN), Version 1.2*, Available from http://www.topquadrant.com/spin/tutorial/SPARQLRulesTutorial.pdf, Retrived on 22/02/2012

VDI 4499 Blatt 1 (2008). Digitale Fabrik - Grundlagen. VDI-Handbuch Materialfluss und Fördertechnik, Bd. 8, Gründruck. Berlin: Beuth

Volz, R. (2008). *Web Ontology Reasoning with Logic Databases*, PhD thesis, 17 Februar 2008, Universität Karlsruhe, Germany

Wollschläger M.; Braune A.; Runde S.; Topp U.; Mühlhause M.; Drumm O.; Thomalla C.; Saboc A. & Lindemann L. (2009), Semantische Integration im Lebenszyklus der Automation, Tagungsband zum Automation 2009 –Kongress Baden-Baden, VDI-Berichte 2067, pp. 167-170, VDI-Verlag, Düsseldorf, Germany

Wünsch, G. (2008). *Methoden für die Virtuelle Inbetriebnahme automatisierter Produktionssysteme*, PhD thesis, Technische Universität München, Herbert Utz Verlag, ISBN 978-3-8316-0795-2, München, Germany

Zäh, M. F.; Wünsch, G. (2005). Schnelle Inbetriebnahme von Produktionssystemen, *wt Werkstattstechnik 95*, Vol. 9 (2005), pp. 699-704

8

Introduction to the Computer Modeling of the Plague Epizootic Process

Vladimir Dubyanskiy[1], Leonid Burdelov[2] and J. L. Barkley[3]
[1]*Stavropol Plague Control Research Institute,*
[2]*M. Aikimbaev's Kazakh Science Center of Quarantine and Zoonotic Diseases,*
[3]*Virginia Polytechnic Institute and State University*
[1]*Russian Federation*
[2]*Republic of Kazakhstan*
[3]*USA*

1. Introduction

Bubonic plague, caused by the bacteria, Yersinia pestis, persists as a public health problem in many parts of the world, including Central Asia and Kazakhstan. Plague is a vector-borne disease, i.e. the disease is spread by arthropod vectors that live and feed on hosts (Gage and Kosoy, 2005). Great gerbil (Rhombomys opimus Liht., 1823, Rodentia, Cricetidae) is a significant plague host in Central Asia and Kazakhstan region. Transmission happens through bites of infected fleas that earlier fed on infected hosts. This plague epizootic process study can help future plague control work.

Modeling the epizootic process of plague in great gerbil settlements allows for quantitative analyses of epizootic characteristics, which we are unable to control for in nature (Soldatkin et al, 1973). Only one such attempt of this type of modeling is known (Soldatkin et al, 1966; Soldatkin and Rudenchik, 1967; Soldatkin et al, 1973). In this study, the authors demonstrated the high potential for this model, yet it possessed many limitations and had a small work space (4,096 holes). The development of high-performance workstations, and integration of geographic information systems (GIS) and remote sensing now makes it possible to model complex, epizootic processes of plague.

We created a probabilistic, cellular automaton model of the epizootic process of plague by using all of the above-named technologies. A probabilistic, cellular automaton with Monte-Carlo, as the chosen statistical method, is the basic mechanism of the model (Etkins, 1987; Grabovskiy, 1995). It is the SEIR (susceptible, exposed, infected and removed) model with specific additions.

2. Description of the model

The model was created using Microsoft Excel 2003 with Visual Basic as the code language. Microsoft Excel has 255 columns and 65,000 rows. A user-defined, fixed size of the modeling area depends on the density of the gerbil colonies. A probabilistic, cellular automaton with Monte-Carlo, as the chosen statistical method, is the basic mechanism of the model

(Dubyanskiy V.M., Burdelov L.A. , 2010, Dubyanskiy V.M, 2010). The choice for this model type was rationalized by the following:

- Commonly-used mathematical models that rely on differential equations require input from many data sources. While, at the same time, the availability of plague surveillance source data is rather insufficient.
- Temporal models, (based on differential equations or other methods), are being used for modeling without consideration of spatial distribution. In these models, it is assumed that all events are happening at an abstract point rather than considering plague focus, landscape region, or primary square.
- If plague simulates nature, it analyzes only a static structure of plague focus or its parts.

The probabilistic, cellular automaton simulation allows the user to combine all of these rationales. The epizootic process of plague in Great gerbil settlements was a good subject for this experiment. The gerbil's burrow system represents a time-spatial, discrete unit of an epizootic process (Sedin, 1985). The data on burrow systems distribution within a plague focus was determined by GIS and remote sensing. Satellite images of the burrow systems provide a coherent picture of the colony structure (Burdelov et al., 2007; Addink et al., 2010) at many plague foci (Fig 1.).

The cell/row structure of the model's workspace allowed us to observe the proportion between burrow systems and settlements and density of colonies. The density of colonies was the main factor approximating the number of rodents. Both the density of colonies and the number of rodents are significant factors for plague prediction (Klassovskiy et al., 1978; Davis et al., 2004, 2008).

Several quantitative and qualitative relationships and characteristics between components of the plague's parasitic system were identified early on (Burdelov et al., 1984; Begon et al., 2006). Analyzing these numerous, character-combinations required that we use Monte-Carlo simulation, rather than differential equation models, to describe each specific time stamp in question. The analysis of burrow systems distribution on satellite images demonstrated that the number of nearest neighbors in the burrow system often exceeded four (von Neumann neighborhood). Hence, our model used the square (eight-cell) Moore neighborhood (Fig 2.).

The state of colonies can be categorized as "infected", "infecting", "immune", or "readied" for reinfecting. An infected colony is one that is infected, (in host, in vector or in substratum), but has not spread the plague microbe (latent period). An infecting colony is one that is infected and is spreading the plague microbe among other colonies. A colony that has been cleared of the plague microbe and is not infectious is called an immune colony. A readied colony is one which can be infected again after the immune period.

Data on the vector's and host's infection were used to the extent that it was necessary for calculating the model's parameters. An infecting agent can be transmitted to other colonies independent of the main mechanism of "host-vector-host" transmission. These processes define whether a colony is infected or not and constitutes the "black box" of the model. Nuances of an epizootic process which might be connected to specific mechanisms of plague transmission have not been reviewed.

The cycle of infection is a duration when an infected colony is being converted to an infecting colony and can spread the plague microbe among other colonies. The obligatory scan of the model's workspace occurs at the end of each cycle of infection. The probability of transfer success is defined as one attempt of the plague microbe to transfer from an infected colony to a non-infected colony. The number of attempts required to infect neighboring colonies from one "infecting" colony during one cycle of infection is termed the coefficient of colony infection. This term was proposed by Litvin et al., (1980). However, this term, in our case, unites the coefficient of infection of a colony by Litvin and the coefficient of transmission by Soldatkin et al., (1973). The rate (or the value) of epizootic contact is the product of the successive transfer probability and the coefficient of a colony infection.

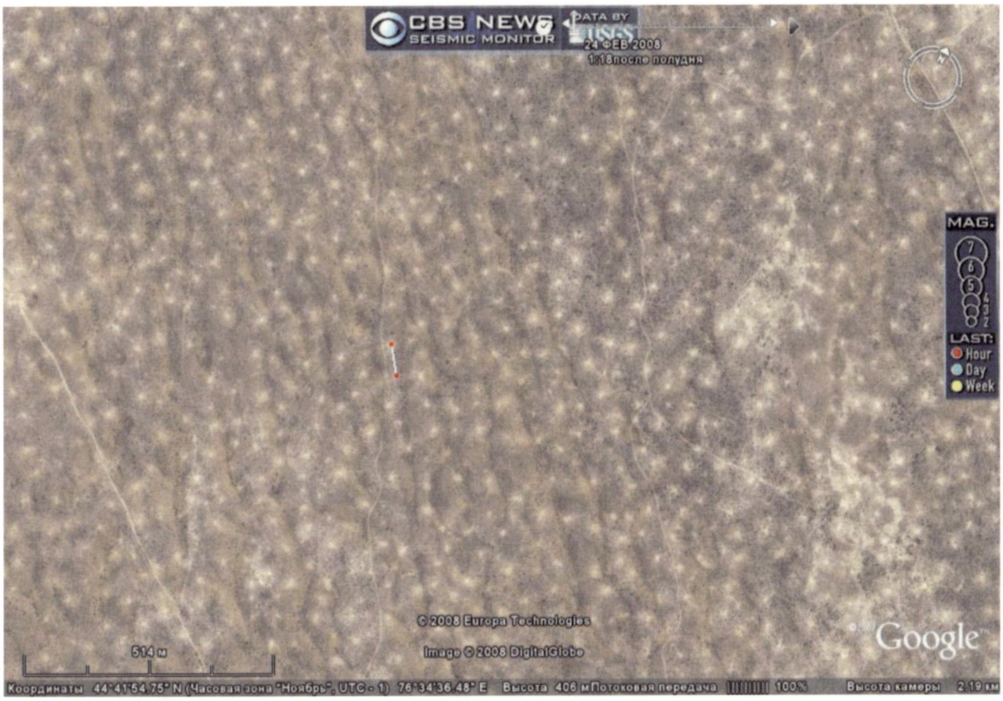

Fig. 1. Colonies of Great Gerbil in a satellite image. Each bright disc represents colonies 10-40m in diameter. The image was captured using the publicly available software Google Earth (http://earth.google.com/). Copyright 2008 DigitalGlobe; Europa Technologies.

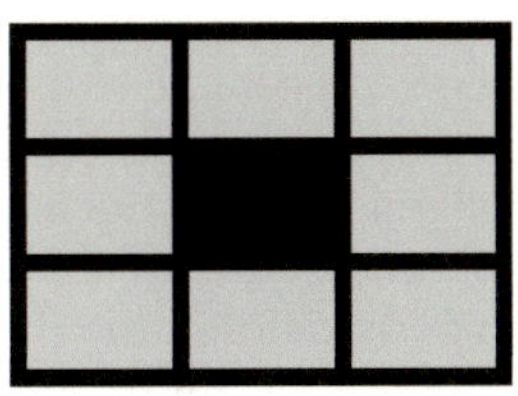

Fig. 2. Moore's environment allows using all cells in model for its greater efficiency.

$$L = C_i * C_t \tag{1}$$

where L is a rate (or the value) of epizootic contact, C_i the coefficient of colony infection, C_t the probability of transfer success.

The relative remote drift is the possibility of more or less remote colonies infecting without nearer colonies infecting. This parameter can be regulated by the change of plague drift distance as well as by the change of plague drift realization probability. It also allows changing the speed of epizootic spread.

The structure of great gerbil settlements was reflected by altering colony-cells and non colony-cells. This allowed working with different configurations of settlements and with different densities of burrow systems. The scheme of burrow system's spatial location could be integrated into the work space of the model from a map or from a satellite image if necessary (Fig. 3). If the epizootic process was imitated, then the spatial location of burrow systems would be defined by a random-number generator. Initial results were published earlier (Dubyanskiy, Burdelov, 2008).

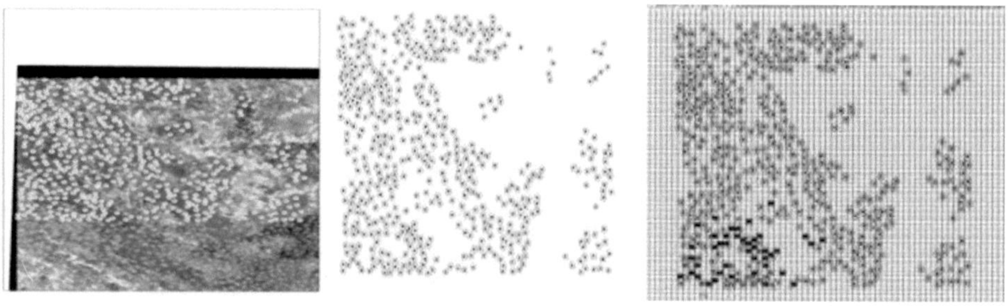

A B C

Fig. 3. Carry of a real configuration of settlement of Great Gerbil from the satellite image adhered in GIS in working space of model. A - Allocated by means of GIS colonies of Great Gerbil. B - Layer of GIS ready for carry to working space of model. C - Working space of model with developing of plague epizooty.

There are two ways of spreading the plague microbe. The "infected" colony could attempt to infect the other colony-cell as well as non colony-cell. In the latter case, the attempt of infecting is useless. However, the tendency of gerbils to migrate among colonies will spread the plague microbe to occupied colonies. The probability of transfer success is a probability that a sick flea (or sick gerbil) will be transferred from an "infected" colony to a non-infected colony and a microbiocenose of colony will be involved in the epizootic process. The coefficient of a colony infection demonstrates how often transfers occurred (i.e. a gerbil can visit non-infected colony two times or 10 times and so on).

The basic quantitative and qualitative characteristics were entered into the parameter sheet. The value of parameters can be changed before modeling, although the dynamic changing of parameters is possible. The parameters could be, for example, quantities of the "cycle of infection" for one season (or session of modeling), the duration of the cycle of infection, the density of colonies, and the probability of infecting neighbor colonies. The technical conditions of the model work were also defined. The built-in menu allows comfortably managing the modeling options.

Cellular flattened models have the edge effect. Naturally, this effect took place in our model too. It was decided that all transmission of plague microbes outside the model workspace was not consequential. In this case the model was made simpler. The model settlement looked like an isolated island of the gerbil's settlement in nature.

2.1 Basic parameters of the model

Even with the seeming simplicity of the model, the real quantity of fulfilled interrelations is rather great. For example, "infection" of a colony depends on: the duration of cycle of infection; the duration of the colony's status as infected; infectious quotient and immunity; the probability of success of transfer; the coefficient of infection of a colony; the density and distribution of colonies in the workspace of the model; and the number and arrangement of infected colonies. Considering that further behavior of a cell depends on probability and status of surrounding cells at PCA, to give the optimum and full analysis of influence of such quantities of the interconnected parameters is very difficult. Therefore, initial selection of the basic working parameters of the model were carried out in a literature search of data, including results of direct or physical modeling of epizootic process using radioactive isotopes in real settlements of the great gerbils. It is necessary to make a stipulation that the analysis of literature has shown an absence of any unequivocal quantitative characteristic of epizootic process in a required aspect.

After infection by the plague microbe, a colony becomes "infected". This period in the model lasts for 10 days where three days are the time of bacteremia and seven days are the time of flea's prohibition (block of the proventriculus) if the flea fed on a sick animal. The colony has the possibility to infect neighbors from day 11 to 50 from the beginning of its own infection. It is not infectious from day 51 to 90, and it is immune (Novikova et al., 1971; Naumov et al., 1972). The colony can be infected again after 90 days (Samsonovich et al., 1971; Rothschild et al., 1975; Rothschild, 1978).

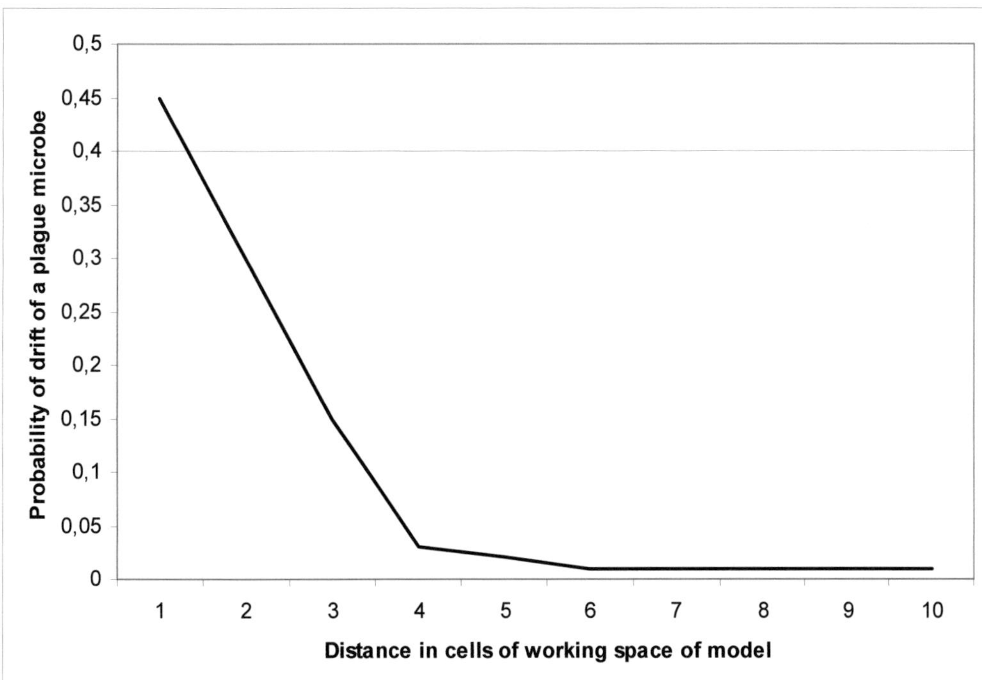

Fig. 4. Probability of drift of a plague microbe from colony to colony in working space of model during one cycle of infection.

The probabilities distribution of plague microbe drift approximately corresponds to data of direct modeling of drift of the fleas in the neighboring holes (Soldatkin, Rudenchik, 1971), and to data on movement distances of gerbils to the neighbor colonies (Alekseev, 1974; Davis et al., 2008) (Fig. 4). The probability of transfer success could be modified by the user within the limits from zero to one. The coefficient of a colony infection could be varied from one up to experimentally received or any other logically admissible value. The density, distribution, occupation, and number of a plague microbe hosts expressed through occupation were given as cell-colonies percentage, from quantity of cells in all working spaces of the model and could be changed from zero to 100. It would be desirable to emphasize that it was possible to change any parameters in the model.

The maximum drift of plague microbe in model can mark as l. Every infecting colony can infect the neighbours in a square with diagonal $2l$.The infecting colony is situated in the center of this square. The probability of contact with any colonies inside this square defined as

$$Pz = \frac{N_l - (i_1 + i_2 + R)}{N_l} \tag{2}$$

where Pz is the probability of contact with the neighbour colony, N_l is the general number of colonies in the square with $2l$ diagonal, i_1 is the infected colony, i_2 is the infecting colony, R is the immune colony.

The probability of colony infect is defined as

$$Pi = Pz * Ct \quad Ci \tag{3}$$

where Pi is the probability of colony infect, Ct is the probability of transfer success (input from experimentalist), Ci is the coefficient of colony infection (input from experimentalist).

Taking into consideration the formula (3)

$$P_l = P_Z * L \tag{4}$$

The intensity of model's epizooty is defined as

$$Y = (i_1 + i_2) / B * 100 \tag{5}$$

where Y is the intensity of epizooty, B is the general number of colonies.

The quantity of infected colonies are increasing in the beginning as Fibonacci sequence

$$F_{n+2} = F_n + F_{n+1}, n \in N \tag{6}$$

However when immune colonies appear, the complexity of the epizootic system increases. The description of epizooty by differential equations are not possibe.

The basic parameters described above would be for a case studying the influence of spatial parameters for epizootic processes. It is possible to establish a reverse problem to study the influence of temporary parameter changes at a static, spatial configuration of the settlement. In an ideal case, data can be accumulated on epizootic processes in nature and in experiments on models.

For this study, the plague spread in various settlement types of great gerbils used the following items: 1. The probability of transfer success was equal to one, 2. the coefficient of a colony infection could be varied from one up to 10, and 3. The coefficient of a colony infection was always one.

A two dimensional array of cells was used for the modeling world. The modeling square was equal to 900 ha. The experiment was successful if the epizootic process continued during 18 epochs (model's half year) 11 times without a break. It was a 0.9999 probability that the experiment was nonrandom (Rayfa, 1977). The start value of the coefficient of a colony infection was equal to one. The start density of colonies were 1.33 per ha. The density of colonies declined for 0.11 colonies per ha after each successful experiment for continuous settlements and it was increasing the distance among colonies per one cell on vertical and horizontal lines for narrow-band settlements. As a result, the density of the colonies declined exponentially. If the experiment was not successful the coefficient of a colony infection increase up per unit and recurred again.

3. Results

We gave an example of developmental epizooty in the feigned settlement of the great gerbil (Fig. 5). The goal of the experiment was an examination of qualitative likeness between modeling and natural epizootic processes. The size of the model's working field was 2x2 km. It corresponded to an area of 400 hectares with density of colonies at 2.4 per hectare. A conditional start date of epizooty was March 1st. The epizootic process was developed as a result of the distribution of a plague microbe from one infected and random established colony. The probability of transfer success and the coefficient of a colony infection were constant and equal, 0.8 and 1 respectively.

The epizooty developed slowly at the initial stage. The analysis of the modeled epizootic process showed that the critical time was the initial transfer of a plague microbe between colonies. This was also noted by Soldatkin et al. (1971). The epizootic process could proceed indefinably long if the "infected" colony transferred the plague microbe even to one "not infected" colony during its "infecting" ability. The random hit of a primary "infected" colony on a site of settlement where transfer of a plague microbe would occur raises the probability of epizootic fade-out. The epizooty will end if the transfers of plague microbe do not provide even short-term constant increase of number in the "infected" colonies (1.1 - 2 times during a minimum of 3 cycles of infection). If the start of epizooty was successful, after three months, small foci, consisting of 4 - 8 infected colonies would have been formed (Fig. 5a). The number of the infected colonies increased exponentially. On reaching a certain level of "infected" colonies, the intensity of epizootic process started varying cyclically (Fig. 6). The exponential increase of the infected colonies number was terminated by virtue of two causes. The first cause is the probabilistic character of the plague microbe transfer success from colony to colony. If two or more colonies are infected by an infected colony, the epizootic process intensity will exponentially increase. At the same time, an infected colony can be a source of infection for all neighboring colonies or infect nothing. The second cause is the relatively remote drift of the plague microbe that forms many small foci, consisting of the infected colonies (Fig. 5b). These foci are dissemination centers of epizooty. The spatial structure of the epizooty site is a complicating factor. It begins as a saturated epizooty site of "infected" and "immune" colonies, whereby the plague microbe transfers to other "infected" and "immune" colonies in greater numbers, decreasing the probability of transferring to "non infected" colonies.

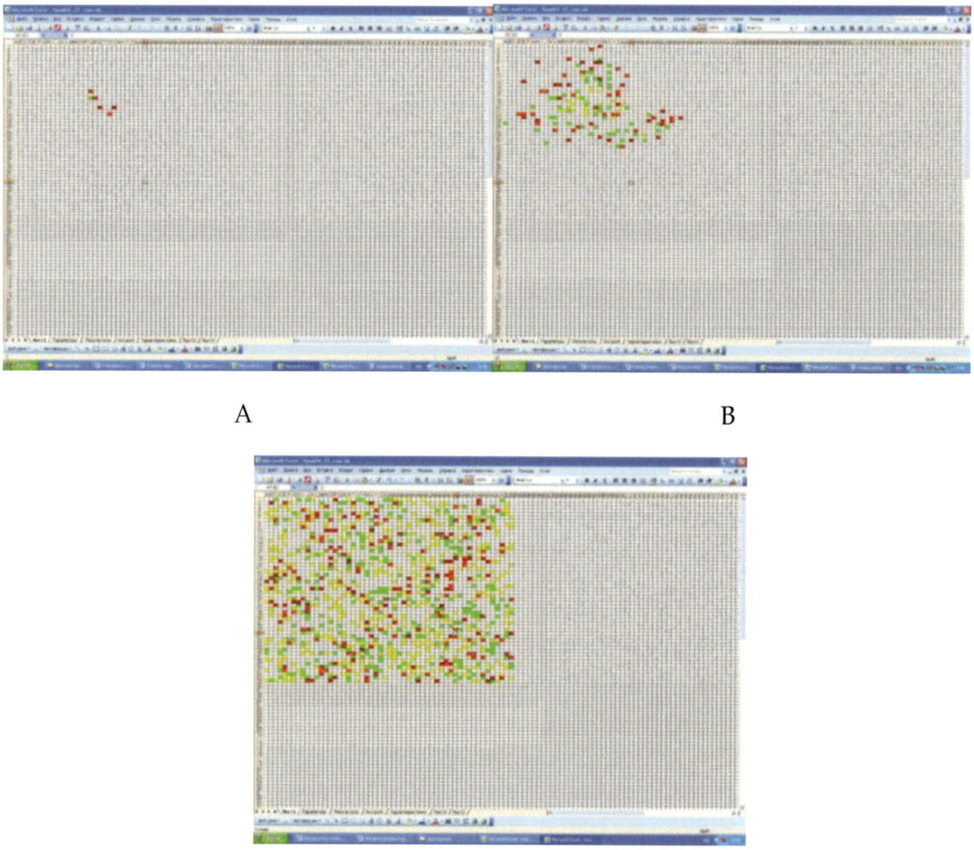

A B

C

Fig. 5. The visual representation of epizootic process had been developed at model. A) The initial stage of the epizooty development (90 days). B) Remote drift and formation of small foci (210 days). Quantity of the "infected" colonies had started an exponential increase. C) A stage of "saturation" of the epizooty site and transition to cyclic fluctuations

Typically, two peaks of epizooty activity are registered during the year, in the spring and in the fall (21). The epizootic contact rate increases during the rut in early spring and at the end of the spring, as young gerbils explore their territory. Epizootic contact rate and process intensity decreases during the high summer temperatures as the gerbils spend less time on the surface. This also coincides with a generational replacement of the *Xenopsilla* fleas. Contact increases again during the fall when the gerbils become hyperactive, resulting in a second peak of epizooty activity. In autumn, the gerbils cache food and redistribute it among the colonies before winter. The younger generations of fleas are actively feeding, possibly transferring the plague microbe among the animals. This complex mechanism is approximated through the rate of epizootic contact within the model. The exact data about probability of "infecting" colonies in different seasons of the year are absent in literature. Thus, for a qualitative look at the model and nature, the following conclusions were garnered from the model: the probability of a colony infection in the summer is half of that in the spring; a colony infection in the autumn is 0.15% less than in the spring; and a colony infection in the winter is three times below that in the autumn.

Numerical values of the probability of transfer success and the coefficient of colony infections were calculated earlier (Soldatkin et al, 1966, 1971). The probability of transfer success was equal to 0.8 and the coefficient of a colony infection was equal to 1.0, for spring. After entering these parameters, the picture of epizootic process in the model was appreciably changed and looked much like a natural occurrence (Fig. 7).

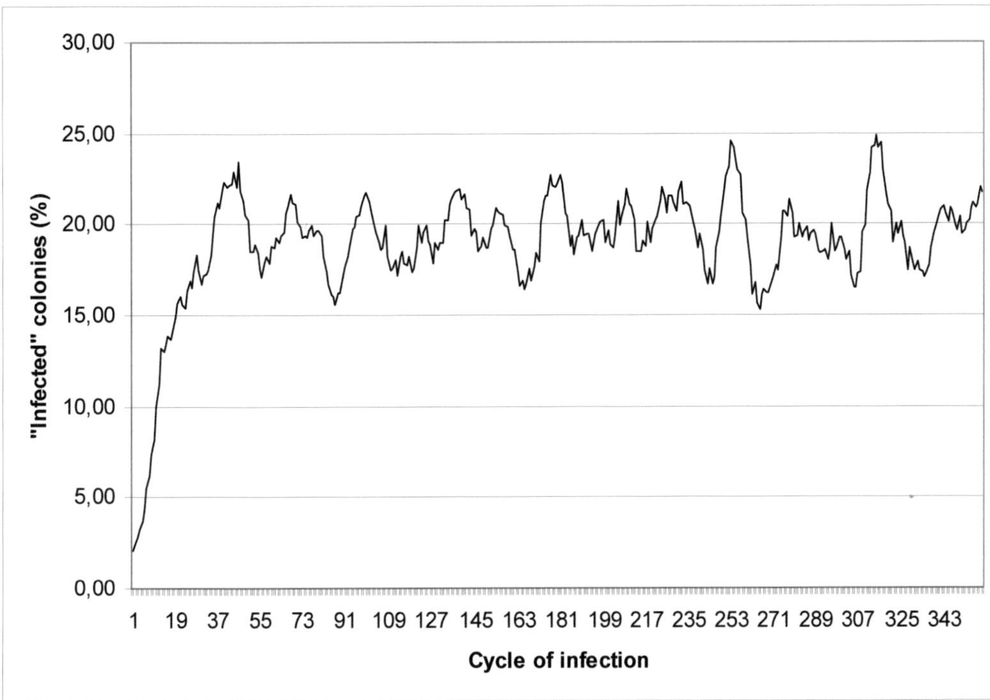

Fig. 6. Intensity of epizootic process in model

There were rare, separate "infected" colonies present in the model's space after the winter period. Those colonies were in the initial stages of the annual cycle of epizooty (Fig. 7a). There were many "immune" colonies that remained after the last epizooty in autumn. The epizooty became sharper at the end of spring when the number of plagued colonies increased, but the number of "immune" colonies decreased (Fig. 7b). In summer, the epizooty recessed and the number of "immune" colonies increased again (Fig. 7c). In autumn, the epizooty raised again as the number of plagued colonies increased and the number of "immune" colonies decreased (Fig. 7d). Thus, the annual epizootic process intensity had qualitative correspondence to a classic, twice-peak epizooty (Fig. 8) as in some natural, plague foci.

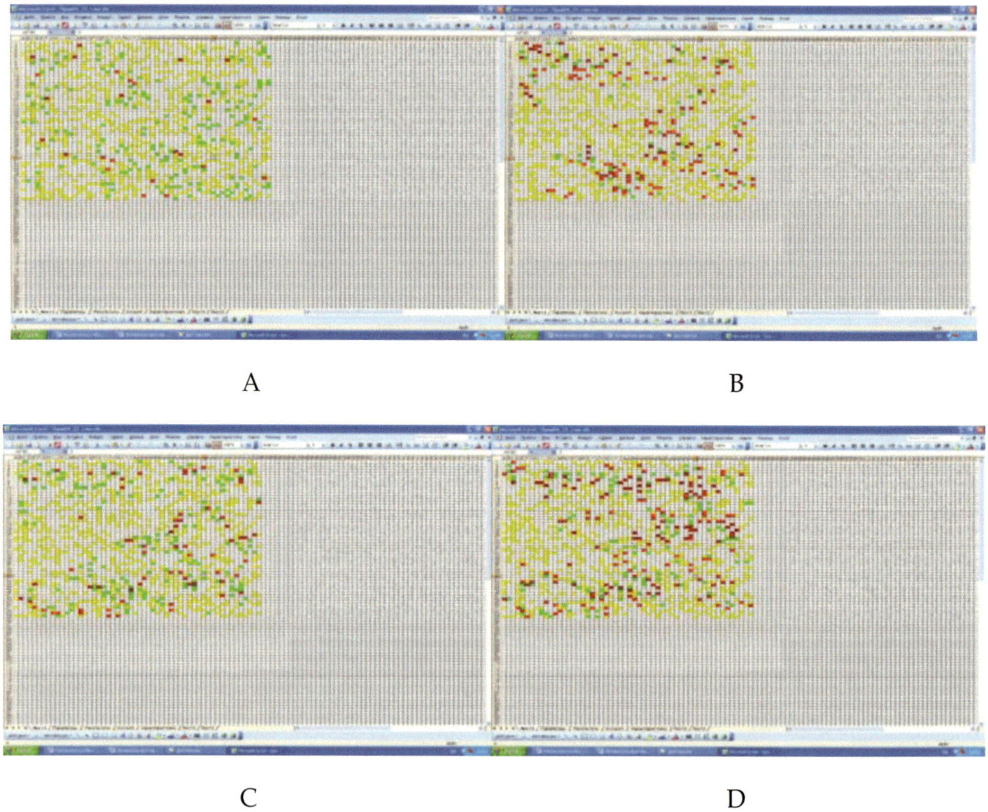

Fig. 7. The epizootic process was developing with a changeable rate of epizootic contact. Intensity of epizooty in the end: A) winter; B) spring; C) summer; D) autumn

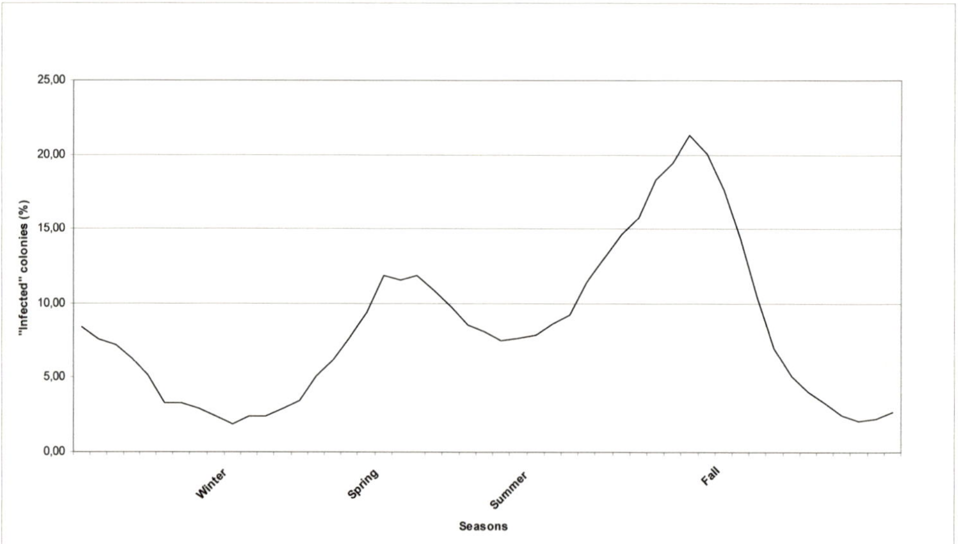

Fig. 8. The intensity of epizootic process modeling within a year

It is interesting, that when a changeable rate of epizootic contact was entered into the model's parameters, the average number of the "infected" and "immune" colonies decreased from 18.92 to 9.31 ($p < 0.001$, Mann-Whitney U test). It was caused by the increase in colonies which were ready for a new infection. As a result, much sharper epizooty could develop. We modeled the sharper epizooty. The coefficient of colony infection was increased to two for this. The picture of epizooty is shown on Fig. 9 and Fig. 10. The epizooty intensity increased almost double in spring and in the fall. However, the number of "infected" colonies was on par with the pre-epizootic level in the winter-summer recession of epizootic activity. The speed of the epizootic spreading fluctuated from 640 to 920 meters per month in these experiments.

It is known that the colonies of the great gerbil have a non-uniform distribution on the surface. According to the distribution pattern, the groups of colonies - the settlement, may be visible as a narrow-band, a wide-band, a large patch, a small patch and continuous settlement (fig 11, 12). Some authors assumed that spatial distribution of the great gerbil settlements played a big role in the epizootic process functioning (Pole, Bibicov, 1991; Dubyanskiy et al., 1992). However that supposition was described as a quality parameter without any quantity values and was quantitatively checked up by modeling. The narrow-band settlements were presented as single colonies, arranged on an identical distance on the vertical and horizontal lines (fig 13).

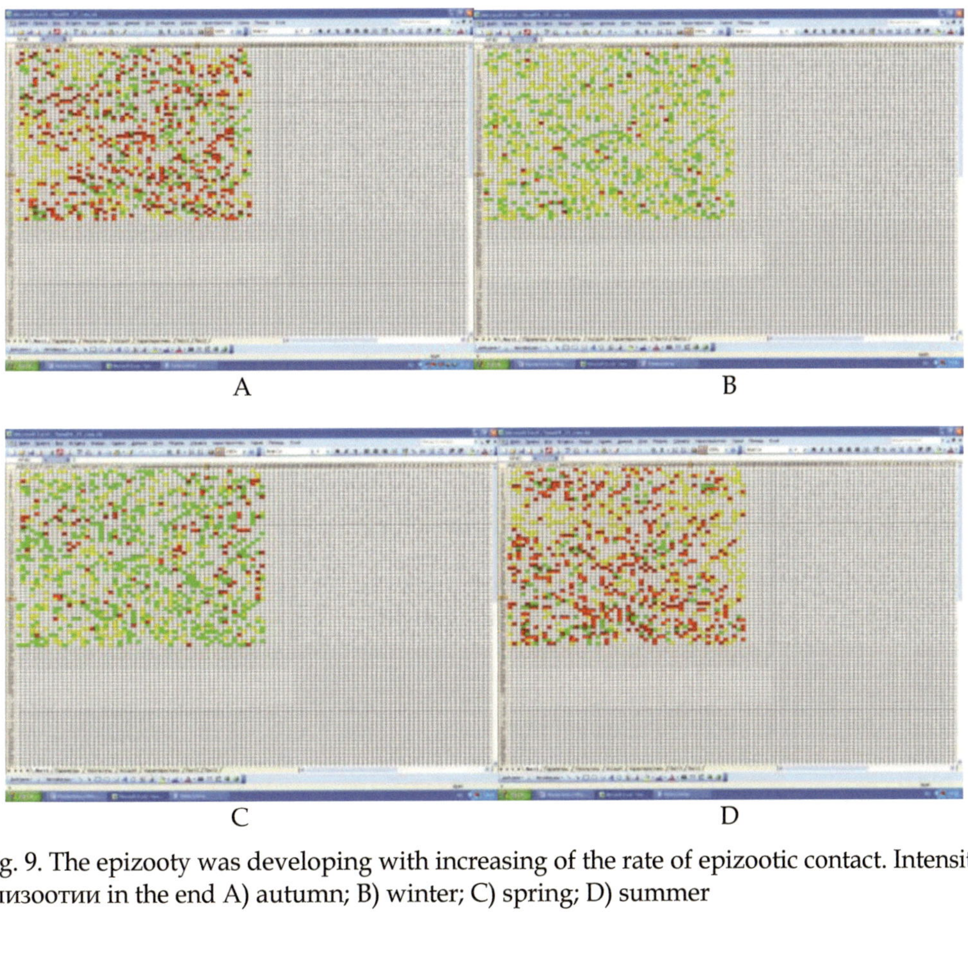

A B

C D

Fig. 9. The epizooty was developing with increasing of the rate of epizootic contact. Intensity эпизоотии in the end A) autumn; B) winter; C) spring; D) summer

First experimental results demonstrated a significant difference of plague spread into continuous and narrow-band settlements (table 1). When the coefficient of a colony infection is one (it is the minimum value of model) the plague can be spread from a density of 1.33 colonies per ha into the continuous settlement. At the same time a density of 0.44 colonies per ha is enough for plague epizootic into the narrow-band settlement. The difference of densities equals three. As the density declines the coefficient of a colony infection proportionally increases into the continuous settlement. A critical threshold is achieved from a density of 0.78 colonies per ha. At the following step of density (0.67 colonies per ha) the epizootic process is recommencing only from value of the coefficient of a colony infection equal 10.

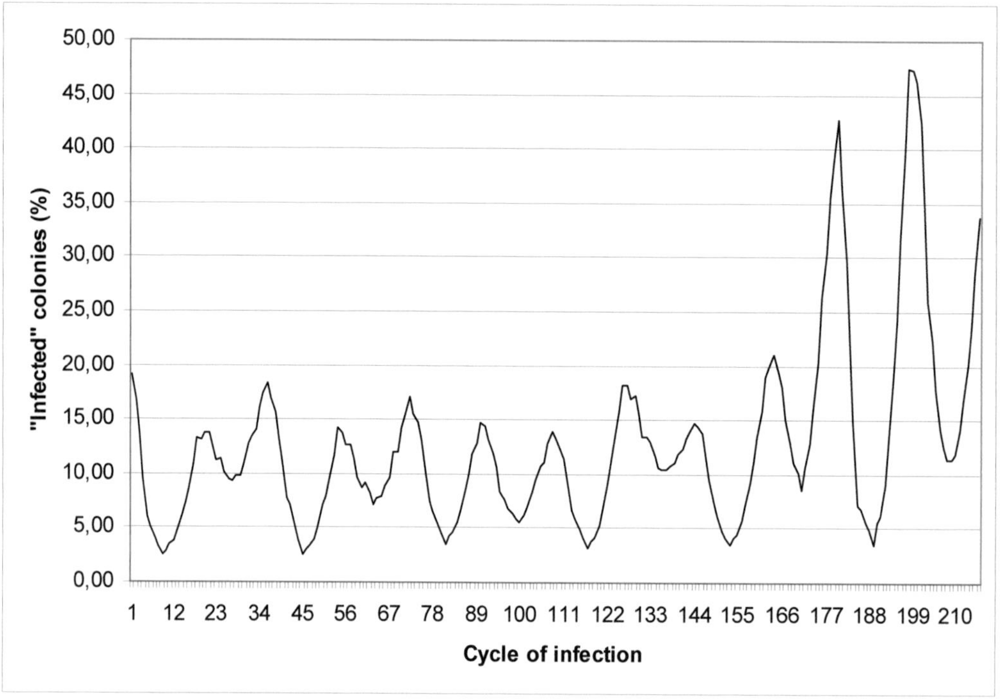

Fig. 10. The intensity of epizootic process which developed at the model with increasing of the rate of epizootic contact (the right part of curve after cycle of infection 171).

In the narrow-band settlement, the minimum limit of density was achieved at 0.11 colonies per ha, and the value of the coefficient of the colony infection was equal to eight. It is not possible to reduce the density of colonies more because of the model limits in the plague microbe drift.

Previous explanation of this phenomenon follows. As monitoring of the model shows, the colonies in the continuous settlement were aggregated to a few groups. The plague epizootic could develop under certain conditions:

1. if the number of colonies in separate groups are enough for plague circulating only in this group;

2. if the distance among groups of colonies is not more than a limit of the relatively remote drift;
3. if the movement activity of animals is enough for connection among the groups of the colonies;
4. if the rate of epizootic contact is not the highest at the given density of colonies.

These conditions were not often established in the continuous settlement. The plague epizootic was faded by the Allee principle (Odum, 1975). It is possible to check up these items by the statistical data from table 1. The critical threshold of the colonies' density appeared when the average distance among the colonies had exceeded the maximum possibility of the plague drift. At the same time the mode of spacing demonstrated that the circulation plague microbe inside the groups of colonies was not limited by the distance. The rare connections among the colonies and groups of colonies which were situated in the plague microbe drift distance were more stable on the highest level of epizootic contact. It could be about 25% of all colonies at 0.67 colonies per ha density. It demonstrated the first quartile of the distance among colonies. Thus the plague microbe could circulate among 150 colonies in the model's space.

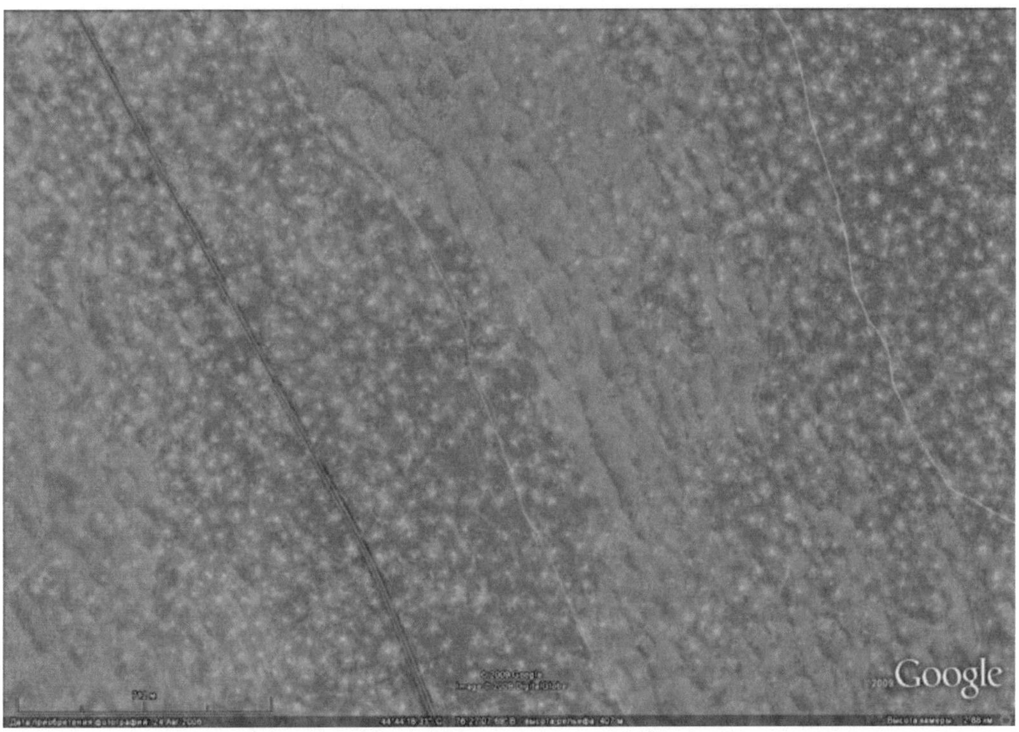

Fig. 11. The continuous settlement of the great gerbil. Colonies of Great Gerbil in a satellite image. Each bright disc represents colonies 10-40m in diameter. The image was captured using the publicly available software Google Earth (http://earth.google.com/). Copyright 2008 DigitalGlobe; Europa Technologies.

Fig. 12. The narrow-band settlement of the great gerbil. Colonies of Great Gerbil in a satellite image. Each bright disc represents colonies 10-40m in diameter. The image was captured using the publicly available software Google Earth (http://earth.google.com/). Copyright 2008 DigitalGlobe; Europa Technologies.

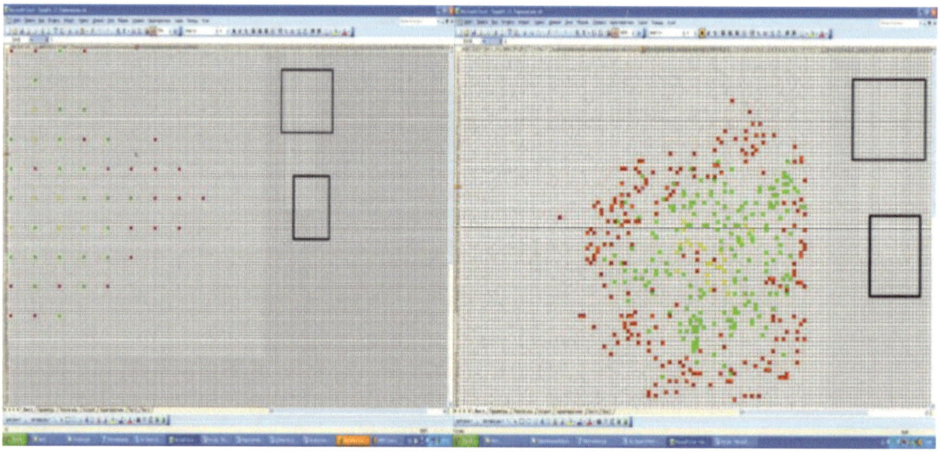

A B

- plague colony, - immune colony, - susceptible colony

Fig. 13. The development of plague epizooty into narrow-band settlement (A) and into continuous settlement (B).

The rate of epizootic contact	Density per ha		The statistical characteristics of distance among colonies (in cells)			
	Continuous settlement	Narrow-band settlement	Average	1 quartile	Mode	Median
1	1.33	0.44	7,33	2	1	5
2	1.11	0.23	8,09	2	0	0
3	1.00	0,17	8,99	2	0	6
4	0.89	0.14	10,11	2	0	7
5	0.78	0,11	10,11	2	0	7
6	0,78	0,11	10,11	2	0	7
7	0,78	0,11	10,11	2	0	7
8	0,78	0.11	10,11	2	0	7
10	0.67	0,11	15,61	4	2	11

Table 1. Dependence of plague spread from colonies density

In the narrow-band settlement the groups of colonies were absent. The colonies were arranged in the form of a lattice and the colonies stood in the vertexes of this lattice. The distance among the colonies was not more than the limit of the relatively remote drift. Each surrounding colony contained enough numbers of colonies for supporting plague epizootic. The colonies were situated in the better places for digging holes in nature (Dubyanskiy, 1970). The migration routes of rodents were not voluntary, and arrangement depended on the settlement structure (Naumov et. al., 1972). The rodents moved along migration routes despite low density, for searching mates, dissemination of young animals, etc., increasing the epizootic contact rate (the dynamic density (Rall, 1965)). The bird-view (fig. 12) demonstrated the limited routes of migration of the animals. These routes are lying along verges of lattice. The similar direction of migration plays a big role in supporting of the plague epizootic.

There is a great difference in the intensity of epizootic processes between the narrow-band and the continuous settlements (fig. 14). The epizooty in the narrow-band settlement is more intensive as in the continuous settlement. The significant difference is represented according to Mann-Whitney U test $p<=0.01$.

If the epizootic contact rate was variable, the intensity of the epizootic process changed according to the density of the colonies. As the epizootic process was more stable in the narrow-band settlement, this type was used for experiments in the model. When the rate of epizootic contact was to equal one the epizootic process was registered by the minimum experimental density of colonies - 0.11 per ha. However the probability of long-term epizootic process was fewer less - 0.9-0.99.

Is persistence of plague possible when the number of great gerbil is in depression? It is an important question for a problem of an interepizootic plague save solution.

It is known that great gerbils' depression, inhabited colonies are often saved as groups which were situated into better ecological niches (Marin et al., 1970; Burdelov L.A. & Burdelov V.A., 1981; Dubrovskiy et. al., 1989). It was interesting that the number of colonies in separate groups were necessary for persistent plague epizooty. The colonies were arranged as a united group in the model space. The spacing among the colonies and the level (rate) of epizootic contact changed during experiments. Results demonstrated in the table 2.

The minimum number of colonies necessary for plague persistent during the modeling year was equal to 50 with maximum value of the probability of transfer success, the coefficient of the colony infection and the spacing among the colonies. After decreasing the spacing among the colonies to eight cells, the epizootic process was developed inside a group consisting of 40 colonies, but after decreasing the spacing to seven cells, 60 colonies were required. Each decrease in the spacing among the colonies (and increasing of plague contact) the number of colonies necessary for supporting plague epizooty increased. The spatial structure of epizootic process was reminiscent of a "layer cake" (fig. 15) or “wave” of infected colonies. The external layer is a layer of infected coloniesfollowed by a layer of immune colonies.

The "readied" colonies were arranged around the first infected colony inside the "Layer cake". The drift of plague microbe from infected to "readied" colonies was difficult. The irregularity of epizootic process, which is one of the main conditions of long existing epizooty, disappeared. Thus the epizootic process degenerated. If the rate of contact decreases, the effect of "wave” is neutralized. In total the plague can persist in a group of 25 colonies when epizootic contact rate is equal to one. The biology base of this effect was confirmed in nature. When great gerbils' density was in deep depression, the movement of rodents decreased and the level of epizootic contacts decreased too (Naumov et al., 1972).

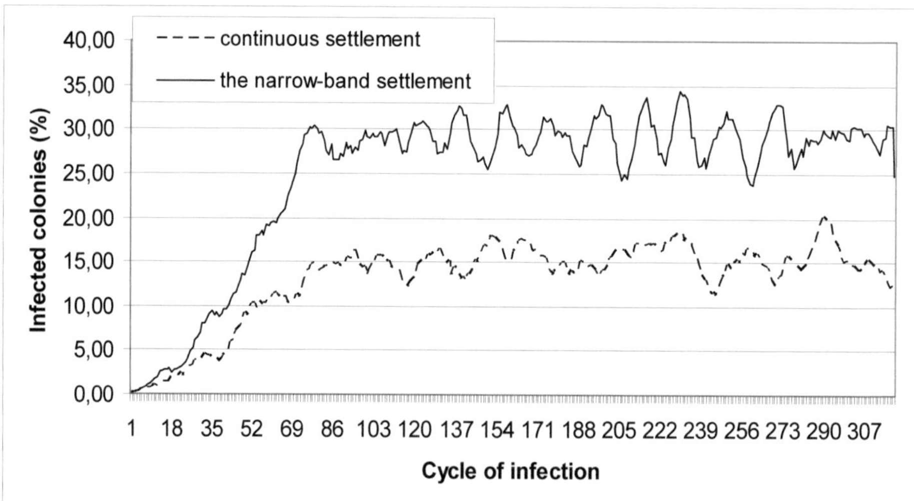

Fig. 14. The intensity of epizootic process in the different type of great gerbils' settlements in model. The density of narrow-band settlement is 0.44 colonies per ha, the density of the continuous settlement is 1.33 colonies per ha.

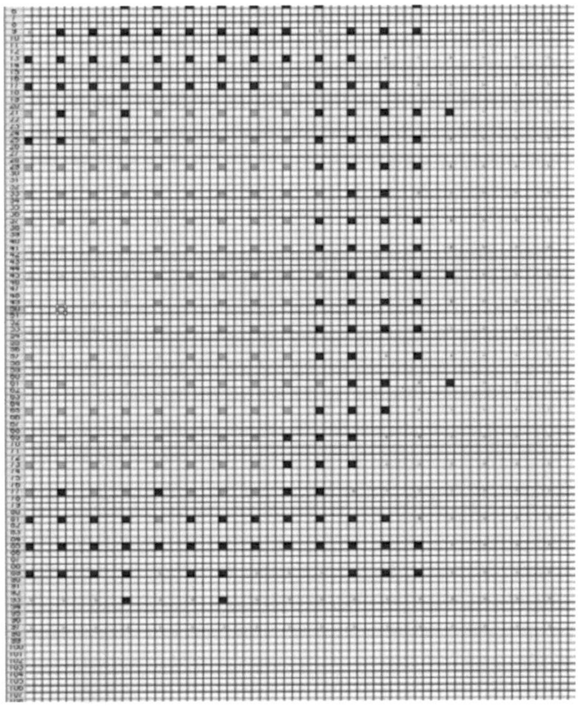

- immune colonies, - infected colonies

Fig. 15. The spatial development of epizooty at the high density of colonies and the highest epizootic contact rate.

The modelling showed that 25 colonies were enough for the plague to persist for a long time. The epizootic contact level had non-line (threshold) dependence from density and number of colonies.

The level of epizootic contact	The colonies quantity	The distance among colonies (cells)
10	50	10
10	40	8
10	60	7
10	50	6
10	70	5
10	400	4
6,48	60	4
1,2	40	2
1,14	30	2
1,0	25	0

Table 2. The level of epizootic contact, the distance among colonies and minimum of colonies quantity, when the plague persistence during 1 year period

4. Discussion

It is important to remember the following for a quality assessment of the modeling and real epizootic process. The investigation of the plague's spatial structure had shown that the small foci of plagued colonies represented pools of the neighbor colonies where there had been living sick or immune gerbils and their infested ectoparasites (Dubyanskiy, 1963; Klassovskiy et al., 1978; Rothschild, 1978; Rivkus et al., 1985). Each pool was irregularly located on the surface and separated by significant distance from other identical pools. It allowed for observing all phases of the epizootic process in a small focus at one time. Sick rodents were observed in some parts of colonies, while an increased fraction of immune rodents were observed in other parts of the colonies. Some colonies were without the plague and these were reserved for epizooty prolongation. All authors agreed on the irregularity of epizootic process in the small focus and permanent moving of epizooty in the territory of the plague focus.

The model-based process was analogous to the described-above epizooty at the small focus or at the pool of the small foci (so-called "epizooty spot"). The analogous types of colonies were present in the model. The well-defined pools of colonies (small foci) were separated from each other and non-uniformity distributed in the working space of the model. The speed of the model-based epizooty spreading coordinated well with field experiments through the marking of animals with radioisotopes and the direct observation of epizooty in nature. The cyclicity arising in the model simulated fluctuations of natural epizootic process intensities (Dubyanskiy and Burdelov, 2010) that confirmed an adequacy of the model-based process to real epizooty.

We have shown the possibility of modeling an annual epizootic cycle by changing the season's epizootic contact rate. This demonstrated the adequacy of the model in transmission theory framework. At the same time, the qualitative picture of epizootic cycle phases was simulated in the model, shown by an increase in the immune colonies' fraction after autumn or spring epizooty peaks. Even a minor increase (up to two) in the single parameter, coefficient of colony infection, was an important factor, as it led to epizooty outbreak. A doubling in the coefficient of colony infection could indicate a gerbil family's contact rate with other families as two times in 10 days instead of only one time. Our thinking, based on literature data, was that both values of the coefficient of colony infection (1 or 2) are usual for the density of a colony, which we used in the model. It demonstrates how minor fluctuations of the epizootic contact rate can lead to the strong changes in the epizootic situation. It is necessary to note that the endless epizootic process can develop even in a small model area. It can be considered as indirect acknowledgement of high survivability of a similar probabilistic process. It was correct for the real epizootic process at natural foci and it was often proved in practice.

It was quantitatively demonstrated in the big role of spatial distribution of great gerbils for plague epizootic. The various kinds of burrow systems constituting settlements, was marked by M. A. Dubyanskiy (1963, 1963a, 1970). He showed that lines of separate kinds of burrow systems played a different role in the plague spread. The modeling demonstrated spatial arrangement is very significant for plague epizootic too. It was known some plague foci had "band" settlements. These are "the foci from North desert subzone". The settlements from these foci are arranged along dry riverbeds, ravines, gullies. The plague in these foci was more intense and had more rare interepizootic periods as in the foci with the "continuous" settlements (Aikimbaev et al., 1987). We believe the spatial kind of great gerbil settlements was a significant influence to this phenomenon.

The background epizooty of low density in the occupancy colonies have proved the opinion that a negative result of surveillance can result not only from a real absence of plague in the focus, but also from insufficient intensity and extent of the surveillance (Dubyanskiy et al., 2007; Dubyanskiy, 2008). Culling surveillance area of plague foci is possible provided 0.1 - 0.6 colonies per ha.

When the great gerbils' density was set in the deep depression, the plague microbe was not detected. However, the plague epizooties were restarted after the gerbils' density recovered. Such "disappearing" of the plague microbe, resulted in a long discussion about routes of plague microbe conservation during interepizootic period. It was believed that the plague microbe requires specific mechanisms for conservation as a soil plague and the transmission mechanism by fleas was not effective (Akiev, 1970; Burdelov, 2001; Popov, 2002). Special works were made in Mangyshlak (Marin et al., 1970) in the South Balhash area (Burdelov L.A. & Burdelov V.A., 1981) and in Kizilkum (Dubrovskiy et al., 1989) it was shown that the occupancy colonies were arranged individually as a group, from five colonies to several. It is enough for plague microbes circulating for a long time. At the same time the most difficult to detect plague by unspecific surveillance because the groups of colonies were rare arranged on the biggest (more over 1000 square kilometers) area. Moreover the plague microbe was not conserved for every colony group. It is a cause, possibly, unregistering of epizooty and existing interepizootic periods. The modeling experiments demonstrated that the classical transmission theory can explain a complexity of plague detected during depression of gerbils and a beginnings of plague at a normal density of rodents.

The modeling allows searching numeric parameters of the epizootic process of plague. However, the simulation may appear a perspective method of forecasting plague epizootics, if applied along with permanent comparing and correcting of modeling results with the real surveillance situation.

The number of great gerbil burrow systems was nearest to natural data about quantity and distribution colonies during deep depressions of great gerbil's number. Thus, the permanent plague epizootics could occur even during deep depressions of great gerbils' number.

5. Conclusion

The model can be useful for studying the general appropriateness of the plague's epizootic process and for obtaining formerly inaccessible quantitative characteristics. It is the result of model simplicity, availability of high-performance computing workstations, model parameter input by anyone, and a large display area in the model. The model allows displaying a real and modeled structure of the rodent's settlement. Thus, it was possible to study the spatial regulation of the epizootic process depending on the density of colonies, distribution of colonies, intrasettlement migration, and so forth. Remote sensing methods and GIS can be used as a base for modeling. The basic characteristics of simulated epizootic processes qualitatively also are quantitatively close to natural analogues. It is important to emphasize the model is not constrained by species-specificity. It can be used for modeling epizootic processes among different species and for different vector-borne diseases.

6. References

Addink E.A., S.M. de Jong, S.A. Davis, V. Dubyanskiy, L.A. Burdelov & H. Leirs. (2010). The use of high-resolution remote sensing for plague surveillance in Kazakhstan. *Remote Sensing of Environment* No. 114, pp. 674-681

Akiev A.K. (1970). Condition of a question on studying the mechanism of preservation of the plague in interepizootic years. (Review). *Problems of especially dangerous infections.* Vol 4 (14), pp. 13-33

Aikimbaev A.M., Aubakirov S.A., Burdelov A.S., Klassovskiy L.N., Serjanov O.S. *The Nature desert plague focus of Middle Asia.* Alma-Ata. Nauka. 1987. 207 p.

Alekseev A. F. (1974). The long of live and peculiarity of moving of Great Gerbils in North-West Kizilkum. *The material VIII scientist conference plagues control establishments of Central Asia and Kazakhstan* pp. 220-222, Alma-Ata

Begon M., Klassovskiy N., Ageyev V., Suleimenov B., Atshabar B. et al.. (2006). Epizootiological parameters for plague (Yersinia pestis infection) in a natural reservoir in Kazakhstan. *Emerging Infectious Diseases.* Vol. 12, Issue 2, pp. 268-273

Burdelov L. A. (2001). Sources, the reasons and consequences of crisis in plague epizootology. *Quarantine and zoonotic infections in Kazakhstan,* Vol. 3, pp. 20-25.

Burdelov L. A., Burdelov A. S., Bondar E. P., Zubov V. V., Maslennikova Z. P., Rudenchik N. F. (1984). Hole using of great gerbil (Rhombomys opimus, Rodentia, Cricetidae) and epizootically significance empty colonies in the Central Asia foci. *Zoology journal,* No. LXIII, Vol. 12, pp. 1848-1858.

Burdelov L. A., Burdelov V. A. (1981). *The optimum ecotope and ways of great gerbils survive in unfavorable conditions. The ecology and medical significance gerbils in the USSR fauna, Proceedings of national meeting by ecology and medical significance gerbils as importance rodents of arid zone,* pp 129-130, Moscow, Russia.

Burdelov L. A., Davis S., Dubyanskiy V.M., Ageyev V. S., Pole D. S., Meka-Mechenko V. G., Heier L., Sapojnikov V. I. (2007) The prospect use remote sensing in monitoring of epidemiology of plague. *Quarantine and zoonotic infections in Kazakhstan,* Vol. 1-2 (15-16), pp. 41-48

Davis S., Trapman P., Leirs H., Begon M., Heesterbeek J.A.P. (2008) The abundance threshold for plague as a critical perculation phenomenon. *Nature.* Vol. 454, pp. 634-637

Davis S, Begon M, De Bruyn L, Ageyev VS, Klassovskiy NL, Pole SB, Viljugrein H, Stenseth NC, Leirs H. (2004). Predictive thresholds for plague in Kazakhstan. *Science,* Vol. 304, pp. 736-738. (doi:10.1126/ science.10.95854).

Rivkus J.Z.,Mitropolskiy O.V.,Urmanov R.A.,Belyaeva S.I. (1985). Features of epizootic plague among rodents Kyzylkum, In: *Fauna and ecology of the rodents,* ol. 16. pp. 5-106. Publishing house of Moscow University

Dubrovskiy J. A., Mitropolskiy O.V., Urmanov R.A., Tretjakov G.P., Zakirov R.H., Sarjinskiy B.A., Fotteler E.R. (1989). The natural laws of great gerbil arrangement during perennial depression of its number in Central Kizilkum. *The gerbils - importance rodents of USSR arid zone. Fan, Tashkent,* pp. 42-44.

Dubyanskiy M. A. (1963). About exterior of plague epizooty on a different phases in settlements of great gerbil. *Materials of scientist conference,* pp. 76-78. Alma-Ata

Dubyanskiy M. A. (1970). *The ecology structure of great gerbil's setllements in Priaral Karakum.* Abstract of Ph. D. thesis. Alma-Ata

Dubyanskiy M. A. (1963). The settlement types of great gerbil and theirs epizotology significant in Priaral Karakum *Zoology journal*, Vol. 42, No. 1, pp. 103-113

Dubyanskiy M. A., Kenjebaev A., Stepanov V. V., Asenov G. A., Dubyanskaja L. D. (1992). *Forecasting epizootic activity of a plague in Near- Aral region and Kizilkum.* Nukus: Karakalpakstan.

Dubyanskiy V.M., Burdelov L.A. (2008). A possibility of mathematic modeling of plague epizootic process in GIS-space. *«Modern technologies in the implementation of the global strategy of combating infectious diseases over the territories of the CIS member states». Release IX international. science.-pract. conf. CIS member states,* pp. 195-197. Volgograd

Dubyanskiy V.M., Burdelov L.A. (2010). The computer model of plague's epizootic process in Rhombomys opimus settlements and some results of its researching. *Zoology journal*, Vol 89, No 1, pp. 79–87

Dubyanskiy, V. M. (2008). Interval evaluation of intensity of plague epizootics. *Quarantinable and zoonotic infections in Kazakhstan. Almaty,* Vol 1-2 (17-18), pp. 52-57

Dubyanskiy, V. M., Pole S. B., Burdelov L.A., Klassovskaja E.. V., Sapojnikov V. I. (2007). Depending on plague detection in Bakanas plain on completeness of epizootologival survey. *Quarantinable and zoonotic infections in Kazakhstan.* Vol. 1-2 (15-16), pp.62-66.

Dubyanskiy V. M. (2010). Modelling plague spread in the different kind settlements of Great Gerbil (Rhombomys opimus Liht., 1823, Rodentia, Cricetidae), In: *Current issuaes of zoonotic diseases. International conference. 29 March 2010.* Ulaanbaatar, pp. 92-99

Etkins P. (1987). *Order and disorder in Nature.* M.: Mir.

Grabovskiy V. I. (1995). Cellular automata as a simple model of complication system. *Progress of modern biology*, No. 115, pp. 412 - 418.

Gage, K. L., & Kosoy, M. Y. (2005). Natural history of plague: Perspectives from more than a century of research. *Annual Review of Entomology*, No. 50, pp. 505–528

Klassovskij L. N, Kunitskiy V. N., Gauzshtejn D. M, Burdelov A. S., Aikimbaev M. A., Dubovizkiy N. M., Novikov G. S., Rasin B. V., Savelov U.V., (1978). To a question on short-term forecasting epizootic situations in Ili-Raratal country between two rivers. *The condition and prospects prophilaxis of plagues: Thesis on USSR's conference plague control institute "Microbe"*, pp. 69-70. Saratov

Litvin V. U., Karulin B. E., Vodomorin N. A., Ohotskij J. V. (1980). Radioisotope modeling epizootic situations and stochastic model of rabbit-fever epizooty in stack, In: *Fauna and ecolgy of rodents.* Vol. 14, pp. 63-84. Publishing house of Moscow University

Marin S.N., Kamnev P.I., Korinfskiy A.N., Nikitenko G.I. (1970). The depression of great gerbils' number on Mangishlak. *Problems of particularly dangerous infections*, Vol. 4 (14), pp. 149-157.

Naumov N. P., Lobachev V. S., Dmitriev P. P., Smirin V. M. (1972). *The nature plague focus in Priaral Karakum.* Moscow. MGU.

Novikova T. N., Pershin I. B., Gubaydullina G. S., Reshetnikova P. I., Sviridov G. G., Kunitskiy V. N. (1971). Results of serology study of Great Gerbils after its infected by plague fleas. *The material V scientist. conference plague control establishments of Central Asia and Kazakhstan,* pp. 148-151. Alma-Ata

Odum E. (1975). *Fundamentals of ecology.* Moscow. Mir.

Pole S. B., Bibicov D. I.. (1991). The dynamic of structure and the mechanisms supporting the optimal density of grey marmot population. *The structure of marmot's population*, pp. 148-171. Moscow

Popov N.V. (2002). *Discrete character – the basic spatial-temporal peculiarity of plague manifestations in natural foci of souslic type.* Saratov University Press.

Rall J.M. (1965). *Natural nidality and epizootology of plague.* Moscow. Medicina

Rayfa G. (1977). *Decision analysis.* Moscow. Nayka

Rothschild E.V. (1978). *Spatial structure of the natural foci of a plague and methods of its studying.* Moscow. The Moscow State University.

Rothschild E.B., Postnikov G.B., Dankov S.S, Kosarev V.P. (1975). Development of groups of plague holes at epizooty among Great Gerbils. *Problems of particularly dangerous infections*, Vol. 3-4 (43-44), pp. 70-77.

Samsonovich L. G., Peysahis L. A., Gordienko O. J. (1971). Susceptibility, sensitiveness and intensity immunity of Great Gerbils after infected it by leucine-depended strain of plague microbe from different foci. *The material V scientist. conference plague control establishments of Central Asia and Kazakhstan*, pp. 169-172. Alma-Ata

Sedin V.I. (1985). Hole's of Great Gerbils as elements of structure plague epizooty in the Central and the Western Kara-Kum. *XII interrepublican scientist conference plague control establishments of Central Asia and Kazakhstan on prophylaxis of a plague*, pp. 141-143. Alma-Ata

Soldatkin I. S, Rodnikovskiy V. B., Rudenchik J. V. (1973). Experience of statistical modeling of plague epizootic process. *Zoology journal*, Vol. 52, pp. 751-756.

Soldatkin I. S., Rudenchik J. V. (1967). About correlation between model and nature plague epizooty process. *The material V scientist. conference plague control establishments of Central Asia and Kazakhstan, to the 50 anniversary of the Great October Socialist revolution*, pp 70-71. Alma-Ata

Soldatkin I. S., Rudenchik J. V., 1971. Some questions a plague enzooty as a self-regulating system "rodent-flea-causative agent", In: *Fauna and ecology of the rodents*, Vol. 10. pp. 5 – 29. Publishing house of Moscow University

Soldatkin I. S., Rudenchik J. V., Ostrovsky I. B., Levoshina A. I. (1966). Quantitative characteristics of condition development of plague epizooty in Great Gerbil's settlements. *Zoology journal*, Vol. 14, pp. 481-485

9

Learning Automation to Teach Mathematics

Josep Ferrer[1], Marta Peña[1] and Carmen Ortiz-Caraballo[2]
[1]*Universitat Politecnica de Catalunya*
[2]*Universidad de Extremadura*
Spain

1. Introduction

One of the recurrent controversies at our Engineering Faculty concerns the orientation of first year basic courses, particularly the subject area of mathematics, considering its role as an essential tool in technological disciplines. In order to provide the basic courses with technological applications, a mathematical engineering seminar was held at the Engineering Faculty of Barcelona. Sessions were each devoted to one technological discipline and aimed at identifying the most frequently used mathematical tools with the collaboration of guest speakers from mathematics and technology departments.

In parallel, the European Space of Higher Education process is presented as an excellent opportunity to substitute the traditional teaching-learning model with another where students play a more active role. In this case, we can use the Problem-Based Learning (PBL) method. This environment is a really useful tool to increase student involvement as well as multidisciplinarity. With PBL, before students increase their knowledge of the topic, they are given a real situation-based problem which will drive the learning process. Students will discover what they need to learn in order to solve the problem, either individually or in groups, using tools provided by the teacher or 'facilitator', or found by themselves.

Therefore, a collection of exercises and problems has been designed to be used in the PBL session of the first course. These exercises include the applications identified in the seminar sessions and would be considered as the real situation-based problems to introduce the different mathematic topics. Two conditions are imposed: availability for first year students and emphasis on the use of mathematical tools in technical subjects in later academic years. As additional material, guidelines for each technological area addressed to faculties without engineering backgrounds are defined.

Some material on Electrical Engineering has been already published ((Ferrer et al., 2010)). Here we focus on control and automation. The guidelines and some exercises will be presented in detail later on. As general references on linear algebra see for example (Puerta, 1976) and on system theory (Kalman et al., 1974) and (Chen, 1984). For other applications, see (Lay, 2007).

Here we describe some of the items regarding control and automation that are presented in the guideline (Section 2).

(1) The input-output description: black box, input-output signals, impulse response, linearity, causality, relaxedness, time invariance, transfer-function matrix, time domain, frequency domain, Laplace transform, Fourier transform, gain, phase, poles, zeros, Bode diagram, filters, resonances.

(2) The state-variable description: the concept of state, state equation, output equation, transfer-function matrix, linear changes, feedbacks, realizations, stability, reachable states, controllability, control canonical form, pole assignment, observability, Kalman decomposition.

The exercises (Section 3) related to this area cover the following topics:

- Matrices. Determinant. Rank: Composition of Systems (ex. 1), Controllability Matrix, Controllable Systems and Controllability Indices (ex. 2), Realizations (ex. 3).
- Vector Spaces. Bases. Coordinates: States in Discrete Systems (ex. 4), Control Functions in Discrete Systems (ex. 4).
- Vector Subspaces: Reachable States for one or several Controls and Sum and Intersection of these Subspaces (ex. 5).
- Linear Maps: Changes of Bases in the System Equations and Invariance of the Transfer-Function Matrix (ex. 6), Controllability Subspace and Unobservable Subspace (ex. 7), Kalman Decomposition (ex. 8).
- Diagonalization. Eigenvectors, Eigenvalues: Invariant Subspaces and Restriction to an Invariant Subspace (ex. 7), Controllable Subsystem (ex. 10), Poles and Pole assignment (ex. 9).
- Non-Diagonalizable Matrices: Control Canonical Form (ex. 11)

2. Guideline for teachers

(1) External system description

Systems are considered as "black boxes" in which each input $u(t)$ (input, control, cause,...) causes an output $y(t)$ (output, effect,...), both multidimensional vectors, in general. We consider only known inputs, ignoring other ones like, for example, disturbance, noise...

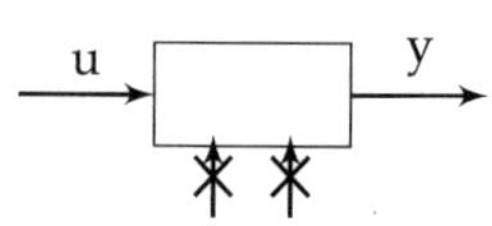

The most usual inputs will be piecewise continuous functions, built from the elementary functions or standard "signals" (impulse or delta, step, ramp, sawtooth, periodics...). Simple systems are adders, gain blocks, integrators, pure delays, filters...

In general, the aim is to analyse their behaviour without looking inside. Indeed, if we consider the "impulse response" $g(t;\tau)$ (that is, the output when the input is an "impulse" in τ, δ_τ) and the system is linear, it results

$$y(t) = \int_{-\infty}^{+\infty} g(t;\tau)u(\tau)d\tau.$$

The upper integration limit will be t if we assume that the system is "causal" (that is, if the current response does not depend on the future inputs, as in all physical systems) and the lower will be t_0 (or simply 0) if it is "relaxed" at t_0 (that is, $y(t) = 0$ for $t \geq t_0$, when $u(t) = 0$ for $t \geq t_0$). Finally, if it is "time-invariant", instead of $g(t;\tau)$ we can write $g(t-\tau)$, resulting in the following expression of $y(t)$ as a convolution product of $g(t)$ and $u(t)$

$$y(t) = \int_{-\infty}^{+\infty} g(t-\tau)u(\tau)d\tau = g(t) * u(t),$$

which is the general system representation in the "time domain".

Applying Laplace transform we get the general representation in the "s-domain"

$$\hat{y}(s) = G(s)\hat{u}(s),$$

where $G(s) = \hat{g}(s)$ is called "transfer function matrix". Indeed, it is the focus of study in this external representation.

If we do the change of variable $s = j\omega$ (imaginary axis of the "complex plane") we obtain the representation in the "frequency domain", (more usual in engineering)

$$y(\omega) = G(\omega)u(\omega),$$

where $G(\omega)$ is called "isochronous transfer function matrix". It can be obtained directly as a "Fourier" transform when appropriate hypotheses hold. This allows the use of tools such as Fourier transform, Parseval theorem... if we have some basic knowledge of functional analysis and complex variable: function spaces, norms, Hilbert spaces, integral transforms...

Its denomination shows that $G(\omega)$ indicates the system behaviour for each frequency. So, if $u(t) = A\sin(\omega t)$, then $y(t) = B\sin(\omega t + \varphi)$, being the "gain" B/A the module of $G(\omega)$, and the "phase" φ is its argument. A widely used tool in engineering is the "Bode diagrams" which represent these magnitudes on the ordinate (usually, the gain is in logarithmic scale, in decibels: $dB = 20log|G|$) as a function of the frequency as abscissa (also in log scale).

In generic conditions (see next section) the coefficients of G are "proper rational fractions", so that the system's behaviour is largely determined by its "degrees", "zeros" and "poles". So:

- The relative degree (denominator degree minus numerator degree) gives the "order of differentiability" at the origin of the response to the input step signal.
- As already mentioned, this difference must be strictly positive (if zero, the Parseval theorem would give infinite energy in the output signal; if negative, it would contradict the causality), so that the gain tends to 0 for high frequencies.
- Roughly speaking, the zeros indicate filtered frequencies (of gain 0, or in practice far below), so that in the Bode diagram place the "inverted comb pas" as "filters". On the contrary, the poles indicate dangerous frequencies because of resonance (infinite gain).

(2) Internal system description

In addition, state variables $x(t)$ (not univocally defined) are considered. They characterize the state in the sense that they accumulate all the information from the past, that is, future outputs are determined by the current state and the future inputs. Typically, the derivative $\dot{x}(t)$

functionally depends on $x(t)$, $u(t)$ and t (state equation), and $y(t)$ as well (output equation), although in this case we may obviate the dependence on $u(t)$. In the linear case:

$$\dot{x}(t) = A(t)x(t) + B(t)u(t); \quad y(t) = C(t)x(t).$$

From elementary theory of ordinary differential equations, it holds that for every continuous (or piecewise continuous) control $u(t)$, there exists a unique "solution" $x(t)$ for every "initial condition" $x_0 = x(0)$.

In particular, it can be applied a "feedback control" by means of a matrix F

$$u(t) = Fx(t).$$

One of the first historical (non linear) examples is the Watt regulator which controlled the velocity of a steam engine acting on the admission valve in the function of the centrifugal force created in the regulator balls by this velocity. Nowadays this "automatic regulation" can be found in simple situations such as thermostats.

When this kind of feedback is applied, we obtain an autonomous dynamical system

$$\dot{x}(t) = Ax(t) + BFx(t) = (A + BF)x(t)$$

called a "closed loop" system. It is natural to consider if we can adequately choose F such that this system has suitable dynamic properties. For example, for being "stable", that is, that the real parts of the eigenvalues of $A + BF$ are negative. Or more in general, that these eigenvalues have some prefixed values. As we will see later, one of the main results is that this feedback "pole assignment" is possible if the initial system is "controllable".

If it is time-invariant (that is, A, B, and C are constant) and we assume $x(0) = 0$, the Laplace transform gives

$$\hat{y}(t) = G(s)\hat{u}(s); \quad G(s) = C(sI - A)^{-1}B$$

recovering the previous transfer matrix. Reciprocally, the "realization" theory constructs triples (A, B, C) giving a prefixed G, formed by proper rational fractions: it can be seen that it is always possible; the uniqueness conditions will be seen later.

It is a simple exercise to check that when introducing a "linear change" S in the state variables, the system matrices become $S^{-1}AS$, $S^{-1}B$ and CS, respectively, and $G(s)$ do not change.

In this description it is clear that the coefficients of $G(s)$ are proper rational fractions and in particular its poles are the eigenvalues of A.

One of the main results is that the set of "reachable states" (the possible $x(t)$ starting from the origin, when varying the controls $u(t)$) is the image subspace $\operatorname{Im} K(A, B)$ of the so-called "controllability matrix"

$$\operatorname{Im} K(A, B); \quad K(A, B) = \begin{pmatrix} B & AB & \ldots & A^{n-1}B \end{pmatrix},$$

that is, the subspace spanned by the columns of B and successive images for A, called "controllability subspace". It is an interesting exercise to justify that we can truncate in $n - 1$, since the columns of higher powers are linearly dependent with the previous ones. In fact we can consider each control individually (each column of B separately), in such a way that the "sum and intersection of subspaces" give the reachable states when using different controls at the same time or for each one of them.

The system is "controllable" if all the states are reachable, that is, if and only if $K(A,B)$ has full "rank". Hence, it is a generic condition (that is, the subset of controllable pairs (A,B) is open and dense).

In the single input case (A,b), it can be seen that if the system is controllable it can be transformed, by means of a suitable basis change, in the so-called "control canonical form"

$$\bar{A} = \begin{pmatrix} 0\,1\;\;0\;\;\ldots\;0\;0 \\ 0\,0\;\;1\;\;\ldots\;0\;0 \\ \ldots\ldots \\ 0\,0\;\;0\;\;\ldots\;0\;1 \\ *\;*\;\;*\;\;\ldots\;*\;* \end{pmatrix}, \quad \bar{b} = \begin{pmatrix} 0 \\ 0 \\ \vdots \\ 0 \\ 1 \end{pmatrix}.$$

Observe that $\bar{A}$ is a "companion" (or Sylvester) matrix. For these kinds of matrices, it can be easily seen that the coefficients of the characteristic polynomial are the ones of the last row (with opposite sign and in reverse order), and they are "non derogatory matrices", that is, for each eigenvalue there is a unique linearly independent eigenvector, and therefore only one Jordan block. Hence, if some eigenvalue is multiple, the matrix is non diagonalizable.

In the multi input case, another reduced form is used (Brunovsky, or Kronecker form), which is determined by the so-called "controllability indices" that can be computed as a conjugated partition from the one of the ranks of: B, (B,AB),...,$K(A,B)$.

From these reduced forms it is easy to prove that the pole assignment is feasible, as well as to compute the suitable feedbacks.

If the system is not controllable, it is easy to see that the subspace $\operatorname{Im} K(A,B)$ is "A-invariant", and that in any "adapted basis" (of the state space) the matrices of the system are of the form:

$$\begin{pmatrix} A_c & * \\ 0 & * \end{pmatrix}, \quad \begin{pmatrix} B_c \\ 0 \end{pmatrix},$$

being (A_c, B_c) controllable. It is also easy to deduce that if $x(0)$ belongs to $\operatorname{Im} K(A,B)$, $x(t)$ also belongs, for all time t and all control $u(t)$, and that its trajectory is determined for the pair (A_c, B_c), which enables considering the "restriction" of the system (A,B) to $\operatorname{Im} K(A,B)$, called the "controllable subsystem" of the initial one.

In the same direction, the "Kalman decomposition" is obtained by considering "(Grassman) adapted bases to the pair of subspaces" $\operatorname{Im} K(A,B)$ and $\operatorname{Ker} L(C,A)$, being

$$L(C,A) = \begin{pmatrix} C \\ CA \\ \ldots \\ CA^{n-1} \end{pmatrix}$$

that is, the "transposed matrix" of $K(A^t, C^t)$. In fact there are interesting properties of "duality" between the systems (A,B,C) and (A^t, C^t, B^t).

We remark the equivalence between the controllability of (A, B, C) and the "observability" of (A^t, C^t, B^t), in the sense that the initial conditions are computable if the outputs for determined inputs are known.

On the other hand it is interesting to note that the transfer matrix of the initial system is the same as that of the controllable subsystem, as well as that of the "complete subsystem" (that is, controllable and observable) obtained by means of the Kalman decomposition.

This means that, given any transfer matrix (formed by proper rational fractions), not only it is possible to find realizations, but also "controllable realizations" and even complete ones. In fact it can be seen that all the complete realizations are equivalent. In particular, they have the same number of state variables. Moreover, a realization is complete if and only if it is a "minimal realization", in the sense that there are not realizations with a smaller number of state variables. This minimal number of state variables of the realizations is called "McMillan degree", which coincides with the dimension of the complete realizations.

3. Exercises for students

We will present here the guideline of proposed exercises for the students and their solutions:

3.1 Proposed exercises

1. Composition of systems

A control system Σ

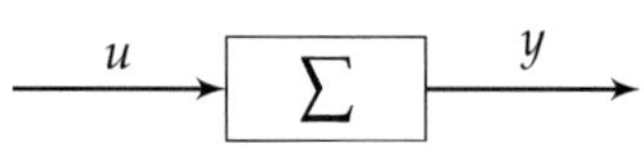

is defined by the equations

$$\dot{x}(t) = Ax(t) + Bu(t); \qquad y(t) = Cx(t),$$

or simply by the "triple of matrices" (A, B, C). This triple determines the "transfer matrix"

$$G(s) = C(sI - A)^{-1}B$$

which relates the Laplace transforms of u, y:

$$\hat{y}(s) = G(s)\hat{u}(s).$$

Given two systems:

$$\Sigma_1: \quad \dot{x}_1 = A_1x_1 + B_1u_1; \quad y_1 = C_1x_1$$
$$\Sigma_2: \quad \dot{x}_2 = A_2x_2 + B_2u_2; \quad y_2 = C_2x_2$$

they can be composed of different ways to obtain a new system. For example in:

(i) Series

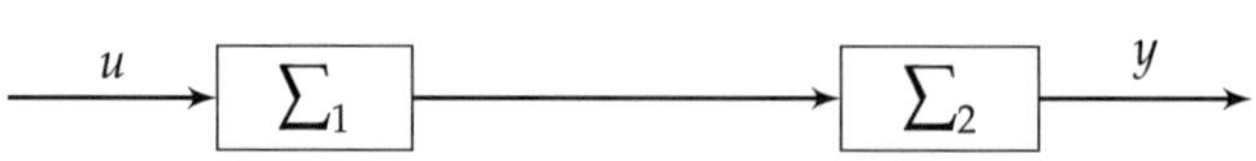

(ii) Parallel

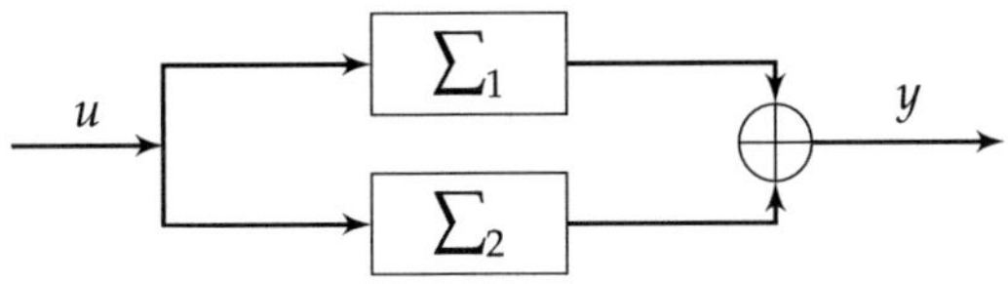

(iii) Feedback

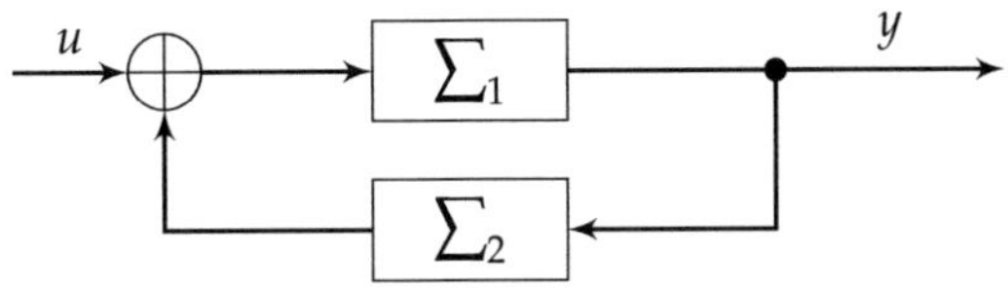

In all these cases the new state variables are:

$$x = \begin{pmatrix} x_1 \\ x_2 \end{pmatrix}$$

(1) Deduce the following relations between the different input and output variables:

(1.i) $u = u_1; y_1 = u_2; y_2 = y$

(1.ii) $u = u_1 = u_2; y = y_1 + y_2$

(1.iii) $u + y_2 = u_1; y_1 = y = u_2$

(2) Deduce the expression of the triple of matrices of each composed system, in terms of $A_1, B_1, C_1, A_2, B_2, C_2$.

(3) Deduce the expression of the transfer matrix of each composed system, in terms of $G_1(s)$ and $G_2(s)$.

2. Controllable systems; controllability indices

A control system

$$\dot{x}(t) = Ax(t) + Bu(t), \quad A \in M_n, \quad B \in M_{n,m}$$

is "controllable" (that is, any change in the values of x is possible by means of a suitable control $u(t)$), if and only if the rank of the so-called "controllability matrix" is full

$$K(A,B) = (B, AB, A^2B, \ldots, A^{n-1}B).$$

(a) Discuss for which values of $\alpha, \beta \in \mathbb{R}$ the system defined by

$$A = \begin{pmatrix} 0 & 1 & 0 \\ -2 & 1 & 0 \\ \alpha & 0 & 1 \end{pmatrix}, \quad B = \begin{pmatrix} 1 & 0 \\ 0 & \beta \\ 0 & 1 \end{pmatrix}$$

is controllable.

(b) Discuss for which values of $\alpha, \beta \in \mathbb{R}$ the system is controllable with only the second control, that is, when instead of the initial matrix B we consider only the column matrix

$$b_2 = \begin{pmatrix} 0 \\ \beta \\ 1 \end{pmatrix}.$$

(c) In general, the "controllability indices" are determined by the rank of the matrices

$$B, \quad (B, AB), \quad (B, AB, A^2B), \ldots, \quad K(A, B).$$

Compute these ranks, in terms of parameters $\alpha, \beta \in \mathbb{R}$, for the system defined by

$$A = \left(\begin{array}{ccccc|cc} 0 & 1 & & & & & \\ & 0 & 1 & & & & \\ & & 0 & 1 & & & \\ & & & 0 & 1 & 0 & \\ & & & & 0 & & \\ \hline & & 0 & & & 0 & 1 \\ & & & & & & 0 \end{array}\right), \quad B = \left(\begin{array}{c|c} 0 & 0 \\ 0 & 0 \\ 0 & \gamma \\ 0 & \delta \\ 1 & 0 \\ \hline 0 & 0 \\ 0 & 1 \end{array}\right).$$

3. Realizations

Given a matrix $G(s)$ of proper rational fractions in the variable s, it is called a "realization" any linear control system

$$\dot{x}(t) = Ax(t) + Bu(t); \quad y(t) = Cx(t)$$

which has $G(s)$ as "transfer matrix", that is,

$$G(s) = C(sI - A)^{-1}B.$$

In particular, the so-called "standard controllable realization" is obtained in the following way, assuming that $G(s)$ has p rows and m columns.

(i) We determine the least common multiple polynomial of the denominators

$$P(s) = p_0 + p_1 s + \ldots + p_r s^r.$$

(ii) Then, $G(s)$ can be written

$$G(s) = \frac{1}{P(s)} \begin{pmatrix} G_{11}(s) & \cdots & G_{1m}(s) \\ \cdots & \cdots & \cdots \\ G_{p1}(s) & \cdots & G_{pm}(s) \end{pmatrix},$$

where $G_{ij}(s)$ are polynomials with degree strictly lower than $r (= \deg P(s))$.

(iii) Grouping the terms of the same degree we can write $G(s)$ in the form

$$G(s) = \frac{1}{P(s)} (R_0 + R_1 s + \ldots + R_{r-1} s^{r-1}),$$

where $R_0, \ldots, R_{r-1} \in M_{p,m}(\mathbb{R})$.

(iv) Then, the standard controllable realization is given by the triple of matrices (where 0_m is the null matrix of $M_m(\mathbb{R})$):

$$A = \begin{pmatrix} 0_m & I_m & 0_m \dots & 0_m \\ 0_m & 0_m & I_m \dots & 0_m \\ & \dots & \dots \dots & \\ 0_m & 0_m & 0_m \dots & I_m \\ -p_0 I_m & -p_1 I_m & \dots \dots & -p_{r-1} I_m \end{pmatrix} \in M_{mr}(\mathbb{R}), \quad B = \begin{pmatrix} 0_m \\ \vdots \\ 0_m \\ I_m \end{pmatrix} \in M_{mr,m}(\mathbb{R}),$$

$$C = \begin{pmatrix} R_0 \dots R_{r-1} \end{pmatrix} \in M_{p,mr}(\mathbb{R}).$$

We consider, for example,

$$G(s) = \begin{pmatrix} 1/s \\ 1/(s-1) \end{pmatrix}.$$

(1) Following the above paragraphs, compute the triple of matrices (A, B, C) which give the standard controllable realization.

(2) Check that it is controllable, that is, that the matrix

$$K(A,B) = (B, AB, \dots, A^{mr-1}B)$$

has full rank.

(3) Check that it is a realization of $G(s)$, that is,

$$\begin{pmatrix} 1/s \\ 1/(s-1) \end{pmatrix} = C(sI - A)^{-1}B.$$

4. Reachable states; control functions

Given a linear control system

$$x(k+1) = Ax(k) + Bu(k); \quad A \in M_n(\mathbb{R}), \quad B \in M_{n,m}(\mathbb{R}),$$

the h-step "reachable states", from $x(0)$, are:

$$x(h) = A^h x(0) + A^{h-1}Bu(0) + \dots + ABu(h-2) + Bu(h-1)$$

where all possible "control functions" $u(0), u(1), \dots$ are considered. More explicitly, if we write $B = (b_1 \dots b_m)$ and $u(k) = (u_1(k), \dots, u_m(k))$, we have that:

$$\begin{aligned} x(h) &= A^h x(0) \\ &+ (A^{h-1}b_1u_1(0) + \dots + A^{h-1}b_m u_m(0)) + \dots \\ &+ (Ab_1u_1(h-2) + \dots + Ab_m u_m(h-2)) + (b_1u_1(h-1) + \dots + b_m u_m(h-1)) \end{aligned}$$

Let

$$A = \begin{pmatrix} 2 & -1 & 1 \\ -2 & 1 & -1 \\ 2 & 1 & 3 \end{pmatrix}, \quad B = \begin{pmatrix} -1 & 1 \\ 1 & 0 \end{pmatrix}.$$

(1) Assume that only the first control acts, that is: $u_2(0) = u_2(1) = \ldots = 0$. Show that in this case, the state $x = (-1, 2, 1)$ is not 3-step reachable from the origin.

(2) Assume now that only the second control acts.
 (2.1) Prove that $x = (-1, 2, 1)$ is 2-step reachable from the origin.
 (2.2) Compute the corresponding control function.

(3) Assume that both controls act.
 (3.1) Determine the control functions set to reach $x = (-1, 2, 1)$ from the origin, at second step.
 (3.2) In particular, check if it is possible to choose positive controls.
 (3.3) Idem for $u_1(0) = u_2(0)$.

5. Subspaces of reachable states

Given a linear control system

$$x(k+1) = Ax(k) + Bu(k); \quad A \in M_n(\mathbb{R}), \quad B \in M_{n,m}(\mathbb{R}),$$

the h-step "reachable states", from $x(0)$, are

$$x(h) = A^h x(0) + A^{h-1}Bu(0) + \ldots + ABu(h-2) + Bu(h-1),$$

where all possible control functions $u(0), u(1), \ldots$ are considered.

(1) We write $K(h)$ the set of these states when $x(0) = 0$. Show that:
 (1.1) $K(h) = [B, AB, \ldots, A^{h-1}B] \subset \mathbb{R}^n$
 (1.2) $K(1) \subset K(2) \subset \ldots \subset K(j) \subset \ldots$
 (1.3) $K(h) = K(h+1) \Rightarrow K(h+1) = K(h+2) = \ldots$
 (1.4) $K(n) = K(n+1) = \ldots$
 This maximal subspace of the chain is called "subspace of reachable states":

$$K = [B, AB, \ldots, A^{n-1}B] \subset \mathbb{R}^n$$

(2) Analogous results hold when only the control $u_i(k)$ acts. In particular, the subspace of reachable states, from the origin, with only this control is:

$$K_i = [b_i, Ab_i, \ldots, A^{n-1}b_i] \subset \mathbb{R}^n; \quad 1 \leq i \leq m$$

 where $B = (b_1, \ldots, b_m)$.
 (2.1) Reason that the reachable states by acting the controls $u_i(k)$ and $u_j(k)$ are $K_i + K_j$.
 (2.2) Reason that $K_i \cap K_j$ is the subspace of reachable states by acting any of the controls $u_i(k)$ or $u_j(k)$.

(3) Let us consider the linear control system defined by the matrices

$$A = \begin{pmatrix} 2 & -1 & 1 \\ -2 & 1 & -1 \\ 2 & 1 & 3 \end{pmatrix}, \quad B = \begin{pmatrix} 0 & 0 \\ -1 & 1 \\ 1 & 0 \end{pmatrix}.$$

(3.1) Determine the subspaces K, K_1 and K_2, and construct a basis of each one.
(3.2) Idem for $K_1 + K_2, K_1 \cap K_2$.

6. Change of state variables in control systems

In the linear control system

$$\dot{x}(t) = Ax(t) + Bu(t); \quad y(t) = Cx(t)$$

$$A \in M_n, \quad B \in M_{n,m}, \quad C \in M_{p,n},$$

we consider a linear change in the state variables given by:

$$\bar{x} = S^{-1}x.$$

(1) Prove that in the new variables the equations of the system are:

$$\dot{\bar{x}}(t) = \bar{A}\bar{x}(t) + \bar{B}\bar{u}(t); \quad y(t) = \bar{C}\bar{x}(t)$$

$$\bar{A} = S^{-1}AS, \quad \bar{B} = S^{-1}B, \quad \bar{C} = CS.$$

(2) The "controllability indices" of the system are computed from the ranks:

$$\text{rank}(B, AB, \ldots, A^h B), \quad h = 1, 2, 3, \ldots$$

Deduce from (1) that they are invariant under linear changes in the state variables.

(3) The "transfer matrix" of the system is:

$$G(s) = C(sI - A)^{-1}B.$$

Deduce from (1) that it is invariant under linear changes in the state variables.

7. Controllability subspaces and unobservability subspaces

Given a linear control system

$$\dot{x}(t) = Ax(t) + Bu(t); \quad y(t) = Cx(t)$$

$$A \in M_n(\mathbb{R}), \quad B \in M_{n,m}(\mathbb{R}), \quad C \in M_{p,n}(\mathbb{R}),$$

the following subspaces are called "controllability subspace" and "unobservability subspace", respectively:

$$K = \text{Im}\begin{pmatrix} B & AB & \ldots & A^{n-1}B \end{pmatrix}$$

$$L = \text{Ker}\begin{pmatrix} C \\ CA \\ \ldots \\ CA^{n-1} \end{pmatrix}.$$

(1) Show that they are A-invariant subspaces.

(2) Let us consider

$$A = \begin{pmatrix} 1 & 1 & 0 & 0 \\ 0 & 1 & 0 & 0 \\ 0 & 0 & -1 & 1 \\ 0 & 0 & 0 & -1 \end{pmatrix}, \quad B = \begin{pmatrix} 1 \\ 0 \\ 0 \\ 1 \end{pmatrix}, \quad C = \begin{pmatrix} 0 & 1 & 0 & 1 \end{pmatrix}.$$

(2.1) Compute the dimensions of K and L.
(2.2) Construct a basis of each one.
(2.3) Obtain the matrices, in these bases, of the restrictions $A|_K$ and $A|_L$.
(2.4) Idem for the subspace $K \cap L$.

8. Kalman decomposition

Given a linear control system

$$\dot{x}(t) = Ax(t) + Bu(t); \quad y(t) = Cx(t)$$

$$A \in M_n(\mathbb{R}), \quad B \in M_{n,m}(\mathbb{R}), \quad C \in M_{p,n}(\mathbb{R}),$$

we consider the subspaces

$$K = \operatorname{Im} K(A,B), \quad L = \operatorname{Ker} L(C,A),$$

where

$$K(A,B) = \begin{pmatrix} B & AB & \dots & A^{n-1}B \end{pmatrix},$$

$$L(C,A) = \begin{pmatrix} C \\ CA \\ \dots \\ CA^{n-1} \end{pmatrix}.$$

A (Grassman) adapted basis to both subspaces is called a "Kalman basis". More specifically, the basis change matrix is of the form

$$\begin{aligned} & S = \begin{pmatrix} S_1 & S_2 & S_3 & S_4 \end{pmatrix} \\ & S_2 : \text{ basis of } K \cap L \\ & (S_1 \quad S_2) : \text{ basis of } K \\ & (S_2 \quad S_4) : \text{ basis of } L \end{aligned}$$

where some of the submatrices $S_1, \dots, S_4$ can be empty.

(1) Prove that with a basis change of this form, the matrices of the system become of the form

$$\bar{A} = S^{-1}AS = \begin{pmatrix} A_{11} & 0 & A_{13} & 0 \\ A_{21} & A_{22} & A_{23} & A_{24} \\ 0 & 0 & A_{33} & 0 \\ 0 & 0 & A_{43} & A_{44} \end{pmatrix}$$

$$\bar{B} = S^{-1}B = \begin{pmatrix} B_1 \\ B_2 \\ 0 \\ 0 \end{pmatrix}$$

$$\bar{C} = CS = \begin{pmatrix} C_1 & 0 & C_3 & 0 \end{pmatrix}.$$

It is called "Kalman decomposition" of the given system.

(2) Consider the system given by

$$A = \begin{pmatrix} 1 & 1 & 0 & 0 \\ 0 & 1 & 0 & 0 \\ 0 & 0 & -1 & 1 \\ 0 & 0 & 0 & -1 \end{pmatrix}, \quad B = \begin{pmatrix} 1 \\ 0 \\ 0 \\ 1 \end{pmatrix}, \quad C = \begin{pmatrix} 0 & 1 & 0 & 1 \end{pmatrix}.$$

(2.1) Determine a Kalman basis.

(2.2) Determine a Kalman decomposition.

(2.3) Check that

(A_{11}, B_1, C_1) is controllable and observable,

$\left(\begin{pmatrix} A_{11} & 0 \\ A_{21} & A_{22} \end{pmatrix}, \begin{pmatrix} B_1 \\ B_2 \end{pmatrix}, \begin{pmatrix} C_1 \\ 0 \end{pmatrix}\right)$ is controllable,

$\left(\begin{pmatrix} A_{11} & A_{13} \\ 0 & A_{33} \end{pmatrix}, \begin{pmatrix} B_1 \\ 0 \end{pmatrix}, \begin{pmatrix} C_1 \\ C_3 \end{pmatrix}\right)$ is observable,

that is, that the following matrices have full rank:

$$K(A_{11}, B_1) = \left(B_1 \; A_{11}B_1 \; \dots \; (A_{11})^{n-1}B_1 \right),$$

$$(L(C_1, A_{11}))^t = \left(C_1^t \; A_{11}^t C_1^t \; \dots \; (A_{11}^t)^{n-1}C_1^t \right),$$

$$K\left(\begin{pmatrix} A_{11} & 0 \\ A_{21} & A_{22} \end{pmatrix}, \begin{pmatrix} B_1 \\ B_2 \end{pmatrix}\right) = \left(\begin{pmatrix} B_1 \\ B_2 \end{pmatrix} \cdots \begin{pmatrix} A_{11} & 0 \\ A_{21} & A_{22} \end{pmatrix}^{n-1} \begin{pmatrix} B_1 \\ B_2 \end{pmatrix}\right),$$

$$\left(L\left(\begin{pmatrix} C_1 & C_3 \end{pmatrix}, \begin{pmatrix} A_{11} & A_{13} \\ 0 & A_{33} \end{pmatrix}\right)\right)^t = \left(\begin{pmatrix} C_1^t \\ C_3^t \end{pmatrix} \cdots \begin{pmatrix} A_{11}^t & 0 \\ A_{13}^t & A_{33}^t \end{pmatrix}^{n-1} \begin{pmatrix} C_1^t \\ C_3^t \end{pmatrix}\right).$$

9. Pole assignment by state feedback

Given a linear control system

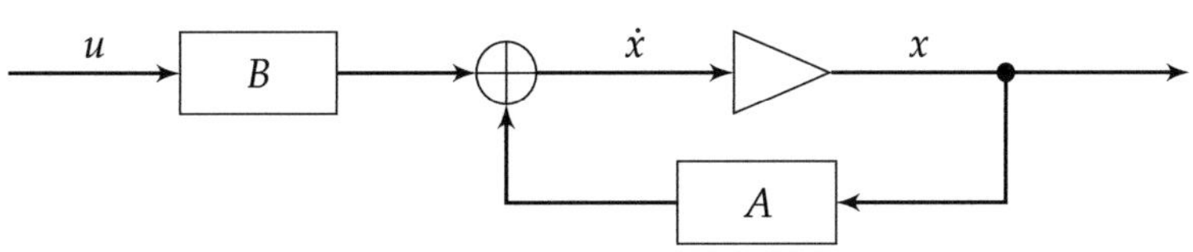

$$\dot{x}(t) = Ax(t) + Bu(t); \quad A \in M_n, \quad B \in M_{n,m},$$

the eigenvalues of the matrix A are called the "poles" of the system and play an important role in its dynamic behaviour. For example, they are the resonance frequencies; the system is "BIBO stable" if and only if the real part of its poles is negative.
If we apply an automatic control by means of a "feedback" F (for example, a thermostat, a Watt regulator,...)

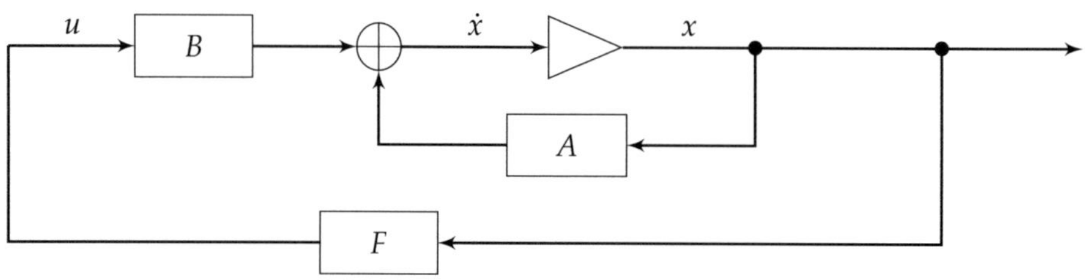

we obtain an autonomous dynamic system (called a "closed loop" system):

$$\dot{x}(t) = Ax(t) + BFx(t) = (A + BF)x(t)$$

with matrix $A + BF$ instead of the initial matrix A. We ask if it is possible to choose adequately F such that the new poles (that is, the eigenvalues of $A + BF$) have some prefixed desired values. For example, with negative real part, so that the automatic control is stable.

A main result of control theory ensures that this pole assignment by feedback is possible if the initial system is controllable, that is, if $\text{rank}(B, AB, A^2B, \dots, A^{n-1}B) = n$.

We will prove and apply that in the one-parameter case ($m = 1$). We will do it by transforming the initial system in the so-called "control canonical form" by means of a suitable basis change. In this form, the feedback computation is trivial, and finally it will only be necessary to undo the transformation.

(1) Consider a one-parameter control system

$$\dot{x}(t) = Ax(t) + bu(t); \quad A \in M_n, \quad b \in M_{n,1},$$

and let $\mu_1, \dots, \mu_n$ be the desired eigenvalues for the feedback system (each one repeated so many times as its algebraic multiplicity). The main hypothesis is that the system is controllable, that is, $\text{rank}(b, Ab, A^2b, \dots, A^{n-1}b) = n$.
Then it is possible to find a linear change S in the state variables such that the new matrices of the system have the so-called "control canonical form":

$$A_c = S^{-1}AS = \begin{pmatrix} 0 & 1 & 0 & \dots & 0 & 0 \\ 0 & 0 & 1 & \dots & 0 & 0 \\ & \dots & \dots & & & \\ 0 & 0 & 1 & \dots & 0 & 1 \\ a_n & a_{n-1} & a_{n-2} & \dots & a_2 & a_1 \end{pmatrix}, \quad b_c = S^{-1}b = \begin{pmatrix} 0 \\ \vdots \\ 0 \\ 1 \end{pmatrix}$$

for certain coefficients $a_1, \dots, a_n$.

(1.1) Check that these coefficients are the same (with opposite sign) than the ones of the characteristic polynomial:

$$Q(t) = (-1)^n(t^n - a_1t^{n-1} - \ldots - a_{n-1}t - a_n).$$

(1.2) Deduce that it is straightforward to find F_c such that:

$$\text{eigenvalues}(A_c + bcF_c) = \{\mu_1, \ldots, \mu_n\}.$$

(1.3) Prove that $F = F_cS^{-1}$ is the sought feedback, that is, that:

$$\text{eigenvalues}(A + bF) = \{\mu_1, \ldots, \mu_n\}.$$

(2) Consider the particular case:

$$A = \begin{pmatrix} 0 & 0 & 1 \\ 0 & 1 & -1 \\ 1 & 0 & -1 \end{pmatrix}, \quad b = \begin{pmatrix} 1 \\ 1 \\ 1 \end{pmatrix}.$$

(2.1) Check that the basis change

$$S = \begin{pmatrix} -2 & 1 & 1 \\ -2 & 0 & 1 \\ -1 & 0 & 1 \end{pmatrix}$$

transform the initial matrices into the control canonical form.

(2.2) Compute a feedback F such that:

$$\text{eigenvalues}(A + bF) = \{\mu_1, \mu_2, \mu_3\}.$$

10. Controllable subsystem

Given a linear control system

$$\begin{gathered} \dot{x}(t) = Ax(t) + Bu(t); \quad y(t) = Cx(t) \\ A \in M_n, \quad B \in M_{n,m}, \quad C \in M_{p,n}, \end{gathered} \tag{1}$$

the following matrix is called its "controllability matrix"

$$K(A,B) = \begin{pmatrix} B & AB & A^2B & \ldots & A^{n-1}B \end{pmatrix}.$$

It can be seen that the subspace spanned by its columns $K = \operatorname{Im} K(A,B)$ is the set of reachable states from the origin, and it is called "controllability subspace". We denote $d = \dim K$.

(1) Prove that K is an A-invariant subspace.

We consider a basis change S in the state space adapted to K, and we denote by x_c the first d new coordinates and by x_{uc} the remaining ones:

$$\bar{x} = S^{-1}x = \begin{pmatrix} x_c \\ x_{uc} \end{pmatrix}, \quad x_c \in M_{d,1}.$$

(2) Check that:

$$x \in K \Leftrightarrow x_{uc} = 0$$

that is:

$$x \in K \Leftrightarrow \bar{x} = S^{-1}x = \begin{pmatrix} x_c \\ 0 \end{pmatrix}.$$

(3) Prove that the equations of the system in the new variables are:

$$\begin{pmatrix} \dot{x}_c(t) \\ \dot{x}_{uc}(t) \end{pmatrix} = \bar{A} \begin{pmatrix} x_c(t) \\ x_{uc}(t) \end{pmatrix} + \bar{B}u(t)$$

$$\bar{A} = S^{-1}AS = \begin{pmatrix} A_c & * \\ 0 & * \end{pmatrix}, \quad A_c \in M_d$$

$$\bar{B} = S^{-1}B = \begin{pmatrix} B_c \\ 0 \end{pmatrix}, \quad B_c \in M_{d,m}$$

$$\bar{C} = CS = \begin{pmatrix} C_c \\ C_{uc} \end{pmatrix}, \quad C_c \in M_{p,d}.$$

(4) Deduce that if the initial state belongs to K, it also belongs all the trajectory, for any applied control $u(t)$:

$$x(0) \in K \Rightarrow x(t) \in K, \quad \forall t, \quad \forall u(t).$$

Therefore, it makes sense to consider the "restriction" to K of the initial system:

$$\dot{x}_c(t) = A_c x_c(t) + B_c u(t); \quad y_c(t) = C_c x_c(t). \tag{2}$$

The following paragraphs justify the interest of this subsystem and its denomination as "controllable subsystem" of the initial one.

(5) Justify that the trajectories of system (1) in K can be computed by means of equations (2) of the subsystem and relation (2).

(6) Prove that this subsystem is "controllable", that is, that the following matrix has full rank:

$$\left(B_c \; A_c B_c \; \ldots \; A_c^{d-1} B_c \right).$$

(7) Prove that the controllable subsystem has the same "transfer matrix" (which, we recall, reflects the input/output behaviour) than the initial system, that is, that:

$$C(sI - A)^{-1}B = C_c(sI - A_c)^{-1}B_c.$$

11. Control canonical form

For controllable systems, the so-called "control canonical form" simplifies the computations, for example, for the pole assignment by feedback. We are going to obtain it for the one-parameter case.

(1) Consider the system

$$\dot{x}(t) = Ax(t) + bu(t)$$

$$A = \begin{pmatrix} 0 & 0 & -1 \\ 0 & 1 & -1 \\ 1 & 0 & 1 \end{pmatrix}, \quad b = \begin{pmatrix} 1 \\ 1 \\ 1 \end{pmatrix}.$$

(1.1) Check that it is controllable, that is, that:

$$\operatorname{rank} \left(b \; Ab \; A^2 b \right) = 3.$$

(1.2) Check that doing the basis change

$$S_1 = \left(A^2 b \; Ab \; b \right)$$

we get

$$\bar{A} = S_1^{-1} A S_1 = \begin{pmatrix} 2 & 1 & 0 \\ 0 & 0 & 1 \\ -1 & 0 & 0 \end{pmatrix}, \quad \bar{b} = S^{-1} b = \begin{pmatrix} 0 \\ 0 \\ 1 \end{pmatrix}.$$

(1.3) Check that with the additional change

$$S_2 = \begin{pmatrix} 1 & 0 & 0 \\ -2 & 1 & 0 \\ 0 & -2 & 1 \end{pmatrix}$$

we get the control canonical form

$$A_c = S_2^{-1} \bar{A} S_2 = \begin{pmatrix} 0 & 1 & 0 \\ 0 & 0 & 1 \\ -1 & 2 & 0 \end{pmatrix}, \quad b_c = S_2^{-1} \bar{b} = \begin{pmatrix} 0 \\ 0 \\ 1 \end{pmatrix}.$$

(2) Consider the system

$$\dot{x}(t) = Ax(t) + bu(t); \quad A \in M_n, \quad b \in M_{n,1}$$

which is assumed to be controllable:

$$\operatorname{rank} \left(b \; Ab \; \dots \; A^{n-1} b \right) = n$$

and let $Q(t)$ be the characteristic polynomial of A

$$Q(t) = (-1)^n (t^n - a_1 t^{n-1} - \dots - a_{n-1} t - a_n).$$

(2.1) Check that doing the basis change

$$S_1 = \left(A^{n-1} b \dots Ab \; b \right)$$

we get

$$\bar{A} = S_1^{-1} A S_1 = \begin{pmatrix} a_1 & 1 & 0 & \dots & 0 \\ a_2 & 0 & 1 & \dots & 0 \\ & & \dots & & \\ a_{n-1} & 0 & 0 & \dots & 1 \\ a_n & 0 & 0 & \dots & 0 \end{pmatrix}, \quad \bar{b} = S_1^{-1} b = \begin{pmatrix} 0 \\ \vdots \\ 0 \\ 1 \end{pmatrix}.$$

(2.2) Check that with the additional change

$$S_2 = \begin{pmatrix} 1 & 0 & \dots & 0 & 0 \\ -a_1 & 1 & \dots & 0 & 0 \\ -a_2 & -a_1 & \dots & 0 & 0 \\ & & \dots & & \\ -a_{n-1} & -a_{n-2} & \dots & -a_1 & 1 \end{pmatrix}$$

we get the control canonical form

$$A_c = S_2^{-1}\bar{A}S_2 = \begin{pmatrix} 0 & 1 & 0 & \dots & 0 & 0 \\ 0 & 0 & 1 & \dots & 0 & 0 \\ & & & \dots & & \\ 0 & 0 & 0 & \dots & 0 & 1 \\ a_n & a_{n-1} & a_{n-2} & \dots & a_2 & a_1 \end{pmatrix}, \quad b_c = S_2^{-1}\bar{b} = \begin{pmatrix} 0 \\ \vdots \\ 0 \\ 1 \end{pmatrix}.$$

(3) As an application, we use the control canonical form to check that A is non-derogatory:

(3.1) Prove that if λ_i is an eigenvalue of A, then:

$$\text{rank}(A_C - \lambda_i I) = n - 1.$$

(3.2) Deduce that A is non-derogatory, and so it does not diagonalize if some of its eigenvalues are multiple.

3.2 Solutions

1. Solution

(1) It follows immediately from the observation of the diagrams.

(2.i)

$$\begin{aligned} \dot{x}_1 &= A_1x_1 + B_1u_1 = A_1x_1 + B_1u \\ \dot{x}_2 &= A_2x_2 + B_2u_2 = A_2x_2 + B_2y_1 = A_2x_2 + B_2C_1x_1 \\ y &= y_2 = C_2x_2 \end{aligned}$$

Hence:

$$\dot{x} = \begin{pmatrix} \dot{x}_1 \\ \dot{x}_2 \end{pmatrix} = \begin{pmatrix} A_1 & 0 \\ B_2C_1 & A_2 \end{pmatrix} \begin{pmatrix} x_1 \\ x_2 \end{pmatrix} + \begin{pmatrix} B_1 \\ 0 \end{pmatrix} u$$

$$y = \begin{pmatrix} 0 & C_2 \end{pmatrix} \begin{pmatrix} x_1 \\ x_2 \end{pmatrix}$$

(2.ii) Reasoning in the same way, it results:

$$\begin{pmatrix} A_1 & 0 \\ 0 & A_2 \end{pmatrix}, \quad \begin{pmatrix} B_1 \\ B_2 \end{pmatrix}, \quad \begin{pmatrix} C_1 & C_2 \end{pmatrix}$$

(2.iii) Analogously:

$$\begin{pmatrix} A_1 & B_1C_2 \\ B_2C_1 & A_2 \end{pmatrix}, \quad \begin{pmatrix} B_1 \\ 0 \end{pmatrix}, \quad \begin{pmatrix} C_1 & 0 \end{pmatrix}$$

(3.i) $\hat{y} = \hat{y}_2 = G_2\hat{u}_2 = G_2\hat{y}_1 = G_2G_1\hat{u}_1 = G_2G_1\hat{u}$. Hence, the transfer matrix is:

$$G_2G_1$$

(3.ii) Analogously:

$$G_1 + G_2$$

(3.iii) $\hat{y} = \hat{y}_1 = G_1\hat{u}_1 = G_1(\hat{u} + \hat{y}_2) = G_1\hat{u} + G_1G_2\hat{u}_2 = G_1\hat{u} + G_1G_2\hat{y}$

$(I - G_1G_2)\hat{y} = G_1\hat{u}$

$\hat{y} = (I - G_1G_2)^{-1}G_1\hat{u}$

2. Solution

(a) The controllability matrix is

$$K(A,B) = \begin{pmatrix} 1 & 1 & 0 & \beta & -2 & \beta \\ 0 & \beta & -2 & \beta & -2 & -\beta \\ 0 & 1 & \alpha & 1 & \alpha & \alpha\beta + 1 \end{pmatrix},$$

which clearly has full rank for all α, β. Hence, the given system is controllable for all α, β.

(b) When only the second control acts, that is to say, when $u_1(t) = 0$ for all t, the system can be written

$$\dot{x}(t) = Ax(t) + b_2u_2(t).$$

Then, the controllability matrix is reduced to

$$K(A,b_2) = \begin{pmatrix} 1 & \beta & \beta \\ \beta & \beta & -\beta \\ 1 & 1 & \alpha\beta + 1 \end{pmatrix},$$

which has full rank for $\beta \neq 0, \alpha \neq \frac{-2}{\beta}$.

(c) Now one has

$$K(A,B) = \left(\begin{array}{cc|cc|cc|cc|cc|c|c} 0 & 0 & 0 & 0 & 0 & \gamma & 0 & \delta & 1 & 0 & & \\ 0 & 0 & 0 & \gamma & 0 & \delta & 1 & 0 & 0 & 0 & & \\ 0 & \gamma & 0 & \delta & 1 & 0 & 0 & 0 & 0 & 0 & 0 & 0 \\ 0 & \delta & 1 & 0 & 0 & 0 & 0 & 0 & 0 & 0 & & \\ 1 & 0 & 0 & 0 & 0 & 0 & 0 & 0 & 0 & 0 & & \\ \hline 0 & 0 & 0 & 1 & 0 & 0 & 0 & 0 & 0 & 0 & & \\ 0 & 1 & 0 & 0 & 0 & 0 & 0 & 0 & 0 & 0 & & \end{array}\right).$$

The "controllability indices" can be computed from the ranks of the $2, 4, 6, \ldots$ first columns:

$$\text{rank}(B) = 2, \text{for all } \gamma, \delta$$
$$\text{rank}(B, AB) = 4, \text{for all } \gamma, \delta$$
$$\text{rank}(B, AB, A^2B) = 5 \text{ if } \gamma = \delta = 0; \quad \text{rank}(B, AB, A^2B) = 6 \text{ otherwise}$$
$$\text{rank}(B, \ldots, A^3B) = 6 \text{ if } \gamma = \delta = 0; \quad \text{rank}(B, \ldots, A^3B) = 7 \text{ otherwise}$$
$$\text{rank}(B, \ldots, A^4B) = 7$$

3. Solution

(1) Let

$$G(s) = \begin{pmatrix} 1/s \\ 1/(s-1) \end{pmatrix}.$$

Then: $p = 2, m = 1$. Following the given pattern:
(i) $P(s) = s(s-1) = -s + s^2$
$r = 2, p_0 = 0, p_1 = -1, p_2 = 1$

(ii) $G(s) = \frac{1}{-s+s^2} \begin{pmatrix} s-1 \\ s \end{pmatrix}$

(iii) $G(s) = \frac{1}{-s+s^2} \left(\begin{pmatrix} -1 \\ 0 \end{pmatrix} + \begin{pmatrix} 1 \\ 1 \end{pmatrix} s \right)$

$R_0 = \begin{pmatrix} -1 \\ 0 \end{pmatrix}, R_1 = \begin{pmatrix} 1 \\ 1 \end{pmatrix}$

(iv) $A = \begin{pmatrix} 0 & 1 \\ 0 & 1 \end{pmatrix}, \quad B = \begin{pmatrix} 0 \\ 1 \end{pmatrix}, \quad C = \begin{pmatrix} -1 & 1 \\ 0 & 1 \end{pmatrix}$

(2) $\text{rank}\, K(A, B) = \text{rank} \begin{pmatrix} 0 & 1 \\ 1 & 1 \end{pmatrix} = 2$

(3) $(sI - A)^{-1} = \left(s \begin{pmatrix} 1 & 0 \\ 0 & 1 \end{pmatrix} - \begin{pmatrix} 0 & 1 \\ 0 & 1 \end{pmatrix} \right)^{-1} = \begin{pmatrix} s & -1 \\ 0 & s-1 \end{pmatrix}^{-1} = \frac{1}{s^2-s} \begin{pmatrix} s-1 & 1 \\ 0 & s \end{pmatrix}$

$$\begin{aligned} C(sI - A)^{-1}B &= \begin{pmatrix} -1 & 1 \\ 0 & 1 \end{pmatrix} \frac{1}{s^2 - s} \begin{pmatrix} s-1 & 1 \\ 0 & s \end{pmatrix} \begin{pmatrix} 0 \\ 1 \end{pmatrix} = \frac{1}{s^2 - s} \begin{pmatrix} -1 & 1 \\ 0 & 1 \end{pmatrix} \begin{pmatrix} 1 \\ s \end{pmatrix} \\ &= \frac{1}{s^2 - s} \begin{pmatrix} -1 + s \\ s \end{pmatrix} = G(s) \end{aligned}$$

4. Solution

(1) $x = (-1, 2, 1)$ is 3-step reachable, from the origin, if there are $u_2(0), u_2(1), u_2(2)$ such that:

$$\begin{aligned} x &= A^2 b_1 u_1(0) + A b_1 u_1(1) + b_1 u_1(2) \\ &= \begin{pmatrix} 10 \\ -8 \\ 8 \end{pmatrix} u_1(0) + \begin{pmatrix} 2 \\ -2 \\ 2 \end{pmatrix} u_1(1) + \begin{pmatrix} 0 \\ -1 \\ 1 \end{pmatrix} u_1(2) \end{aligned}$$

or equivalently::

$$\begin{pmatrix} -1 \\ 2 \\ 1 \end{pmatrix} = \begin{pmatrix} 10 & -2 & 0 \\ -8 & -2 & -1 \\ 8 & 2 & 1 \end{pmatrix} \begin{pmatrix} u_1(0) \\ u_1(1) \\ u_1(2) \end{pmatrix}$$

No solutions exist because

$$\text{rank} \begin{pmatrix} 10 & -2 & 0 \\ -8 & -2 & -1 \\ 8 & 2 & 1 \end{pmatrix} = 2 \qquad \text{rank} \begin{pmatrix} 10 & -2 & 0 & -1 \\ -8 & -2 & -1 & 2 \\ 8 & 2 & 1 & 1 \end{pmatrix} = 3$$

(2) Analogously:

$$\begin{pmatrix} -1 \\ 2 \\ 1 \end{pmatrix} = \begin{pmatrix} -1 & 0 \\ 1 & 1 \\ 1 & 0 \end{pmatrix} \begin{pmatrix} u_2(0) \\ u_2(1) \end{pmatrix}$$

whose solution is: $u_2(0) = 1, u_2(1) = 1$.

(3.1) Analogously:

$$\begin{pmatrix} -1 \\ 2 \\ 1 \end{pmatrix} = \begin{pmatrix} 2 & -1 & 0 & 0 \\ -2 & 1 & -1 & 1 \\ 2 & 1 & 1 & 0 \end{pmatrix} \begin{pmatrix} u_2(0) \\ u_2(0) \\ u_1(1) \\ u_2(1) \end{pmatrix}$$

The solutions can be parameterized by $u_1(0)$ as follows:

$$\begin{aligned} u_2(0) &= 1 + u_1(0) \\ u_1(1) &= 1 - 2u_1(0) - u_2(0) = -4u_1(0) \\ u_2(1) &= \ldots = 1 - 4u_1(0) \end{aligned}$$

(3.2) It is not possible, because $u_1(0)$ and $u_1(1)$ have opposite signs.

(3.3) It is possible $u_1(0) = u_2(0) = -1$. Then $u_1(1) = 4, u_2(1) = 5$.

5. Solution

(1.1) When $x(0) = 0$, one has:

$$\begin{aligned} x(k) = {} & (A^{k-1}b_1u_1(0) + \ldots + A^{k-1}b_mu_m(0)) \\ & \vdots \\ & + (Ab_1u_1(k-2) + \ldots + Ab_mu_m(k-2)) \\ & + (b_1u_1(k-1) + \ldots + b_mu_m(k-1)) \end{aligned}$$

where $B = (b_1, \ldots, b_m)$ and $u(0), \ldots, u(k-1)$ run over all possible control functions. Therefore

$$K(h) = [A^{k-1}b_1, \ldots, A^{k-1}b_m, \ldots, Ab_1, \ldots, Ab_m, b_1, \ldots, b_m] = [B, AB, \ldots, A^{k-1}B].$$

(1.2) It is obvious from (1.1).

(1.3) $K(h) = K(h+1)$ if and only if $A^h b_1, \dots, A^h b_m \in K(h)$.
Then $A^{h+1} b_1, \dots, A^{h+1} b_m \in K(h+1)$, so that $K(h+2) = K(h+1)$.

(1.4) As $\dim K(h) \leq n$, the length of the increasing chain is n at most.

(2)

$$\begin{aligned} K_i + K_j &= [b_i, Ab_i, \dots, A^{n-1} b_i] + [b_j, Ab_j, \dots, A^{n-1} b_j] \\ &= [(b_i, b_j), A(b_i, b_j), \dots, A^{n-1}(b_i, b_j)] \end{aligned}$$

(3.1) $K = \mathrm{Im} \begin{pmatrix} 0 & 0 & 2 & -1 & 10 & -2 \\ -1 & 1 & -2 & 1 & -8 & 2 \\ 1 & 0 & 2 & 1 & 8 & 2 \end{pmatrix} = \mathbb{R}^3$

$K_1 = \mathrm{Im} \begin{pmatrix} 0 & 2 & 10 \\ -1 & -2 & -8 \\ 1 & 2 & 8 \end{pmatrix} = \mathrm{Im} \begin{pmatrix} 0 & 2 \\ -1 & -2 \\ 1 & 2 \end{pmatrix}$

A basis of K_1: $(0, -1, 1)$, $(1, 0, 0)$

$K_2 = \mathrm{Im} \begin{pmatrix} 0 & -1 & -2 \\ 1 & 1 & 2 \\ 0 & 1 & 2 \end{pmatrix} = \mathrm{Im} \begin{pmatrix} 0 & -1 \\ 1 & 1 \\ 0 & 1 \end{pmatrix}$

A basis of K_2: $(0, 1, 0)$, $(-1, 0, 1)$

(3.2) $K_1 + K_2 = \mathrm{Im} \begin{pmatrix} 0 & 1 & 0 & -1 \\ -1 & 0 & 1 & 0 \\ 1 & 0 & 0 & 1 \end{pmatrix} = \mathbb{R}^3$

$K_1 = \{y + z = 0\}$
$K_2 = \{x + z = 0\}$
$K_1 \cap K_2 = \{y + z = x + z = 0\}$
A basis of $K_1 \cap K_2$: $(1, 1, -1)$

6. Solution

(1) If $\bar{x} = S^{-1} x$, then:

$$\begin{aligned} \dot{\bar{x}} &= S^{-1} \dot{x} = S^{-1}(Ax + Bu) = S^{-1} A S \bar{x} + S^{-1} B u \\ y &= Cx = CS\bar{x} \end{aligned}$$

(2) $\mathrm{rank}(\bar{B}, \bar{A}\bar{B}, \dots, \bar{A}^h \bar{B}) = \mathrm{rank}(S^{-1}B, S^{-1}ASS^{-1}B, \dots, (S^{-1}AS)^h S^{-1}B)$
$= \mathrm{rank}\, S^{-1}(B, AB, \dots, A^h B) = \mathrm{rank}(B, AB, \dots, A^h B)$

(3) $\bar{G}(s) = \bar{C}(sI - \bar{A})^{-1}\bar{B} = CS(sI - S^{-1}AS)^{-1}S^{-1}B = CS(S^{-1}(sI - A)S)^{-1}S^{-1}B$
$= CSS^{-1}(sI - A)^{-1}SS^{-1}B = G(s)$

7. Solution

(1) From Cayley-Hamilton theorem:

$$A^n = a_0 I + a_1 A + \dots + a_{n-1} A^{n-1}$$

Therefore, if $x \in \operatorname{Im}\left(B\ AB \ldots A^{n-1}B\right)$, then

$$Ax \in \operatorname{Im}\left(AB\ A^2B \ldots A^nB\right) \subset \operatorname{Im}\left(B\ AB \ldots A^nB\right),$$

and

$$\begin{aligned}\operatorname{Im}\left(B\ AB \ldots A^nB\right) &= \operatorname{Im}\left(B\ AB \ldots (a_0B + a_1AB + \ldots + a_{n-1}A^{n-1}B)\right)\\ &= \operatorname{Im}\left(B\ AB \ldots A^{n-1}B\right)\\ &= K.\end{aligned}$$

If $x \in L$, then:

$$Cx = CAx = \ldots = CA^{n-1}x = 0$$

Clearly:

$$C(Ax) = \ldots = CA^{n-2}(Ax) = 0$$

It is sufficient to prove that $CA^{n-1}(Ax) = 0$, but

$$\begin{aligned}CA^{n-1}(Ax) &= CA^nx\\ &= C(a_0I + a_1A + \ldots + a_{n-1}A^{n-1})x\\ &= a_0Cx + a_1CAx + \ldots + a_{n-1}CA^{n-1}x = 0\end{aligned}$$

(2.1) $K = \operatorname{Im}\begin{pmatrix} 1 & 1 & 1 & 1\\ 0 & 0 & 0 & 0\\ 0 & 1 & -2 & 3\\ 1 & -1 & 1 & -1\end{pmatrix}; \quad \dim K = 3$

$L = \operatorname{Ker}\begin{pmatrix} 0 & 1 & 0 & 1\\ 0 & 1 & 0 & -1\\ 0 & 1 & 0 & 1\\ 0 & 1 & 0 & -1\end{pmatrix}; \quad \dim L = 4 - 2 = 2$

(2.2) Basis of K: (u_1, u_2, u_3)
$u_1 = (1,0,0,1), \quad u_2 = (1,0,1,-1), \quad u_3 = (1,0,-2,1)$
Basis of L: (v_1, v_2)
$v_1 = (1,0,0,0), v_2 = (0,0,1,0)$

(2.3) $Au_1 = u_2; \quad Au_2 = u_3; \quad Au_3 = \begin{pmatrix}1\\0\\3\\-1\end{pmatrix} = u_1 + u_2 - u_3$

$\operatorname{Mat} A|_K = \begin{pmatrix} 0 & 0 & 1\\ 1 & 0 & 1\\ 0 & 1 & -1\end{pmatrix}$

$Av_1 = v_1; \quad Av_2 = -v_2$

$\operatorname{Mat} A|_L = \begin{pmatrix} 1 & 0\\ 0 & -1\end{pmatrix}$

(2.4) $K = \{(x_1, x_2, x_3, x_4) : x_2 = 0\}$
$L = \{(x_1, x_2, x_3, x_4) : x_2 = x_4 = 0\}$
$K \cap L = L$

8. Solution

(1) We recall (see ex. 7) that K and L are A-invariant subspaces. Therefore:

$A(S_2) \subset [S_2]$

$A(S_1) \subset [S_1, S_2]$

$A(S_4) \subset [S_2, S_4]$

so that $\bar{A} = S^{-1}AS$ has the stated form.
Moreover Im $B \subset K$. Hence, $\bar{B} = S^{-1}B \subset [S_1, S_2]$.
Finally, as $L \subset \operatorname{Ker} C$, we have $C(S_2) = C(S_4) = 0$.

(2.1) According to the solution of ex. 7, we can take:

$$S_2 = \begin{pmatrix} 1 & 0 \\ 0 & 0 \\ 0 & 1 \\ 0 & 0 \end{pmatrix} \quad S_1 = \begin{pmatrix} 0 \\ 0 \\ 0 \\ 1 \end{pmatrix} \quad S_3 = \begin{pmatrix} 0 \\ 1 \\ 0 \\ 0 \end{pmatrix}$$

and S_4 is empty. Hence:

$$S = \begin{pmatrix} 0 & 1 & 0 & 0 \\ 0 & 0 & 0 & 1 \\ 0 & 0 & 1 & 0 \\ 1 & 0 & 0 & 0 \end{pmatrix}$$

(2.2)

$$\bar{A} = S^{-1}AS = \left(\begin{array}{c|cc|c} -1 & 0 & 0 & 0 \\ \hline 0 & 1 & 0 & 1 \\ 1 & 0 & -1 & 0 \\ \hline 0 & 0 & 0 & 1 \end{array}\right)$$

$$\bar{B} = S^{-1}B = \begin{pmatrix} 1 \\ \hline 1 \\ 0 \\ \hline 0 \end{pmatrix} \qquad \bar{C} = CS = (1 | 0\ 0 | 1)$$

That is to say:

$$A_{11} = (-1) \quad A_{13} = (0\ 0) \quad A_{21} = \begin{pmatrix} 0 \\ 1 \end{pmatrix} \quad A_{22} = \begin{pmatrix} 1 & 0 \\ 0 & -1 \end{pmatrix} \quad A_{23} = \begin{pmatrix} 1 \\ 0 \end{pmatrix}$$

$$A_{33} = (1) \quad B_1 = (1) \quad B_2 = \begin{pmatrix} 1 \\ 0 \end{pmatrix} \quad C_1 = (1) \quad C_3 = (1)$$

(2.3) $\operatorname{rank} K(A_{11}, B_1) = \operatorname{rank} (1\ -1) = 1$
$\operatorname{rank}(L(C_1, A_{11}))^t = \operatorname{rank} (1\ -1) = 1$

$$\operatorname{rank} K\left(\begin{pmatrix} A_{11} & 0 \\ A_{21} & A_{22} \end{pmatrix}, \begin{pmatrix} B_1 \\ B_2 \end{pmatrix}\right) = \operatorname{rank}\begin{pmatrix} 1 & -1 & 1 \\ 1 & 1 & 1 \\ 0 & 1 & -2 \end{pmatrix} = 3$$

$$\operatorname{rank}\left(L\left(\begin{pmatrix} C_1 & C_3 \end{pmatrix}, \begin{pmatrix} A_{11} & 0 \\ A_{21} & A_{33} \end{pmatrix}\right)\right)^t = \operatorname{rank}\begin{pmatrix} 1 & -1 \\ 1 & 1 \end{pmatrix} = 2$$

9. Solution

(1.1) It is easy to check that there is only one main minor of each sign. Then:

$$a_1 = \operatorname{tr} A_c, \quad a_2 = -\det\begin{pmatrix} 0 & 1 \\ a_2 & a_1 \end{pmatrix}, \quad a_3 = \det\begin{pmatrix} 0 & 1 & 0 \\ 0 & 0 & 1 \\ a_3 & a_2 & a_1 \end{pmatrix}, \ldots, \quad a_n = \det A_c.$$

(1.2) If $F_c = \begin{pmatrix} f_1 & \ldots & f_n \end{pmatrix}$, then

$$A_c + b_c F_c = \begin{pmatrix} 0 & 1 & \ldots & 0 & 0 \\ 0 & 0 & \ldots & 0 & 0 \\ & & \ldots & & \\ 0 & 0 & \ldots & 0 & 1 \\ a_n + f_1 & a_{n-1} + f_2 & \ldots & a_2 + f_{n-1} & a_1 + f_n \end{pmatrix}$$

Therefore, $f_1, \ldots, f_n$ are the solutions of:

$$t^n - (a_1 + f_n)t^{n-1} - \ldots - (a_{n-1} + f_2)t - (a_n + f_1) = (t - \mu_1)(t - \mu_2)\cdots(t - \mu_n).$$

(1.3)

$$\begin{aligned} \text{eigenvalues}(A + bF) &= \text{eigenvalues}(SA_cS^{-1} + Sb_cF_cS^{-1}) \\ &= \text{eigenvalues}(S(A_c + b_cF_c)S^{-1}) \\ &= \text{eigenvalues}(A_c + b_cF_c) = \{\mu_1, \ldots, \mu_n\}. \end{aligned}$$

(2.1) It is straightforward to check that:

$$S^{-1}AS = \begin{pmatrix} 0 & 1 & 0 \\ 0 & 0 & 1 \\ -1 & 2 & 0 \end{pmatrix}, \quad S^{-1}b = \begin{pmatrix} 0 \\ 0 \\ 1 \end{pmatrix}.$$

(2.2) First, we look for $F_c = \begin{pmatrix} f_1 & f_2 & f_3 \end{pmatrix}$ such that:

$$t - f_3t^2 - (2 + f_2) - (-1 + f_1) = (t - \mu_1)(t - \mu_2)(t - \mu_3)$$

Therefore:

$$\begin{aligned} f_3 &= \mu_1 + \mu_2 + \mu_3 \\ f_2 &= \mu_1\mu_2 + \mu_2\mu_3 + \mu_3\mu_1 - 2 \\ f_1 &= \mu_1\mu_2\mu_3 + 1 \end{aligned}$$

Finally: $F = F_c S^{-1}$.

10. Solution

(1) From Cayley-Hamilton theorem, if the characteristic polynomial of A is

$$Q(t) = (-1)^n (t^n - a_1 t^{n-1} - \ldots - a_{n-1} t - a_n),$$

then

$$A^n = a_n I + a_{n-1} A + \ldots + a_1 A^{n-1}.$$

Therefore, if $x \in K = \text{Im}\left(B \; AB \; \ldots \; A^{n-1}B \right)$, then

$$Ax \in \text{Im}\left(AB \; A^2B \; \ldots \; A^n B \right)$$

$$\subset \text{Im}\left(B \; AB \; \ldots \; A^{n-1}B \; (a_0 B + \ldots + a_{n-1} A^{n-1} B) \right) = K$$

(2) If $(u_1, \ldots, u_d, \ldots, u_m)$ is a basis of the state space adapted to K, then a state x belongs to K if and only if the last $n - d$ coordinates are 0.

(3) In the conditions of (2), the d first columns of $\bar{A}$ are $Au_1, \ldots, Au_d$, which belong to K (see (1)). Hence, again from (2), their last $n - d$ coordinates are 0. The same argument works for B, because its columns belong to K.

(4) From (3), it is clear that $x_{uc}(0) = 0$ implies $x_{uc}(t) = 0$ for any control $u(t)$.

(5) Again from (3), if $x_{uc}(t) = 0$, the remainder coordinates $x_c(t)$ are determined by system (2).

(6) By hypothesis

$$\text{rank}\left(B \; AB \; \ldots \; A^{n-1}B \right) = d.$$

The rank is preserved under changes of bases. Hence:

$$d = \text{rank}\left(\bar{B} \; \bar{A}\bar{B} \; \ldots \; \bar{A}^{n-1}\bar{B} \right) = \text{rank}\left(\begin{pmatrix} B_c \\ 0 \end{pmatrix} \begin{pmatrix} A_c B_c \\ 0 \end{pmatrix} \cdots \begin{pmatrix} A_c^{n-1} B_c \\ 0 \end{pmatrix} \right)$$

$$= \text{rank}\left(B_c \; A_c B_c \; \ldots \; A_c^{n-1} B_c \right) = \text{rank}\left(B_c \; A_c B_c \; \ldots \; A_c^{d-1} B_c \right)$$

where the last equality follows from Cayley-Hamilton theorem.

(7) Recall that the transfer matrix is preserved under changes of bases:

$$C(sI - A)^{-1}B = \bar{C}S^{-1}(sI - S\bar{A}S^{-1})^{-1}S\bar{B} = \bar{C}S^{-1}(S(sI - \bar{A})S^{-1})^{-1}S\bar{B} =$$

$$= \bar{C}S^{-1}(S^{-1})^{-1}(sI - \bar{A})S^{-1}S\bar{B} = \bar{C}(sI - \bar{A})\bar{B}$$

From (2):

$$(sI - \bar{A})^{-1} = \left(sI - \begin{pmatrix} A_c & * \\ * & * \end{pmatrix} \right)^{-1} = \begin{pmatrix} sI_d - A_c & * \\ 0 & * \end{pmatrix}^{-1} = \begin{pmatrix} (sI_d - A_c)^{-1} & * \\ 0 & * \end{pmatrix}$$

where it does not matter the form of the blocks *:

$$\bar{C}(sI - \bar{A})^{-1}\bar{B} = \begin{pmatrix} C_c & C_{uc} \end{pmatrix} \begin{pmatrix} (sI_d - A_c)^{-1} & * \\ 0 & * \end{pmatrix} \begin{pmatrix} B_c \\ 0 \end{pmatrix}$$

$$= \begin{pmatrix} C_c & C_{uc} \end{pmatrix} \begin{pmatrix} (sI_d - A_c)^{-1}B_c \\ 0 \end{pmatrix}$$

$$= C_c(sI_d - A_c)^{-1}B_c$$

11. Solution

(1)

$$A = \begin{pmatrix} 0 & 0 & -1 \\ 0 & 1 & -1 \\ 1 & 0 & 1 \end{pmatrix}; \quad b = \begin{pmatrix} 1 \\ 1 \\ 1 \end{pmatrix}$$

(1.1)

$$\text{rank}\begin{pmatrix} b & Ab & A^2b \end{pmatrix} = \text{rank}\begin{pmatrix} 1 & 1 & 2 \\ 1 & 0 & -2 \\ 1 & 2 & 3 \end{pmatrix} = 3$$

(1.2)

$$S_1 = \begin{pmatrix} 2 & 1 & 1 \\ -2 & 0 & 1 \\ 3 & 2 & 1 \end{pmatrix}; \quad S_1^{-1} = \frac{1}{3}\begin{pmatrix} 2 & -1 & -1 \\ -5 & 1 & 4 \\ 4 & 1 & -2 \end{pmatrix}$$

$$\bar{A} = \frac{1}{3}\begin{pmatrix} 6 & 3 & 0 \\ 0 & 0 & 3 \\ -3 & 0 & 0 \end{pmatrix} \quad \bar{b} = \frac{1}{3}\begin{pmatrix} 2 & -1 & -1 \\ -5 & 1 & 4 \\ 4 & 1 & -2 \end{pmatrix}\begin{pmatrix} 1 \\ 1 \\ 1 \end{pmatrix} = \frac{1}{3}\begin{pmatrix} 0 \\ 0 \\ 3 \end{pmatrix}$$

(1.3)

$$S_2 = \begin{pmatrix} 1 & 0 & 0 \\ -2 & 1 & 0 \\ 0 & -2 & 1 \end{pmatrix}; \quad S_2^{-1} = \begin{pmatrix} 1 & 0 & 0 \\ 2 & 1 & 0 \\ 4 & 2 & 1 \end{pmatrix}$$

$$A_c = \begin{pmatrix} 0 & 1 & 0 \\ 0 & 0 & 1 \\ -1 & 0 & 2 \end{pmatrix}; \quad b_c = \begin{pmatrix} 1 & 0 & 0 \\ 2 & 1 & 0 \\ 4 & 2 & 1 \end{pmatrix}\begin{pmatrix} 0 \\ 0 \\ 1 \end{pmatrix} = \begin{pmatrix} 0 \\ 0 \\ 1 \end{pmatrix}$$

(2) The computations are analogous to those in (1).

(3.1) If λ_i is an eigenvalue of A_c, then:

$$\text{rank}(A_c - \lambda_i I) < n.$$

On the other hand,

$$\text{rank}(A_c - \lambda_i I) = \text{rank}\begin{pmatrix} -\lambda_i & 1 & 0 \dots & 0 & 0 \\ 0 & -\lambda_i & 1 \dots & 0 & 0 \\ & \dots & \dots & & \\ 0 & 0 & 0 \dots & -\lambda_i & 1 \\ * & * & * \dots & * & * \end{pmatrix} \geq n - 1.$$

(3.2) If λ_i is an eigenvalue of A, then λ_i is also en eigenvalue of A_c. Then, from (3.1):

$$\dim \mathrm{Ker}(A - \lambda_i I) = n - \mathrm{rank}(A - \lambda_i I) = n - \mathrm{rank}(A_c - \lambda_i I) = n - (n-1) = 1.$$

Therefore, A is non-diagonalizable if the algebraic multiplicity of λ_i is greater than 1.

4. References

Chen, Chi-Tsong (1984). *Linear System Theory and Design*, Holt-Saunders International Editions, Japan. ISBN: 4-8337-0191-X.

Ferrer, J.; Ortiz-Caraballo, C. & Peña, M. (2010). Learning engineering to teach mathematics, *40th ASEE/IEEE Frontiers in Education Conference*

Kalman, R. E.; Falb, P. L. & Arbib, M. A. (1974). *Topics in mathematical system theory*, PTMH, India, McGraw-Hill

Lay, D.C (2007). *Algebra lineal y sus aplicaciones*, Ed. Pearson, Mexico. ISBN: 978-970-26-0906-3

Puerta Sales, F. (1976). *Algebra Lineal*, Universidad Politecnica de Barcelona. ETS Ingenieros Industriales de Barcelona in collaboration with Marcombo Boixareu Ed. ISBN: 84-600-6834-X

10

Rapid Start-Up of the Steam Boiler, Considering the Allowable Rate of Temperature Changes

Jan Taler and Piotr Harchut
Department of Power Installations, Cracow University of Technology
Poland

1. Introduction

The development of wind turbine engineering, which can be characterized by large inconsistencies in the amount of delivered electricity over time, generated new problems regarding the regulation of power engineering systems. At high and low wind speeds, the energy production falls rapidly, meaning that the conventional heat or steam-gas blocks have to be activated very quickly. Steam boilers in blocks have to be designed in a way that allows for a start-up of the block within a few dozen minutes. The main elements limiting the quick activation of steam boilers are thick-walled structural elements, in which heat induced stresses can occur during the start-up cycle.

The start-up and the shut-down processes shall be conducted in a way that ensures that the stresses at the places of stress concentration do not exceed the allowable values.

To be able to ensure the appropriate durability and safety of the blocks adapted for a rapid start up, a precise analysis of the flow-heat processes, together with the strength analysis, is required. Such analysis is covered in this paper.

German boiler regulations, TRD - 301 and the European Standard, EN-12952-3 allow determining two allowable rates for heating the pressure element:

- v_{T1} the pressure p_1 at the beginning of the start-up process,

- v_{T2} at the pressure p_2 at the end of the start-up process.

Heating rates for intermediate pressures shall be determined using the method of linear interpolation (Fig. 1).

To be able to heat the boiler evaporator at the maximum allowable rate, heat flow rate $\dot{Q}_k$ shall be delivered to the evaporator in the boiler furnace chamber (Fig. 2). The value of $\dot{Q}_k$ is a function of many parameters, but for the boiler of a specific construction, it depends mainly on the rate of changes of the saturation pressure dp_s / dt and the mass of the saturated steam mass flow rate.

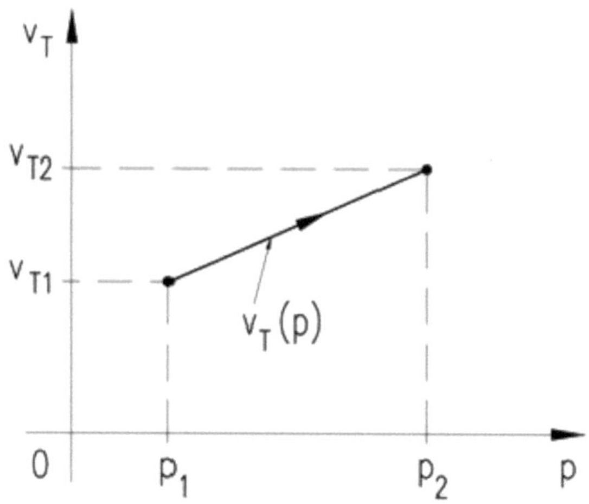

Fig. 1. Heating rate of the pressure boiler element as a pressure function, determined in accordance with TRD-301 regulations.

2. Pressure and temperature changes in the evaporator during the boiler heating process

In accordance with TRD 301 regulations (Fig. 1), the changes of the heating rate $v_T = dT_n/dt$ in the pressure function are interpolated using a straight line.

$$\frac{dT_n}{dt} = \frac{p_2 v_{T1} - p_1 v_{T2}}{p_2 - p_1} + \frac{v_{T2} - v_{T1}}{p_2 - p_1} p_n(T_n) \tag{1}$$

The water saturation pressure $p_n(T_n)$ from Eq. (1) can be expressed in the following way in the temperature function:

$$p_n = \exp(-19{,}710662 + 4{,}2367548 T_n) \tag{2}$$

Where pressure p_n is expressed in bars and temperature T_n is expressed in °C.

Eq. (1) has been integrated using the Runge-Kutty method of the 4-th rank at the following initial condition:

$$T_n\big|_{t=0} = T_n(p_1) \tag{3}$$

After determining, by solving the initial condition (1-3), the flow of the saturation temperature in a function of time, that is $T_n(t)$, the pressure changes $p_n(t)$ can be determined, e.g., from function (2). The pressure changes rate dp_n/dt and the saturation temperature changes rate dT_n/dt in a function of time shall be determined on the basis of the assumed run $v_T(p_n)$ (Fig. 1).

In this paper, the dynamics of the boiler evaporator of the OP-210M boiler will be analysed.

OP-210M is a fine coal fuelled boiler with natural circulation. The $T_n(t)$ and $p_n(t)$ flows, determined for p_1=0 bar, v_{T1}=2 K/min and p_2=108,7 bar, v_{T2}=5 K/min are presented in Fig. 2.

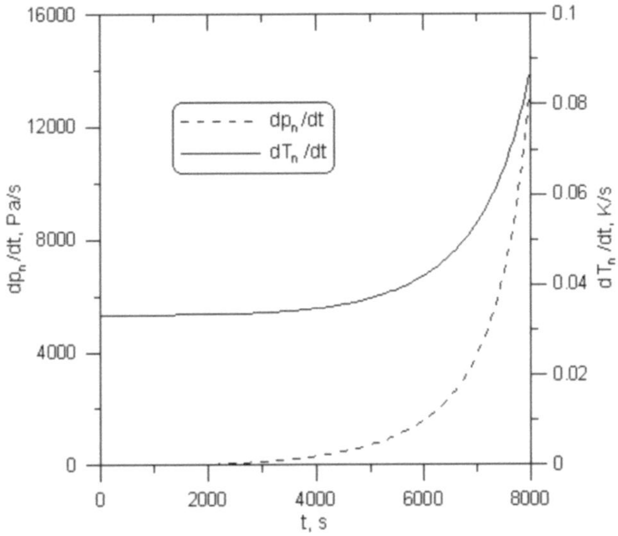

Fig. 2. Temperature and saturation pressure changes in the boiler evaporator during the heating process.

After determining the $T_n(t)$ and dT_n/dt time run, it is possible to determine the rate of pressure changes dp_n/dt using the following formula:

$$\frac{dp_n}{dt} = \frac{dp_n}{dT_n}\frac{dT_n}{dt} \tag{4}$$

Where functions dp_n/dT_n for saturated steam can be determined by the integration of formula (2).

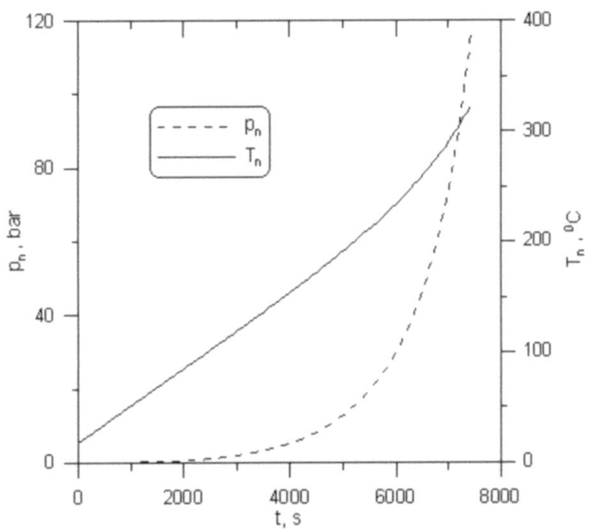

Fig. 3. Rates of the pressure dp_n/dt and temperature dT_n/dt changes, during the evaporator heating process.

The pressure dp_n/dT_n changes rate can also be calculated using an approximate formula:

$$\frac{dp_n}{dt} = \frac{p_n(t+\Delta t) - p_n(t-\Delta t)}{2\Delta t} \tag{5}$$

Where: Δt is the time step.

For the calculations, it can be assumed, for example, that $\Delta t = 1$ s. Temperature dT_n/dt and pressure dp_n/dt changing rates are presented in Fig. 3.

3. Mass and energy balance for the boiler evaporator

Modelling of the transient state effects occurring in the boiler evaporator is usually done while assuming that it is an object having a concentrated mass and heat capacity [4]. The start point for the determination of the heat flow rate $\dot{Q}_k$, which assures that the evaporator is heated at the desired rate vT(t), are the mass and energy balance equations for the evaporator (Fig. 4):

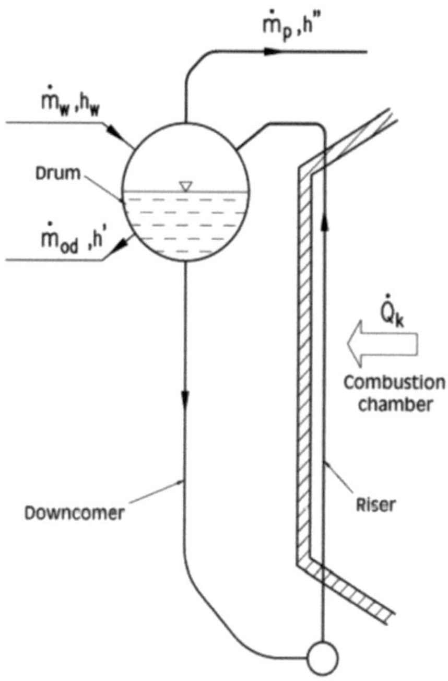

Fig. 4. Diagram of the boiler evaporator.

$$\frac{d(V'\rho' + V''\rho'')}{d\tau} = \dot{m}_w - \dot{m}_p - \dot{m}_{od} \tag{6}$$

$$\frac{d(V'\rho' + V''\rho'')}{d\tau} = \dot{m}_w - \dot{m}_p - \dot{m}_{od} \tag{7}$$

After transforming formulas (6) and (7) we obtain:

$$\dot{Q}_k = \dot{m}_{od}\frac{\rho''}{\rho'-\rho''}(h''-h') - \dot{m}_w(h_w - \frac{\rho'h'-\rho''h''}{\rho'-\rho''}) + \dot{m}_p\frac{\rho'(h''-h')}{\rho'-\rho''} +$$
$$[V'(\rho'\frac{dh'}{dp} + \frac{\rho''(h''-h')}{\rho'-\rho''}\frac{d\rho'}{dp} - 1) + V''(\rho''\frac{dh'}{dp} + \frac{\rho(h''-h')}{\rho'-\rho''}\frac{d\rho''}{dp} - 1) + m_m c_m \frac{dT_n}{dp_n}]\frac{dp_n}{dt} \quad (8)$$

Formula (8) allows determining the heat flow rate $\dot{Q}_k$, which shall be delivered to the evaporator from the furnace chamber in order to ensure the assumed rate of the pressure change dp_n/dt. Changes of the $d\rho'/dp$ and $d\rho''/dp$ functions and the dh'/dp and dh''/dp functions, appearing in formula (8) are presented in Fig. 5 and Fig. 6.

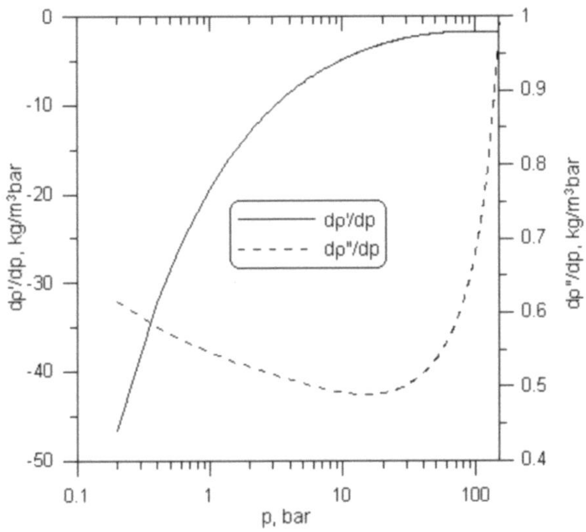

Fig. 5. Changes of the $d\rho'/dp$ and $d\rho''/dp$ derivatives as a function of pressure.

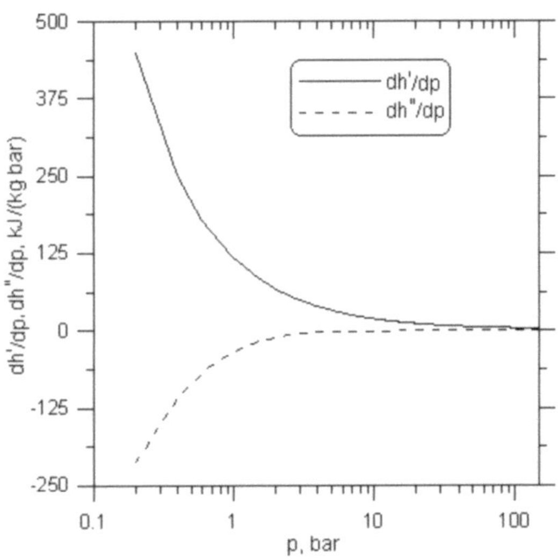

Fig. 6. Changes of the dh'/dp i dh''/dp derivatives as a function of pressure.

The precision in determining those functions is very important. The best way is to determine those functions through analytical differentiation of the $\rho'(p)$ and $\rho''(p)$ functions, obtained earlier using the least square methods on the basis of the pair tables.

4. Heat flow rate absorbed in combustion chamber

From the conducted calculations, we can see, that the heat flow rate $\dot{Q}_k$ delivered to the evaporator can be lowered significantly by reducing the mass of the steam mass flow rate $\dot{m}_p$ delivered from the boiler drum. The minimum steam mass flow rate $\dot{m}_p$, which shall be produced in the boiler during the start-up process can be determined from the condition, stating that the maximum allowable temperatures for the particular stages of the heater shall not be exceeded. The furnace oil mass flow rate $\dot{m}_{pal}$, delivered to the burners during the boiler start up process, which is necessary to assure the heat flow rate $\dot{Q}_k$ shall be determined on the basis of the calculations of the boiler furnace chamber.

The heat flow rate $\dot{Q}_{k,kom}$ absorbed by the walls of the boiler furnace chamber, expressed in W, can be calculated from the following formula:

$$\dot{Q}_{k,kom} = \dot{m}_{pal} W_d + \dot{m}_{pow}\, c_{p,pow}\Big|_0^{T_{pow}} T_{pow} - \dot{m}_{sp}\, c_{p,sp}\Big|_0^{T''_{sp}} T''_{sp} \tag{9}$$

$\dot{m}_{pal}$	- fuel mass flow rate, kg/s,
W_d	- fuel calorific value, J/kg,
$\dot{m}_{pow}$	- air mass flow rate, kg/s,
$c_{p,pow}\Big\|_0^{T_{pow}}$	- average specific heat of air, at constant pressure and at temperature from 0 to T_{pow} [°C], J/(kg·K)
T_{pow}	- temperature of air delivered to the furnace chamber, °C
$\dot{m}_{sp}$	- flue gas mass flow rate, kg/s
$c_{p,sp}\Big\|_0^{T''_{sp}}$	- average specific heat of flue gas, at constant pressure and at temperature from 0 to T''_{sp} [°C], J/(kg·K)
T''_{sp}	- flue gas temperature at the exit of the furnace chamber, °C

The temperature of flue gases at the exit of the furnace chamber, T''_{sp} is calculated from the equation:

$$T''_{sp} = \frac{T_{ad} + 273,15}{M\left(\dfrac{a_p}{Bo}\right)^{0,6} + 1} - 273,15 \tag{10}$$

Where Bo is the *Boltzmann number*, determined by the formula:

$$Bo = \frac{\dot{m}_{sp}\,\overline{c}_{p,sp}}{\sigma \psi A_k T_{ad}^3} \tag{11}$$

$\overline{c}_{p,sp}$ is the average specific heat of flue gases in J/(kg·K), at temperature ranging from T''_{sp} to T_{ad}

$$\overline{c}_{p,sp} = c_{p,sp}\Big|_{T_{sp}-273,15}^{T_{ad}-273,15} \tag{12}$$

- a_p – emissivity of the furnace chamber,
- M – parameter characterizing the area, in which the flame temperature in the chamber is highest,
- A_k – area of the walls of the boiler furnace chamber, m²
- T_{ad} – theoretical (adiabatic) combustion temperature, °C
- σ – Stefan-Boltzmann constant, $\sigma = 5{,}67 \cdot 10^{-8}$ W/(m² · K⁴)
- ψ – thermal efficiency coefficient of the screens, being the quotient of the heat flux absorbed by the screen and the incident flux.

The calculation of the heat flow rate $\dot{Q}_{k,kom}$ absorbed by the walls of the furnace chamber of the OP-210M boiler has been calculated using (9). In the function of the fuel mass flow rate $\dot{m}_{pal}$ and the air excess coefficient λ, the calorific value of the oil was assumed to be: W_d = 41060 kJ/kg.

The results of the calculations are presented in Fig. 7.

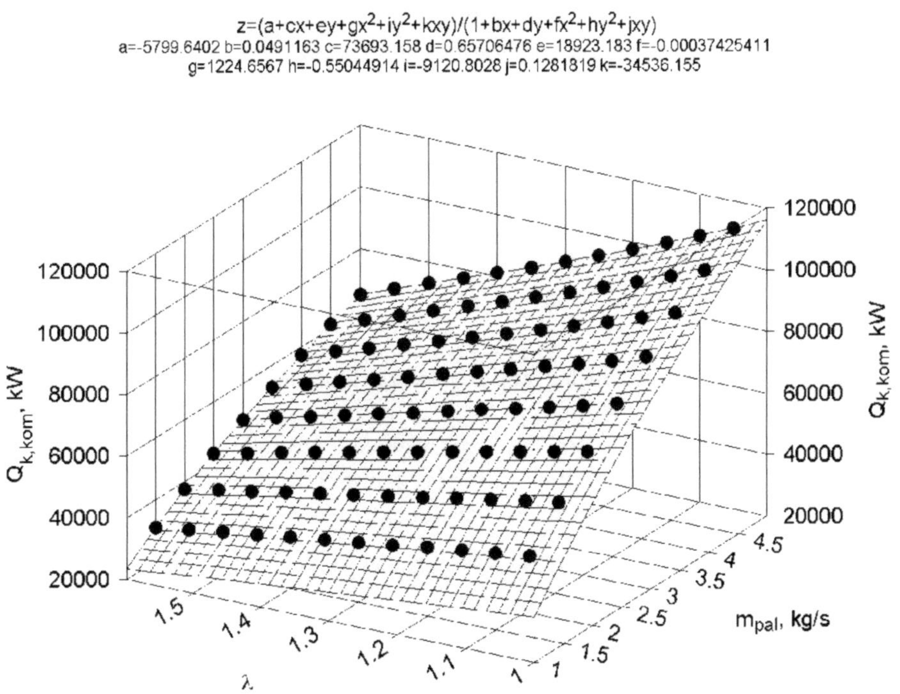

Fig. 7. Heat flow rate $\dot{Q}_{k,kom}$, in kW, delivered to the evaporator of the OP-210M boiler, as a function of the fuel mass flow rate $\dot{m}_{pal}$ in kg/s and the air excess coefficient λ.

It was proven that the increase of the air excess coefficient λ at the assumed fuel mass flow rate $\dot{m}_{pal}$ leads to the reduction of the heat flux delivered from flue gases in the furnace chamber to the evaporator. This is due to the fact that the temperature of the flue gases in the furnace chamber T_{pl} was lowered significantly, which in turn led to a significant decrease of the heat flow rate $\dot{Q}_{k,kom}$, as the heat flux is proportional to temperature difference between the flame temperature and the temperature of the chamber walls raised to the fourth power, that is $\left(T_{pl}^4 - T_{sc}^4\right)$ when λ increases $\dot{Q}_{k,kom}$ is reduced, and the heat flux absorbed by the steam over heaters increases. Increasing the fuel mass flow rate $\dot{m}_{pal}$ at the assumed air excess coefficient λ leads to the increase of the heat flow rate $\dot{Q}_{k,kom}$ absorbed by the walls of the furnace chamber.

Next, the fuel mass flow rate $\dot{m}_{pal}$ delivered to the boiler during the start-up (heating) process was determined. After determining from (8) the heat flow rate $\dot{Q}_k(t)$, the non-linear algebraic equation is solved.

$$\dot{Q}_k(t) = \dot{Q}_{k,kom}\left(\dot{m}_{pal}, \lambda\right) \tag{13}$$

$\dot{m}_p$ $\dot{m}_{pal}$ The fuel mass flow rate, at the assumed value of the air excess coefficient λ, is determined from (13). Changes of the mass flow rate in a time function, during the start-up process of the OP-210M boiler, determined from (13), for λ = 1,1 is presented in Fig. 9a and 9b. The calculations were conducted for different steam mass flows produced in the boiler evaporator. The mass of the fuel oil used for the boiler start up process from time $t=0$ to $t=t_k$ can be calculated using the formula:

$$m_{pal} = \int_0^{t_k} \dot{m}_{pal} dt \tag{14}$$

The integral (14) was calculated numerically, using the trapezium rule for integration.

5. Calculation results

The calculation has been completed for the evaporator of the OP-210M boiler, for the pressure change rate dp_n/dt presented in Fig. 3. The temperature of the incoming water T_w during the evaporator heating process is at a given moment lower by 10K than the saturation temperature Tn. The temperature run of the incoming water T_w, the saturation temperature T_n and the pressure in the evaporator p_n are presented in Fig. 8.

The calculations have been conducted for the following data:

$\dot{m}_{ods} = 0{,}51$ kg/s, $\dot{m}_w = 17{,}1$ kg/s, $\dot{m}_p = 16{,}57$ kg/s, m_m=171900 kg, c_m = 511 J/(kg·K), V' = 43,6 m^3, V'' = 15,9 m^3.

The results of the $\dot{Q}_k$ calculations, determined for various saturated steam mass flow rates, are presented in Fig. 9a and Fig. 9b.

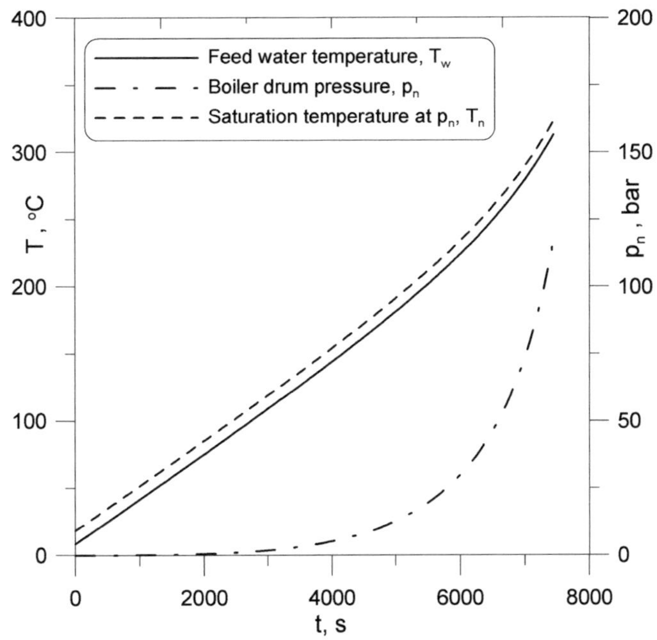

Fig. 8. Temperature run of the incoming water temperature T_w, saturation temperature T_n and pressure in the evaporator p_n during the boiler start up cycle.

After analysing the results shown in Fig. 9a we can see that if the mass capacity of the evaporator $\dot{m}_p$ equals zero, then the flow rate of the heat flow rate $\dot{Q}_k$ delivered to the evaporator shall increase. In this case, all heat delivered from the furnace chamber is used mainly for heating water contained in the boiler evaporator. This is due to the increase of the specific heat of water c_w, which increases in line with the pressure increase. For example, at pressure p_n =1 MPa water specific heat equals: c_w = 4405 J/(kg·K), and at the pressure p_n =10 MPa water specific heat equals: c_w = 6127 J/(kg·K). Moreover, at the end of the start-up process a large increase of the rate of temperature changes in the evaporator was noticed (Fig. 3), which required increased heat flow rate $\dot{Q}_k$ to be delivered to the evaporator.

In larger mass capacities $\dot{m}_p$ the heat flow rate $\dot{Q}_k$, which shall be delivered from the furnace chamber to the evaporator, increases significantly, because there is a large power demand on water evaporation (Fig. 9). For the evaporator efficiencies greater than 90 t/h (Fig. 9b), however, it has been noted that for the assumed flux $\dot{m}_p$, the reduction of the heat flow rate $\dot{Q}_k$ occurred during the end phase of the heating process, despite the increase of the rate of evaporator heating (Fig. 3). This is due to the fact that the water evaporation heat, $r = h'' - h'$ decreases in line with the pressure increase in the evaporator. For example, at the pressure p_n=1 MPa, water evaporation heat equals: r = 2014,4 J/(kg·K), and at the pressure p_n=10 MPa, the water evaporation heat is reduced to: r = 1317,7 J/(kg·K).

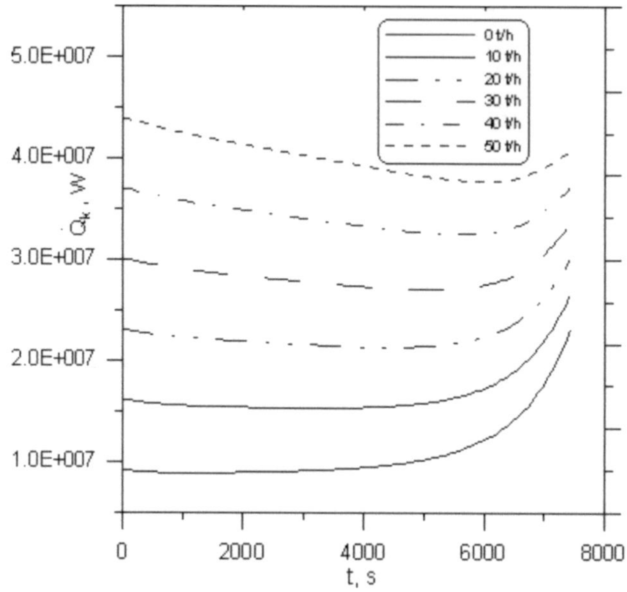

a)

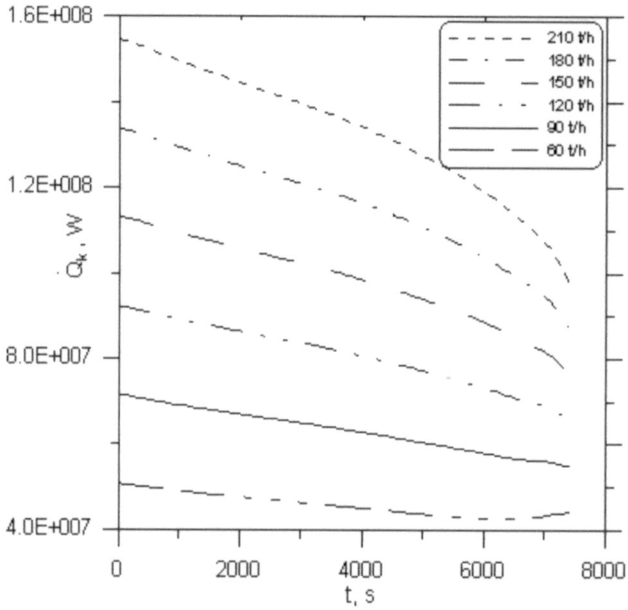

b)

Fig. 9. Heat flow rate $\dot{Q}_k$ delivered to the evaporator of the boiler in the furnace chamber for various steam fluxes $\dot{m}_p$.

Fuel mass flow rate $\dot{m}_{pal}$ during the start-up of the OP-210M boiler obtained from Eq. (13) for the excess air number λ=1,1 are shown in Figs. 10a and 10b.

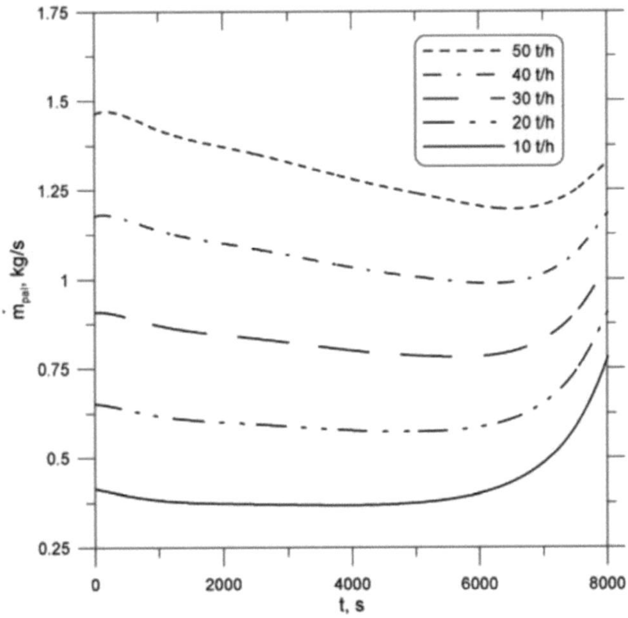

a)

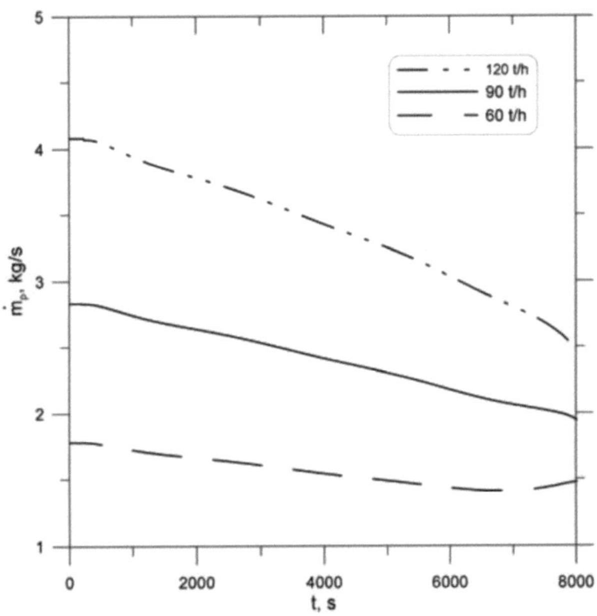

b)

Fig. 10. Changes of the fuel mass flow rate $\dot{m}_{pal}$ in kg/s, during the start-up of the OP-210M boiler, depending on the steam mass flow rate $\dot{m}_p$ in t/h, produced in the boiler evaporator.

The calculations are carried out for various steam flow rates $\dot{m}_p$ generated in the evaporator.

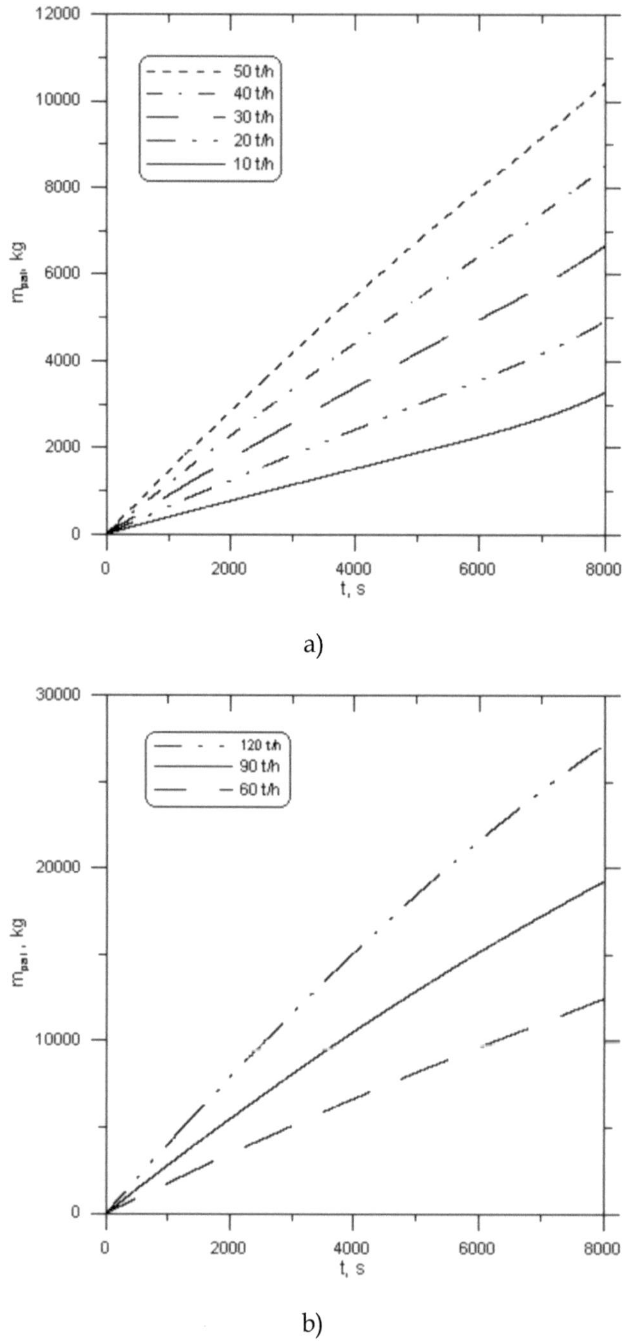

Fig. 11. Fuel consumption m_{pal}, in kg, during the start-up process of the OP-210M boiler, depending on the steam mass flow rate $\dot{m}_p$, in t/h, produced in the boiler evaporator.

The results of the calculations of m_{pal} corresponding to the runs of the fuel mass flow rates $\dot{m}_{pal}$ from Fig. 10 are presented in Fig. 11.

From the analysis of the results shown in Fig. 11 we can see that the fuel consumption depends mainly on the steam mass flow rate $\dot{m}_p$ removed from the boiler drum. During the start-up process, lasting 8000s, the fuel consumption varies from m_{pal} = 3308 kg, at the steam mass flow rate $\dot{m}_p$ = 10 t/h, to m_{pal} = 27207 kg, at the steam mass flow $\dot{m}_p$ = 120 t/h. Due to the high price of the fuel oil, the boiler start up process shall be conducted at the minimum steam mass flow rate $\dot{m}_p$, ensuring the appropriate cooling of the pipes of the over heaters.

Strength analyses show that temperature can change at the beginning of warm-up by leaps and bounds, without exceeding the allowable stress.

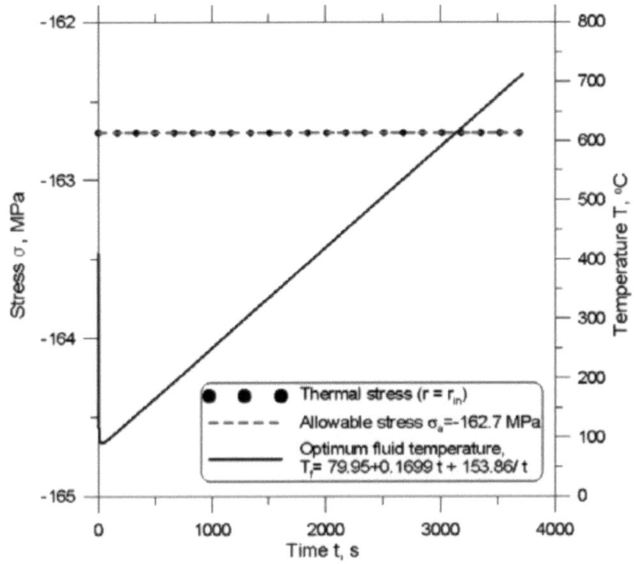

Fig. 12. The allowable process of the temperature due to the thermal stresses on the inner surface of the boiler drum.

The allowable temperature of the process due to the thermal stresses on the inner surface of the drum can be approximated by using:

$$T_n(t) = a + bt + \frac{c}{t^2} \tag{15}$$

where the constants a, b and c are determined from the condition:

$$\sigma_\varphi(r_{in}, t_i) \cong \sigma_a \quad t_i = (i-1)\Delta t \quad i = 1,2,3,\ldots,n \tag{16}$$

Condition (16) means that for optimal temperature of the process defined by formula (15) constants *a*, *b* and *c* should be chosen so that the thermal stress on the inner circumferential surface of the drum was, for n time points, equal to the allowable stress. In the case of the drum analysed in this paper, for allowable stress $\sigma_a = -162{,}7$ MPa, the following constants were obtained: a=79,95°C, b=0,1699 °C/s, c=153,86 s. Starting with an optimal start-up boiler water temperature change in the drum is very difficult, because the optimal temperature is very high at the beginning of start-up and then very quickly reduced to the minimum value.

Therefore, during start-up an approximation of optimal temperature change will be implemented by the step up temperature increase at the beginning and then heating at a constant speed. When the pressure in the drum is higher than the atmospheric pressure, after the jump of temperature, the speed of temperature change is equal v_{T1} (at the beginning) and v_{T2} (at the end of start-up). Determination of the heating or cooling rate of the drum for any pressure p is the same as in TRD 301 regulations or European Standard BS EN 12952-3.

Fig. 13a and fig. 13b show the changes in temperature and pressure of water in the drum at different initial temperature jumps.

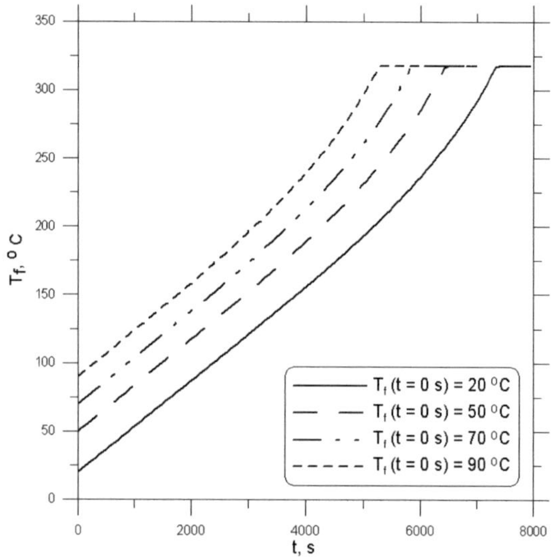

a)

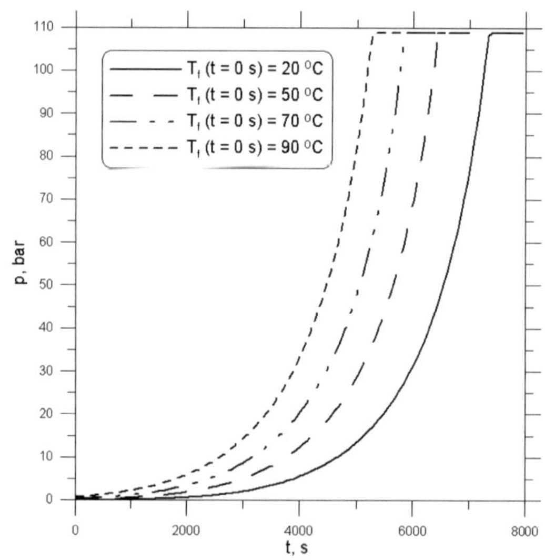

b)

Fig. 13. Temperature and pressure of the water in the drum during boiler start-up.

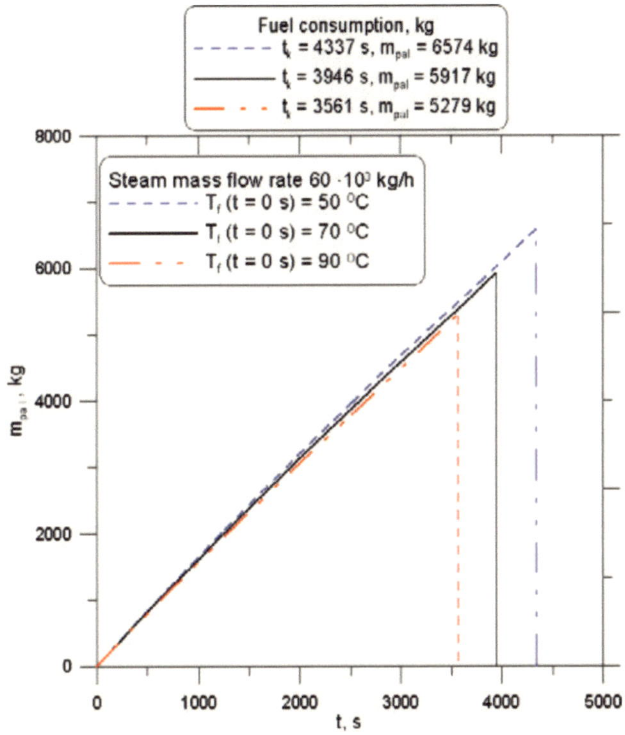

Fig. 14. Fuel consumption and boiler start-up time at different initial jumping changes in water temperature in the evaporator boiler.

An analysis of the results in Fig. 14 shows that the release of the initial jump in the temperature of the boiler start-up reduces the time and reduces fuel consumption.

6. Conclusions

The model of the dynamics of the boiler evaporator, presented in this paper, can be used for the determination of the heat flux, flowing from flue gases in the furnace chamber to the boiler evaporator, necessary to warm up the boiler at the assumed rate of the medium temperature change in the boiler evaporator. The fuel mass flow rate $\dot{m}_{pal}$, for which the heat flux delivered to the evaporator equals $\dot{Q}_k$, is determined on the basis of the calculations of the boiler furnace chamber. Fuel consumption during the boiler start up process, necessary for heating the boiler evaporator to the assumed temperature, from the assumed nominal start temperature, assuming that the heating of the boiler drum is conducted at the maximum allowable speed, due to the limitations caused by heat stresses at the boiler drum/down comers intersection, has been determined. In order to limit the consumption of the expensive fuel oil during the boiler start up process, the process shall be conducted at the minimum steam mass flow rate, ensuring the appropriate cooling of the over heaters and at the minimum allowable air coefficient. The dependencies derived in this paper can be used for the preparation of the optimal technology of the boiler start up process.

It is shown that allowing the rapid increase of water temperature at the start of commissioning of the boiler reduces the boiler start-up time and helps to reduce fuel oil consumption.

7. References

TRD 301 (2001), Zylinderschalen unter innerem Überdruck. *Technische Regeln für Dampfkessel* (TRD), pp. 143-185, Heymanns Beuth Köln - Berlin, Germany.

Staff Report, (2007). Dealing with cycling: TRD 301 and the Euro Norm compared, *Modern Power Systems*, Vol 27, No 5, p 33-38.

European Standard, EN 12952-3, *Water-tube boilers and auxiliary installations – Part 3: Design and calculation for pressure parts* , CEN - European Committee for Standardization, rue de Stassart 36, B-1050 Brussels, 25. July 2001.

Profos, P., (1962). *Die Regelung von Dampfanlagen,* Springer, Berlin, Germany.

Styrykowicz, M. A., Katkowskaja, K. J., Sierow, E. P. (1959). *Kotielnyje agregaty.* Gosenergoizdat, Moscow, Russia.

Sierow, E.P., Korolkow, B. P. (1981). *Dinamika parogienieratorow.* Energija, Moscow, Russia.

Taler, J., Dzierwa, P. (2010): A new method for optimum heating of steam boiler pressure components. *International Journal of Energy Research,* Volume 35, Issue 10, pages 897–908, Wiley, August 2011.

11

A Computer-Aided Control and Diagnosis of Distributed Drive Systems

Jerzy Świder and Mariusz Hetmańczyk
The Silesian University of Technology, Poland

1. Introduction

Modern systems of the industrial processes automation most often are composed of several basic components, among others: controllers (PLCs, regulators, industrial computers, etc.), actuators (electric, pneumatic, hydraulic and their combinations), control and measuring equipment (sensors and detectors), industrial networks and SCADA/HMI systems (Świder & Hetmańczyk, 2009c). Functional subassemblies have been divided into several groups (mechanical elements, electrical and electronic parts, control equipment and software packets). All of the enumerated components are susceptible to failures, therefore, the main problem is an estimation of the trouble-free service period. In the case of individual components the described task is easy, but an analysis of distributed applications defined as a collection of dependent subsystems the reliability function is dependent on all reliability factors assigned to all components (Dhillon, 2002).

Sequential distributed systems include most often: automated machining centres, assembly lines, conveyors etc. Specified examples (for safety reasons) involve extended monitoring procedures strictly connected with requirements of the employed maintenance model.

The preventive maintenance model includes several meaningful assumptions, among other things: risk reduction, failure avoidance, the continuous improving of the reliability of the considered system, defects elimination (with attention focused on minimization of the time needed for the detection of error sources and their elimination), maximization of a productivity combined with decreasing of the total production costs, with continuous improvement of the quality level. An elaboration of the method that fulfils all the specified requirements with enumerated factors will be the most important task for engineers derived from different domains. A utilization of more advanced devices allows for a combination of diagnostic and control functions at various levels. Furthermore, in the most of contemporary applications, an integration was omitted for many reasons (economic premises, time and work efforts, the complexity of the algorithm etc.).

2. Problems connected with the maintenance of distributed systems

The maintenance of distributed systems is a very important task in many engineering applications (Ryabinin, 1976). Industrial control systems are described by several qualities:

control system organization (centralized or distributed), application size, type of measuring equipment (sensors and detectors), technical orientation, etc.

Industrial productive systems run at a risk of impact by many external factors. Among meaningful impacts occur two main groups in the form of: command signals (purposefully dispatched by users) and disturbances. Progress at a complexity of manufacturing departments, productive systems, as well as machines, causes a continuous increase in the number of failure sources (De Silva, 2005). Nowadays the diagnosis of automated industrial productive systems is a significant issue.

Design, implementation and operation of machines without taking into consideration aspects related to the monitoring of operating conditions and planning of the duty are not possible in applications taking advantage of mechatronics devices.

The described problem is especially apparent in the case of the monitoring and control of the states of electric drives equipped with AC asynchronous motors that are subjected to extensive investigation in many research institutes. However, tests on a single drive treated as an autonomous machine constitute a predominating tendency nowadays.

Such an approach does not meet the needs of operational and maintenance engineers, as industrial conditions compel the requirements of complying with several rules, including: fast detection and removal of failures, minimization of the number of configuration procedures run on the diagnostic equipment (a universality securing the operation of numerous communication protocols and methods of access to the diagnosed object) as well as a wide range of available tools (Bloch & Geitner, 2005).

Up to this moment, devising a concept of diagnosing distributed drive systems operated by frequency converters coupled with the ProfiBus industrial network (Mackay & Wright, 2004) and a centralized PLC control has not been developed.

2.1 A structural analysis of the electric drive unit

Electric motors dominate in industrial drive systems of modern machines. The reasons for such wide spread popularity are their functional features such as: high efficiency of the individual drive unit, wide ranges of torque and speed adjustments, small dimensions, ease of a power supply, good prices regarding functional features.

The availability of a wide selection of the components belonging to the group of the automatic control equipment has made it possible to construct new technical devices that have not previously corresponded to the traditional definition of a machine (Fig. 1a). The determination of the state of the machine on the grounds of control, energy, disturbances and mass fluxes does not consider many factors to ensure a proper course of work of distributed units. The problems of control and identification of failures of units that have a different degree of technological advancement are still important tasks to be addressed.

Automation involves the necessity of providing man-free operation of the duty, for which a two-directional information flow is required (enabling both control of the parameters and change in the states).

The specification of the components of the machine block (Fig. 1b) enables a complex identification of the internal information fluxes (signals exchanged at the level of individual control of drives by means of frequency inverters) and external ones (a composite flux, corresponding to the operational state of the machine as a whole).

Each drive subsystem has features that predispose it to a group of autonomous machines. In each subsystem were extracted an electric drive, extended by a reduction gear and a control subsystem (in the form of a frequency inverter or an I/O module). Detailed relationships between functional subsystems of the drive system are shown in Figure 1c.

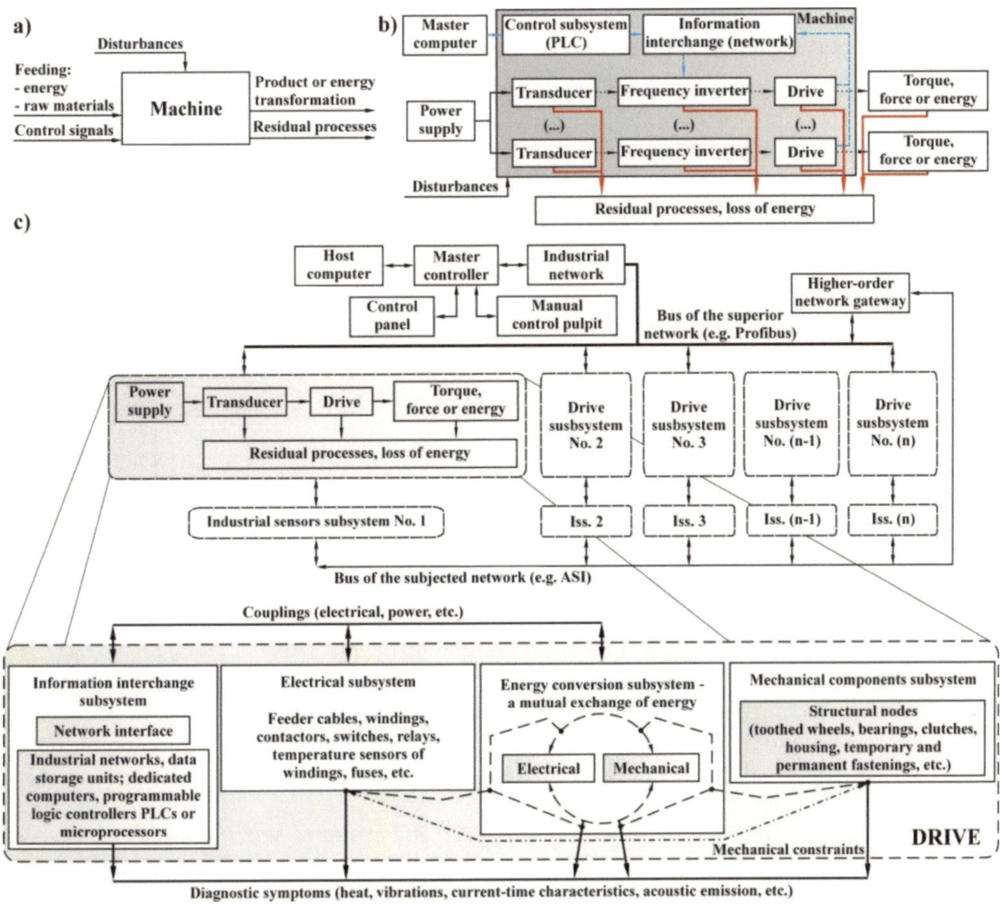

Fig. 1. Definitions of the machine: a) a traditional approach (Korbicz et al., 2002), b) the approach adjusted to the requirements of the analysed systems, c) simplified view of the drive subsystem in the form of coupled components collection

The proposed structure refers to, among others, conveyors, production lines treated as a one cell, in which the proper course of the work cycle depends on all the components.

Distributed systems with the appropriately formed information flux may be subjected to an analysis in view of the system parameters at different levels (from the values recorded by sensors, through the state of the control elements, to production data).

2.2 An overview of the most often used approaches for the monitoring of distributed systems

The supervision (Boyer, 2004) over industrial processes and the change of the parameters of the states of distributed systems are most frequently performed by three means (Fig. 2).

Portable analysers (Fig. 2a) make it possible to make measurements of the operational parameters of machines at a work place, yet, in view of the nature of their construction (specialized units enabling the determination of parameters, mostly for the one type of phenomenon), the measurements are conducted only at planned time intervals or in the case of visible symptoms of damage. An additional disadvantage of such an approach is the need to provide the service using experienced diagnostics staff.

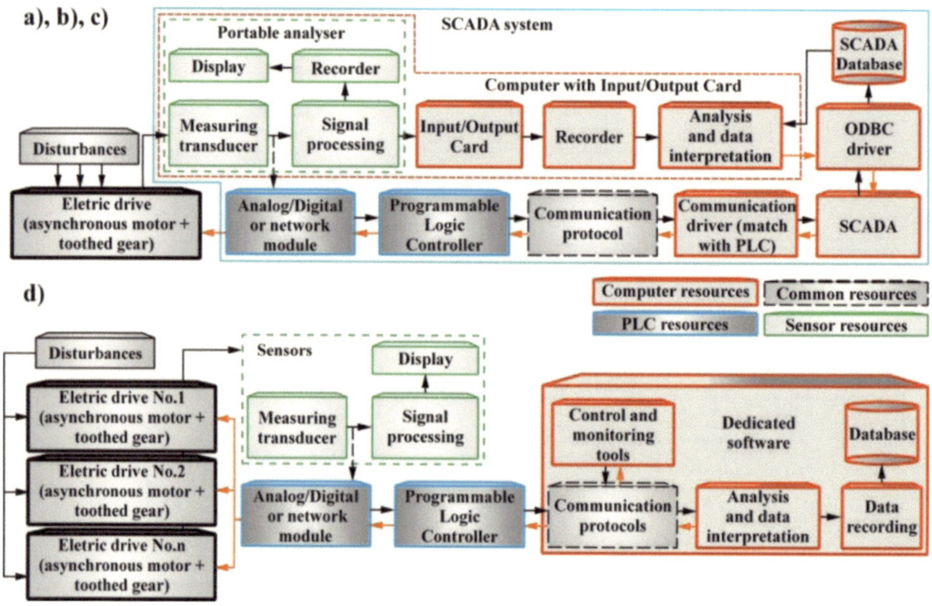

Fig. 2. Methods of controlling the states of technical systems: a) analysis of the machine under operation made by the analyser, b) use of the input/output card, c) implementation of SCADA system and data collector, d) a concept of the customized computer environment

Computer input/output cards (Fig. 2b) are used in systems with a small number of measuring or executive devices. SCADA systems (Fig. 2c) do not facilitate access to special functions of the supervised equipment and are, at the same time, relatively expensive (Boyer, 2004). Furthermore, their satisfactory recording resolution requires the configuration of an optional collector. The remaining two means have many limitations and are most frequently used as elements supporting exact diagnostic activities.

The last of the concepts (Fig. 3d) is based on the application of customized software, which combines benefits of presented supervisory and control systems, expanding simultaneously opportunities for additional features such as: dedicated diagnostic and control blocks, utilization of the ProfiBus main line as a carrier of the diagnostic information, advanced features of the ProfiBus protocol. The main advantage of the proposed concept is the ability of the implementation in any process (or machine) operated by the PLC controller.

3. Integration of diagnostic and control functions

The developed method is characterized by some degree of universality (all applications of control and diagnostic activities of electric drives supported by PLC controllers), but at the design stage the authors have adopted significant limitations related to the organization of hardware and software structures (Kwaśniewski, 1999).

Usage of programmable logic controllers and distributed drives equipped in industrial network interfaces is a common case in industrial applications. PLC's popularity results from a fact that exchanged data most often have a status of control commands (Legierski, 1998).

In the majority of cases diagnostic and control functions are separated. The traditional structure of an organization of control algorithms is oriented to the failure-free execution of the course of work, but the diagnostic information is most often omitted. The main cause of this situation arises from a substantial complexity of the control algorithm extended by diagnostic functions. Integrated diagnostic and control functions compound specified functional qualities, which combine presented domains.

A set of diagnostic and control functions creates a program for realizing self-diagnostic functions of the controlled system. A fault detection function should be prepared for every unit (Świder & Hetmańczyk, 2010). A superior level function is performed by the computer system, which consists of dedicated screens that allow interpretation of current and historical data. To primary parameters belonged, among other things current, minimum and maximum bus cycles, diagnostic information, a current work status, error codes, reactive and active powers, current status of I/Os, etc.

3.1 The model of the information flow

The proposed solution is intended for the real time analysis in the case of industrial applications that use automatically controlled equipment and are mainly operated by means of customized controllers with properly selected modules and communication standards.

A preventive maintenance of programmable logic controllers contains several simple actions. On the basis of this statement the PLC unit could be assigned to the group of the most reliable elements of automated productive systems (Prasad, 2004).

The reliability of constituent elements and a whole system depends on a fast assessment of operating conditions and prediction of failures on the grounds of the current state.

So that execution of detailed results could be undertaken, research objects were restricted to Mitsubishi PLC controllers. A functional extension can be obtained on most PLCs available in engineering applications.

In the diagram of a distributed system shown in Fig. 3a, the system has been divided into particular functional subsystems. The PLC holds a superior function as it enables the execution of the control parameters for all drives (Świder & Hetmańczyk, 2009d).

The basic assumption of the decentralized system, in the form of distributed actuators and centrally located PLC (or separated control systems connected by the communication main line, facilitating access to the data in the entire system) leads to the conclusion that the system components are connected hierarchically. An elaborated method in the form of Computer Control and Diagnostic System of Distributed Drives (CCaDSoDD) enables the control of the states of the PLCs, actuators and industrial sensors without interferences into the operation of the distributed system, and a simultaneous running of other software connected with the actuators or control units (Świder & Hetmańczyk, 2009a). The structure of the described system, based on connection relationships, conforms to an automation hierarchy (Fig. 3):

- electric drives and frequency inverters are placed in the lower level (a primary technology level),
- a Programmable Logic Controller performs the role of a supervised unit (within the algorithm defined in the PLC memory) and is placed at the group control level (Świder & Hetmańczyk, 2009b),
- the elaborated expert system is located on the highest level (the PLC and the ProfiBus DP interface mediate at the stage of data exchange between frequency inverters and the expert system),
- work of the CCaDSoDD does not interfere with an operation of the PLC controller, but it is possible to enforce certain conditions of drives (performance of active diagnostic tests).

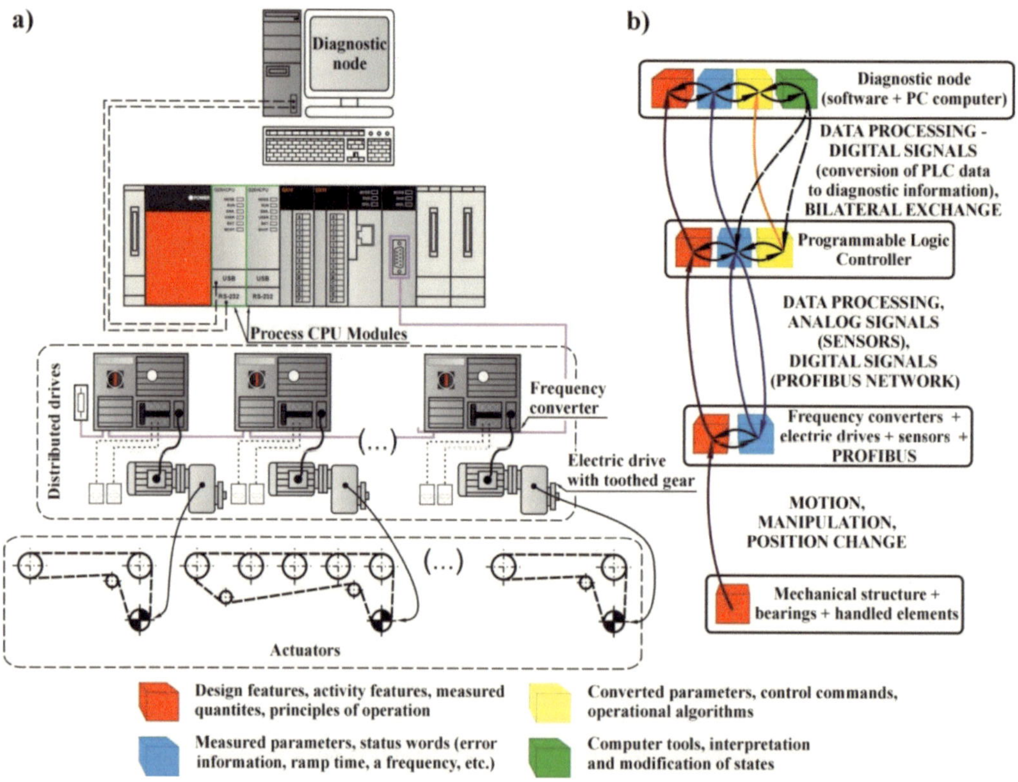

Fig. 3. Schematic diagrams of: a) the considered system, b) a data flow

Despite a precise division, a problem with an inspection of distributed mechatronic systems still exists. The maintenance of big systems requires utilization of advanced tools.

The proposed approach is based on making direct or indirect connections by means of a coupled network. The topology of the connection depends on the type of the communication network.

The choice of Mitsubishi controllers has been made in view of the ease of connection with selected logical units at each level of the defined hierarchy of the system (Fig. 4).

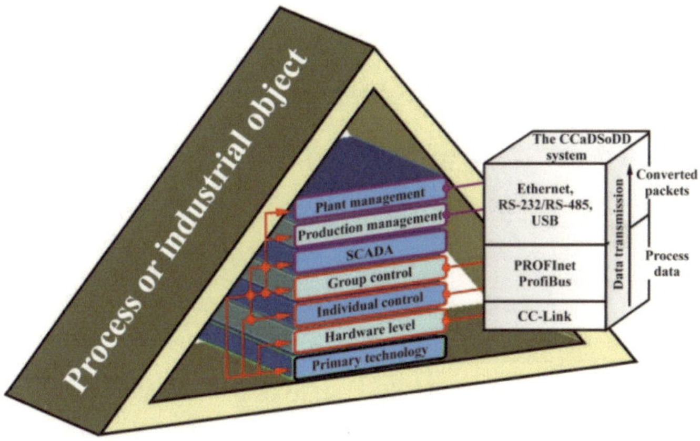

Fig. 4. A schematic diagram of dependences of the considered system taking into account the automation hierarchy

The acquisition of data from remote units is possible by means of the access via gate function (for example, the HMI panel - Fig. 5a).

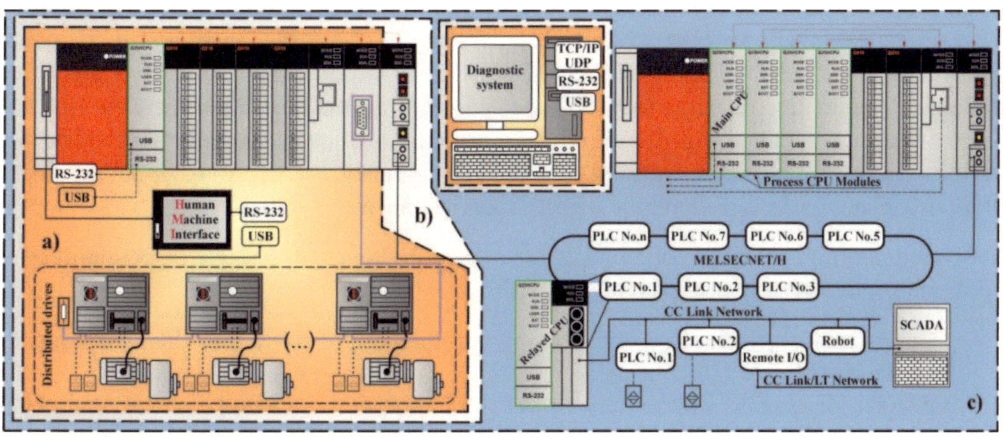

Fig. 5. Manners of access to the discussed distributed system: a) 1:1 connection (direct or by means of the operator's panel), b) and c) by means of 1:n connection

The software provided by producers must be connected directly (Fig. 6) to the drive section (frequency converter) by means of a network coupling the entire manufacturing company (the Industrial Ethernet is very popular). It is also possible to use the communication socket and the USB port of a PC, however this is still 1:1 type connection. Moreover, the contents of the data file are limited to control signals of the electric motor.

The classification of industrial networks in view of their possible applications and the configuration of the system in accordance with the assumed hierarchy of a control enables access from any level (Fig. 6). Such a solution offers many advantages, as it facilitates the configuration of the connection in a coupled or in a direct mode using any unit extended by a selected network module.

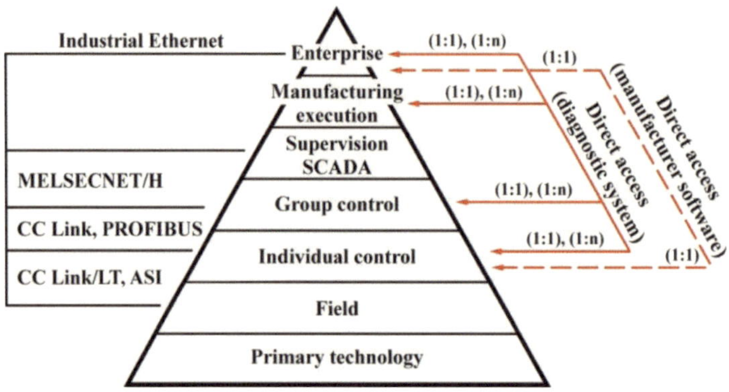

Fig. 6. The pyramid of access methods to the resources of control units

Knowing the address or the relations of the connections among the system elements, it is possible to utilize a direct access to the resource (a coupled network), both for the ascending graph trend (from lower to higher level networks) and for the descending one.

3.2 Integrated control and diagnostic blocks

In the discussed case, the communication and control of the operational parameters of each drive are performed by means of a control algorithm, saved in a PLC memory (the sequence of operating conditions). Correct course of work results from a faultless control algorithm definition. Elimination of errors leads to acquiring a faultless control algorithm, but is rather time consuming. A better approach is elimination of roots of problems and that method has been used in elaboration of integrated control and diagnostic blocks. An essential drawback of industrial networks is difficulty in direct access to transmitted data.

Integrated control and diagnostic blocks have been divided, at an angle of possibilities of applications, into several groups (Fig. 7):

- primary control blocks (for: the Movimot frequency inverter, the Movifit frequency inverter, distributed I/O modules),
- universal diagnostic blocks (of: the SLAVE unit, an interpretation of diagnostic information, the interpretation of extended diagnostic information, the interpretation of alarm information, a distribution of control data).

Basic blocks are necessary for correct system operation and are required (in case of the elementary configuration) to proper realization of the control process, however, maintenance of diagnostic functions is unfeasible. Achievement of the complete functionality requires utilization of optional blocks. Error elimination, based on internal algorithms, permits a programmer to abandon of the PLC memory allocation and algorithm testing stages (Świder & Hetmańczyk, 2010, 2009c).

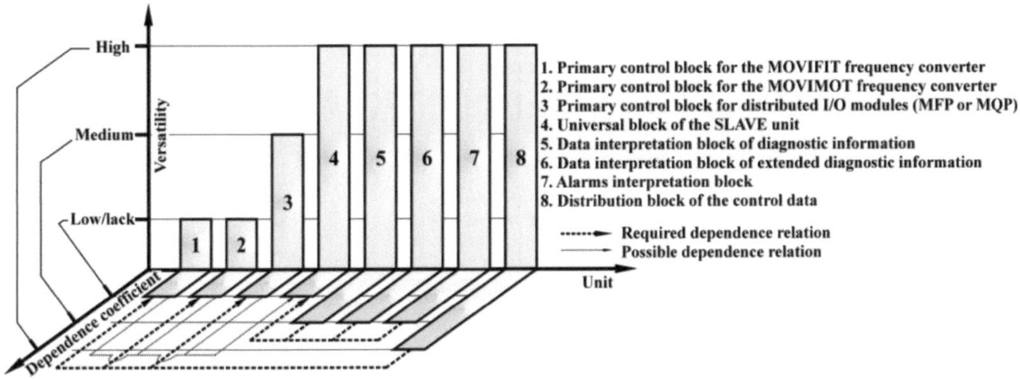

Fig. 7. A schematic representation of relationships between types of blocks, degrees of universality and the signal flow dependency

Moreover the algorithm structure for each SLAVE unit is unequivocally defined, without possibilities of mistakes. Each presented block has different tasks in the control algorithm.

An elaboration of integrated control and diagnostic blocks has been based on characteristic features of Programmable Logic Controllers. Principal assumptions of the presented method can be characterized in several points:

- each of the SLAVE unit is identified by an unique ordinal number (dependent on the ProfiBus network address),
- for proper system operation, the configuration of minimum three primary control blocks is necessary,
- additional functions (extended diagnostic functions, alarm control, etc.) demand an extension at the level of diagnostic blocks,
- exists the possibility of programming without the knowledge about a network configuration and a system structure,
- presented universal blocks have been characterized by general-purpose internal structure (implementation to control functions of all SLAVE devices with ProfiBus interfaces),
- control blocks have been elaborated at an angle of individual units.

The presented scheme of the signal flow and blocks arrangement is universal for all units (not only for the presented devices). Accommodation for an arbitrary configuration requires an implementation of the control block dedicated to the concrete unit.

Up to now, the method of traditional programming with algorithms acquiring the faultless data exchange between PLCs networks modules and SLAVE units requires:

- a continuous inspection of addresses defined in options of a network module and addresses assigned by a programmer,
- multiple testing connected with the task of a proper memory allocation,
- a definition of the amount of input and output words of each SLAVE unit (most often on the basis of manuals),
- realization of time consuming additional operations (in the case of requirements concerning diagnostic information).

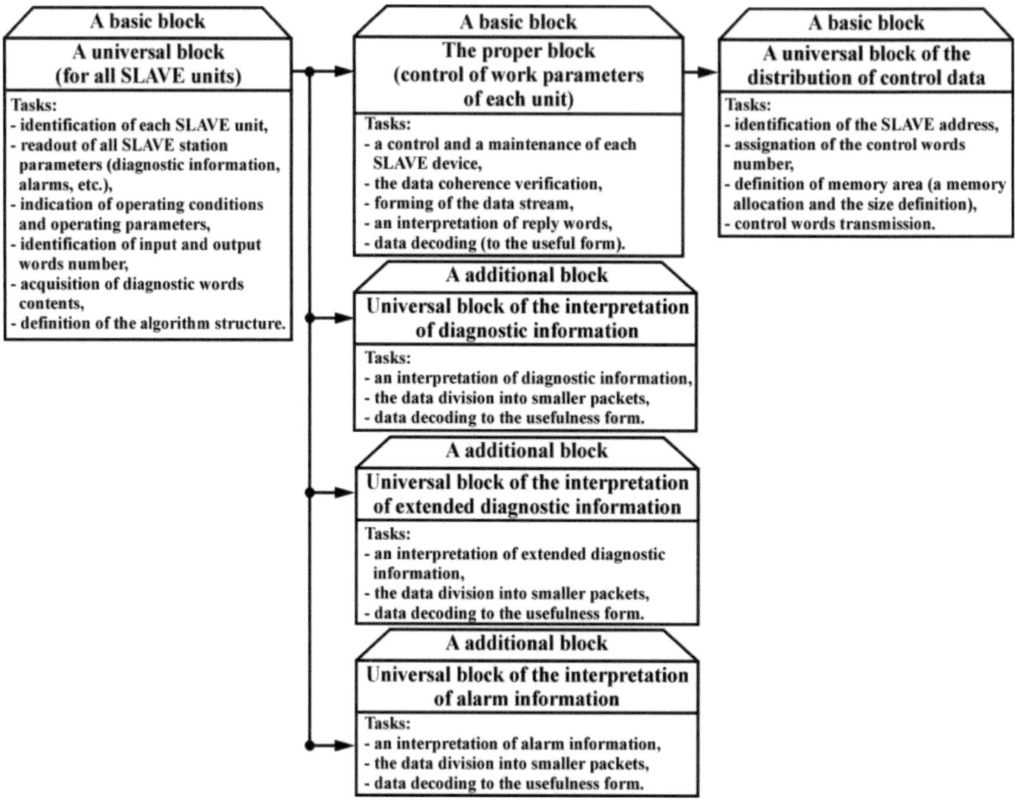

Fig. 8. A scheme of the signal flow and the blocks configuration for the single Slave unit

The proposed solution is based on the memory stack (Fig. 9) which is the primary principle of operation of all devices that convert digital data. The main problem arises from difficulties with an identification of registers which store corresponding data.

The first step is checking of statistical data of a concerned system, at an angle of number of devices, number of input (control) words and output words (diagnostic words). On the grounds of those information a programmer is able to identify the type of the SLAVE unit, without a knowledge of the functional structure of a distributed system. All described functions are performed by the universal block (Fig. 8).

The main difference between a traditional programming method and the proposed solution might be defined as a direct data exchange between the PLC processor and a special network module.

An internal algorithm of the universal block is presented in the Figure 11. The described program block belongs to the group of main elements of the control algorithm. Basic features of this block are presented in Figure 8, but the most important task is the dispatching of reply words from the SLAVE unit. A contrary task characterizes a distribution block of the control data. The block located at the centre (Fig. 8) is dependent on two described blocks and must be matched to the individual features of each SLAVE unit.

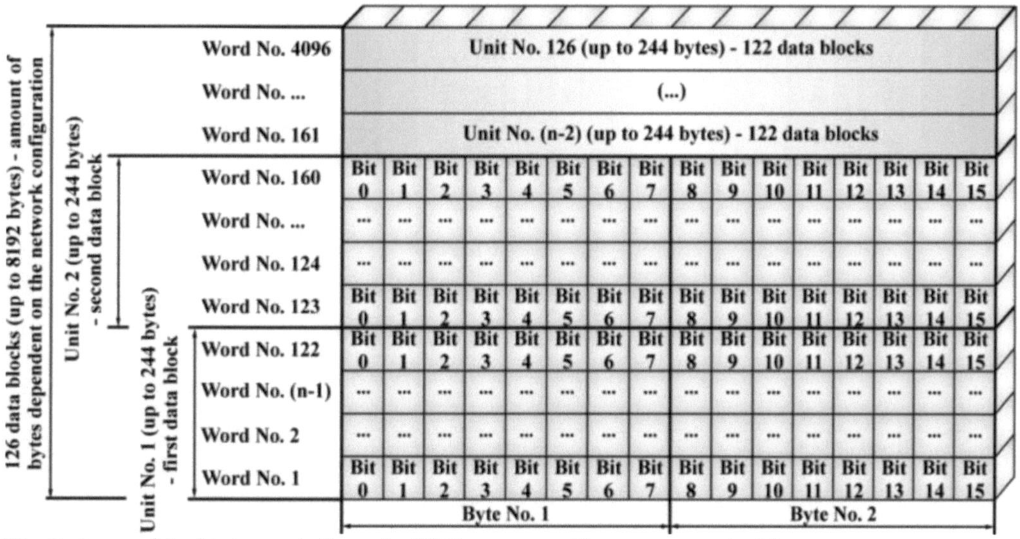

Fig. 9. A graphical interpretation of a PLC memory (the memory stack)

An internal algorithm formulates contents of control words dispatched to a block of distribution of the control data. Due to required organization of each control and diagnostic blocks, the set must be arranged in accordance with the presented scheme.

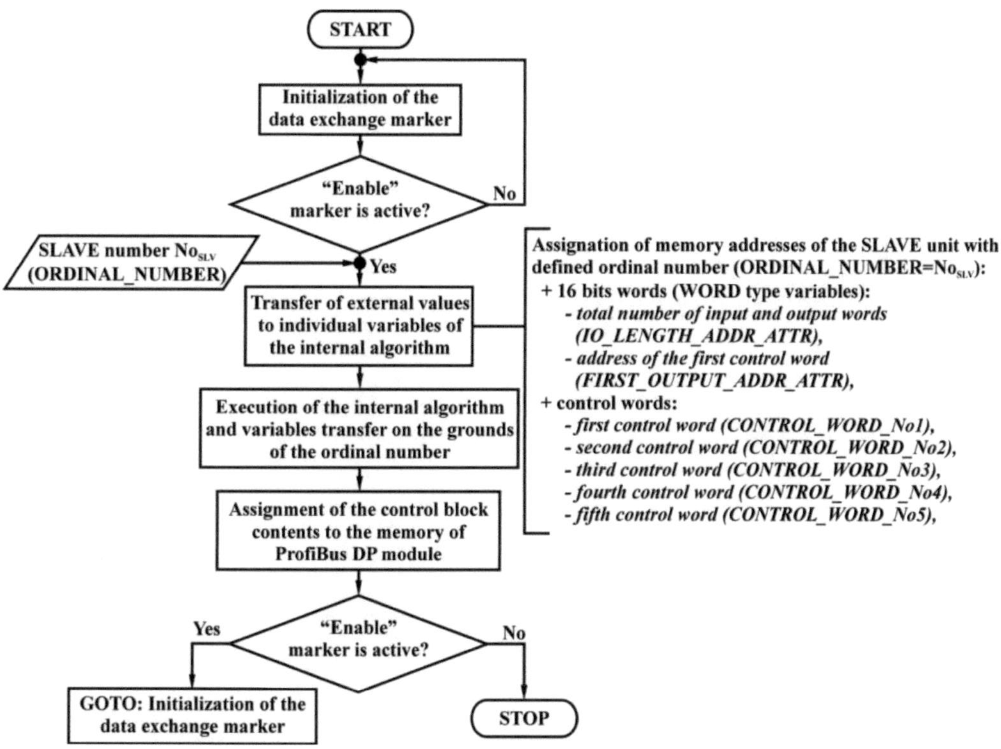

Fig. 10. A block diagram of the block of a distribution of the control data

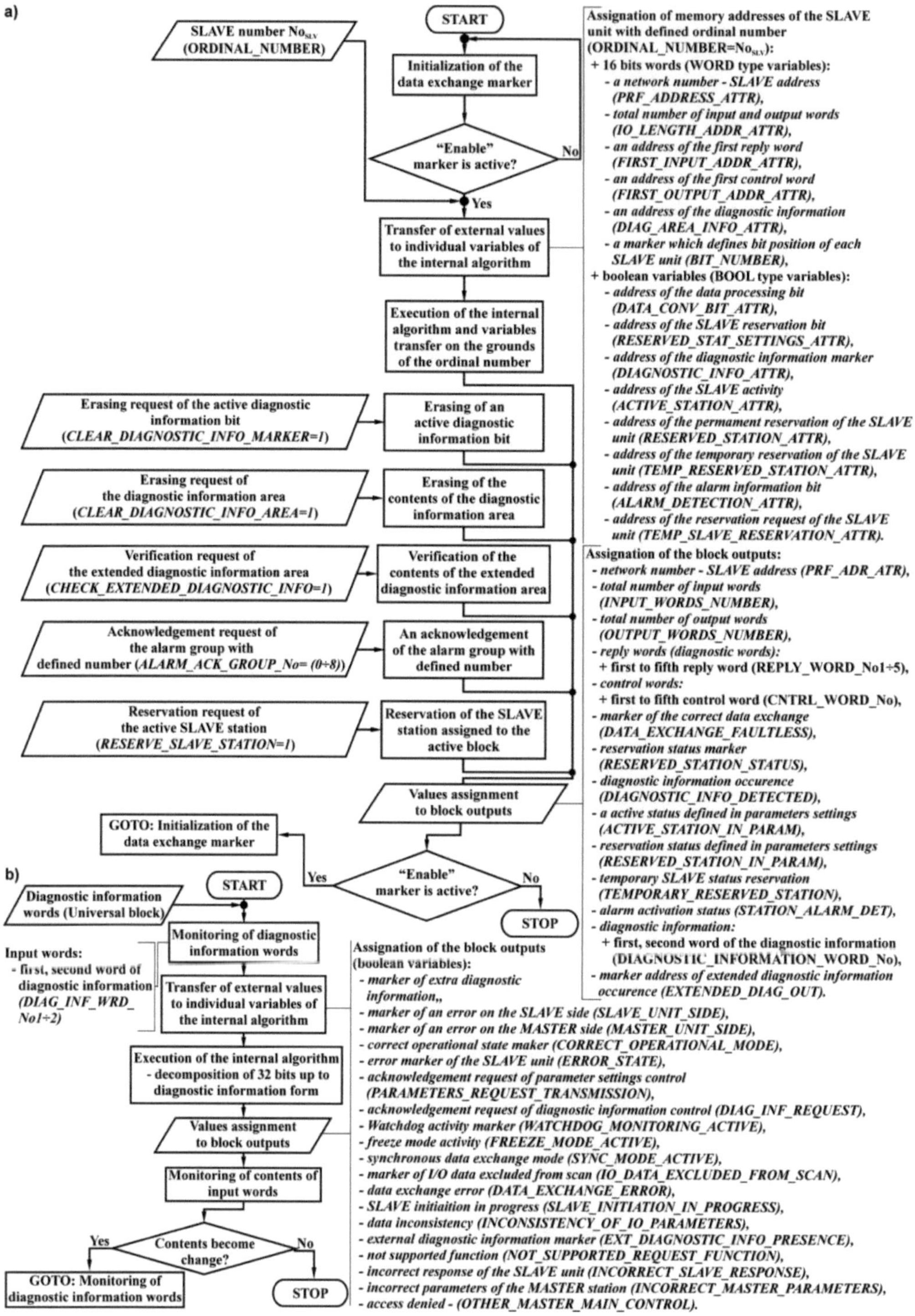

Fig. 11. Block diagrams of: a) the main universal block, b) the data interpretation block of the diagnostic information

The principle of an operation of other blocks is similar to the data interpretation block of diagnostic information and its description has been neglected.

4. Embedded functions of identification and interpretation of errors

4.1 The binary states vector method – a manual definition of failure states

A common malfunction of systems with implemented industrial networks is a loss of the communication. In consequence, often parameter reading of the inverters (or I/O modules) becomes impossible, resulting mostly from improper configuration, electromagnetic interferences, interruptions or damages in bus structures, etc.

The CCaDSoDD system allows a definition of the binary diagnostic vector (Świder & Hetmańczyk, 2011). The vector is defined on the basis of damage or failure symptoms indicated by external elements of control devices (status diodes placed on front plates).

In the case of integrated frequency inverter of the MOVIFIT unit, contents of the state's vector are defined according to the following formula:

$$W_i=[\ S_t\ \ T_r\ \ K_{zi}\ \ K_{zl}\ \ K_c\ \ P_r\ \ S_{rp}\ \ P_{dk}\ \ P_{zp}\ \ K_r2x\ \ K_r3x\ \ K_r4x\ \ K_r5x\ \ K_r6x\ ] \qquad (1)$$

where:

i - the unique ordinal number of the diagnosed unit,
S_t - the current state of a status diode (0 - passive, 1 - active),
T_r - a work mode (0 - a continuous signal, 1 - the pulse signal),
K_{zi}, K_{zl}, K_c - colours (accordingly: green, yellow or red; marker equals logical true corresponds to the current colour),
P_r, S_{rp}, P_{dk}, P_{zp} - accordingly: a uniform pulsation, the fast uniform pulsation, the pulsation in two colours, the pulsation with a pause,
K_r - times of pulsation (3, 4, 5, 6-times)

Each failure has been assigned to the unequivocally defined diagnostic vector, which is the pattern used for comparison of functions that define errors.

In the case of determination of bus work parameters, the described vector has been divided into components, with the aim of easier identification of damaged states:

$$W_i=[\ [W_{RUN}]\ [W_{BUS}]\ \ [W_{SYS}]\ \ ] \qquad (2)$$

where:

W_{RUN} - a six-elements vector of the hardware components diagnosis,
W_{BUS} - the three-elements vector of the communication status diagnosis,
W_{SYS} - the five-elements vector of the configuration diagnosis, a data exchange management program, etc.

Table 1 shows forms of vectors constituting the main vector. The result of the submission is the fourteen-elements diagnostic vector, whose components were determined as:

- S_{tA}, S_{tP}, S_{tpls} - current states, accordingly: active, passive or pulse,
- K_{zi}, K_{zl}, K_c - as previously.

Vector denotation	Contents
W_{RUN}	W_{RUN}=[S_{tA} K_{zi} K_{zl} K_c S_{tp} S_{tpls}]
W_{BUS}	W_{BUS}=[S_{tA} S_{tp} S_{tpls}]
W_{SYS}	W_{RUN}=[S_{tA} S_{tp} S_{tpls} K_c K_{zl}]

Table 1. Vectors corresponded to isolated components

The encoded state's vector is unreadable for the end user. The transformation of the vector content W_i into a type of damage U_n is realized through the use of rules of an inference in the form of a simple conjunction, for example, the first vector:

$$\text{If } (S_t=s_j) \cap (T_r=s_j) \cap (K_{zi}=s_j) \cap \ldots \cap (K_r6x=s_j) \text{ Then} \cap U_n) \tag{3}$$

where:

s_j - a current state (0 or 1, j=2),
U_n- n-th damage (defined as a unique tag of a data record).

An error description is identified on the basis of the unique damage marker Un (syntax consistent with the equation 3). The selection function of corrective actions is defined in the following form:

$$\text{If } ZU=U_n \text{ Then } [Z_{AKR}= k[z]] \tag{4}$$

where:

ZU - a damage marker,
Z_{AKR} - the z-th element of the corrective actions vector.

The correct definition of the vector can lead to two situations: determined as:

- the corrective actions means the vector is strictly assigned to one type of damage,
- a given state vector corresponds to n failures.

In the second case the selection function of corrective actions takes the form of a logical alternative and its form of notation can be presented in the following form:

$$\text{If } ZU=U_n \text{ Then } [Z_{AKR1}= k[z] \text{ or } Z_{AKR2}= k[s] \text{ or } \ldots \text{ or } Z_{AKRm}= k[p]] \tag{5}$$

A size of corrective action vectors is variable and depends on the type of error, impact degree at the diagnosed system and number of steps required in order to restore the correct status.

It is worth mentioning that the presented system is complete (contains all possible combinations of fault states of frequency inverters). An implementation of the method was performed using a high order language. Figure 12 shows a screen of Computer Control and Diagnostic System of Distributed Drives used to determine the status of the integrated frequency inverter of a MOVIFIT unit. Representation of the diagnostic vector is realized by a state's definition in subsequent dialog windows (Fig. 13).

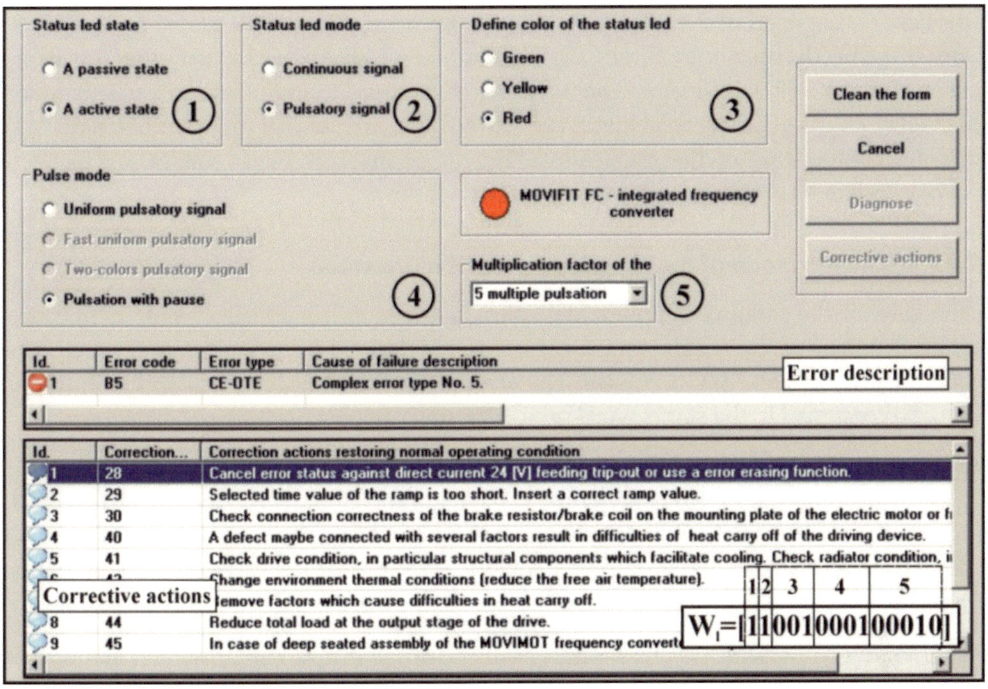

Fig. 12. A view of the diagnosis window of an integrated frequency converter of the MOVIFIT unit

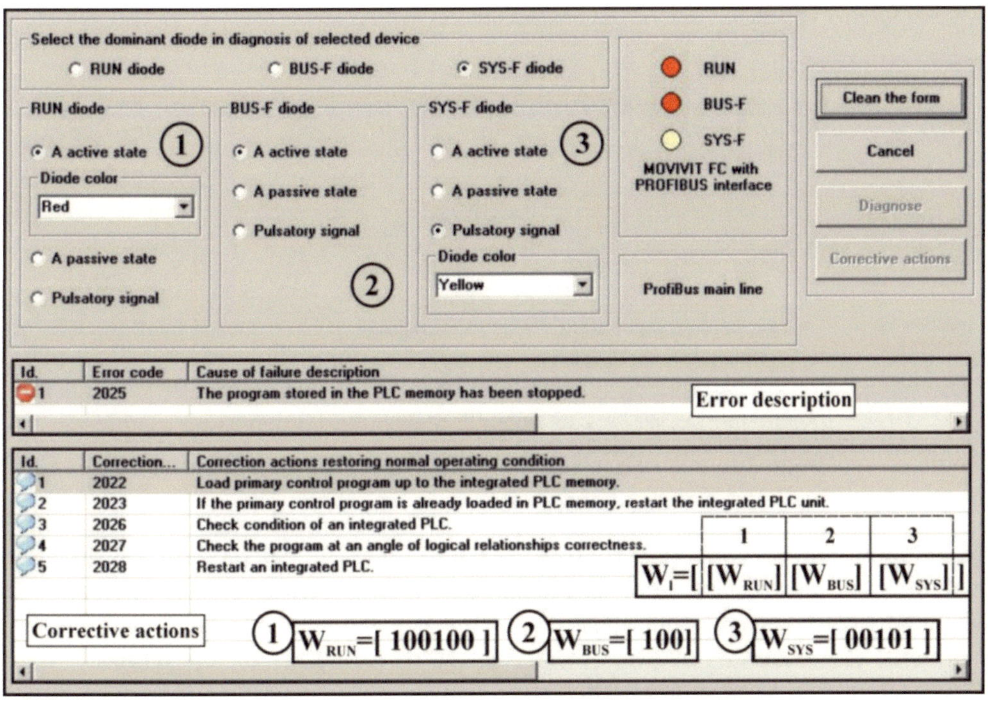

Fig. 13. A view of the MOVIFIT communication bus diagnosis window

In the case of diagnosis of the MOVIFIT bus, the exemplary diagnostic vector is presented in Figure 13. The division into three sub-vectors were introduced by defining the weight marker of diagnosed subsystems. The weight is introduced to define the subsystem, among which exist an irregularity. Selecting a tag "RUN LED" will search for solutions among the hardware components of the application. The other markers work similarly. The whole advisory system has been elaborated in fourteen windows.

4.2 An automatic mode of the definition of the failure states

In the case of the automatic failure detection, a phase of diagnostic vector definition is omitted. Content of the failure vector U_n is directly defined by using internal functions of frequency inverters and I/O modules. The error is identified on the basis of the individual 8-bit code contained in the status word of SLAVE units.

The primary disadvantage of this approach is dependence on the correct condition of the PLC controller and the ProfiBus network. The communication disturbance or incorrect PLC operational status is indicated by the diagnosing system of the PLC controller. The structure of the CCaDSoDD is hierarchical; that allows a reciprocal supplementation of defined subsystems.

It is necessary to present the main assumption, which confirms the presented stand of the authors, in the following form: entrusting to the advanced mechatronics devices of the control functions (which require a high level of reliability) allows functionality extension at additional functions and data connected with diagnostic functions without loss of the accepted quality factor.

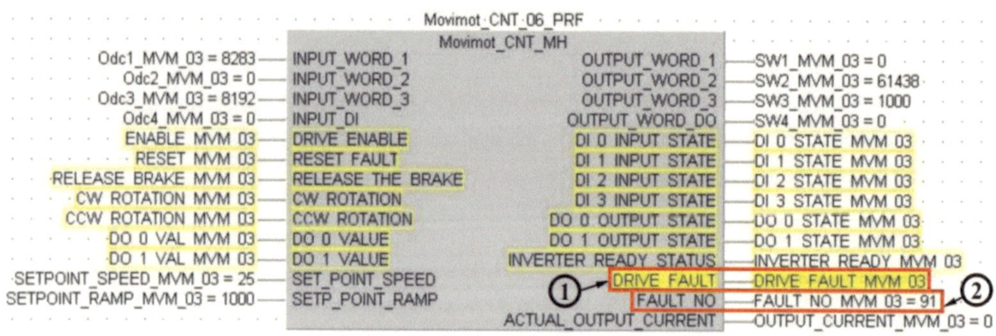

Fig. 14. A general view of the MOVIMOT functional block, dedicated to the fault identification of frequency inverter, where: 1 – a fault marker, 2 – a number of the identified error

The group of frequency inverters coupled by the ProfiBus DP network are controlled and monitored via integrated control and diagnostic blocks (placed in PLC memory) and elaborated expert system, at an angle of functional features of integrated self-diagnostic algorithms of frequency inverters (acquiring error codes) which are utilized in the CCaDSoDD system (a set of implication rules that converts numerical codes to text strings with user directions).

4.3 Self-diagnostic functions of frequency inverters

The presented example of frequency inverters allows the identification of a domain which could be used for computer-aided failures detection and their elimination. It is worth stressing that modern mechatronic devices offer very reliable and expanded functions connected with monitoring of internal subsystems.

Utilization of inner self-diagnostic algorithms coverage by specialized electronic circuits, integrated inside mechatronic devices, brings many benefits, among others:

- a separation of system functions (discharging of resources),
- shortening of time assigned to detection and elimination of identified failures,
- increasing application reliability.

A result of self-diagnostic functions is given in the form of individual error codes. Error codes contain 8-bits and are easy to gather by the PLC connected with the frequency inverters by the ProfiBus DP network. Additionally, the PLC performs several functions at the same time:

- processing of the control algorithm determined by dependences defined in accordance with the cyclogram,
- gathering the diagnostic data from frequency inverters circuits,
- checking current parameters of additional elements (like sensors, electric drives, frequency inverters, etc.) within the defined dependences (relations between states of constituent elements) determining the current state.

Ordinal number	Assigned error set	Description of sources
1.	AE-PTE	• incorrect configuration of hardware resources, • causes generated in the stage of loading of programs, settings or parameters, • an overflow of a memory stack, • disturbances of the beginning or an execution of program actions, • abnormalities of data conversion, settings or an active mode change.
2.	EE-HTE	• the ProfiBus DP bus failures, • incorrect matching of the hardware layer, • an abnormal work of hardware devices (reading errors, incorrect hardware configurations, etc.), • diagnostic information, • failures of hardware components like electronic circuits, feeding interfaces, • abnormal conditions of supervised units (i.e. decay of feeding voltage, phases, etc.).
3.	CE-OTE	• complex relationships between identified failures belonged to the AE-PTE and EE-HTE collections

Table 2. Types of identified failure states with description of their sources

All identified failure conditions were grouped by the authors into three main domains:

- AE-PTE - application errors (connected with programming activities, configuration parameters, etc.),
- EE-HTE - hardware errors,
- CE-OTE - complex errors (coupled by a conjunction of previously specified groups).

The end user has not got the possibility of any influence over the form and size of the error set notified by the internal resources of frequency inverters, because access to the embedded circuits and functions in many cases is blocked by several methods (lack of sufficiently documentation, resource interlocking and simplification of accessible functions).

A complete exploitation of advantages of AE-PTE and EE-THE collections enables continuous monitoring of values stored in variables included in the control algorithm.

A designer of the control algorithm decides how to use codes and how to define mutual dependences which define complex errors CE-OTE. Error codes are collected and gathered in the PLC memory. On the basis of their current values and conditions of other variables, a user realizes the diagnostic inference. Each of the frequency inverter error codes is treated like an independent diagnostic premise.

An identification of complex errors is performed by a comparison of two (or more) simultaneously identified errors belonging to different groups within the defined relations placed in the statements of the expert system of the CCaDSoDD application. Data gathered in the CCaDSoDD expert system are transmitted in on a package in every cycle of the PLC controller.

The identified error code is sent to a PLC controller in the form of one byte, transformed up to the hexadecimal number. For the considered groups of frequency inverters produced by the SEW Eurodrive group, the authors isolated more than 70 failures identified only by embedded self-diagnostic functions.

Taking into consideration functional constraints of considered devices, the final results are satisfactory, because a defined number describes attributes referred to the hardware subsystems (a group of diagnostic premises identified directly). The identification of residual conditions could be realized as a result of an elaboration of decision rules matched to the concrete application of distributed drives. A diagnostic inference is possible in two independent modes:

- directly on the basis of premises delivered in the form of frequency inverters error codes,
- indirectly by connection of frequency inverters error codes and additional statements implemented in the expert system.

Isolated errors (used in the direct diagnostic inference) expanded by relations connected with monitored values of the control algorithm allow defining a group of corrective actions with a quantity exceeding 270. The accepted inference method is based on dynamic checking of a diagnosed subsystem and a failures localization model at an assumption supported by the serial inference algorithm.

The method of identification and definition of reciprocal dependences of complex errors were defined in accordance with the following assumptions:

- a complex error is implicated by two (or more) errors belonging to the set of AE-PTE or EE-HTE errors coupled with a conjunction condition (definition of used implications shows Figure 15),
- in the case of the occurrence of two or more errors from the same set (AE-PTE or EE-HTE), errors are treated as individual diagnostic premises (adequately $u_{AE\text{-}PTE}$ and $u_{EE\text{-}HTE}$); occurrence of many errors in the domain of one collection do not implicate a complex error (identified diagnostic premises take part in the inference process in accordance with defined rules in the CCaDSoDD system),
- each of the errors (AE-PTE or EE-HTE) can belong to many rules implicating complex errors.

The mediate form of the inference has a lot of advantages, among others:

- a possibility of an unbounded forming of statements, fitted to the concrete applications,
- shortening of time connected with an elaboration of expert systems,
- an open structure and feasibility of development.

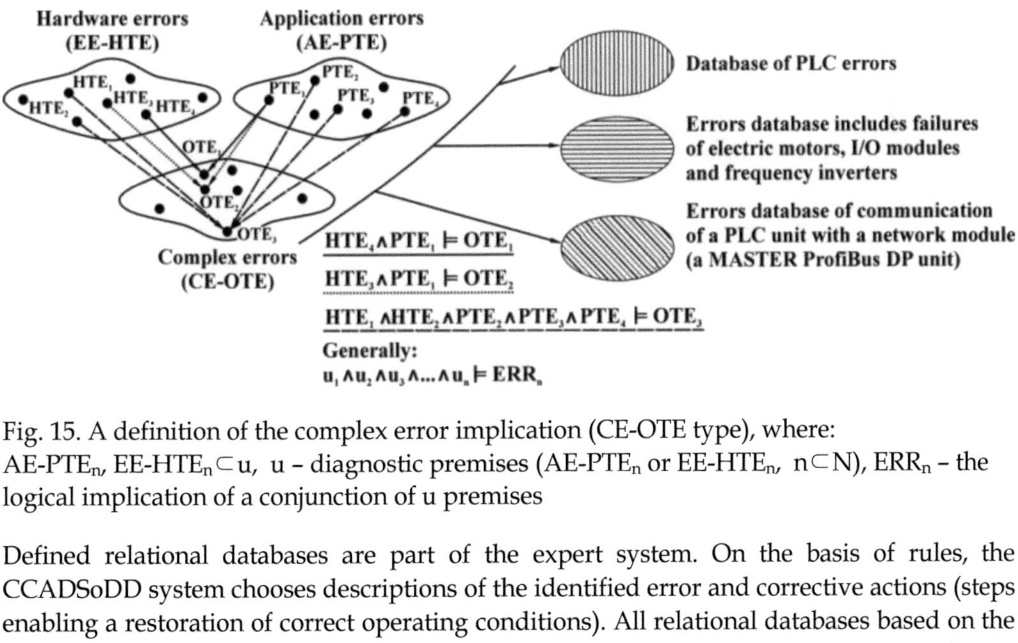

Fig. 15. A definition of the complex error implication (CE-OTE type), where: $AE\text{-}PTE_n$, $EE\text{-}HTE_n \subset u$, u – diagnostic premises ($AE\text{-}PTE_n$ or $EE\text{-}HTE_n$, $n \subset N$), ERR_n – the logical implication of a conjunction of u premises

Defined relational databases are part of the expert system. On the basis of rules, the CCADSoDD system chooses descriptions of the identified error and corrective actions (steps enabling a restoration of correct operating conditions). All relational databases based on the SQL language (Structured Query Language) were divided into tables (the main key is equivalent to the error code).

At the stage of an error identification, the authors accepted a few meaningful assumptions:

- the monitoring and collection of all data are realized within PLC tasks,
- maximal time of a symptom generation is equal to the PLC cycle, increased at the time connected with ProfiBus DP bus refreshing, within the equation (Korbicz et al., 2002):

$$\theta_j = \max_{k:f_k \in F(s_j)} \left\{ \theta_{jk}^2 \right\} \tag{6}$$

where:

$F(s_j)$ - a failures set detected by the diagnostic signal s_j,

θ_{jk}^{2} - maximal time from the moment of occurrence of the k failure to generation of the j symptom.

- a set of diagnostic tests does not include elements corresponding to symptoms with time occurrence exceeding the time predicted to a diagnosis period,
- protective functions (safety actions) of the analysed case are executed in the time equal to the value of one PLC cycle,
- dynamic properties of the diagnostic symptoms formation is dependent on the PLC cycle,
- in the elaborated expert system, the authors do not improve a states detection on the basis of the time and detection sequence of diagnostic symptoms (a packet block transfer).

Implication rules are referred to all groups of diagnosed units and databases of the CCaDSoDD system. Recording of error values are stored within a defined standard (Fig. 16).

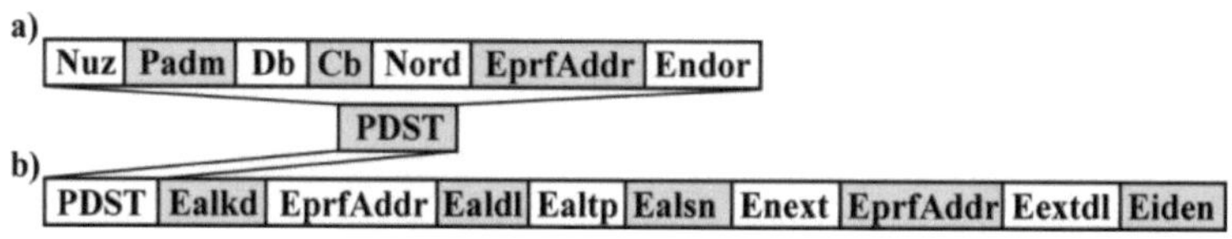

Fig. 16. A view of entry error patterns of the frequency inverter: a) basic, b) extended, where: Nuz - a user name, Padm - a authorization level, Db - date, Cb - current time, Nord - the ordinal number of the SLAVE unit, EprfAddr - network address error, Endor - ordinal number error of the SLAVE unit, Ealkd - alarm code, Ealdl - length of the alarm information, Ealtp- alarm type, Ealsn - socket number with an active error state, Enext - extended diagnostic information code, Eextdl - data length of the extended diagnostic information, Eiden - identification number of SLAVE unit

The pattern of error entries has been divided into two main subgroups:

- basic - main errors of the SLAVE (frequency inverter) unit,
- extended - includes all data of the basic pattern widen of diagnostic information of the ProfiBus DP network and hardware devices.

5. An investigation of the impact of volume and syntax of the algorithm at the performance of the PLC controller

From the standpoint of the operation of the control unit, a main parameter is refresh time values stored in registers of the control device. The assumption of the entrusting of control and diagnostic tasks to one device requires a high reliability factor of the data processing element (in this case the PLC controller).

During the study, the influence of the complexity of the integrated control and diagnostic algorithm at processing parameters of the PLC was specified (Świder & Hetmańczyk, 2009d). Table 3 summarizes the configuration parameters and their numerical values, adapted to the verification needs of the PLC algorithm.

Networks configuration parameters	Configuration	Physical unit
Data transmission type	Digital differential signal compatible with RS485 standard, NRZ coding type	-
Data rate	variable	kbits
Data protection type	HD=4, parity bit, start/end transmission separator	-
Data carrier	Two-wire twisted shielded pair (A type)	-
Power supply type	With additional cable	-
Network topology	Main line	-
Bus length	12	[m]
Number of stations	9 (MASTER + 8 x SLAVE)	-

Table 3. Basic configuration parameters of the tested network

Refresh time values of the Profibus DP bus were recorded on the basis of parameters recorded by the MASTER QJ71PB92V_ContDiagBLK dedicated block. A set of variable parameters of realized verification contains the data transfer rate of the ProfiBus DP bus and organization of the syntax of the control and diagnostic algorithm.

Time values of maximum and minimum cycles were subjected to an exceptional reading, while the current time is calculated as the arithmetic mean of samples, obtained within 10-minutes cycles of observation (recorded with a sampling resolution equal to the PLC cycle, labelled as TC_PLC).

The presented graphs also show values of the PLC cycle time. It should be noted that the numerical value of the first cycle of the PLC controller is not known. The only value registered during the first cycle is the maximum refresh time value of the bus network, given in the second cycle of the controller.

The first series of measurements were performed using the following assumptions:

- program blocks were placed in the main program,
- control blocks of drives are supported from the calling subroutine placed in the main program,
- additional blocks of a control algorithm (determining a course of work of controlled objects) were positioned inside individually called subprograms.

Figure 17 shows measured values of PLC refresh times, with a characteristic of work specified in the form of assumptions defined with respect to the first measurement cycle and configuration of slave units shown in Table 3.

The second of the considered cases assumed placement of all blocks in the main program window (Fig. 18). This configuration of the algorithm affects the growth of the recorded maximal value of the refresh time of the ProfiBus network bus. The identified phenomenon is related to the need for processing a larger number of variables, their support and assignment of parameters contained in the main program.

SLAVE/MASTER unit type	Category	Network address	Unit configuration
QJ71PB92V	M	0	-
-	NU	1÷2	-
MFP/MOVIMOT	S	3	[000] 3PD+DI/DO (MFP2x)
MFP	S	4	[000] 0PD+DI/DO (MFP2x/3x)
MFP/MOVIMOT	S	5	[000] 3PD+DI (MFP2x/3x)
MFP	S	6	[000] 0PD+DI (MFP2x/3x)
MFP/MOVIMOT	S	7	[000] 3PD+DI/DO (MFP2x)
MFP	S	8	[000] 0PD+DI/DO (MFP2x/3x)
MFP/MOVIMOT	S	9	[000] 3PD+DI (MFP2x/3x)
MOVIFIT	S	10	[000] Slot unused [001] MOVIFIT Status [002] FC 3PD [003] Slot unused [004] FC/SC: 6/8 DI [005] 2/4 D0 [006÷008] Slot unused

Table 4. A matching of units and configuration parameters of the considered application

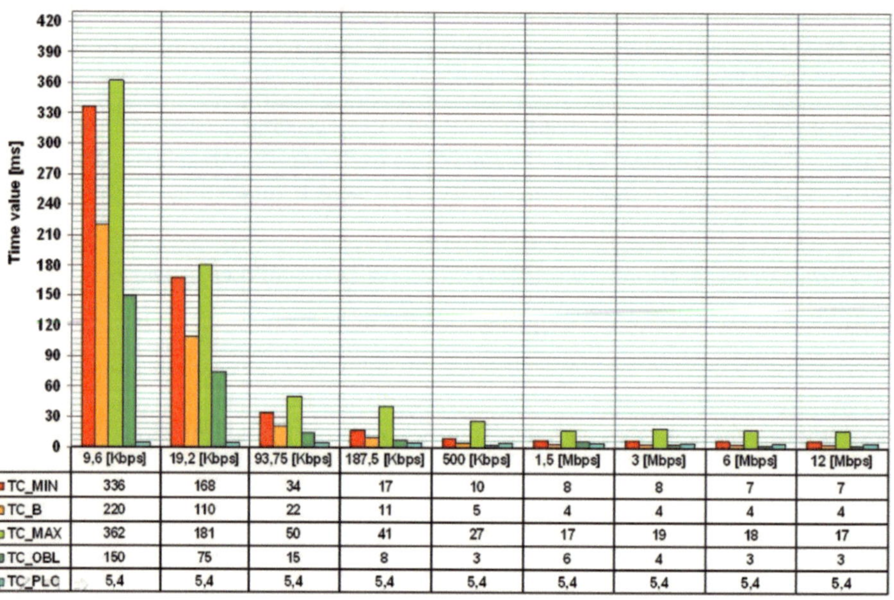

	9,6 [Kbps]	19,2 [Kbps]	93,75 [Kbps]	187,5 [Kbps]	500 [Kbps]	1,5 [Mbps]	3 [Mbps]	6 [Mbps]	12 [Mbps]
TC_MIN	336	168	34	17	10	8	8	7	7
TC_B	220	110	22	11	5	4	4	4	4
TC_MAX	362	181	50	41	27	17	19	18	17
TC_OBL	150	75	15	8	3	6	4	3	3
TC_PLC	5,4	5,4	5,4	5,4	5,4	5,4	5,4	5,4	5,4

Fig. 17. The dependence of characteristic time periods of the ProfiBus DP network (case of the complete configuration including all diagnostic and control blocks called from subroutines), where: TC_MIN - minimum value [ms], TC_B - current value [ms], TC_MAX - maximum value, TC_OBL - calculated value (predicted value), TC_PLC - PLC cycle time (onefold time of an algorithm processing)

In the first configuration case a deficiency of variable that initiate subroutine execution causes an omission of the algorithm execution contained in its structure, which significantly limits the data amount subjected to processing in the first cycle of the driver.

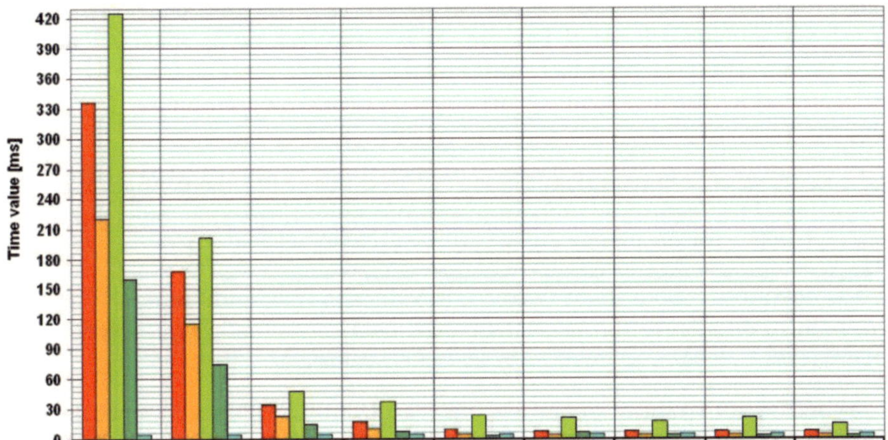

	9,6 [Kbps]	19,2 [Kbps]	93,75 [Kbps]	187,5 [Kbps]	500 [Kbps]	1,5 [Mbps]	3 [Mbps]	6 [Mbps]	12 [Mbps]
TC_MIN	336	168	34	17	10	8	8	8	7
TC_B	220	115	23	11	5	4	4	4	4
TC_MAX	426	201	48	38	24	22	17	21	15
TC_OBL	160	75	15	8	3	6	4	3	3
TC_PLC	5,4	5,4	5,4	5,4	5,4	5,4	5,4	5,4	5,4

Fig. 18. The dependence of characteristic time periods of the ProfiBus DP network (case of the complete configuration including all diagnostic and control blocks placed in the main program)

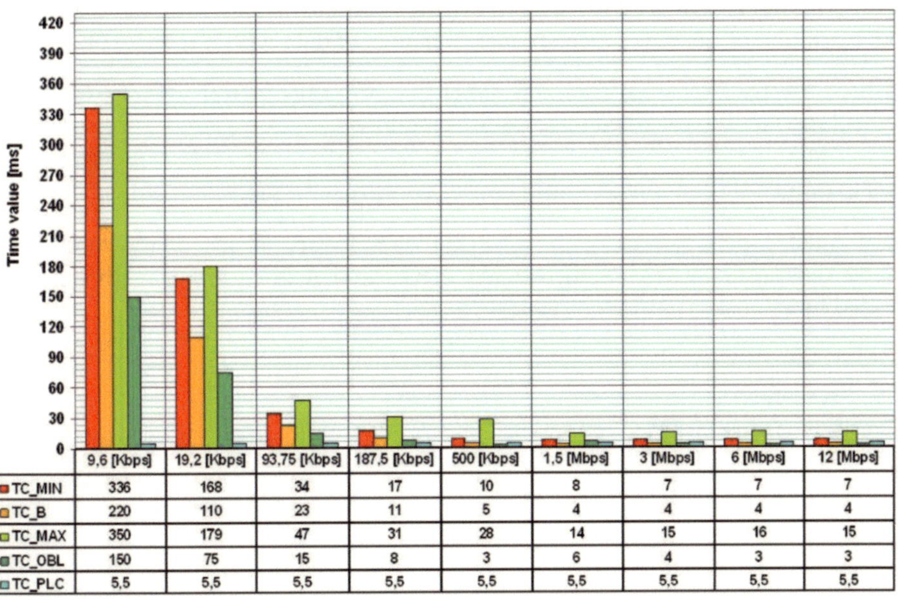

	9,6 [Kbps]	19,2 [Kbps]	93,75 [Kbps]	187,5 [Kbps]	500 [Kbps]	1,5 [Mbps]	3 [Mbps]	6 [Mbps]	12 [Mbps]
TC_MIN	336	168	34	17	10	8	7	7	7
TC_B	220	110	23	11	5	4	4	4	4
TC_MAX	350	179	47	31	28	14	15	16	15
TC_OBL	150	75	15	8	3	6	4	3	3
TC_PLC	5,5	5,5	5,5	5,5	5,5	5,5	5,5	5,5	5,5

Fig. 19. The dependence of characteristic time periods of the ProfiBus DP network (case of the configuration based on main blocks)

Because of an application of the rapid CPU unit (Q25HCPU) there were no significant differences in the recorded values of current bus cycle times and refresh times of the PLC resources.

Estimated (computed) values of the refresh time of the bus network, obtained from the MELSOFT GX CONFIGURATOR DP, are on average two times smaller than the actual recorded values. The presented fact must be taken into account at the design stage of the industrial network structure (step of matching network module and a data transfer rate choice).

A third case of measurements (Fig. 19) was carried out using the configuration shown in Table 4 and the PLC algorithm, without loading of the control code (only control blocks of the data exchange, i.e. QJ71PB92V_ContDiagBLK and PROFIBUS_NET).

In the described test cycle a slight decrease in values of maximum refresh time of bus was noticed, both in relation to the previously described configuration. However, current bus refresh times oscillate around previously identified values.

It can be concluded that the current cycle time of a ProfiBus DP bus does not depend on the configuration architecture of a control and diagnostic algorithm, which confirms the usefulness of the application in complex appliances extended at Fieldbus network buses.

Increasing data transmission speed of the network bus does not cause significant problems in the recorded times, despite changes in the structure of the configuration of the control and diagnostic algorithm.

The proposed form of the structure organization of the developed control and diagnostic system is configuration organized in subroutines called from the main program.

The configuration is beneficial for reasons arising from the need to fulfil the requirements of smooth operation of the PLC processor and exchange of data with remote units.

Small time values of the current bus cycle (TC_B) confirm the validity of the correct implementation of the described diagnostic functions.

6. Conclusions

The presented expert system of the CCaDSoDD software is one of the modules of the computer diagnostic system of distributed drives and it is equipped with the following embedded functions:

- on-line diagnosis,
- off-line diagnosis,
- preliminary diagnosis.

The computer environment is convenient to use, but an elaboration of the expert system is rather complicated. Significant difficulties can arise for the following reasons:

- the process of data gathering involves participation of high-class engineers,
- a database designer must identify all possible failures,
- the expert system requires performing of the additional database with stored corrective actions,

- the readable user interface is necessary,
- work stability depends on the specified protocol.

Self-diagnostic functions are frequently used in removal of states of failures by qualified staff, but an intention of the authors was to turn the attention to the utilization of embedded diagnostic functions at the stage of design of the final form of applications included in modern mechatronic devices.

Advantages of the presented approach are meaningful and can be summarized by several points:

- increasing reliability of the considered application,
- a possibility of preparation of fault detection functions for every mechatronic unit (taking into account its functional constraints and embedded functions),
- utilization of more advanced mechatronic units allowing a widening of the amount of possible identified operating conditions by embedded subsystems,
- to primary parameters belong, among other things: current, minimal and maximal bus cycles (detected by a industrial network module of the PLC unit), the diagnostic information, a current work status, error codes, reactive and active powers, etc.

Beyond the specified advantages a group of functional constraints also exist, resulting in an enlargement of expenditure connected with:

- a preparation of the control algorithm widened by memory cells which include values of diagnostic premises,
- an elaboration of the expert system and inference rules (diversified at an angle of the concrete application).

The application of developed control and diagnostic blocks with the proposed structure of the algorithm allows for:

- reduction of the time needed for an elaboration of the complex control and diagnostic system,
- resignation of the phase of copying the contents of global memory addresses to memory buffers (a direct manipulation on memory buffer registers),
- the implementation of different configuration types:
  - with an implementation of the function corresponding to the control algorithm,
  - using diagnostic functions,
  - combining these functions into the one system extended with additional capabilities of an expert system.

7. References

Ryabinin, I. (1976). *Reliability of engineering systems*, MIR Publishers, ISBN: 0714709492, Moscow

Legierski, T. (1998). *PLC Controllers Programming*, Jacek Skalmierski Publishing, ISBN: 838664416-8, Gliwice

Kwaśniewski, J. (1999). *Industrial programmable controllers in control systems*, AGH Publishers, ISBN: 83-86320-45-1, Kraków

Dhillon, B.S. (2002). *Engineering Maintenance - A Modern Approach,* CRC Press, ISBN 978-1-58716-142-1, New York

Korbicz, J. & Kościelny, J. & Kowalczuk, Z. & Cholewa, W. (2002) *Processes diagnostic: models, methods of the artificial intelligence, applications,* WNT, ISBN 83-204-2734-7, Warsaw

Boyer, S. (2004). *SCADA: Supervisory Control and Data Acquisition,* ISA, ISBN: 1556178778, 2004, New York Fraden J. (2004) *Handbook of modern sensors –physics, design, applications,* Springer, ISBN 1441964657, New York

Prasad, K.V. (2004). *Principles of Digital Communication Systems and Computer Networks,* Charles River Media, ISBN 158450-3297, New York

Mackay, St. & Wright, E. (2004). *Practical Industrial Data Networks: Design, Installation and Troubleshooting,* Elsevier Linacre House, ISBN 07506-5807X , Oxford

Bloch, H.P. & Geitner, F. K. (2005). *Practical Machinery Management for Process Plants – Machinery Component Maintenance and Repair,* Elsevier Linacre House, ISBN 0750677260, Oxford

De Silva, C.W. (2005). *Mechatronics: an integrated approach,* CRC Press, ISBN 0849312744, New York

Świder, J. & Hetmańczyk, M. (2009). Hardware and software integration of mechatronic systems for an example measurement path for temperature sensors, *Solid State Phenomena,* Vol. 147-149, 2009, pp. 676-681, ISSN 1012-0394

Świder, J. & Hetmańczyk, M. (2009). Method of indirect states monitoring of dispersed electric drives, [in:] Proceedings of the The Sixth International Conference on Condition Monitoring and Machinery Failure Prevention Technologies, Dublin, (June 2009), pp. 1171-1179

Świder, J. & Hetmańczyk, M. (2009). The visualization of discrete sequential systems, *Journal of Achievements in Materials and Manufacturing Engineering,* Vol. 34, No. 2, (April 2009), pp. 196-203, ISSN 1734-8412

Świder J. & Hetmańczyk, M. (2009) Computer-aided diagnosis of frequency inverter states, *Problems of Working Machines,* pp. 57-66, Vol. 33, ISSN 1232-9304

Świder, J. & Hetmańczyk, M. (2010). Adaptation of the expert system in diagnosis of the connection of the PLC user interface system and field level, *Solid State Phenomena,* Vol. 164, 2010, ISSN 1012-0394

Świder, J. & Hetmańczyk, M. (2011). *The computer integrated system of control and diagnosis of distributed drives,* Silesian Technical University Publishing, ISBN: 978-83-7335-844-7, Poland

12

Optical Interference Signal Processing in Precision Dimension Measurement

Haijiang Hu[1,*], Fengdeng Zhang[2], Juju Hu[1] and Yinghua Ji[1]
[1]*Jiangxi Normal University, Nanchang,*
[2]*University of Shanghai for Science and Technology, Shanghai,*
China

1. Introduction

The optical techniques are widely used in the automatic measurement and test. In the precision dimension measurement, there are some optical instruments that are widely applied for the microscale or nanoscale dimension measurement, such as the grating, the homodyne interferometer, and the heterodyne interferometer. An optical instrument system is often divided into two main subsystems: the optical subsystem and the electrical subsystem. Moreover, some optical instrument systems also have the mechanical subsystem and the computer subsystems. The optical subsystem includes optical source and other optical elements, such as the polarizing beam splitter, the quarter-wave plate, and so on. The electronic subsystem turns the received optical signals into the electrical signals by the photoelectric receiver, and uses the analog signal system and the digital signal system to process the received signals. A typical structure of the optical instruments is shown as Fig.1.

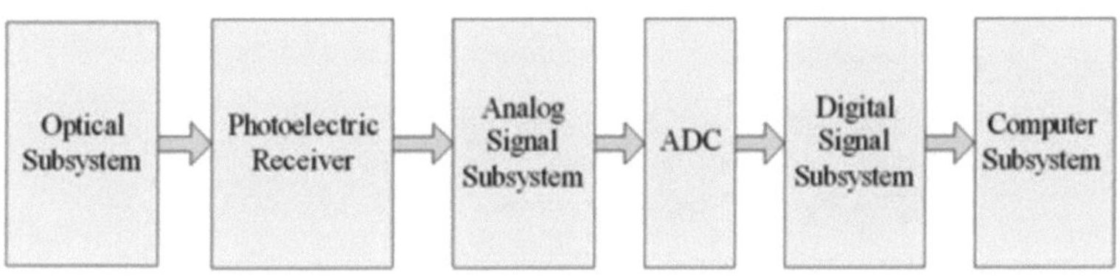

Fig. 1. A typical structure of optical instruments: ADC, analog-digital converter.

In the performance evaluation of the optical measurement instrument, the resolution and the precision (JCGM, 2008) are two important parameters. The resolution describes the ability of a measurement system to resolve detail in the object that is measured. The precision of a measurement system, also called reproducibility or repeatability, is the degree to show the same results when the measurement process is repeated under the unchanged conditions. This chapter mainly discusses two topics: the signal processing method for the improvement of measurement resolution and the elimination errors for the improvement of measurement precision. Generally speaking, the optical source, the design of optical path, and the signal processing are the main factors to improve the measurement resolution. For

* Corresponding Author

example, in a heterodyne Michelson interferometer with double path optical difference that is shown as Fig.2, the traveling distance of the micro-displacement platform L is as follows:

$$L = \frac{\lambda}{2}\int_0^T \Delta f dt, \tag{1}$$

where λ is the wavelength of laser, and Δf is the Doppler frequency shift. Obviously, λ is smaller, the measurement resolution is higher.

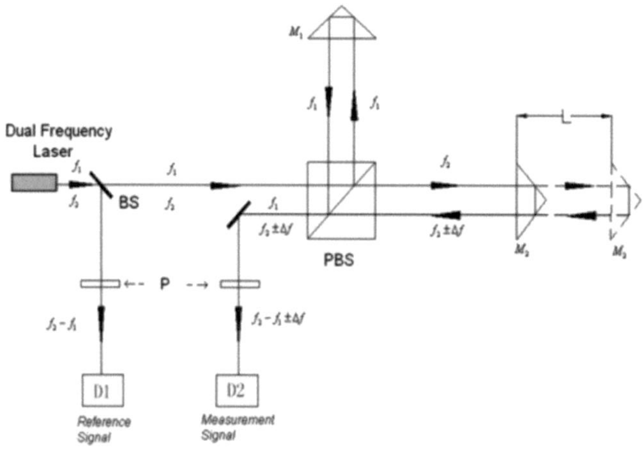

Fig. 2. A heterodyne Michelson interferometer with double path optical difference: BS, beam splitter; PBS, polarizing beam splitter; P, polarization analyzer.

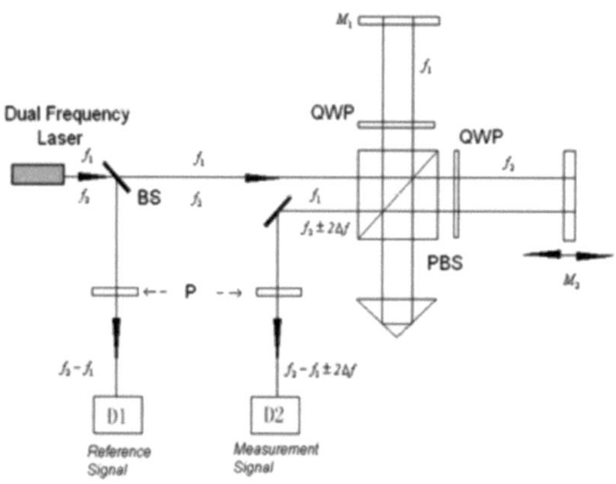

Fig. 3. A heterodyne Michelson interferometer with fourfold path optical difference: BS, beam splitter; PBS, polarizing beam splitter; P, polarization analyzer; QWP, quarter-wave plate.

The multiple reflection of optical path, also called the optical subdivision, is the second factor to influence the measurement resolution. A heterodyne Michelson interferometer with fourfold path optical difference is shown as Fig.3. When the micro-displacement platform M_2 moves the same distance L in Fig.2 and Fig.3, the Doppler frequency shift Δf in Fig.3 is twice as large as Δf in Fig.2. In the same configuration of laser source and electrical signal processing system, the measurement resolution in Fig.3 is twice as large as the measurement resolution in Fig.2. Therefore, the optical subdivision relies on the optical path design of optical system and can be realized by the multiple reflections in the optical path (Cheng et al., 2006).

The third method is the electrical signal processing. Because of the light attenuation in optical parts, the optical subdivision is hard to realize the high subdivision number. Moreover, the measurement error from the nonideal characters and arrangement of optical parts is gradually increased when the number of optical subdivision is added. So the electrical signal processing becomes the most useful method to improve the measurement resolution.

Moreover, the detection and elimination of errors is the important research targets for the improvement of measurement precision. These errors have very serious influence on the measurement precision. In this chapter we discuss electrical signal processing of orthogonal signals (fringes), which is widely used in the optical gratings, the homodyne interferometers, and other optical instruments.

2. Signal preprocessing

In the optical gratings and the homodyne interferometers, the photoelectric converter turns the intensity of light into the electrical signal. Because of the requirement of the direction recognition when the measured object moves, the orthogonal signals whose phase difference is $\pi/2$ are used in these instruments. In order to obtain the orthogonal signals, 4-channel signal receiving systems are designed in the optical systems. Fig.4 shows a typical 4-channel signal receiving systems (Keem et al., 2004).

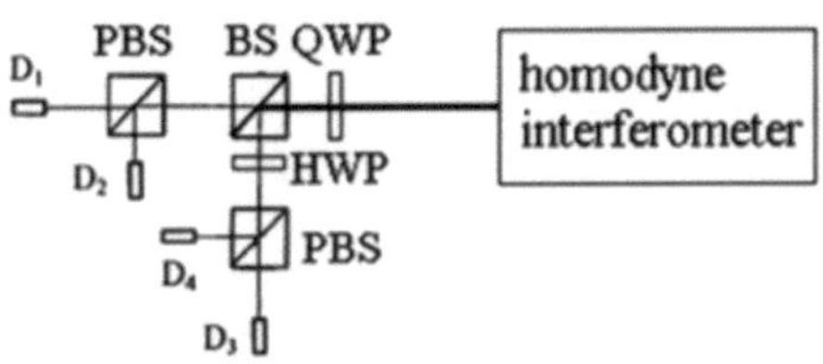

Fig. 4. 4-channel signal receiving systems: BS, beam splitter; PBS, polarizing beam splitter; QWP: quarter-wave plate; HWP: half-wave plate;

When the measured object moves forward, the 4-channel electrical signals are collected from the photosensors and shown as follows:

$$\begin{cases} I_1 = A_1 + B_1 \sin\theta \\ I_2 = A_2 - B_2 \cos\theta \\ I_3 = A_3 - B_3 \sin\theta \\ I_4 = A_4 + B_4 \cos\theta \end{cases}, \tag{2}$$

where A_n is the DC component of channel n, B_n is the amplitude of AC component of channel n, and θ is the change of phase that is related to the displacement ΔL when the measured object moves, respectively. In the homodyne interferometer with two path optical differences, the relationship between θ and ΔL is shown as follows:

$$\theta = \frac{4\pi}{\lambda}\Delta L\,, \tag{3}$$

where λ is the wavelength of laser.

Two signals I_x and I_y can be obtained by subtracting:

$$\begin{cases} I_y = I_1 - I_3 \\ I_x = I_4 - I_2 \end{cases}. \tag{4}$$

From Eqs.2 and Eqs. 4,

$$\begin{cases} I_y = A_1 - A_3 + (B_1 + B_3)\sin\theta \\ I_x = A_4 - A_2 + (B_2 + B_4)\cos\theta \end{cases}. \tag{5}$$

Under the ideal condition, A_2=A_4, A_1=A_3 and B_2+B_4 = B_1+B_3. So, two orthogonal signals are obtained as follows when the measured object moves forward:

$$\begin{cases} I_y = B\sin\theta \\ I_x = B\cos\theta \end{cases}, \tag{6}$$

where B=B_1+B_3=B_2+B_4.

Similarly, the 4-channel signals are shown as follows when the measured object moves backward:

$$\begin{cases} I_1 = A_1 + B_1\sin\theta \\ I_2 = A_2 + B_2\cos\theta \\ I_3 = A_3 - B_3\sin\theta \\ I_4 = A_4 - B_4\cos\theta \end{cases}, \tag{7}$$

where the parameters of Eqs.7 are the same as the parameters of Eqs.2. Similarly, from Eqs.4 and Eqs.7, two orthogonal signals are obtained as follows when the measured object moves backward:

$$\begin{cases} I_y = B\sin\theta \\ I_x = -B\cos\theta \end{cases}, \tag{8}$$

where B=B_1+B_3=B_2+B_4.

So, ΔL is obtained by using the half periodic counting of orthogonal signals and the calculation of θ as follows:

$$\Delta L = \frac{\lambda}{2} N + \varepsilon , \qquad (9)$$

with

$$\varepsilon = \frac{\Delta\theta}{4\pi}\lambda ,$$

where λ is the wavelength of laser, N is the counting value of half period, and ε is the additional displacement value that is shorter than $\lambda/2$ and can be calculated from the orthogonal signals, respectively. For the improvement of measurement resolution, some bidirectional subdivision methods are studied to raise the resolution of N and ε.

3. Bidirectional subdivision

3.1 λ/4 bidirectional subdivision

The simplest bidirectional subdivision method (Yoshizawa, 2005) is shown as Fig.5. This method is applied on both the analog orthogonal signals and the digital orthogonal signals. The principle of this method is shown in the right part in Fig.5. During the measured object moves forward, when signal u_2 crosses zero from positive to negative, signal u_1 is positive; when signal u_1 crosses zero from positive to negative, signal u_2 is negative; when signal u_2 crosses zero from negative to positive, signal u_1 is negative; when signal u_1 crosses zero from negative to positive, signal u_{2f} is positive. Therefore, the counting of four times in a period for forward movement is realized.

Similarly, during the measured object moves backward, when signal u_2 crosses zero from negative to positive, signal u_1 is positive; when signal u_1 crosses zero from positive to negative, signal u_2 is positive; when signal u_2 crosses zero from positive to negative, signal u_1 is negative; when signal u_1 crosses zero from negative to positive, signal u_2 is negative. From the above detection, the counting of four times in a period for bidirectional movement is realized.

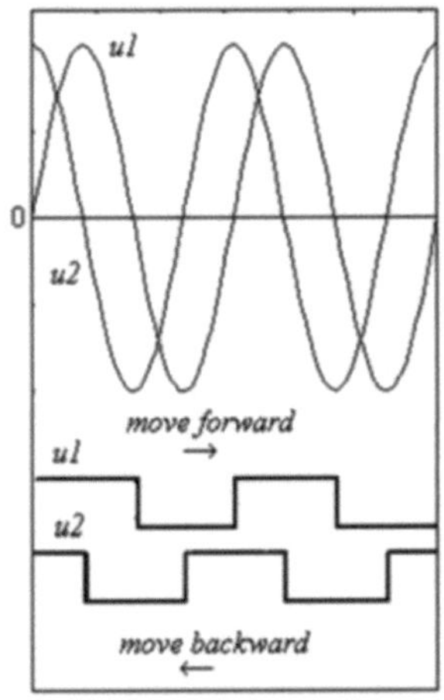

Detected signal	Cross-zero detection	Sign of signals u_1	Sign of signals u_2	Counting value
u_1	+ → -	n/a	-	+1
u_1	+ → -	n/a	+	-1
u_1	- → +	n/a	+	+1
u_1	- → +	n/a	-	-1
u_2	+ → -	+	n/a	+1
u_2	+ → -	-	n/a	-1
u_2	- → +	-	n/a	+1
u_2	- → +	+	n/a	-1

Fig. 5. $\lambda/4$ bidirectional subdivision

3.2 λ/8 bidirectional subdivision

In order to realize the counting of more than four times in a period, some methods are studied. The first method uses two thresholds to realize λ/8 bidirectional subdivision by (Chu et al., 2003). The method defines two thresholds T_h and T_l . T_h is the amplitude of normalized signal in the $\pi/4$ phase while T_l is the negative of T_h. When u_1 cross across T_h from the bottom to the top, the movement direction is positive if u_2 is positive at the same time or negative if u_2 is negative at the same time. Similarly, when u_1 cross across T_l from the bottom to the top, the movement direction is positive if u_2 is positive at the same time or negative if u_2 is negative at the same time. By combination between λ/4 bidirectional subdivision and two thresholds detection, the λ/8 bidirectional subdivision is realized.

(Cui et al., 2000) proposes a method to realize λ/8 bidirectional subdivision by the constructed function. The constructed function is as follows:

$$f = |u_1| - |u_2| . \tag{10}$$

Fig.6 shows the function images of u_1, u_2 and f. According to the change of sign of u_1, u_2 and f, λ/8 bidirectional subdivision is achieved. The principle of this method is as shown in Fig.7.

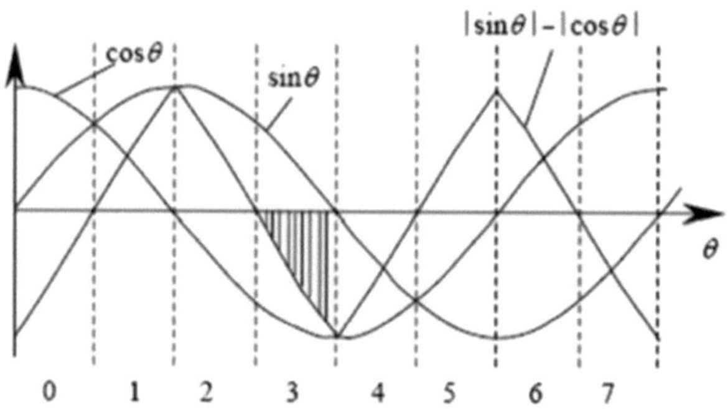

Fig. 6. Function images of u_1, u_2 and f

sinθ	cosθ	*f*	*code*	*status*
> 0	> 0	< 0	110	A
> 0	> 0	> 0	000	B
> 0	< 0	> 0	101	C
> 0	< 0	< 0	100	D
< 0	< 0	< 0	000	E
< 0	< 0	> 0	001	F
< 0	> 0	> 0	011	G
< 0	> 0	< 0	010	H

move forward
A->B->C->D->
E->F->G->H->A

move backward
A->H->G->F->
E->D->C->B->A

Fig. 7. Principle of λ/8 bidirectional subdivision by Eq.10

3.3 λ/16 bidirectional subdivision

Similar to (Cui et al., 2000), (Chen et al., 2005) proposes another constructed function and status code method to achieve $\lambda/16$ bidirectional subdivision. The difference from (Cui et al., 2000), (Chen et al., 2005) uses 4 bits code as status parameter.

3.4 λ/2^{n+2} bidirectional subdivision

(Hu et al., 2009) and (Hu & Zhang, 2012) proposes a method of $\lambda/2^{n+2}$ bidirectional subdivision that uses both the constructed function and the cross-zero detection. The $\lambda/8$ and $\lambda/16$ bidirectional subdivision methods are simply introduced at first. Commonly, two orthogonal signals are defined as follows when the measured object moves forward:

$$\begin{cases} u_{1f} = U_0 \sin\theta \\ u_{2f} = U_0 \cos\theta \end{cases}, \tag{11}$$

where U_0 is the amplitude of the orthogonal signals, and θ is the change of phase that is related to the displacement when the measured object moves.

The following two relative functions are constructed to be the reference signals in order to achieve the bidirectional subdivision:

$$\begin{cases} r_{11f} = U_0 \sin\theta + U_0 \cos\theta \\ r_{12f} = U_0 \sin\theta - U_0 \cos\theta \end{cases}. \tag{12}$$

Reference signals r_{11f} and r_{12f} can be acquired by the digital adder. Therefore, the $\lambda/8$ subdivision of the two orthogonal signals can be realized easily by the zero-cross detection which consists of the orthogonal signals and the reference signals.

Based on the $\lambda/4$ bidirectional subdivision When the positive voltage drives the measured object to move forward, reference signal r_{12f} crosses zero from negative to positive, signals u_{1f} and u_{2f} are positive; when reference signal r_{11f} crosses zero from positive to negative, signal u_{1f} is positive and signal u_{2f} is negative; when reference signal r_{12f} crosses zero from positive to negative, signals u_{1f} and u_{2f} are negative; when reference signal r_{11f} crosses zero from negative to positive; signal u_{1f} is negative and signal u_{2f} is positive; Therefore, the counting of eight times in a period for the forward movement is realized.

The same to the forward counting, it's easy to realize the counting of eight times in a period for the backward movement. When the measured object moves backward, the two orthogonal signals are defined as follows:

$$\begin{cases} u_{1b} = U_0 \sin\theta \\ u_{2b} = -U_0 \cos\theta \end{cases}. \tag{13}$$

The same to Eqs.12, the two relative functions are constructed to be the reference signals:

$$\begin{cases} r_{11b} = U_0 \sin\theta - U_0 \cos\theta \\ r_{12b} = U_0 \sin\theta + U_0 \cos\theta \end{cases}. \tag{14}$$

Similarly, the counting of eight times in a period for the backward movement is also achieved. So, a set of reference signals are only built as follows:

$$\begin{cases} r_{11} = u_1 + u_2 \\ r_{12} = u_1 - u_2 \end{cases}. \tag{15}$$

The bidirectional subdivision of $\lambda/8$ can be realized by the zero-cross detection of the orthogonal signals and the reference signals.

On the basis of the $\lambda/8$ subdivision, a group of the reference signals is built which is constituted by the two sets of reference signals. During the time of forward counting, the reference signal sets are defined as follows:

$$\begin{cases} r_{11f} = U_0 \sin\theta + U_0 \cos\theta \\ r_{12f} = U_0 \sin\theta - U_0 \cos\theta \end{cases}, \tag{16}$$

$$\begin{cases} r_{21f} = U_0 \sin 2\theta + U_0 \cos 2\theta \\ r_{22f} = U_0 \sin 2\theta - U_0 \cos 2\theta \end{cases}. \tag{17}$$

In order to calculate the reference signals r_{21f} and r_{22f}, the two orthogonal signals are used to express them. r_{21f} is obtained as follows:

$$r_{21f} = U_0 \sin 2\theta + U_0 \cos 2\theta = U_0 \left[2\cos\theta(\sin\theta + \cos\theta) - 1 \right]. \tag{18}$$

Similarly, r_{22f} is obtained as follows:

$$r_{22f} = U_0 \left[2\sin\theta(\sin\theta + \cos\theta) - 1 \right]. \tag{19}$$

U_0=1 is still assumed. The function images of the two orthogonal signals and reference signals for the $\lambda/16$ subdivision are shown in Fig. 8 and Fig. 9 during the time of the forward counting and the backward counting. The same as the $\lambda/8$ subdivision, the cross-zero detection is also used to realize the bidirectional subdivision of $\lambda/16$. Therefore, on the basis of the $\lambda/8$ subdivision, the bidirectional subdivision of $\lambda/16$ for the orthogonal signals is realized by adding a set of reference signals:

$$\begin{cases} r_{21} = \dfrac{2u_2}{U_0}(u_1 + u_2) - U_0 \\ r_{22} = \dfrac{2u_1}{U_0}(u_1 + u_2) - U_0 \end{cases}. \tag{20}$$

Based on the $\lambda/8$ subdivision and the $\lambda/16$ subdivision, n sets of reference signals can be built to realize the $\lambda/2^{n+2}$ bidirectional subdivision as follows:

$$\begin{cases} r_{n1} = U_0 \sin 2^{n-1}\theta + U_0 \cos 2^{n-1}\theta \\ r_{n2} = U_0 \sin 2^{n-1}\theta - U_0 \cos 2^{n-1}\theta \end{cases}, \tag{21}$$

where n is integer and n>0.

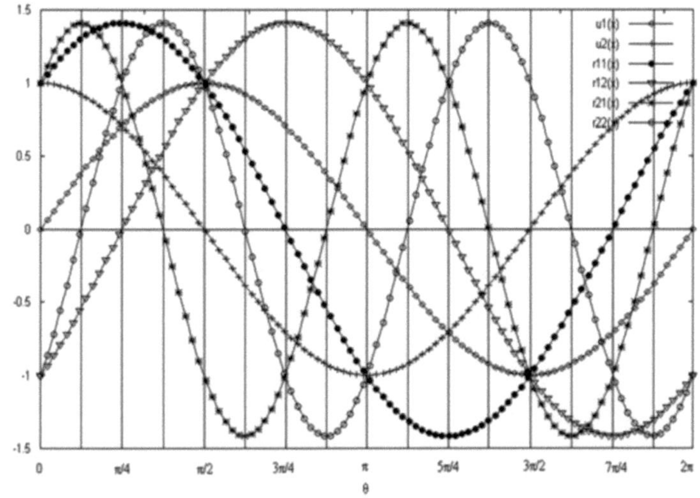

Fig. 8. Forward orthogonal signals and reference signals for $\lambda/16$ subdivision

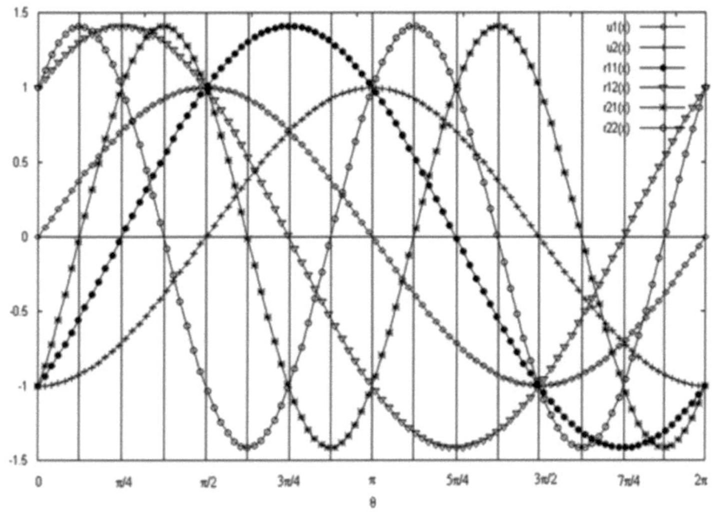

Fig. 9. Backward orthogonal signals and reference signals for $\lambda/16$ subdivision

From Euler's formula:

$$e^{j\theta} = \cos\theta + j\sin\theta\,, \tag{22}$$

the following equation is obtained:

$$e^{jn\theta} = \cos n\theta + j\sin n\theta = (\cos\theta + j\sin\theta)^n\,. \tag{23}$$

Then Eq. 23 is expanded as follows:

$$\cos n\theta + j\sin n\theta = \binom{n}{0}\cos^n\theta + j\binom{n}{1}\cos^{n-1}\theta\sin\theta + j^2\binom{n}{2}\cos^{n-2}\theta\sin^2\theta + \ldots + j^n\binom{n}{n}\sin^n\theta$$

$$= \begin{cases} \binom{n}{0}\cos^n\theta + j\binom{n}{1}\cos^{n-1}\theta\sin\theta - \ldots + \binom{n}{n}\sin^n\theta, n = 4N \\ \binom{n}{0}\cos^n\theta + j\binom{n}{1}\cos^{n-1}\theta\sin\theta - \ldots + j\binom{n}{n}\sin^n\theta, n = 4N+1 \\ \binom{n}{0}\cos^n\theta + j\binom{n}{1}\cos^{n-1}\theta\sin\theta - \ldots - \binom{n}{n}\sin^n\theta, n = 4N+2 \\ \binom{n}{0}\cos^n\theta + j\binom{n}{1}\cos^{n-1}\theta\sin\theta - \ldots - j\binom{n}{n}\sin^n\theta, n = 4N+3 \end{cases}, \quad (24)$$

where N is integer and $N \geq 0$.

From Eqs.24, the following equations are obtained:

$$\begin{cases} \sin 2\theta = 2\sin\theta\cos\theta \\ \cos 2\theta = \cos^2\theta - \sin^2\theta \end{cases} \quad (25)$$

$$\begin{cases} \sin n\theta = \binom{n}{1}\cos^{n-1}\theta\sin\theta - \binom{n}{3}\cos^{n-3}\theta\sin^3\theta + \ldots - \binom{n}{n-1}\cos\theta\sin^{n-1}\theta \\ \cos n\theta = \cos^n\theta - \binom{n}{2}\cos^{n-2}\theta\sin^2\theta + \ldots + \sin^n\theta \end{cases}, n = 4N, \quad (26)$$

where N is integer and $N \geq 1$.

So, from Eqs.16, Eqs.25 and Eqs.26, the following equations are obtained:

$$\begin{cases} r_{11} = U_0\sin\theta + U_0\cos\theta \\ r_{12} = U_0\sin\theta - U_0\cos\theta \end{cases}, n = 1, \quad (27)$$

$$\begin{cases} r_{21} = U_0\left[2\cos\theta(\sin\theta + \cos\theta) - 1\right] \\ r_{22} = U_0\left[2\sin\theta(\sin\theta + \cos\theta) - 1\right] \end{cases}, n = 2, \quad (28)$$

$$\begin{cases} r_{n1} = U_0\left(\cos^{2^{n-1}}\theta + \binom{2^{n-1}}{1}\cos^{2^{n-1}-1}\theta\sin\theta - \binom{2^{n-1}}{2}\cos^{2^{n-1}-2}\theta\sin^2\theta - \ldots + \sin^{2^{n-1}}\theta\right) \\ r_{n2} = U_0\left(-\cos^{2^{n-1}}\theta + \binom{2^{n-1}}{1}\cos^{2^{n-1}-1}\theta\sin\theta + \binom{2^{n-1}}{2}\cos^{2^{n-1}-2}\theta\sin^2\theta - \ldots - \sin^{2^{n-1}}\theta\right) \end{cases}, n > 2 \quad (29)$$

where n is integer.

The same as the $\lambda/8$ subdivision and the $\lambda/16$ subdivision, u_1 and u_2 are used to express r_{n1} and r_{n2}:

$$\begin{cases} r_{11} = u_1 + u_2 \\ r_{12} = u_1 - u_2 \end{cases}, n = 1, \tag{30}$$

$$\begin{cases} r_{21} = \dfrac{2u_2}{U_0}(u_1 + u_2) - U_0 \\ r_{22} = \dfrac{2u_1}{U_0}(u_1 + u_2) - U_0 \end{cases}, n = 2, \tag{31}$$

$$\begin{cases} r_{n1} = \dfrac{1}{U_0^{2^{n-1}-1}}\left(u_2^{2^{n-1}} + \begin{pmatrix} 2^{n-1} \\ 1 \end{pmatrix} u_2^{2^{n-1}-1} u_1 - \begin{pmatrix} 2^{n-1} \\ 2 \end{pmatrix} u_2^{2^{n-1}-2} u_1^2 - \ldots + u_1^{2^{n-1}}\right) \\ r_{n2} = \dfrac{1}{U_0^{2^{n-1}-1}}\left(-u_2^{2^{n-1}} + \begin{pmatrix} 2^{n-1} \\ 1 \end{pmatrix} u_2^{2^{n-1}-1} u_1 + \begin{pmatrix} 2^{n-1} \\ 2 \end{pmatrix} u_2^{2^{n-1}-2} u_1^2 - \ldots - u_1^{2^{n-1}}\right) \end{cases}, n > 2, \tag{32}$$

where n is integer.

For n=1, the number of subdivision times is 8.

For n=2, the number of subdivision times is 16. This is just shown in Fig.9 during the time of the forward counting and Fig.10 during the time of the backward counting.

Similarly, the $\lambda/32$, $\lambda/64$, $\lambda/128$, ..., $\lambda/2^{n+2}$ subdivision can be obtained where n is integer and n>2.

Consequently, a conclusion is obtained as follows:

On the basis of the $\lambda/2^{n+1}$ bidirectional subdivision, the $\lambda/2^{n+2}$ bidirectional subdivision of the orthogonal signals can be realized by adding the two reference signals r_{n1} and r_{n2} and their zero-cross detection. The zero-cross detection uses the original orthogonal signals and the reference signals r_{11}, r_{12}, r_{21}, r_{22}, ..., $r_{(n-1)1}$ and $r_{(n-1)2}$ as the positive-negative judgment items for the forward counting and the backward counting.

4. Error detection and elimination

In the ideal conditions, the phase difference of the orthogonal signals is $\pi/2$. However, because of the imperfect design and manufacture of measurement system, the environmental disturbance and the system noise, the signals often have some errors, such as the nonorthogonality, the non-equality of amplitude, the drift of DC signals, and so on (Heydemann, 1981), which have very serious influence on the precision of the fringe subdivision. The typical orthogonal signals with errors are shown as follows:

$$\begin{cases} u_1^d = u_1 + p \\ u_2^d = \dfrac{1}{r}(u_2 \cos a - u_1 \sin a) + q \end{cases}, \tag{33}$$

where p,q are drift of DC, r is non-equality of amplitude coefficient, a is nonorthogonal error, respectively.

(Heydemann, 1981) proposes a universal method to compensate the errors. From Eqs.33, another equation is obtained as follows:

$$(u_1+p)^2+\left(\frac{1}{r}u_2\cos a-\frac{1}{r}u_1\sin a+q\right)^2=U_0{}^2, \tag{34}$$

and

$$(u_1^d-p)^2+\left[\frac{(u_2^d-q)r+(u_1^d-p)\sin a}{\cos a}\right]^2=U_0{}^2. \tag{35}$$

From Eq.35, the error factors are obtained as follows:

$$Au_1^{d2}+Bu_2^{d2}+Cu_1^du_2^d+Du_1^d+Eu_2^d=1, \tag{36}$$

where

$$\begin{aligned} &A=(R^2\cos^2 a-p^2-r^2q^2-2rpq\sin a)^{-1}\\ &B=Ar^2\\ &C=2Ar\sin a\\ &D=-2A(p+rq\sin a)\\ &E=-2Ar(rq+p\sin a)\end{aligned},$$

respectively. Obviously this is an elliptic function express. The parameter A, B, C, D, and E are called the error factors. Therefore, if the error factors A, B, C, D, and E are obtained, p, q, r, a are calculated as follows:

$$\begin{cases} a=\arcsin C(4AB)^{-\frac{1}{2}}\\ r=(\frac{B}{A})^{\frac{1}{2}}\\ p=\dfrac{2BD-EC}{C^2-4AB}\\ q=\dfrac{2AE-DC}{C^2-4AB}\end{cases}. \tag{37}$$

From Eqs.37, u_1 and u_2 are obtained as follows:

$$\begin{cases} u_1=u_1^d-p\\ u_2=\dfrac{1}{\cos a}\left[(u_1^d-p)\sin a+r(u_2^d-q)\right]\end{cases}. \tag{38}$$

(Heydemann, 1981) sets up a complete method for the error compensation. This method needs at least a period of signals for the calculation of the error factors. Therefore it has the poor real-time performance. (Zumberge et al., 2004, Eom et al., 2001) also calculate error factors for the compensation by least-squares fitting. (Song et al., 2000) uses phase-modulated grating to compensate for nonorthogonal errors. (Keem et al., 2004) uses Jones matrix calculation to investigate the remaining error of a homodyne interferometer with a quadrature detector system. (Cheng et al., 2009) uses the normalized waveforms to eliminate DC offsets and amplitude variation. (Hu & Zhang, 2011) proposes a method for detection and elimination of the nonorthogonal errors based on digital sampled signals as follows when (Hu et al., 2009, Hu & Zhang, 2012) are applied in the fringe subdivision.

In (Birch, 1990), when the measured object moves forward, the two orthogonal signals which only contain nonorthogonality error are shown as follows:

$$\begin{cases} u_1 = U_0 \sin\theta \\ u_{2f} = U_0 \sin(\theta + \frac{\pi}{2} + \Delta\varphi) \end{cases}, \tag{39}$$

where $\Delta\varphi$ is the nonorthogonality error. Similarly, when the measured object moves backward, the original signals which only contain the nonorthogonality error are shown as follows:

$$\begin{cases} u_1 = U_0 \sin\theta \\ u_{2b} = U_0 \sin(\theta - \frac{\pi}{2} - \Delta\varphi) \end{cases}. \tag{40}$$

The detection method of $\Delta\varphi$ is as follows:

At first, a counter is defined as C that records the sample number of the original signal u_2. Then, in the forward movement of the measured object, when the original signal u_1 crosses zero from negative to positive, the counter C resets itself and begins to count. When the original signal u_2 crosses zero, the value of the counter C_e is saved, and the counter C goes on counting. When the original signal u_1 crosses zero from positive to negative, the value of the counter C_{hp} is saved. Consequently, the nonorthogonality error of two original signals can be calculated as follows:

$$\Delta\varphi = \frac{\frac{1}{2}C_{hp} - C_e}{C_{hp}}\pi . \tag{41}$$

Similarly, when the measured object moves backward, the counter C resets itself and begins to count. When the original signal u_2 crosses zero, the value of the counter C_{eb} is saved and the counter C goes on counting. When the original signal u_1 crosses zero from positive to negative, the value of the counter C_{hp} is saved. the nonorthogonality error of two original signals is calculated as follows:

$$\Delta\varphi = \frac{C_{eb} - \frac{1}{2}C_{hp}}{C_{hp}}\pi . \tag{42}$$

The precision of $\Delta\varphi$ is related with the sample number in the half period obviously. When the sample number in the half period is higher, the precision of $\Delta\varphi$ is higher. Moreover, the multiple-averaging technique can be used based on sampling many times to improve the computational accuracy.

If d is defined as the displacement value when the original signal moves a period, the relationship between the displacement L and the phase θ is as follows:

$$L = Nd + \frac{\theta}{2\pi} d, \theta \in [0, 2\pi), \tag{43}$$

where N is the number of periods when the measured object moves, which can be obtained by the cross-zero detection, and θ can be acquired by the method of subdivision and the counting. If the nonorthogonality error does not exist, Eq. 43 can be represented by the $\lambda/8$ bidirectional subdivision from (Hu et al., 2009) as follows:

$$L = \frac{N_8 d}{8} + \frac{4\theta}{\pi} d, \theta \in [0, \frac{\pi}{4}), \tag{44}$$

where N_8 is the number of 1/8 periods when the counting system uses $\lambda/8$ subdivision algorithm, and

$$N = \frac{N_8}{8}. \tag{45}$$

When the nonorthogonality error exists, the algorithm can not divide a signal period into eight subintervals evenly. Therefore Eq.44 is useless if the nonorthogonality error exists. The relationship of the displacement L and θ has to be redefined.

C_p is defined as the sampling value of the signal in a period and can be obtained by means of the cross-zero detection and the counting for the original signals. When the nonorthogonality error does not exist, Eqs.43 and 44 can be shown as follows:

$$L = Nd + \frac{n}{C_p} d, n \in \left(-C_p, C_p\right), \tag{46}$$

and

$$L = \frac{N_8 d}{8} + \frac{n}{C_p} d, n \in \left(-\frac{C_p}{8}, \frac{C_p}{8}\right), \tag{47}$$

where n is positive when the measured object moves forward or negative when the measured object moves backward. In the Eq. 46, n is the count value of the original signal when the original signal does not pass through a complete period. Similarly, in the Eq.47, n is the count value of the original signal when the original signal does not pass through one eighth periods.

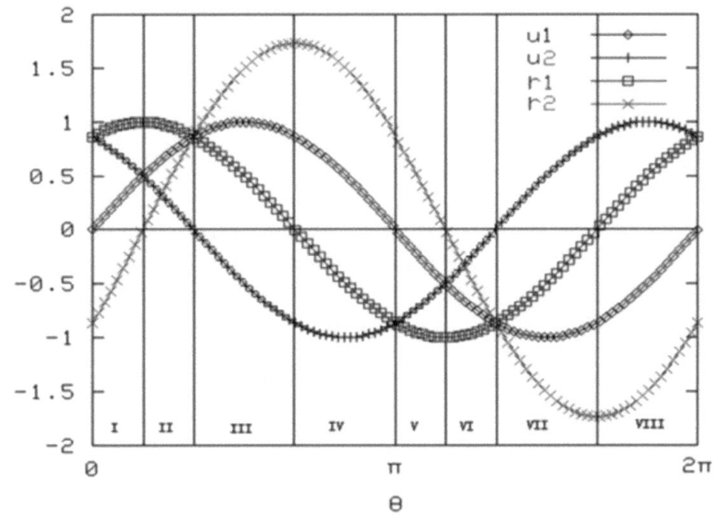

Fig. 10. Subintervals of nonorthogonal signals in $\lambda/8$ subdivision

When the nonorthogonality error exists, eight subintervals can be arranged in two groups. It is shown as Fig. 12. The first group includes the subinterval I, II, V, and VI, while the second group includes the subinterval III, IV, VII, and VIII. Every subinterval in a same group has the same interval length. So the interval length of eight subintervals can be obtained as Table 1:

Number of subinterval	Range of θ in $[0,2\pi)$	Interval length
I	$[0,\frac{\pi}{4}-\frac{\Delta\varphi}{2})$	$\frac{\pi}{4}-\frac{\Delta\varphi}{2}$
II	$[\frac{\pi}{4}-\frac{\Delta\varphi}{2},\frac{\pi}{2}-\Delta\varphi)$	$\frac{\pi}{4}-\frac{\Delta\varphi}{2}$
III	$[\frac{\pi}{2}-\Delta\varphi,\frac{3\pi}{4}-\frac{\Delta\varphi}{2})$	$\frac{\pi}{4}+\frac{\Delta\varphi}{2}$
IV	$[\frac{3\pi}{4}-\frac{\Delta\varphi}{2},\pi)$	$\frac{\pi}{4}+\frac{\Delta\varphi}{2}$
V	$[\pi,\frac{5\pi}{4}-\frac{\Delta\varphi}{2})$	$\frac{\pi}{4}-\frac{\Delta\varphi}{2}$
VI	$[\frac{5\pi}{4}-\frac{\Delta\varphi}{2},\frac{3\pi}{2}-\Delta\varphi)$	$\frac{\pi}{4}-\frac{\Delta\varphi}{2}$
VII	$[\frac{3\pi}{2}-\Delta\varphi,\frac{7\pi}{4}-\frac{\Delta\varphi}{2})$	$\frac{\pi}{4}+\frac{\Delta\varphi}{2}$
VIII	$[\frac{7\pi}{4}-\frac{\Delta\varphi}{2},2\pi)$	$\frac{\pi}{4}+\frac{\Delta\varphi}{2}$

Table 1. The interval length of eight subintervals

So the displacement L can be calculated when the nonorthogonality error exists:

$$L = (\frac{\pi - 2\Delta\varphi}{8\pi} N_{8a} + \frac{\pi + 2\Delta\varphi}{8\pi} N_{8b})d + \frac{n}{C_p} d \,, \tag{48}$$

where N_{8a} is the count value of the subinterval when the original signal passes by the subinterval I, II, V, and VI, N_{8b} is the count value of the subinterval when the original signal passes by the subinterval III, IV, VII, and VIII, and n is the count value of the original signal when the original signal does not pass through a subinterval. N_{8a}, N_{8b}, and n are positive when the measurement platform moves forward or negative when the measurement platform moves backward.

Therefore, without the extra compensation for the nonorthogonality error, the displacement L can be also calculated and the influence of the nonorthogonality error can be eliminated by means of Eq.48.

5. Implementation platform

With the development of electronic and computer technologies, many platforms are widely used in the signal processing for orthogonal signals, such as the hardware (Birch, 1990, Downs & Birch, 1983), FPGA (Hu et al., 2009), the software programming (Zhang et al., 1994, Su et al., 2000), LabVIEW (Yacoot et al., 2001), DSP(Zumberge et al., 2004), and so on. Besides the analog signal processing and the digital signal processing, the image processing is also an important method to realize the fringe analysis (Takeda et al., 1982, Qian, 2004). The image processing system often get data from the CMOS or CCD cameras, and use digital image processing methods to calculate the phase shift. The image processing technique offers many new solutions for the fringe analysis (Patorski & Styk, 2006, Marengo-Rodriguez et al., 2007, Bernini et al., 2009, Larkin et al., 2001, Pokorski & Patorski, 2010, Zhang, 2011) and extends the application fields of fringe analysis (Geng, 2011, Riphagen et al., 2008).

6. Summary

In this chapter the signal processing methods for orthogonal signals are introduced. The orthogonal signals are widely used in the optical grating and the homodyne interferometers. In order to improve the resolution of these optical instruments, some bidirectional subdivision methods are introduced. Besides the subdivision methods, the algorithms for error detection and elimination are also stated for improving the measurement precision. With the development of the hardware technology and the software technology, the precision dimension measurement technology will be further developed and improved on the resolution, the precision, the real-time characteristics, the applicability, and the maneuverability.

7. References

JCGM. (2008). International vocabulary of metrology. Available from: <http://www.bipm.org/utils/common/documents/jcgm/JCGM_200_2008.pdf>.

Cheng, Z. Gao, H. Zhang, Z. Huang, H. & Zhu, J. (2006). Study of a dual-frequency laser interferometer with unique optical subdivision techniques. *Appl. Opt.*, Vol. 45, pp. 2246-2250.

Keem, T. Gonda, S. Misumi, I. Huang, Q. & Kurosawa T. (2004). Removing Nonlinearity of a Homodyne Interferometer by Adjusting the Gains of its Quadrature Detector Systems. *Appl. Opt.*, vol. 43, pp. 2443-2448.

Yoshizawa, T. (2008). *Handbook of Optical Metrology - Principles and Applications*, CRC Press.

Chu, X. Lü, H. & Cao, J. (2003). Research on direction recognizing and subdividing method for Moiré(interference) fringes. *Chin. Opt. Lett.*, vol. 1, pp. 692-694.

Cui, J. Li, H. & Chen, Q. (2000). New digital subdividing snd rester-sensing technique for moiré fringe of grating. *Opt. Tech.*, vol. 26, pp. 591-593.

Chen, B. Luo, J. & Li, D. (2005). Code counting of optical fringes: methodology and realization. *Appl. Opt.*, vol. 44, pp. 217-222.

Hu, H. Qiu, X. Wang, J. Ju, A. & Zhang, Y. (2009). Subdivision and direction recognition of λ/16 of orthogonal fringes for nanometric measurement. *Appl. Opt.*, vol. 48, pp. 6479-6484.

Hu, H. & Zhang, F. (2012). Advanced bidirectional subdivision algorithm for orthogonal fringes in precision optical measurement instruments. *Optik*, in press. Available from <http://www.sciencedirect.com/science/article/pii/S0030402611005973>.

Heydemann, P. L. M. (1981). Determination and correction of quadrature fringe measurement errors in interferometers. *Appl. Opt.*, vol. 20, pp. 3382-3384.

Zumberge, M. A. Berger, J. Dzieciuch, M. A. & Parker R. L. (2004). Resolving quadrature dringes in real time. *Appl. Opt.*, vol. 43, pp. 771-775.

Eom, T. B. Kim, J. Y. & Jeong, K. (2001). The dynamic compensation of nonlinearity in a homodyne laser interferometer. *Meas. Sci. and Technol.*, vol. 12, pp. 1734-1738.

Song, J. H. Kim, K. C. & Kim, S. H. (2000). Reducing tilt errors in moiré linear encoders using phase-modulated grating. *Rev. Sci. Instrum.*, vol. 71, pp. 2296-2300.

Cheng, F. Fei, Y. T. & Fan, G. Z. (2009). New method on real-time signal correction and subdivision for grating-based nanometrology. *Proc. SPIE.*, vol. 7284, 728403.

Birch, K. P. (1990). Optical fringe subdivision with nanometric accuracy. *Precis. Eng.*, vol. 12, pp.195-198.

Downs, M. J. & Birch, K. P. (1983). Bi-directional fringe counting interference refractometer. *Precis. Eng.*, vol. 5, pp. 105-110.

Zhang, G. X. Wang, C. H. & Li, Z. (1994). Improving the accuracy of angle measurement system with optical grating. *CIRP Ann. Manuf. Technol.*, vol. 43, pp. 457-460.

Su, S. Lü, H. Zhou, W. & Wang, G. (2000). A software solution to counting and subdivision of Moiré fringe with wide dynamic range. *Proc. SPIE*, vol.4222, pp. 308-312.

Yacoot, A. Kuetgens, U. Koenders, L. & Weimann, T. (2001). A combined scanning tunnelling microscope and x-ray interferometer. *Meas. Sci. Technol.*, vol. 12, pp. 1660.

Takeda, M. Ina, H. & Kobayashi, S. (1982). Fourier-transform method of fringe-pattern analysis for computer-based topography and interferometry. *J. Opt. Soc. Am.*, vol. 72, pp. 156-160.

Qian, K. (2004).Windowed Fourier Transform for Fringe Pattern Analysis. *Appl. Opt.*, vol. 43, pp. 2695-2702.

Patorski, K & Styk, A. (2006). Interferogram intensity modulation calculations using temporal phase shifting: error analysis. *Opt. Eng.*, vol. 45, 085602.

Marengo-Rodriguez, F. A. Federico, A. & Kaufmann, G. H. (2007). Phase measurement improvement in temporal speckle pattern interferometry using empirical mode decomposition. *Opt. Commun.*, vol. 275, pp. 38-41.

Bernini, M. B. Federico, A. & Kaufmann, G. H. (2009). Normalization of fringe patterns using the bidimensional empirical mode decomposition and the Hilbert transform. *Appl. Opt.*, vol. 48, pp. 6862-6869.

Larkin, K. G. Bone, D. J. & Oldfield, M. A. (2001). Natural demodulation of two-dimensional fringe patterns. I. General background of the spiral phase quadrature transform. *J. Opt. Soc. Am.*, vol. 18, pp.1862–1870.

Pokorski, K. & Patorski, K. (2010). Visualization of additive-type moiré and time-average fringe patterns using the continuous wavelet transform. *Appl. Opt.*, vol. 49, pp. 3640–3651.

Zhang, S. (2011). High-resolution 3-D profilometry with binary phase-shifting methods. *Appl. Opt.*, 50(12), 1753-1757.

Geng, J. (2011). Structured-light 3D surface imaging: a tutorial. *Adv. Opt. Photon.*, vol. 3, pp. 128-160.

Riphagen, J. M. van Neck, J. W. van Adrichem, & L. N. A. (2008). 3D surface imaging in medicine: a review of working principles and implications for imaging the unsedated child. *J. Craniofac. Surg.*, pp. 19, vol. 517-524.

Hu, H. & Zhang, F. (2011). Real-time detection and elimination of nonorthogonality error in interference fringe processing. *Appl. Opt.*, vol. 50, pp. 2140-2144.

13

Land Administration and Automation in Uganda

Nkote N. Isaac
Makerere University Business School, Kampala
Uganda

1. Introduction

This chapter examines automation and land administration in Uganda with the goal of assessing the introduction of automation and whether it has contributed to more efficiency and reliability in the operation of land administration in Uganda. From the findings we make recommendations about the appropriate level of automation that can fully solve the challenges of land administration in Uganda.

Land administration refers to management of land issues that involve keeping custody of the land title records, documentation and ensuring land transactions are secure to promote investments. The land administration system in Uganda prior to automation was paper-based and premised on the Torrens system that guarantees land title by registration. The land administration system was regionally based, providing separate registry offices countrywide and one in Kampala, the capital city of Uganda. The managerial and technical functions were vested in the Land Registrars.

Since land is an important factor in production and the government's desire for more investments to meet the needs of the growing population, this relatively stable paper-based system was deemed incompatible with the complex and increasing number of the land transactions. The solution to these cumbersome and manual procedures was embracing automation of the processes. Automation in this context called for the use of computers, information technology and adopting online service channels that have been a dominant theme in land administration circles worldwide. It is argued that countries that have adopted automation of their land administration have had a complete transformation of the way in which land registries operate. This led Uganda to embrace the initial automation by computerizing some transactions in the 1990s in order to meet modern service level expectations. This computerized land registration system saw the conversion of existing paper-based registers, to being maintained electronically and providing online delivery of title registration functions.

These developments have improved the efficiency of land administration operations and made records more accessible to all stakeholders. However, the system does not fully allow for online transaction links to the key players in land transaction such as lawyers, banks, investors, sellers and buyers. Therefore, future plans will require that land administration in

Uganda is fully automated, with the possibility of online title registration system that enables subscribers to submit land title transfers, and discharge of mortgage for registration in electronic form over the Internet, thereby enhancing/adopting e-dealings access by the licensed law firms, banks, other transacting parties and administrators. It is anticipated that obtaining digital certificates and existing office procedures and document management systems must be aligned to the electronic environment.

2. Automation

Automation has gained widespread usage in recent years in various processes of both public and private organizations. Indeed there is agreement among scholars that automation application is usually pursued judiciously in organizations (Qazi, 2006). Sheridan (2002) offers varied conceptualizations of automation, such as data processing and decision making by computers, while Moray, Inagaki and Itoh (2000) perceive it to be any sensing, information processing, decision making and control action that could be performed by humans, but is actually performed by machines. As Sheridan and Parasuraman (2006), and earlier Kaber and Endsley (2000), contended, human-automation interaction explains the complex and large scale use of automation in various fields. Further, it explains the ability of humans to interact with adaptive automation in information processing, hence enhancing the achievement of optimal performance within an organization.

The literature argues that automated assistance is usually adaptively applied to information acquisition, information analysis, decision making and action implementation aspects, (Lee and Moray, 1994). However, the choice of a framework for analysing and designing automation systems is grounded in the theoretical framework and addresses aspects such as the role of trust, system acceptability and awareness measures (Lee and Moray, 1992). While automation is deemed to perform higher level problem-solving tasks, often the human capacity limitations lead to errors thereby intensifying the challenges of adopting of automation Frank (1998). This constraint subsequently determines the level of automation in allocating the functions to be automated, in particular, the level of the desired autonomy that represents the scale of delegation of tasks to automation and the associated implications for reliability, use and trust (Lee and See, 2004; Lewandowsky et al., 2000).

One driver of adopting automation is its ability to enhance operational efficiency of processes and the associated positive changes in the productivity of the organization. Notwithstanding this increased output rate, automation is often problematic, especially if people fail to rely upon it appropriately (Adam et al., 2003). The support for this contention is that technology is shaped by the social setting on the one side and trust that guides the assurance and reliance of the stakeholders on the other. This is in line with the fact that automation characteristics and cognitive processes affect the appropriateness of trust (Itoh, 2011).

Endsley and Kaber (1999) look at another dimension of automation by arguing that automation is applicable to different aspects of organizations, but to a varying degree, hence creating different levels of task autonomy. Therefore, the debate on automation can only be conclusive if it addresses the issue of whether partial or full automation is the desired goal of the organization. The relevance of full automation achieving the desired performance of that organization is usually at contention, though achieving a reasonable level of automation, especially for those tasks that are performed by human beings, is certainly the aspiration of every organization.

The automation of individual intellectual capabilities would allow accomplishing a higher level of automation in organizations, since within the information systems domain, managerial decision making is of special importance (Kaber and Endsley 2004). The variety of tasks that are performed by managers is immense, hence justifying automation of some, especially those that are simple, complex and repetitive with a distinct application domain while leaving those that seem not to permit automation. The argument for limiting automation is that managers act in an environment that is characterized by ambiguity and risk, hence the creation of room for discretion or judgment.

The above position finds support in the management automation scenarios advanced by Koenig et al. (2009), however, they castigate the failure to understand the effects of automation. Muir (1992) states that this is due to the problem associated with either under- or over-trusting of the automation process. The answer to the problem lies in embracing progressive automation as a way to introduce automation in a manner that quickly leverages the positive effects of automation, while reducing the potential negative effects through a gradual increase in automation. This will help align the various management automation systems to a simple common model, where users will have a better knowledge and more experience when pursuing a high level of automation in the future. Further, one contribution of progressive automation is its ability to have people build up the appropriate amount of trust in the automation system's structure and behaviour, its components and data.

According to Lee and Moray (1992), the decision to rely on automation by the users depends on both trust and self-confidence. Where there is overriding self-confidence in one's ability to perform a task over one's trust in automation, one is most likely to perform the task manually. The reverse is true in incidences where trust in automation is greater, then reliance on automation will dominate. However, Riley (1996) introduces mediating domains of automation reliability and the level of risk associated with the particular situation. The decision to use automation can depend upon different system management strategies (Lee, 1992) and user attitudes (Singh, Molloy & Parasuraman, 1993).

Another distinction in how to use automation is reliance (Meyer, 2001). Reliance refers to the assumption that the system is in a safe state and operates within a normal range, (Dzindolet 2003).Over-reliance is attributed to factors of workload, automation reliability and consistency, and the saliency of automation state indicators (Parasuraman & Riley, 1997). Inappropriate reliance on automation relative to the automation's capabilities may reflect poorly on calibrated trust, automation bias and complacency, and may also reflect failure rate behaviours (Moray, 2003).

Trust as an attitude is a response to knowledge, but other factors do intervene to influence automation usage or non-usage decisions (Muir, 1997). While trust is an important element in those decisions, it is far from the only one. According to Lee and See (2004) humans use alternate routes by which they develop their trust, namely the analytic methods that assume rational decision making on the basis of what is known about the motivations, interests, capabilities and behaviours of the other party. Cialdini (1993) argues that we tend to trust those people and devices that please us more than those that do not. Further, there is realization of the temporal element to trust building that takes time to acquire, whether through experience, training or the experiences of others.

3. Extent of automation in land administration

In setting the stage for automation there is acknowledgement that automation is applied to various organizations, however, the feasibility of any applicability necessitates understanding the nature of the transactions involved and the managers that perform the tasks (McLaughlin, 2001).

Introduced in 1908, the land administration management in Uganda is based on the Torrens system developed in 18thcentury Britain. The tenets of the Torrens system are that the government office is the issuer and the custodian of all original land titles and all original documents registered against them. Further, the government employees in their management tasks examine documents and then guarantee them in terms of accuracy (Barata, 2001).

The Torrens system has three principles: the mirror, curtain and insurance principles. While the mirror principle refers to certificates of title, which accurately and completely reflect the current facts about a person's title, the curtain principle ensures that the current certificate of title contains all the relevant information about the title that creates certainty and offers assurance to the potential purchaser about the dealings on any prior title. The whole trust and confidence in the transactions covered by the insurance principle will guarantee the compensation mechanism for loss of the correct status of the land. These combined principles contribute to secure land transactions and the development of the land market in any given country.

Practitioners in land transactions argue that the Torrens system ensures that the rights in land are transferred cheaply, quickly and with certainty. For manually managed systems, this is conceivable in cases of low volumes of transactions; otherwise, large volumes of transactions necessitate automation. It is argued that the benefits of an automated system will lead to efficiency, accuracy, integrity and cost containment. This will overcome the challenges of retrieval of documents and the inability to manage and store large amounts of data efficiently (Ahene, 2006).

The management tasks in land offices include capturing the precise parcel of land, the owner, limitations of the right of ownership and any right or interest which has been granted or otherwise obtained. Another management task is the cancellation and creation of certificates of title, land notifications and transfers, subdivisions, showing all outstanding registered interest in the land, such as mortgages, caveats and easements

In the management of land transactions there is legal examination of all the associated data entry on documents to make sure that the documents are correct and in compliance with the law affecting land transactions. The objective of the examination is to ensure that the document complies with all applicable law and therefore, this process involves making judgments upon the relevancy of the law and ensuring certainty of the transaction.

Another management task within land transactions are those roles performed by the survey staff which involve reviewing and making the associated data entries, and comparing and interpreting the existing land survey evidence with the new ones to ensure that the land surveyed on the new plan does not encroach upon adjacent lands (Barata, 2001).

In reference to the existing documents, the management task involves processing of searches of the records in the land office that are classified as public records which can be searched by anyone wishing to transact in land matters. Request for searches are usually received from agents, particularly lawyers and bank staff, that seek to verify the documents to conclude the transactions of their clients. To obtain a title search you must know the legal description, land identification numbers or the title number for the property you want to search. The primary purpose of the name search facility is to enable creditors and other parties with statutory rights to determine what interests in land are owned by the person affected by instruments.

While the above management tasks are simple, with increasing volumes of transactions, the land office in Uganda was characterized as inefficient and riddled with unethical behaviour. This outcry by stakeholders led to the adoption of the automation of the processes of the land office in two phases. The first step of automation was initiated by the government through the Ministry of Land and Urban Development. The approach involved capturing of the records in a digital format to create a reliable computer database to allow prompt searches and retrieval of information. The second phase saw the introduction of automation-based decision making. These two phases meant that automation included data entry, indexing and scanning of Mailo land records. These fully automated interventions in the land registry helped generate computer-related information on land ownership, information on Mailo land transactions such as changes in property ownership and encumbrances thereon. The automation of the land office meant all live paper titles were converted from their paper format into an electronic medium. In addition to each certificate of title being assigned a unique title number at the time of its creation, each parcel of land contained within each title was also assigned a unique code number.

A critical review of the documents produced during the process of automation reveal that the conceptual plan included a progressive computerization of the legal and administrative records, and cadastral maps. The aim was to address the shortcomings of the manual system, restore the integrity of the land registry and ensure modernization of land registry operations to meet the needs of a growing economy. It was quite logical to start the process of the actual computerization of the land registry after getting the filing right and re-organizing all the registry records, reconstructing the torn and damaged records and vetting of all records in the manual system. This helped identify and get rid of any forgeries or problem land titles in circulation.

As a result of automation, most titles, all registered documents with a registration number and all plans can now be searched and retrieved electronically, in comparison to the previous practice of only being able to search through a registry agent.

There are many advantages in the automation of the land office over the manual system, meaning that documents can be searched electronically and in a fast and expeditious manner. The documents can also be easily replicated, averting potential loss from natural disasters and lastly the documents are linked with other documents. This chimes well with Maggs' (1970) assertion that automating land systems is only commercially practical where highly formalized rules are applied to highly standardized data. Where rules and data are not of a highly formal nature and so require a large degree of individual judgment, progress can only be achieved through the substitution of new formal rules and data structures.

4. Methodology

The aim of the research for this book chapter was to examine the automation and efficiency of the land transactions management office in Uganda. In analysing the automation of the land transactions in the Ugandan land office, we adopt a cross sectional, qualitative and analytical study design in order to examine automation operations in the three districts of Kampala, Mukono and Mpigi. Primary information was acquired via semi-structured in-depth interviews and focus group discussions (FGDs). The main instrument of data collection was a structured questionnaire for the Key Informant Interviews (KIIs) comprising of lawyers, bankers and staff of the land office.

The questionnaire contained open-ended questions and was developed based on the literature review, as well as drawing from other studies on automation. The key questions that were used related to trust in automation, reliance, usability and confidence. The information obtained formed the basis for the analysis of the findings. The instrument was pilot tested prior to beginning fieldwork to assess the clarity, flow and appropriate context of the questions. Following the pilot test, the questionnaire was refined based on the initial feedback.

While carrying out the study, there should be a population from which a sampling frame can be drawn. The population of the study was drawn from the three district land offices of Kampala, Wakiso and Mpigi that have fully automated their operations. Further, focus was on the institutions that were identified as being key stakeholders to the land transactions, namely the banks, law firms and the administrators of the land offices. Therefore, interviews and focus group discussions comprising lawyers, land office administrators and bankers were conducted to evaluate how the three principle users gauged the automation efforts by the land office.

Using a purposive sampling technique, 11 lawyers were purposively selected from law firms that frequently transact land issues, seven staff from the legal departments of seven banks and 15 staff of the land administration. The purposive sampling was based on the frequency of land transaction by the law firms and the banks in the last six months preceding the study. To capture different perspectives and experiences, the staff of the land offices had varying roles and levels of participation in activities and decision-making processes.

The data collection fieldwork in Kampala commenced in July 2011 and lasted through to August 2011 in the other two districts. To deal with the ethical considerations of the study, the firms and institutions were briefed about the study and appointments were made in order to have meaningful and well-planned discussions without disrupting their normal work. The participants were briefed and verbal consent was sought from all participants. The exercise started with giving the respondents background information about the study.

5. Findings on automation and land management

The proponents of automation argue that when organizations undertake automation it impacts positively on employee productivity and improves efficiency and decision making, besides allowing better management control of operations. At the same time, the opponents of automation are critical of it because it lessens the responsiveness of the bureaucrats to citizens, puts technological elite in charge of processes and shifts power to managers, among others. These diametrically opposed positions are therefore considered when analysing the impact of automation in the organizations that have adopted or intend to adopt it. The findings follow the objective of this study which was to examine automation and the efficiency of land administration in Uganda.

The results of the study show that there has been successful automation in land administration in the three districts following the adoption of a comprehensive computerization, with online access to the system minimizing the need for the paper-based system.

The study sought an assessment of the experiences of the employees with automation of land management. The response was varied as the consequences to the clerical and managerial level staff were differentiated. Most of the clerical staff thought that their own interest in the job had been eroded. However, when responding to how automation has influenced the productivity of the employees, they revealed a registered increase in output. This is quite conceivable since automation released the employees from the clumsy, tedious and dusty manual tasks, thereby allowing them to learn new skills that are ideal for enhancing the production of better quality work at a fast rate. Automation has changed our perception of our jobs which has resulted in improved quality, quantity and accessibility of data and hence, improved services from the land office. This was against the findings that indicated an initial high sensitivity and fear of learning the new systems. The automated institutional infrastructure so far has reduced the internal resistance to operate the computerized system to its full potential. While this has been achieved, there remains the fact that the transactions of the stakeholders are not automated and in this the non-linkage to key offices, such as law and banking firms, is a limitation on the availability of the information. This means they have to come to the land office to undertake transactions instead of doing it online through networked systems.

Secondly automation has, besides improving productivity, impacted on efficiency. As indicated by the responses of participants it has been positive because there has been a substantial reduction in the transaction time from years in some instances to just a month or two from the time of lodging the transaction. This can be attributed to far easier searches and retrieval of land records in digital format through the automated system. The quick, easy information retrieval and search procedure means that the partial automation of the tasks has facilitated the public sector managers to continue making decisions in line with the set procedures and rules. This implies that while full automation has eased the retrieval of information, the decision making processes still require judgement based on the procedures and hierarchy, and this impacts upon the level of adoption of automation and the efficiency of public managers' action. Indeed the respondents reported cases of a similar approach and therefore, no indication of a drastic reduction in red tape. This argument points to what is discussed in the literature about the level of automation in that certain decisions have to be made by the managers hence, limiting the much anticipated benefits of automation.

Thirdly the sceptics of automation doubt the rationale for automation, a fact reiterated by many stakeholders who contributed their insight to the study. They argue that even under the manual system, retrieval personnel were always quick if one complied with their rent seeking motives. By proxing that the manual process was slow, one negates the fact that the motivation of rent seeking created the delays, rather than the inability to get the information. The clerical personnel usually made standard statements to the effect that information was not available, could not be traced or that documents were missing with the stakeholders having no avenue by which to prove otherwise. As one respondent noted: *"the problem hasn't been really the manual system, rather the attitudinal behaviour of the clerical staff that perform their duties upon extra facilitation or inducement by those seeking the services" (comment by one legal respondent).*

The findings point to the fact that notwithstanding the positive effects of automation, the services of the land administration have been depersonalized. While this is good because it is likely to limit the interface between the service providers and the stakeholders and therefore, reduce the rent seeking motives of the managers, on the other hand it erodes the trust, confidence and networks that have been built by the frequent and regular users of the system. Stakeholders continue to express misgivings about automation, including mistrust about the extent to which automation will free them from forgeries and malpractice as experienced in the past. Automation has removed the avenue by which to obtain preferential treatment by the managers, particularly in matters of urgency. The question is, has this reduced the avenue by which to pursue personal or institutional vested interests. While some would answer in the affirmative, others think that even with automation, the respondents revealed that still a number of officials intervene in a single transaction. Hence, quick access and transparency of the land records is not fostering investments in land as it can facilitate the speeding up of processes of transactions, credits, transfers and mortgages.

One of the findings from the public managers indicates their pleasure with the ability to offer easier access to information and services through the use of the web and integrated database. *"With automation we can't reach citizens in different environments and we are now handling public requests for information and service in a timely manner" (remarks by public manager)*. However, there is also a fear in the minds of the people about the potential manipulation of these records.

6. Conclusion

The qualitative findings have been presented and the evidence from the respondents point to increased efficiency, productivity and job satisfaction in land administration. The automation of the land administration has been an achievement given the fact that it had been crippled for years and was unable to cope with the increasing volume of transactions. Secondly, automation of the land transactions offers the society a great opportunity in public access to information that had hitherto been under the sole custody of state functionaries, thereby leading to rent seeking practices on a massive scale. The automation of land administration has facilitated citizens' access to it and, by implication, increased tenure security.

While the public managers have gradually gained trust in automation, the daunting task that remains is to build trust among the stakeholders that the information generated through automation will not be subject to fraud and the risk of loss of property. Many people still insist on the old paper documents as a reassurance to the different market actors and state public managers. Associated with this is the fact that while automation benefits have eased decision making, the findings revealed that in order to restore citizens' confidence in land administration, induced by the questionable ethical behaviour of the managers, internal change is required.

What remains to be seen is whether the new system steers clear of corruption or alternatively aims at grappling with the existing system to change its dynamics by bringing about structural and technological change.

The project in the three districts has served as a testing ground for working towards achieving a more efficient nationally managed system. It will enhance efforts to increase easy public access to land records and the idea of running an automated system parallel to the existing manual system raises the expectation that the automated-based system will demonstrate its superiority over the manual system.

Lastly automation and creating an electronic system containing the possibility for electronic searches, analysis, retrieval and manipulation with an electronic version of the document in a non-revocable electronic form would give the required safeguards.

Like any study, we experienced the shortcoming of unwillingness by respondents to release information due to a fear that it would be revealed. This was overcome by asking the respondents to remain anonymous. Further limitations of the study were the fact that the system is still new and therefore, there is still a lot of old practice, which make assessment of the gains rather difficult. Despite all this the findings reveal there is a general positive attitude to automation.

7. References

Adams B.D, Bruyn L.E, Houde S and Angelopoulos P (2003) Trust in Automated Systems Literature Review Human system incorporated, Toronto

Ahene, R. (2006) Moving from Analysis to Action: Land in the Private Sector Competitiveness Project :World Bank / Private Sector Foundation Uganda.

Barata, K. (2001) Rehabilitating the Records in the Land Registry, Final Report; USAID/SPEED (www.decorg/partners/dex_public/index)

Cialdini, R.B. (1993). *Influence: Science and Practice. 3rd Ed.* New York; Harper Collins.

Dzindolet,M. (2003) The role of trust in automation reliance, *International Journal of Human-Computer Studies,*Vol. 58, Issue 6, Publisher: Academic Press, Inc., pp.697-718

Endsley, M.R.(1996) Automation and systems awareness: *In Parasuraman & M Mouloua (Eds) Automation and Human Performance: Theory and Applications* (pp. 163-181)

Endsley, M.R. & Kaber, D.B. (1999) Level of automation effects on performance, situation awareness and workload in a dynamic control task *Ergonomics,* 42(3), 462-492

Frank U (1998) Increasing the level of automation in organization, remarks on formalization, contingency and social construction of reality. In: The Systemist, Vol.2091998) p, 98-113

Itoh, M. (2011) A model of trust in automation: Why humans over-trust? SICE Annual Conference (SICE), *2011* Proceedings:

Kaber, D B. and Endsley, M (2000) Situation Awareness & Levels of Automation, Hampton, VA: NASA Langley Research Center; SA Technologies -

Kaber, D B. and Endsley, M (2004) The effect of level of automation and adaptive automation on human performance, situation awareness and workload in dynamic control tasks: Theoretical issues in Ergonomics science 5, 113-153

Koenig, R., Wong, E.,& Danoy, G (2009) Progressive automation to gain appropriate trust in management automation systems. Degstuhl Seminar proceedings, Combanatorial scientific computing, http://drops.degstuhl.de/opus/volliexte/2009/2093

Lee, J. (1992). *Trust, self-confidence, and operators' adaptation to automation.* Unpublished doctoral thesis, University of Illinois, Champaign.

Lee, J. & See, K. (2004). Trust in Automation: Designing for Appropriate Reliance. *Human Factors, 46* (1), 50-80.

Lee J &Moray N (1994) Trust, self-confidence and operation's adaptation to automation. *International Journal of Human Computer Studies 40,153-184*

Lee, J.& Moray,N. (1992) Trust, control strategies and allocation of functions in human-machine systems. Ergonomics 35, 1243-1270

Lewandowsky,S., Mundy, M.&Tan, G. (2000) The dynamics of trust: comparing humans to automation.*Journal of experimental Psychology: Applied,* 6(2) 104-123

McLaughlin, J. (2001) Land Information Management: From Records to Citizenship Online

Meyer (2001) Effects of warning validity and proximity on response warnings.*Human Factors* 43(4) 563-572

Maggs, P. (1970) Automation of the land title system

Moray, N. (2003) Monitoring, complacency, scepticism and eutectic behavior : International Journal of Industrial Ergonomics, Volume 31, Number 3, pp. 175-178(4) Publisher: Elsevier

Moray, N., Inagaki, T.&Itoh,M. (2000) Adaptive automation, trust and self-confidence in fault management of time critical tasks, *Journal of Experimental Psychology Applied* 6(2) 44 -58

Muir, B.M. (1997) Trust between humans and machines: the design of decision aids, *International Journal of Man-Machines studies* 27, 527- 539

Muir B.M (1992 Trust in automation: Part I. Theoretical issues in the study of trust and human intervention in automated systems Ergonomics Volume 37, Issue 11, 1994 (1905-1922)

Parasuraman, R. & Riley, V. (1997). Humans and automation: Use, misuse, disuse, abuse. *Human Factors, 39,* 230-253.

Qazi, M. U. (2006) "Computerization of Land Records in Pakistan", LEAD International, Islamabad.

Riley, V. (1996). Operator reliance on automation: theory and data. In R. Parasuraman & M. Mouloua (Eds.), *Automation and human performance: theory and applications* (pp. 19-35). Mahwah, NJ: Lawrence Erlbaum Associates, Inc.

Sheridan, T.M. (2000) Humans and automation: systems design and research issues, *HFES Issues in Human Factors and Ergonomic Series* volume 3 Santa Monica CA; Whiley and Sons Publication Inc.

Sheridan and Parasuraman, 2006. Human-automation interaction. In: Nickerson, R.S. (Ed.), Reviews of Human Factors and Ergonomics, vol. 1.

Singh, I., Molloy, & Parasuraman, (1993). Automation induced complacency: development of complacency potential rating scale: *International Journal of Aviation Psychology*

14

A Graphics Generator for Physics Research

Eliza Consuela Isbăşoiu
Department Informatics, Faculty of Accounting and Finance,
Spiru Haret University,
Romania

1. Introduction

The great quantity of information which emerges from the necessity of an informational society can be mastered in a proper way, due to the perfection of the informatics systems used in all the fields. The pieces of information, due to the fact that they are numerous as concerns their content, become hard to work with, on the whole; thus, only a part of the content is represented as compound and flexible objects, by assuring both their direct usage and re-usage within other conditions and situations. An important factor in the trend of spreading and adapting the application on the web zone is represented by distributed systems and web services.

Generally speaking, the administration of resources and fluxes in distributed systems of great dimensions are confronted with many difficulties. These are provoked by the complexity of the hardware and software platforms and, especially, by the dynamism which is characteristic of these systems' compounds.

These drawbacks can be met through two necessary conditions, based on the understanding of the performing characteristics of the resources of administration systems and of the delivery solutions for a better dynamic adaptation.

This chapter proposes to demonstrate the utility of distributed systems in the design of educational and research software in physics. Educational systems are generally rich in structured knowledge, but not so complex, which does not imply behavioural relations among users.

However, all of the existing architectures being their construction with the experts in the experience of the learning domain, but not from the user's perception, as well as from the study of the materials of the way one constructs the schemes of mental concepts' (Hansen, 2007). One has to underline that there is an intrinsic connection between the results obtained by the users and the method of teaching. Due to the restriction imposed by time and resources, the great part of the implementation is reduced to simple models of knowledge. Thus, the present educational systems do not propose to determine the real state of knowledge, and so the strategies used by the users are omitted as are the specific strategies for each and every domain.

2. Theoretical background

Because of problem which one follows wants to develop consists in realizing web services with numerical calculating methods, I provide a presentation of some points of view connected to the subject. The basic standards, originally proposed by IBM, Microsoft, HP and others, are: SOAP (Simple Object Access Protocol), which offers a standard modality of connecting to the web, UDDI (Universal Description, Discovery and Integration), which represents a standard modality of issuing one's own web services, and WSDL (Web Service Description Language), which provides further information for understanding and using the offered services (Landau 2008). WSIF (Web Services Invocation Framework) represents the invoking of the web services, defined by the standards WSDL and UDDI (Universal Resource Locator). This kind of invoking does not suppose the creation of SOAP XML messages. There are four kinds of transmitting an invoking: one-way - the final point receives an answer; request-answer – the final point receives a message and transmits a correlated message; solicitation-response - the final point transmits a message and receives a correlated message; notification – the final point transmits a message. The most used are one-way and request-answer. The location transparency is realized through link patterns, as a UDDI type. The web services revolute the way in which one application communicates with other applications, thus offering a universal pattern of data which can easily be adapted or transformed. Based on XML, the web services can communicate with platforms and different operating systems, irrespective of the programming language used for their writing. For easy communication among the business parameters: the Java language allows the applications' realization regardless of the platform used. The apparition of the Web services represents a means of communication on a large scale. This can be done by applying some standardized protocols, defined by standardized public organizations. The basic protocols of the Web service are not complete as regards the description and implementer terms, their security, reliability and complexity. It is known that by the fact of using Web technology, the information uses many sources. The portal is the mediator for the access to the information. The portal servers are, in their turn, elementary components in the interchange of information. The importance of these technologies is due to the fact that they emerged as a response to the present tendencies manifested in the software industry: distributed remaking, the development of applications based on the components, the development of the services for companies and the modifying of the Web paradigm, in the sense of the development of applications.

3. Services for numerical calculation

The purpose of the present study is the analysis of the surplus introduced by the Web service interface and its exposure in as simple as possible a variant of the facilities introduced by it. The study uses the book *Numerical Library in Java for Scientists and Engineers*, and so the present work does not re-implement the codes in that book, but only uses those already implemented. The passing from an interface with the user in a command line to an interface based on Web services represents a big advantage because it allows the combining and re-using of some procedures with a prior definition. Many discussions about the algorithms in the speciality literature on the Web omit important details, which can only be uncovered by encoding or be made suitable by reading the code. We also need the real code for object-oriented programming, found in the Numerical Library in Java for Scientists and Engineers (Lau, 2004). An orientation of these solutions is for scientific explanations which are based on numerical calculation, beginning with elementary structures, progressing to complicated ones, as the integration of ordinary differential equations (Lau, 2004). The digital technologies introduced to the market include Web technology; this completely fulfils all of the requests connected with the information cost, stocking and spreading (Petcu, 2008). From these first steps where the sites were using simple visiting cards, iterative processes have been created and developed. Once, with the movement towards an informational society, a novelty is represented by the distributed systems of large dimensions, by which facility is enabled access to a great variety of resources. The above mentioned book offers a general discussion for each subject, a certain amount of mathematical analysis (IBM, 2009), a certain discussion of the algorithm and, most importantly, the implementation of these ideas in a real mode as routines. There is a proper equilibrium among these ingredients for each subject. A starting point is concerned with the construction of the services and the inclusion of one or more methods in implementation is constituted by the analysis of the data structures from the elementary to the complex (Isbasoiu, 2009). The majority of the books on numerical analysis use a specific subject "standard", such as: linear equations, interpolation, extrapolation and ordinary differential equations. Other discussed points are: the functions' evaluation and especially mathematical functions, aleatory numbers, the Monte Carlo method, sorting, optimization, multidimensional methods, the Fourier transformation, the spectral methods and other methods concerning statistical descriptions and data modelling and relaxing methods. The book (Lau, 2004) says that there are seven groups of themes that should be focused on: vectors and matrices, algebraic evaluations, linear algebra, analytical evaluations, analytical problems, special functions, proximity and interpolation. Each group has the name Basic followed by the appropriate number for each theme. A schematic representation of these groupings is presented in Figure 1 (Isbasoiu, 2009):

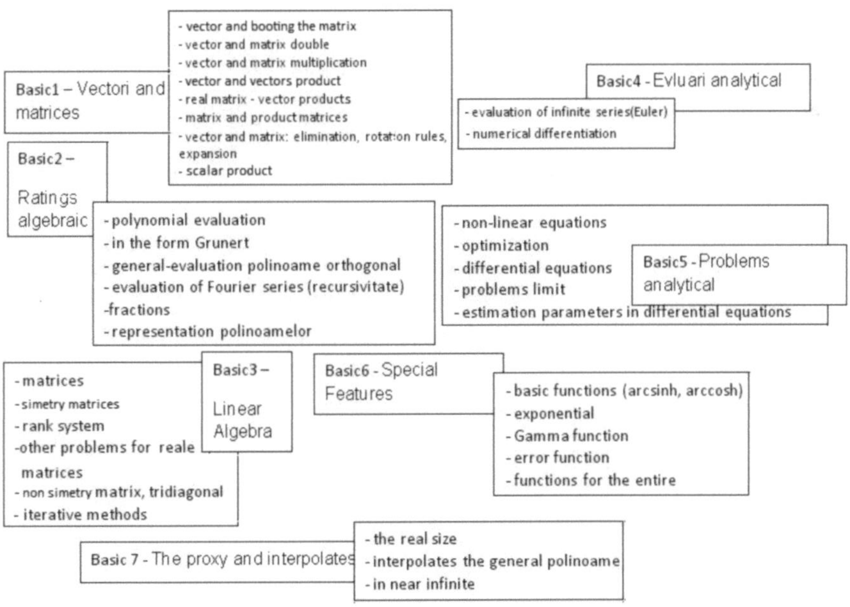

Fig. 1. The basic grouping

Every group contains an impressive number of functions. Through an attentive analysis, we noticed the following: Basic 1 is the base for the construction of the other basics, and some functions of Basic 1 are applied even with the construction of the other functions of this basic; every basic interacts with the other basics; there are functions which do not interact or else do not represent the basis for the construction of other functions. In trying to conclude all of these representations in one, it appears in the following figure:

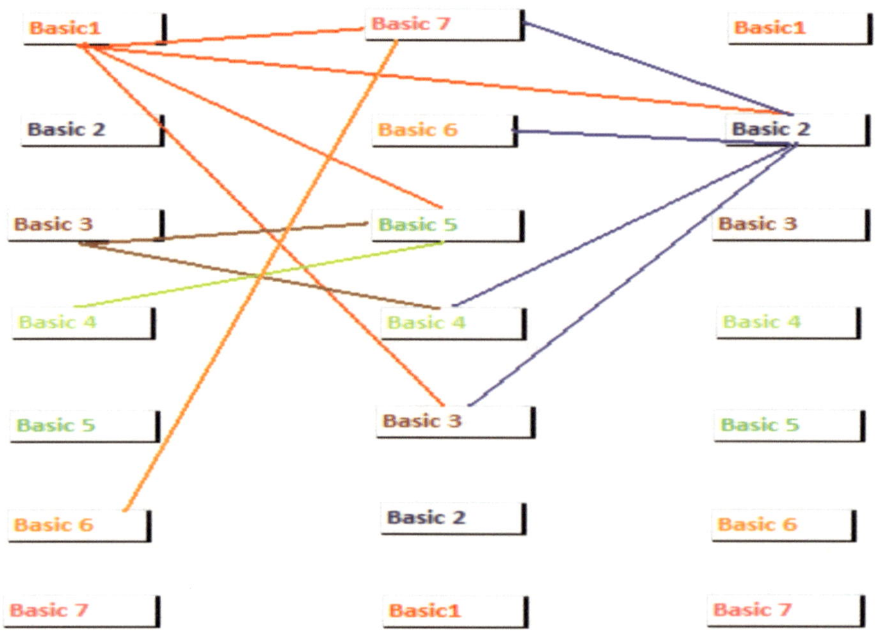

Fig. 2. All dependencies' representations

In the following table is presented the number of times a function is used by other functions:

	Basic 2	Basic 3	Basic 4	Basic 5	Basic 6	Basic 7
Basic 7						
Basic 6						1
Basic 5				1		
Basic 4				1		
Basic 3			2	14		
Basic 2		1	3		4	1
Basic1	1	77		20		2

Table 1. The dependencies' representation

A graphic representation of these dependencies, depending on the number of functions that appear, is realized as follows:

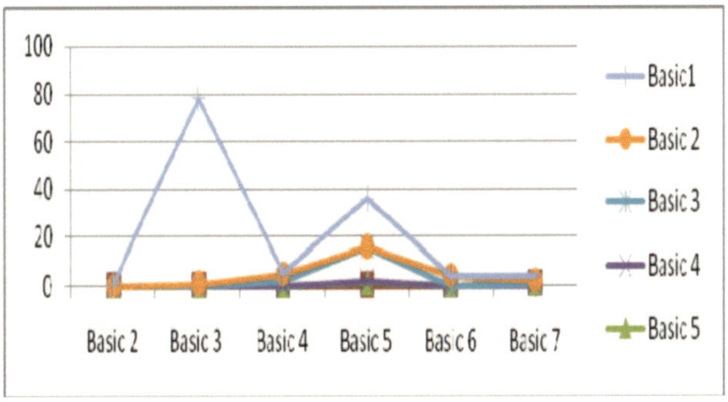

Fig. 3. The graphical representation of the dependencies according to the number of functions

3.1 The functions for numeric calculation – Time comparison

The services were realized using the Eclipse platform. For each function, the response time is specified. We can notice that it varies a lot. We repeat the measurements for more complex input data, both for double type and integer type and vectors and matrices because these units of time depend on this input data.

Function Name	Time of response with Web Service (val/ms) -simple values	Time of response with Web Service (val/ms)-complex values	Function Name
Rnk1min	16	143	Rnk1min
Praxis	11	580	Praxis
Marquardt	13	541	Marquardt
Gssnewton	14	642	Gssnewton
Multistep	10	533	Multistep
Ark	15	570	Ark
Efrk	13	13	Efrk
Efsirk	561	616	Efsirk
Liniger1vs	456	603	Liniger1vs
Liniger 2	473	583	Liniger 2
Gms	605	470	Gms
Impex	562	524	Impex
Peide	628	587	Peide
Minmaxpol	611	552	Minmaxpol

Table 2. Time of response for every function with the Web Service

An elegant graphical representation of the response variations can be seen in the following graphic:

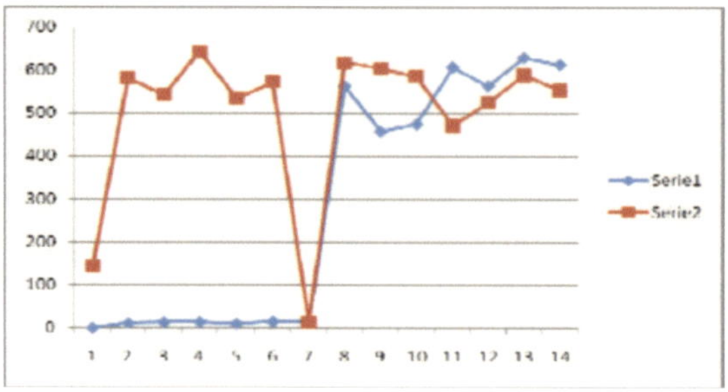

Fig. 4. Graphical representation of the response variations for every function

We repeat the measurements, without the Web Service, for more complex input data, both for double type and integer type and vectors and matrices, because these units of time depend on this input data.

Function Name	Time of response without Web Service (val/ms) - simple values	Time of response without Web Service (val/ms) - complex values	Function Name
Rnk1min	10	36	Rnk1min
Praxis	8	368	Praxis
Marquardt	9	336	Marquardt
Gssnewton	7	398	Gssnewton
Multistep	8	327	Multistep
Ark	11	352	Ark
Efrk	10	7	Efrk
Efsirk	314	326	Efsirk
Liniger1vs	289	318	Liniger1vs
Liniger 2	293	359	Liniger 2
Gms	348	236	Gms
Impex	311	312	Impex
Peide	362	362	Peide
Minmaxpol	353	342	Minmaxpol

Table 3. Time of response for every function without the Web Service

A more elegant representation of time variation can be seen in the following graphic:

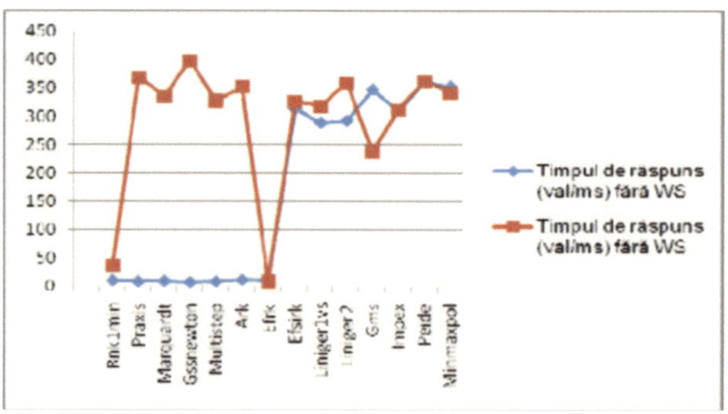

Fig. 5. The graphical representation of time response of each function

In analysing with a focus on the two graphics (Fig. 3 and Fig. 4), we can notice the following:

a. generally, the response times are longer for complex parameters, the main cause stands for the value itself of those parameters
b. the longest time belongs to those functions which resolve determinants, matrices and vectors
c. there are extremely close response times for some functions
d. we can see two groups of functions, corresponding to two types of functions
e. it is clear that the only function which does not modify its time is Efrk, having the value of 13 ms - with the Web Service
f. irrespective of the type of data introduced - or of their complexity - the response time is bigger, but still close
g. by comparing the bits of response time with the Web Service there are significant differences, so with this case it is important to consider the complexity of the input data

3.2 Scenarios

The evaluation of a function and the discovery of its minimum (using the function Rnk1min)

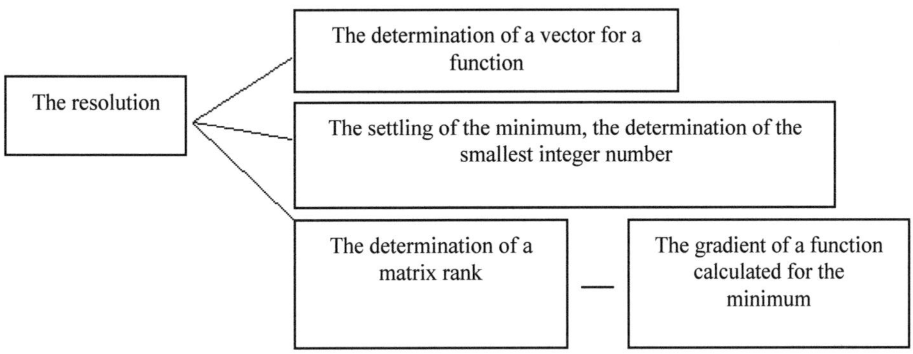

Fig. 6. The graphical representation of a scenario for the evaluation of a function and the discovery of its minimum

Procedures used: vectors product, the determination of a superior triangle matrix.

The determination of the minimum of a function with a minimum number of searches (using the function Praxis):

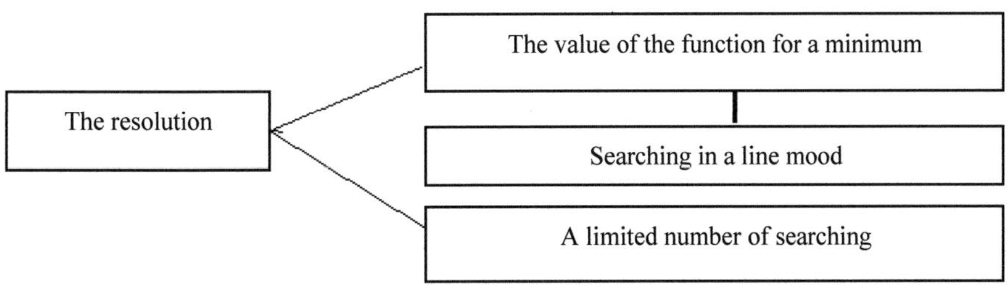

Fig. 7. The graphical representation of the scenario for the determination of the minimum of a function with a minimum number of searches

Procedures used: initialization of the constants, initialization of the sub matrices, matrices' product, matrices' rank.

The resolution[L1] of a non-linear system (using the function Marquardt)

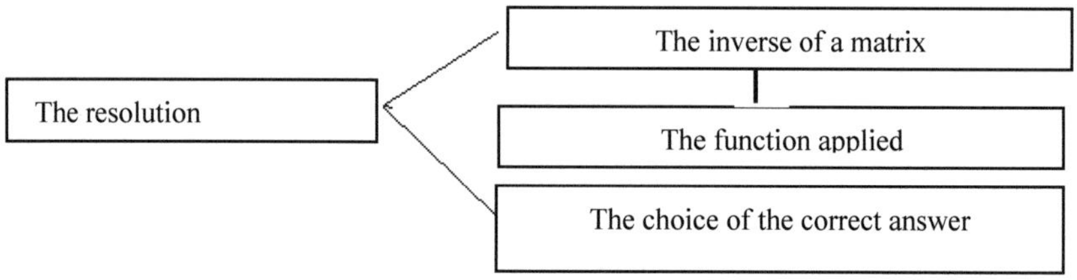

Fig. 8. The graphical representation of the resolution of a non-linear system

Procedures used: the product of two vectors, the product of a vector and a matrix.

The determination of the solutions of a non-linear system (using the function Gssnewton)

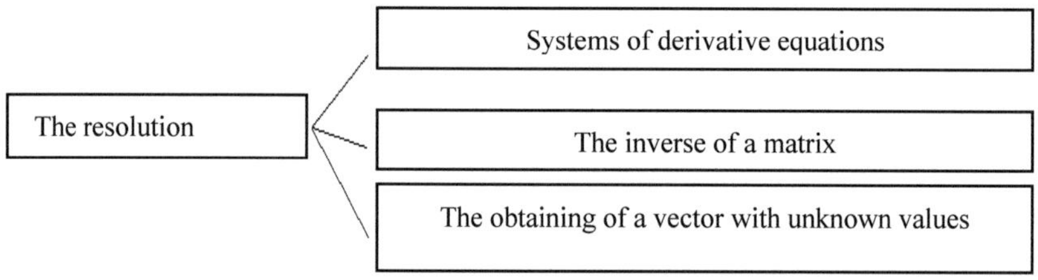

Fig. 9. The graphical representation for the determination of the solutions of a non-linear system

Procedures used: pivot method, the resolution of the systems through successive appeals, the interchanging of lines and columns for the inversion of a matrix, the product of two vectors, adding a multiple constant, systems with the same matrix coefficient, resolved through successive appeals.

The resolution of the systems with differential equations (using the function Multistep)

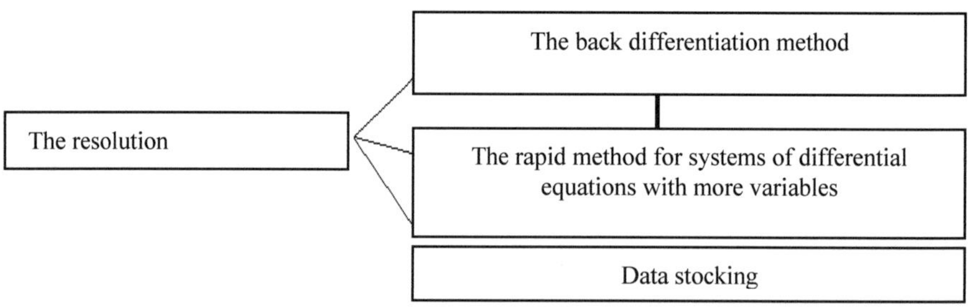

Fig. 10. The graphical representation of the scenario for the resolution of systems with differential equations

Procedures used: the product of two vectors, the product of a vector and a matrix, matrices decomposition, particular cases: a smaller number of variables as against the number of binary numbers in the number of representation, the resolution of the linear systems whose matrix was distorted triangularly.

The resolution of the parabolic and hyperbolic equations (using the function Ark)

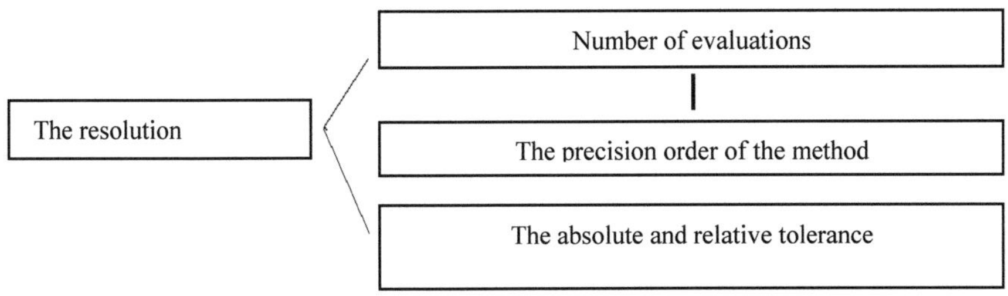

Fig. 11. The graphic representation of the scenario for the resolution of parabolic and hyperbolic equations

Procedures used: the initialization of a vector after certain constants, vectors with multiple setting forms, factors for multiplication, duplicate elements in a vector, the product of an element belonging to a vector and other elements belonging to another vector, adding a multiple constant from a vector to other elements from another vector being in a certain state, the resolution of the systems of linear equations

The resolution of the systems with ordinary differential equations (using the function Efrk)

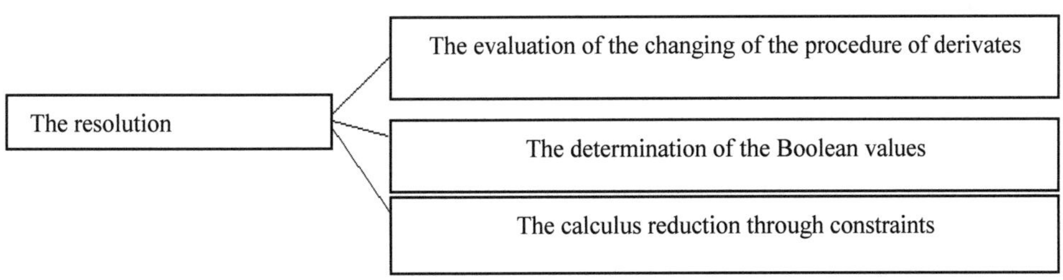

Fig. 12. The graphical representation of the scenario for the resolution of the systems with ordinary differential equations

Procedures used: adding a multiple constant from a vector to an element belonging to another vector which is in a certain state, matrices decomposition, permuting, resolution linear systems.

Resolution the autonomous systems of differential equations (using the function Efsirk)

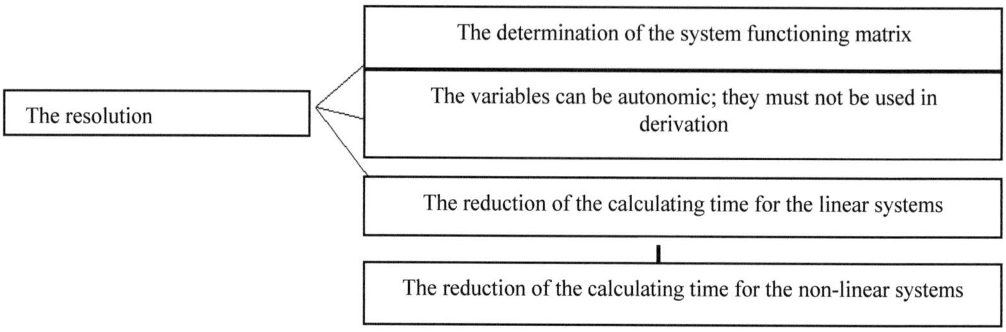

Fig. 13. The graphical representation of the scenario for the resolution of the autonomous systems of differential equations

Procedures used: the product of an element of a vector and another element of another vector, the product of a vector and a matrix, the product of a vector represented by the index of a row and a matrix represented by the index of a column, matrices decomposition, permuting, the use of the partial pivot, factors for multiplication, decomposition of the linear system as the matrices type.

The resolution of the autonomous systems for differential equations (using the function Liniger1vs)

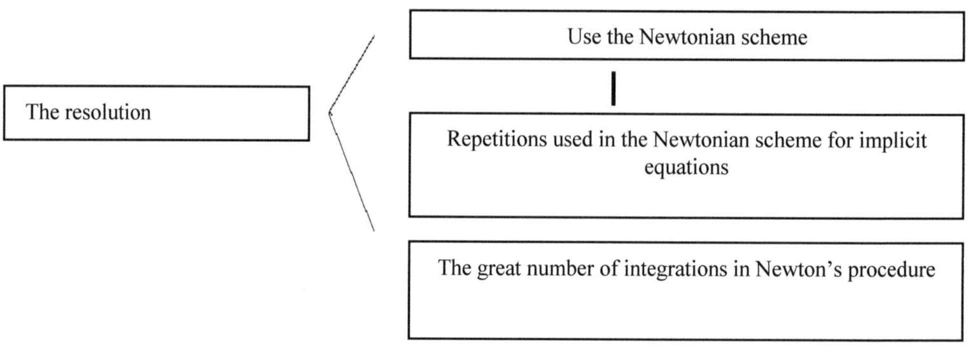

Fig. 14. The graphical representation of the scenario for the resolution of the autonomous systems for differential equations

Procedures used: the initialization of the matrices and constants, factor of multiplication, replacement the elements of a matrix with a succession of elements in a rectangular matrix, duplication of an element in a vector, the product of a vector and a matrix, adding a multiple constant, the product of two vectors, matrix decomposition, permutation of matrices, resolution the linear systems.

The resolution of the autonomous systems for differential equations (using the function Liniger2)

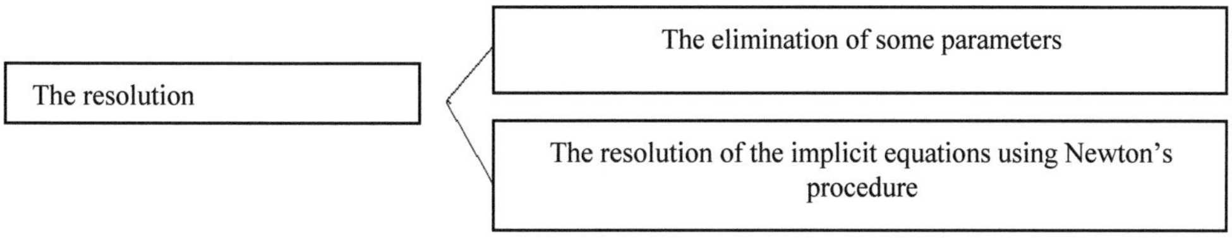

Fig. 15. The graphical representation of the scenario for the resolution of the autonomous systems for differential equations

Procedures used: the product of an element of a vector and another element of another vector, the product of a vector and a matrix, the product of a matrix line and the column of another matrix, matrices decomposition, permuting, the resolution of the linear systems.

The determination of the number of differential equations (using the function Gms)

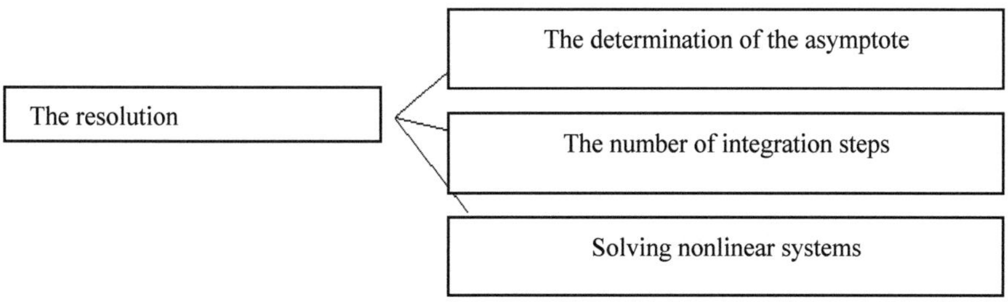

Fig. 16. The graphical representation of the scenario for the determination of the number of differential equations

Procedures used: the product of a vector and another vector, the product of a line in a matrix and a column in another matrix, adding a multiple constant, the resolution of the linear systems, the replacement of a matrix column with constant complex elements and factors of multiplication.

The determination of the number of differential equations (using the function Impex)

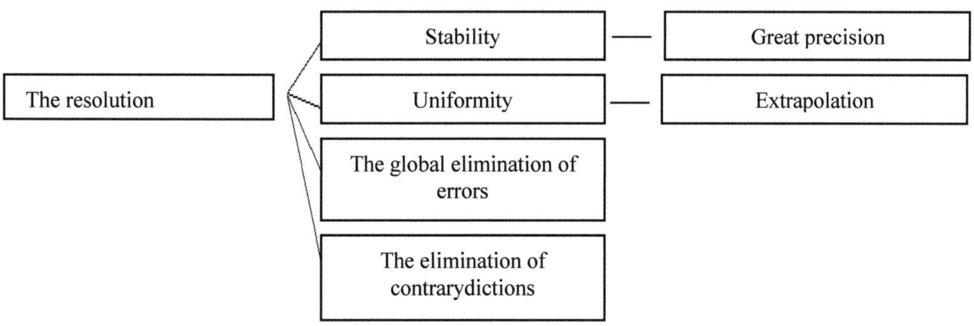

Fig. 17. The graphical representation of the scenario for the determination of the number of differential equations

Procedures used: the initializing of the matrices and constants, the initializing of a sub matrix, the replacement of an element in a vector, the replacement of a column in a matrix with complex constant elements, the product of an element of a vector and another element of other vector, the product of a vector and a matrix, the product of a line in a matrix and the column of another matrix, matrices decomposition, permuting, resolution the linear systems.

Estimations on the unknown variables (using the function Peide)

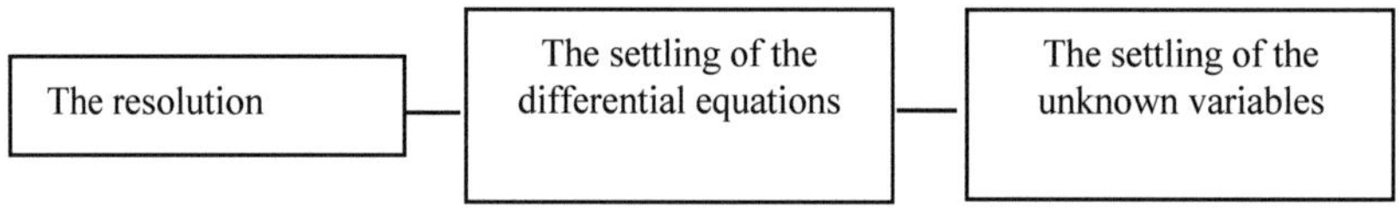

Fig. 18. The graphical representation of the scenario for the estimations on the unknown variables

Procedures used: initializing of the matrices and constants, initializing of a sub-matrix, factor of multiplication, replacement of an element in a vector, replacement of a column in a matrix with complex constant elements, the product of an element of a vector and another element of another vector, the product of a vector and a matrix, the product of a line in a matrix and the column of another matrix, matrix decomposition, permuting, resolution the linear systems, the maximum number of repetitions to be made.

Calculating the polynomial coefficients (using the function Minmaxpol)

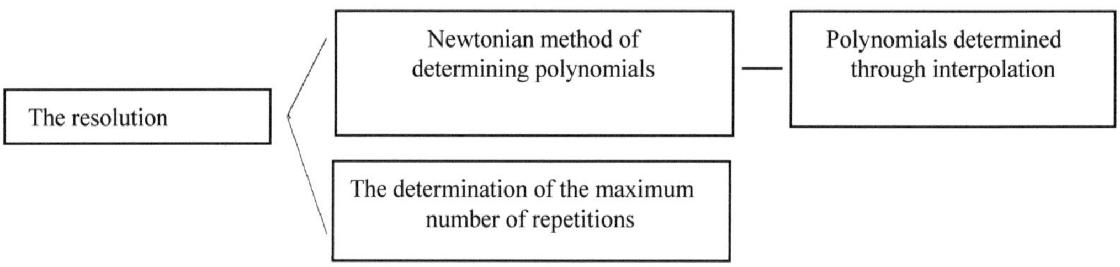

Fig. 19. The graphical representation of the scenario for the calculating polynomial coefficients

Procedures used: adding a multiple constant from a vector into another vector, the duplicate settling of an element of a vector, calculating the Newtonian polynomial coefficients obtained through the interpolation of the corresponding coefficients, calculating the polynomial sum, determination of the polynomial degree.

3.3 Tests realized on the basis of the scenario

I realized a series of tests on the basis of these scenarios. I calculated the sum, the subtraction, the product, the division, the determinant, the inverse, the medium matrices and the covariant, for matrices beginning with 2X2 up to 9X9, both with the Web Service and without the Web Service.

	WS 2x2	Con 2x2	WS 3x3	Con 3x3	WS 4x4	Con 4x4	WS 5x5	Con 5x5	WS 6x6	Con 6x6	WS 7x7	Con 7x7	WS 8x8	Con 8x8	WS 9x9	Con 9x9
Sum 1	124	31760	155	63933	293	104647	317	118323	642	179817	1445	243078	96	286256	-	-
Sum 2	87	32082	122	60776	137	99928	146	112502	188	179686	1517	324743	3160	286083	-	-
Sum 3	89	29685	117	64181	103	100201	135	114665	213	171903	217	379526	263	371584	-	-
Subtraction 1	56	31719	116	63946	128	100474	138	117425	190	180489	210	366958	259	280512	-	-
Subtraction 2	55	32245	173	56843	791	100124	905	114534	1013	179535	1390	364693	2512	285700	-	-
Product	63	31372	167	55817	173	99639	247	116339	426	192731	504	356460	8477	399901	-	-
Division	62	31922	173	62806	118	91329	1420	119942	200	183242	1668	360198	428	416070	-	-
Transposed 1	4	28385	69	48881	91	87872	124	118044	168	180539	5777	352467	2646	405871	-	-
Transposed 2	50	29611	68	62897	105	91625	1494	114164	2137	183603	3280	467799	4870	510209	-	-
Determinant	31	156	124	286	515	3922	6508	4567	24233	23344	76555	255284	967344	826578	-	-
Inverse	286	34208	468	52574	5670	91807	18898	132468	151344	315996	1048074	2301332	8829661	10053879	-	-
Media	50	5362	61	5016	77	8467	124	11956	9	11921	140	19830	157	124004	-	-
Medium matrix	45	6134	57	10812	66	11434	83	11423	794	19435	116	19838	132	34184	-	-
Covariant	75	29850	139	56160	166	87511	1064	112456	432	172651	1210	380630	698	439619	-	-

*) Time expressed in micro seconds
*) WS – Web Service
*) Con – Executed in console

Table 4. The time obtained with testing

Due to the calculation of the inverse the execution time, beginning with 9x9, is bigger than 30 seconds.

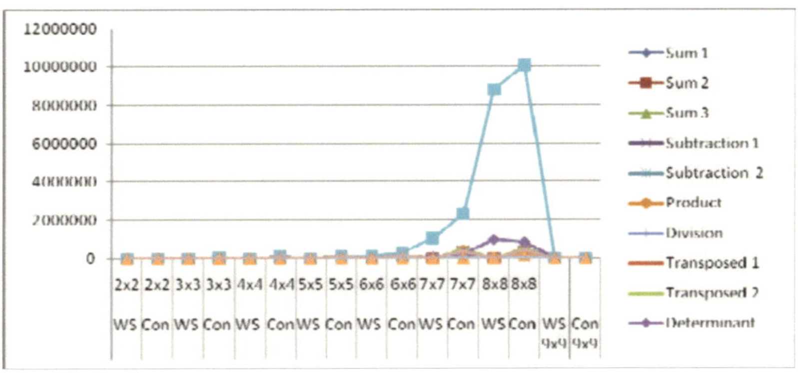

Fig. 20. The graphical representation of the response time for tests, with and without WS

The advantages of the web service use with these scenarios:

a. the possibility of resolution problems of great dimensions which do not go in the client computer's memory
b. the identification of the common costs
c. the reduction of the costs
d. the reduction of the response time
e. the client can be involved with the problem description

4. A graphics generator for physics research

The proper architecture for the web service used for physics simulations is divided into a working area. At first, we realized a series of Web services which grounds the numerical calculation. These facts are oriented towards scientific applications, beginning with elementary structures, up to complex ones, such as the integration of ordinary differential equations. The passing from an interface of the type line command with the user, to an interface based on Web services, represents a great advantage because it allows the combination and use of certain procedures - routines - defined earlier on.

The methods for numerical calculation are practical, efficient and elegant, and the realization of a proper architecture of a Web service will significantly contribute to this process [2].

Beginning here, I would like to develop this idea, applying it to both physics and physics research. By selecting this, I chose the most representative phenomena in physics which allow for applications; next, I followed the interpretation of them from the point of view of numerical calculation.

The chosen phenomena are:

a. A motor with a continuous current;
b. The forces of inertia on the earth's surface;
c. The dynamics of a variable mass point;
d. Admissible resistance;
e. Floating bodies.

In order to set out every phenomenon, they are described physically and are mentioned in terms of the manner in which they are to be found within the created service. The platform is divided into two parts: the former is designated to numerical calculation; the latter is for the realization of graphics. Also, within this service, there is a static zone where I present the selected physical phenomena, as can be seen in the following image (retrieved on http://www.eliza-isbasoiu.muscel.ro 05/10/2009).

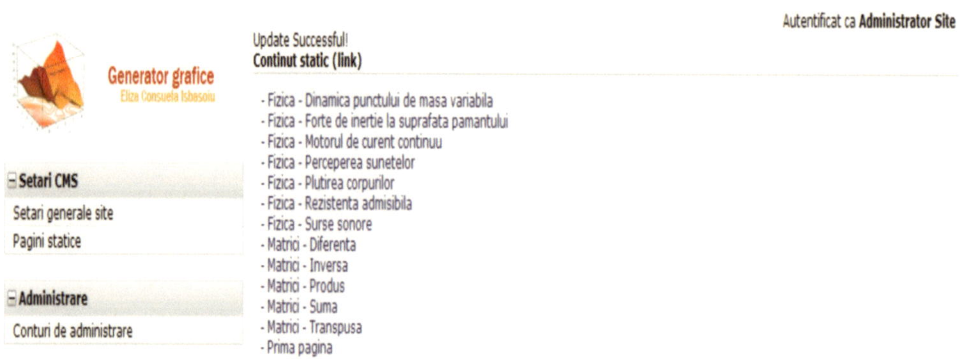

Fig. 21. WS Static content (link-uri)

The user is first welcomed by the image presented beneath: (retrieved on http://www.eliza-isbasoiu.muscel.ro 05/10/2009)

Fig. 22. The starting image of the platform

In order to know the variables of the constants and equations at the basis of every simulation, the user receives pieces of information about the physical phenomena and the possible discussions after the simulation.

A motor with continuous current

One should notice that if we apply a tension to the motor terminals - given the same polarity as where it works as a generator - the rotor spins in the opposite direction as that of the generator (Maxwell, 1892).

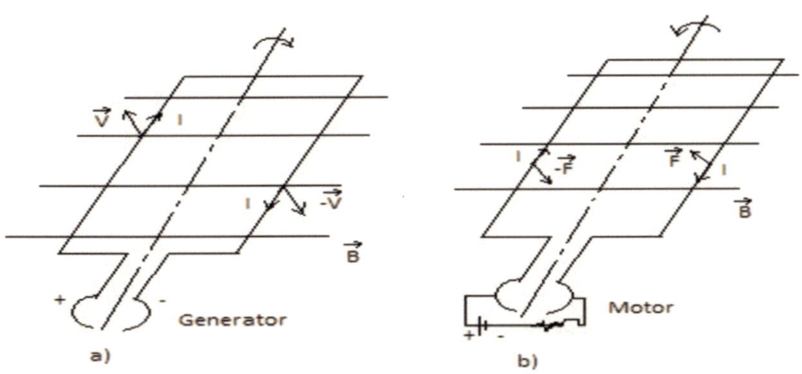

Fig. 23. The graphical representation of the generator and the motor

The parameters that vary are: external voltage, ranging from a minimum value and a maximum, R and r, and contrary[L2] electromotor tension. According to these values, the graph appears. Such findings may be established for efficiency (retrieved on http://www.eliza-isbasoiu.muscel.ro 05/10/2009).

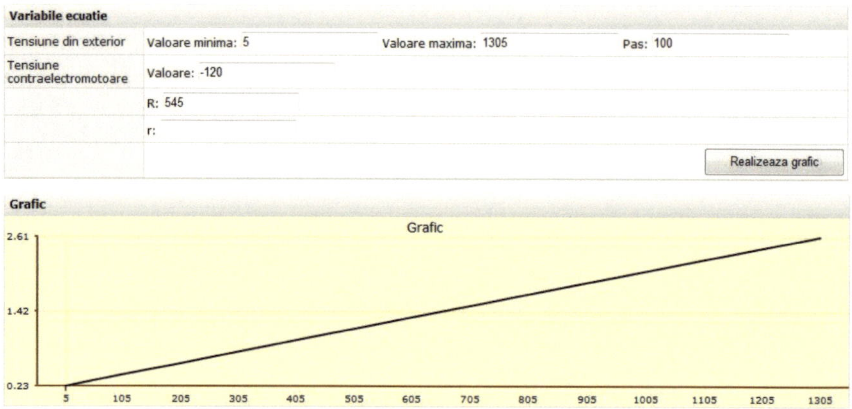

Fig. 24. The graphical representation of the contrary[L3] electromotor tension

Forces of inertia on the earth's surface

The Earth makes a revolutionary motion around the Sun, under the influence of the gravitational forces of different cosmic bodies in our solar system.

We consider a material mass point m, being at rest against the Earth in an A point. This can be seen by the following picture:

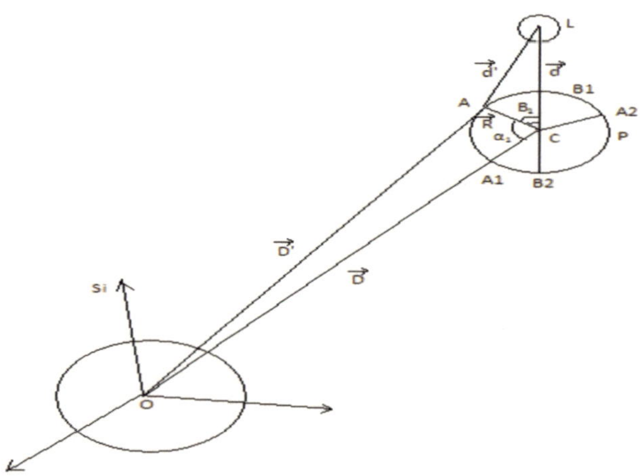

Fig. 25. The graphical representation of the material point as against the Earth

In the inertial system Si against the material points, thus act the gravitational forces of the Sun, Moon and Earth.

The determination of inertial forces on the earth's surface has been, for a long time, a very attractive notion in physics.

Mass: the point of mass is variable. The angle is also variable. (retrieved on http://www.eliza-isbasoiu.muscel.ro 05/10/2009).

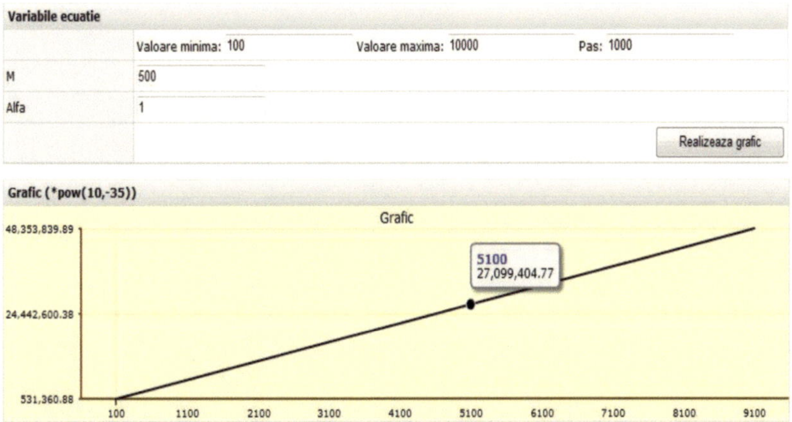

Fig. 26. The graphical representation of the inertial forces on the earth's surface

The dynamics of the variable mass point's applications in the case of detachment.

We consider a material point M, with the mass m, whereby the variability of its mass is given by the joining or detaching of some particles from M. These two situations are, in fact, two cases: the joining case and the detaching case.

The variables in this situation are greater. The time ranges between a maximum and a minimum. The matter the value it receives is v0. The angle, the reaction speed and the gravitational acceleration influence the graph (retrieved on http://www.eliza-isbasoiu.muscel.ro 05/10/2009).

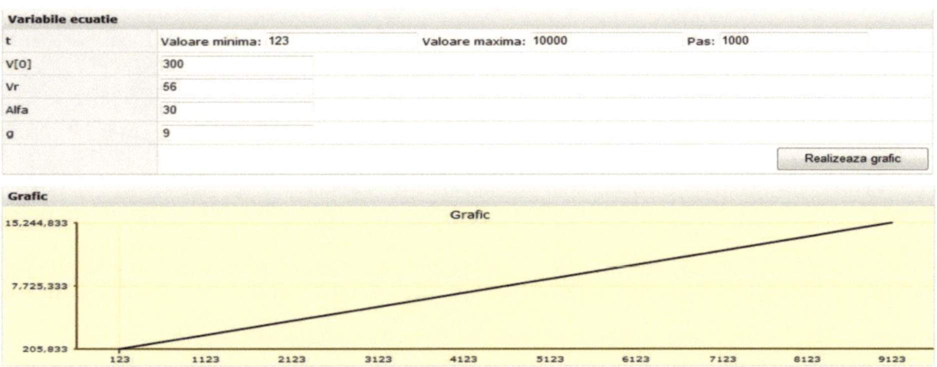

Fig. 27. The graphical representation of a variable mass point speed, after an angle of $\alpha = 30^0$

The floating bodies

Knowledge of the conditions which determine the stability of floating bodies is very important in naval construction techniques. We assume a floating body which has a centre in the same place with the O point, as with the figure below, and that over the superior part of the body outside the liquid on which the F horizontal force acts, by tending to turn the body towards the left (Balibar, 2007).

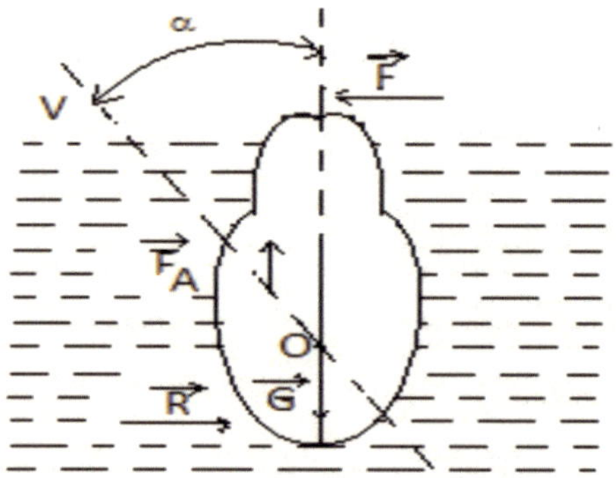

Fig. 28. The graphical representation of the centre of mass an inclined floating body

The factor which influences this phenomenon the most is represented by the surface marked by S. After we introduced the minimum and maximum of the value, we obtained the following graph (retrieved on http://www.eliza-isbasoiu.muscel.ro 05/10/2009).

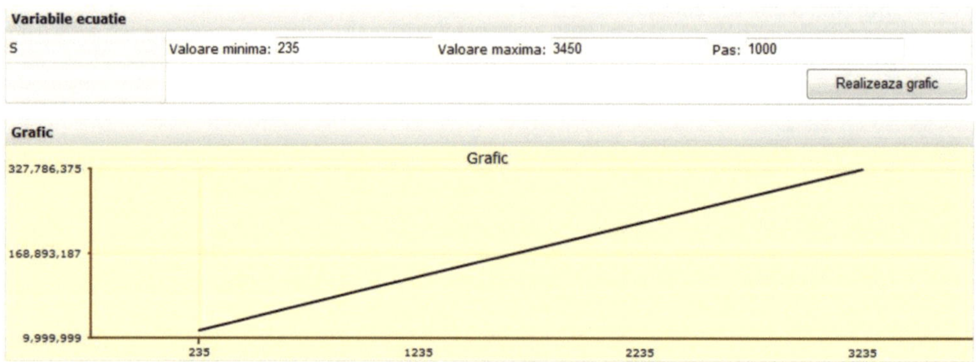

Fig. 29. The graphical representation of the limits between a body can or cannot float for a specific surface S

The graphical interface which the user employs is a simple one but extremely efficient, because it is generalized for every area of study (retrieved on http://www.eliza-isbasoiu.muscel.ro 05/10/2009).

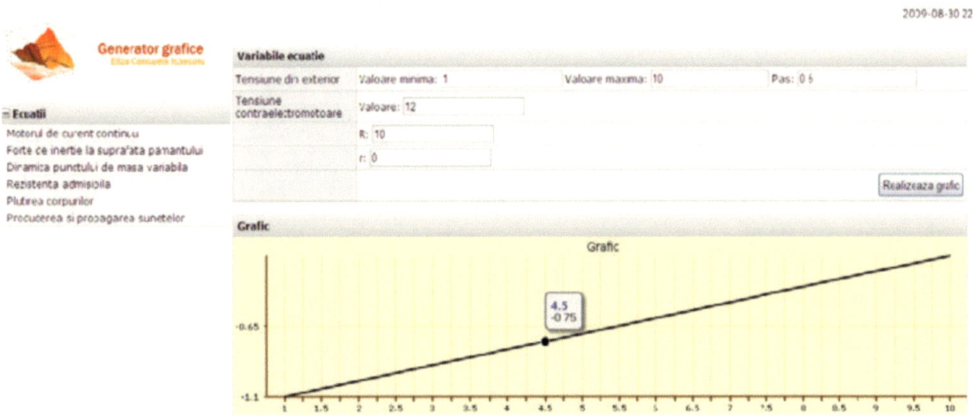

Fig. 30. The image of the graphical interface of the Web service - graphics generator

When we press the button "generate graphs", the user is asked which essential parameter they want as a graphical representation. For example, for control electromotor tension, there are three variants: the exterior variation of U tension, the variable E contro electromotor tension, the R variable.

The following image presents this fact (retrieved on http://www.eliza-isbasoiu.muscel.ro 05/10/2009).

Fig. 31. Select the variant for the graphic

The platform is divided into two working areas: the former is for numerical calculation; the latter is for the graph's realization.

We can represent the sound's intensity; we introduce two values, the minimum and the maximum, the representation being made after a desired pattern (retrieved on http://www.eliza-isbasoiu.muscel.ro 05/10/2009).

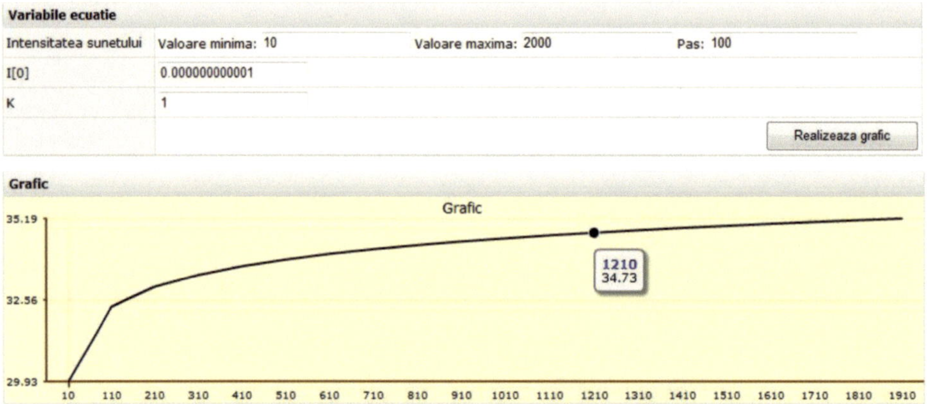

Fig. 32. The graphical representation of the sound's intensity

If we modify the weight, then by increasing it simultaneously with the increasing of the α angle, the graph is as follows: (retrieved on http://www.eliza-isbasoiu.muscel.ro 05/10/2009)

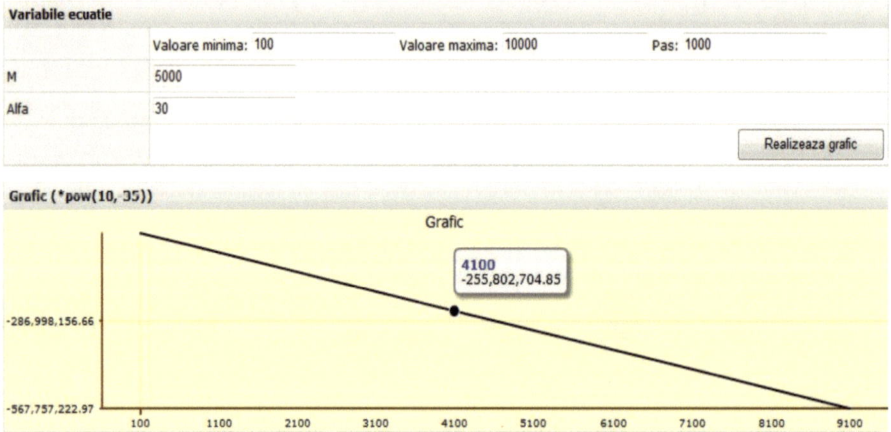

Fig. 33. The graphical representation of the inertial force at the earth surface, for the angle $\alpha = 30^{\circ}$

The above representations suggest the following conclusions:

a. the necessity of knowing of all of the equations, but also the particular cases which represent the basis of the physical phenomena;

b. the possibility of interpretation, depending upon the input values;
c. simulations for finding other possible particular cases;
d. the understanding of the physical phenomena, at a practical level.

All these solutions are oriented towards scientific applications, which are grounded on numerical calculations. They are followed by more complicated constructions.

There are many authors who deeply value such software architectures. Some of them consider it to be the separate profession of the software engineer [5]. Others consider it be independent, separate for the role of the software engineer and, thus, in need separate learning approaches.

The activity can be considered from: psychological points of view; systematic points of view; organizational points of view.

From the psychological point of view, the application is a creative process, which implies knowledge within close fields, such as: software engineering, computer science, logic, cognitive science and programming etc. From systematic point of view, the project is seen as an activity which implies the discovery of optimal solutions for a set of problems, taking into account the balance between obstacles and forces.

The organizational perspective offers the possibility that some software elements are re-used for other products.

The main ideas which result from the above examples are as follows:

a. the necessity of knowing all of the equations and the particular cases which represent the bases of the physical phenomena;
b. the possibility of their interpretation depending upon the input data;
c. simulations for finding other particular cases;
d. the practical understanding of the phenomena of physics;

All of these comparisons between values and interpretations, based upon the graphs, depend upon users' objectives. It is very important that a possible user of the application can use the interface despite different levels of knowledge. The created Web service presents a few physics applications based upon numerical calculations. The selected phenomena can assure the construction of more complicated applications. This method displays ideal cases which are not present in nature, but also specific cases. If one interprets them correctly, one can find optimal solutions to a set of problems; besides this, there is the possibility of the further use of the results obtained. The main advantages of using the Web service are: the possibility of solving problems of great dimensions which exceed the computer's memory; the identification of common costs and their reduction; the reduction of the time response value; the fact that client can be involved in the problems' description.

The graphic must be defined within the text editor under the form **equation (formula,min,max,step,precision).**

a. the formula must be x variable, framed by double inverted commas this *'x'*. The x variable must have a space before it and another space after it in an equation, whenever it appears. For example: "sin(x) + cos(x + 4) + 2 * x "
b. min, max represent the interval margins where the calculation is made;

c. the step represents the incrementation value of the values for which the calculation was made;
d. the precision represents the ranging interval of the graphical representation. It is recommended to use 0.01; one can also use other values in order to see the results.

A few examples of graphs generated by this method are presented below: (retrieved on http://www.eliza-isbasoiu.muscel.ro 21/10/2010).

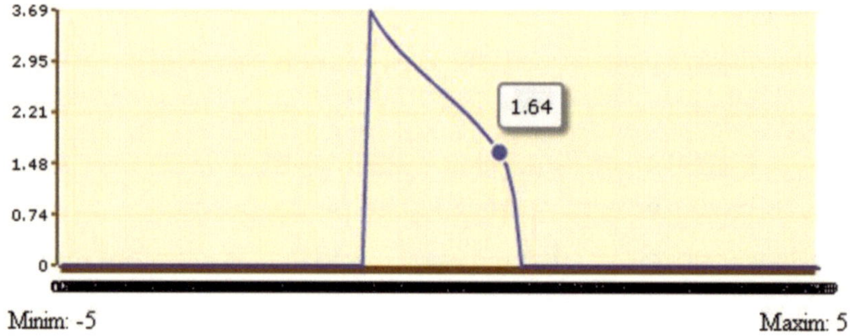

Fig. 34. The movement of a projectile

Figure 32 presents the movement of a projectile according to the equations of the theory depicted in the area "the description of physical phenomenon" (retrieved on http://www.eliza-isbasoiu.muscel.ro 21/10/2010).

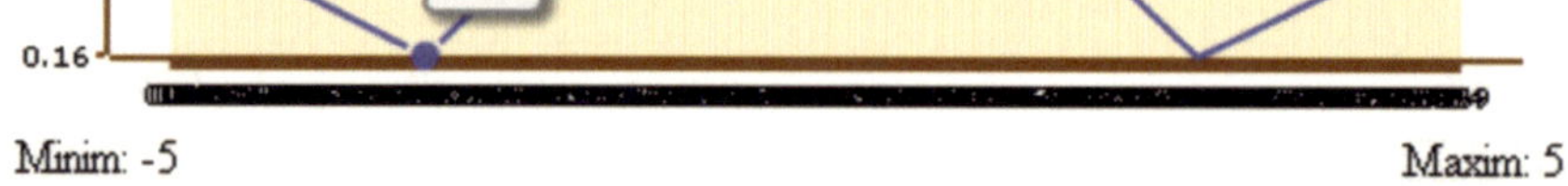

Fig. 35. Newton's Second Law

Figure 33 presents the Second Law of Newton, in conformity with the equations from the theory depicted in the area related to the description of physical phenomena.

5. Conclusion

We presented in this article concepts connected to the architecture of distributed systems, demonstrating the criteria for a proper choice in order to use adequate technology. The problem that we developed was represented by the realization of certain Web Services using methods of numeric calculation. The applications were created on the Eclipse platform; these were realized with the solutions offered by the book *Numerical Library in Java for Scientists and Engineers*. These solutions are oriented towards scientific applications based upon numeric calculation. They begin with elementary structures move on to more complex structures. By grouping and creating services for each and every structure, we realized that many of the functions used for their construction appeal to other functions; we created a table where those functions appear only once. We created a service for every function and followed the response time. We saw that, depending upon the function's complexity, this time differs. By way of comparison, we constructed a graphical representation with the time for both individual functions, with simple input data, and parameters with complex values. The next step would be to realize Web services for education in physics, as well as services for scientific research in physics, using these services.

So long as informatics and communications technology evolve, the implications that they will have on the educational system are difficult to be foresee. There will be always new opportunities and new difficulties, but there will also be results and benefits.

Computer simulations seem to offer the most efficient methods for using computers in physics. There, the processes used in research physics are encouraged: for determining the cause of an effect, in prognosis, and also for the interpretation of research data. As a rule, such simulations develop an inductive and deductive way of thinking, by assuring the capacity of problem solving, the formulation of new hypotheses and the realization of tests. We feel that users should be allowed to realize other experiments as a preliminary to the physics laboratory.

Beginning with these elements and using the Web services for numerical calculation subsequently realized, I selected a series of mathematical functions, namely those useful for the equations used for continuous electricity, the calculation of inertial forces on the earth's surface, the calculation of the dynamics of variable weight points for finding admissible resistance, and in the production and propagation of sounds.

6. References

Balibar F., (2007), *Einstein, bucuria gândirii*, pp. 95 -100, Publisher Univers, ISBN 978-1-60257-025-2, Bucureşti

Hang T.L., (2004), *Numerical Library in Java for Scientists and Engineers*, pp. 38 - 751, ISBN-13: 978-1584884309, Publisher Chapman & Hall/CRC, USA

Isbasoiu E.C. (2009), *Services for Numerical Calculation*, pp. 245 - 252, CSSim - Conference on Computational Intelligence, Modelling and Simulation, ISBN: 978-1-4244-5200-2, Brno, Czech, 2009

Landau R.H. & Bordeianu C.C., (2008), *Computational Physics. A Better Model for Physics Education?*, pp. 22 - 30, Computing in Science & Engineering, ISSN 1521-9615, Oregon State University, USA, 2006

Maxwell J.C., (1892), *A Treatise on Electricity and Magnetism*, pp. 98 - 115, 3rd ed., vol. 2. Publisher Oxford: Clarendon

Petcu D., (2008), *Distributed Systems- Lecture notes*, pp. 35 - 87, Publisher UVT, Retrieved from http://web.info.uvt.ro/~petcu/distrib.htm, Timisoara

http://www.eliza-isbasoiu.muscel.ro retrieved on 05/10/2009

http://www.eliza-isbasoiu.muscel.ro retrieved on 21/10/2010

15

Genetic Algorithm Based Automation Methods for Route Optimization Problems

G. Andal Jayalakshmi
Intel,
Malaysia

1. Introduction

Genetic Algorithms (GA) are robust search techniques that have emerged to be effective for a variety of search and optimization problems. The primary goal of this chapter is to explore various Genetic Algorithm (GA) based automation methods for solving route optimization problems. Three real world problems: Traveling Salesman, Mobile Robot Path-Planning and VLSI global routing are considered here for discussion. All the three problems are *Non-deterministic Polynomial (NP)-complete* problems and require a heuristic algorithm to produce acceptable solutions in a reasonable time.

The basic principles of GAs were first laid down by *Holland*. GAs work with a population of individuals each representing a solution to the problem. The individuals are assigned a fitness value based on the solution's quality and the highly fit individuals are given more opportunities in the reproduction. The reproduction process generates the next generation of individuals by two distinct processes named 'crossover' and 'mutation'. The new individuals generated by crossover share some features from their parents and resemble the current generation whereas the individuals generated by mutation produces new characters, which are different from their parents. The probability of crossover operation is usually very high compared to the probability of mutation operation due to the nature of their operations. The reproduction process is carried out until the population is converged which usually takes hundreds of iterations for complex real world problems. The time taken for convergence is dependent on the problem and it is the progression towards increasing uniformity among the individuals of a population.

A standard GA described by *Goldberg* uses binary encoding for representing the individuals, one-point crossover for reproduction which exchanges two consecutive sets of genes and random mutation which randomly alters the selected gene. The probability for applying crossover typically ranges from 0.6 to 1.0 and the probability for mutation operation is typically 0.001. Generally crossover enables the rapid exploration of the search space and mutation provides a small amount of random search to ensure that no point in the search space is given zero probability of being examined.

Traditional GAs are generally not the successful optimization algorithms for a particular domain as they blindly try to optimize without applying the domain knowledge. *L.Davis* states in the "Handbook of Genetic Algorithms", that the "Traditional genetic algorithms although robust are generally not the most successful optimization algorithms on any particular domain". *Davis* argues that the hybridization will result in superior methods. Hybridizing the genetic algorithm with the optimization method for a particular problem will result in a method which is better than the traditional GA and the particular optimization method. In fact this will produce a more superior method than any of the individual methods. The standard GA can be improved by introducing variations at every level of the GA component including the encoding techniques, the reproduction mechanisms, population initialization techniques, adaptation of genetic parameters and the evolution of the individuals. These thoughts have resulted in a class of genetic algorithms named *'Hybrid Genetic Algorithms'* (HGA). These are the customized genetic algorithms to fit the traditional or simple GA to the problem rather than to fit the problem to the requirements of a genetic algorithm. The HGAs use real valued encodings as opposed to binary encodings and employs recombination operators that may be domain specific.

We will explore further on the use of HGA by discussing a solution to the *Traveling Sales Person* (TSP) problem. The TSP problem has been a typical target for many approaches to combinatorial optimization, including classical local optimization techniques as well as many of the more recent variants on local optimization, such as Simulated Annealing, Tabu Search, Neural Networks, and Genetic Algorithms. This problem is a classic example of *Non deterministic Polynomial hard* problem and is therefore impossible to search for an optimal solution for realistic sizes of *N*. The HGA that is described here is as proposed by *Jayalakshmi* et. al. which combines a variant of an already existing crossover operator with a set of new heuristics. One of the heuristics is for generating the initial population and the other two are applied to the offspring either obtained by crossover or by shuffling. The heuristics applied to the offspring are greedy in nature and hence the method includes proper amount of randomness to prevent getting stuck up at local optimum.

While the hybrid GAs exploit the domain knowledge, in many realistic situations, a priori knowledge of the problem may not be available. In such cases, it is fortunately possible to dynamically adapt aspects of the genetic algorithm's processing to anticipate the environment and improve the solution quality. These are the '*Adaptive GAs*' which are distinguished by their dynamic manipulation of selected parameters or operators during the course of evolving a problem solution. In this chapter we will see an adaptive GA solution to the *mobile robot path planning* problem which generates collision free paths for mobile robots. The problem of generating collision-free paths has attracted considerable interest during the past years. Recently a great deal of research has been done in the area of motion planning for mobile robots as discussed by Choset et al. Traditional planners often assume that the environment is perfectly known and search for the optimal path. On the other hand on-line planners are often purely reactive and do not try to optimize a path. There are also approaches combining offline planers with incremental map building to deal with a partially known environment such that global planning is repeated whenever a new object is sensed and added to the map. The developments in the field of *Evolutionary Computation* (EC) have inspired the emergence of EC-based path planners. However traditional EC-based planners have not incorporated the domain knowledge and were not adaptive and reactive to the changing environments. Recent research has offered EC-based planners for dynamic environments.

The solution to the *mobile robot path planning* problem discussed here is as proposed by *Jayalakshmi* et. al, which incorporates domain knowledge through domain specific operators and uses an initialization heuristics to convert infeasible paths into feasible ones. The fitness of the solution is measured based on the number of fragments, acute edges and the angle between the turns in the path. The algorithm plans the path for the current environment and the robot travels in that direction. If an obstacle is found in its path, the robot senses the presence of the obstacle before the critical time to avoid collision and calls the path planner algorithm again to find the new path from that point onwards.

The CHC genetic algorithm proposed by *Eshelman* has emerged as an alternative to resolve the perennial problem with simple GAs which is the premature convergence. The simple GA allows a sub-optimal individual to take over a population resulting in every individual being extremely alike and thus causing premature convergence. The consequence of premature convergence is a population which does not contain sufficient genetic diversity to evolve further. The CHC genetic algorithm uses crossover using generational elitist selection, heterogeneous recombination by incest prevention and cataclysmic mutation to restart the search when the population starts to converge. The CHC GA has a very aggressive search by using monotonic search through survival of the best and offset the aggressiveness of the search by using highly disruptive operators such as uniform crossover.

In this chapter we will also explore a solution to the *VLSI global routing problem* using CHC GA. One of the most important VLSI Design Approaches is the Macro Cell design. Macro cells are large, irregularly sized parameterized circuit modules that are generated by a silicon compiler as per a designer's selected parameters. Usually the physical design process for macro cells is divided into Floor Planning/Placement, Global Routing and Detailed Routing. Floor Planning/Placement constructs a layout indicating the position of the macro cells. The placement is then followed by routing, which is the process of determining the connection pattern for each net to minimize the overall routing area. Before the global routing process begins, a routing graph is extracted from the given placement and routing is done based on this graph. Computing a global route for a net corresponds to finding a corresponding path in the routing graph. Each edge represents a routing channel and the vertex is the intersection of the two channels. First the vertices that represent the terminal of the net are added to the routing graph and then the shortest route for the net is found. Both the placement and routing problems are known to be NP-complete. Thus it is impossible to find optimal solutions in practice and various heuristics are used to obtain a near optimal solution. There has been a lot of work on optimization for routing, including Simulated Annealing algorithms and Genetic Algorithms.

The details of the genetic algorithm solutions to each of these problems are described in the following sections.

2. Design of a hybrid GA for TSP

A heuristic approach employs some domain knowledge in providing a solution to the problem. A good heuristics can be devised provided one has the knowledge of the problem being solved. In cases, where there is no knowledge of the problem, it is best to use a more general heuristic, often called a meta-heuristic. Meta-heuristics are sometimes also called black-box optimization algorithms or simply, general-purpose optimization algorithms. Coding complex data structures by simple lists of bits or real values leads to the problem that has no one-to-one correspondence between these lists and the problem instances. Hence problem knowledge is necessary either to repair operators to deal with invalid solutions or to design special operators tailored to the problem.

Of the present evolutionary algorithms, hybrid genetic algorithms have received increasing attention and investigation in recent years. This is because of the reason that the hybrid GAs combine the global explorative power of conventional GAs with the local exploitation behaviours of deterministic optimization methods. The hybrid GAs usually outperform the conventional GAs or deterministic methods in practice. To hybridize the genetic algorithm technique and the current algorithm, the following three principles are suggested by *Davis*:

- Use the current algorithm's encoding technique in the hybrid algorithm. This guarantees that the domain expertise embodied in the encoding used by the current algorithm will be preserved.
- Hybridize where possible by incorporating the positive features of the current method in the hybrid algorithm.
- Adapt the genetic operators by creating new crossover and mutation operators for the new type of encoding by analogy with bit string crossover and mutation operators. Incorporate domain based heuristics on operators as well.

Theoretical work as well as practical experience demonstrates the importance to progress from fixed, rigid schemes of genetic algorithms towards a problem specific processing of optimization problems.

This section explores how a HGA is used to solve the TSP problem. The TSP is probably the most studied optimization problems of all times. In the *Travelling Sales Person* problem, given a set $\{c_1, c_2, \ldots, c_n\}$ of cities, the goal is to find an ordering π of the cities that minimizes the quantity

$$\Sigma\, d(c_{\pi(i)}, c_{\pi(i+1)}) + d(c_{\pi(n)}, c_{\pi(1)}) \quad 1 \leq i \leq n-1$$

Where $d(c_i, c_j)$ is the distance associated with each pair of distinct cities $<c_i, c_j>$. This quantity is referred to as the *tour_length,* since it is the length of the tour a salesman would make when visiting the cities in the order specified by the permutation, returning at the end, to the initial city. The *Euclidean Travelling Sales Person* problem involves finding the shortest *Hamiltonian Path* or Cycle in a graph of *N* cities. The distance between the two cities is just the *Euclidean* distance between them.

In a *symmetric* TSP, the distances satisfy $d(c_i, c_j) = d(c_j, c_i)$ for $1 \leq i, j \leq N$. The symmetric traveling salesman problem has many applications, including VLSI chip fabrication X-ray crystallography and many other. It is NP-hard and so any algorithm for finding optimal tours must have a worst-case running time that grows faster than any polynomial. This leaves researchers with two alternatives: either look for heuristics that merely find *near-*optimal tours, but do so quickly, or attempt to develop optimization algorithms that work well on 'real-world' rather than worst-case instances. Because of its simplicity and applicability the TSP has for decades served as an initial proving ground for new ideas related to both these alternatives.

2.1 The hybrid GA solution

The HGA proposed by *Jayalakshmi* et al. to solve the TSP problem use heuristics for initialization of population and improvement of offspring produced by crossover. The

Initialization Heuristics algorithm is used to initialize a part of the population and the remaining part of the population is initialized randomly. The offspring is obtained by crossover between two parents selected randomly. The tour improvement heuristics: *RemoveSharp* and *LocalOpt* are used to bring the offspring to a local minimum. If cost of the tour of the offspring thus obtained is less than the cost of the tour of any one of the parents then the parent with higher cost is removed from the population and the offspring is added to the population. If the cost of the tour of the offspring is greater than that of both of its parents then it is discarded. For shuffling, a random number is generated within one and if it is less than the specified probability of the shuffling operator, a tour is randomly selected and is removed from the population. Its sequence is randomized and then added to the population.

2.1.1 The crossover operator

The initial city is chosen from one of the two parent tours. This is the current city and all the occurrences of this city are removed from the edge map. If the current city has entries in its edgelist then the city with the shortest edge is included in the tour, and this becomes the current city. Any ties are broken randomly. This is repeated until there are no remaining cities. An example is given below:

Let the distance matrix be

0	10	4	15	5	20
10	0	5	25	5	10
4	5	0	13	6	2
15	25	13	0	6	10
5	5	6	6	0	20
20	10	2	10	20	0

Let the *genotype* p_1 be equal to (2,3,4,5,0,1) which encodes the TSP tour (2,3,4,5,0,1,2) and p_2 be equal to (2,3,1,4,0,5) which encodes the TSP tour(2,3,1,4,0,5,2). The combined edge map M_{12} contains the combined edge relationships from both the parents. The first gene value in p1 i.e. 2 is added to the child c1. Then the gene value 2 is removed from the edge map. The combined edge map before and after are given below:

Gene value	Edge map (p_1)	Edge map (p_2)	Combined Edge map M_{12} (before)	Combined Edge map M_{12} (after)
0	5,1	4,5	1,4,5	1,4,5
1	0,2	3,4	0,2,3,4	0,3,4
2	3,1	3,5	1,3,5	1,3,5
3	2,4	2,1	1,2,4	1,4
4	3,5	1,0	0,1,3,5	0,1,3,5
5	4,0	0,2	0,2,4	0,4

Table 1. Combined edge map

Now $|E(2)| = 3$ therefore an edge j is chosen such that $j \varepsilon E(2)$ and $|<2,j,>|$ is minimum and is added to the child c_1. In this example j is 5. Now j is removed from the edge map, and the same procedure is followed until the child c_1 is filled with all the genes. For the example the child c_1 will become (2,5,0,4,1,3).

2.1.2 The Initialization Heuristics

The *Initialization Heuristics* (IH) initializes the population based on a greedy algorithm which arranges the cities depending on their x and y coordinates. The tours are represented in linked-lists. First an initial list is obtained in the input order which is the Input List. The linked-list that is obtained after applying the *Initialization Heuristics* is the "Output List". During the process of applying the *Initialization Heuristics* all the cities in the "Input List" will be moved one by one to the "Output List". Four cities are selected, first one with largest x-coordinate value, second one with least x-coordinate value, third one with largest y-coordinate and fourth one with least y-coordinate value. These are moved from the "Input List" to the "Output List". The sequence of the four cities in the Output List is changed based on minimum cost. The elements in the Input List are randomized and the head element is inserted into the Output List at a position where the increase in the cost of the tour is minimal. This process is repeated until all the elements in the Input List are moved to the Output List.

Figure 1(a) shows a 8-city problem. Figure 1(b) shows the Boundary Tour formed from four extreme cities. Figure 2 (a), (b), (c) & (d) shows the four possible tours that can be formed when city 'E' is moved to the "Output List". It is obvious from the figures that the Tour in Figure 2(a) will result in minimum increase in the cost of the tour in the "Output List". Similarly other cities will be moved one by one to the "Output List".

(a) (b)

Fig. 1. (a) Input cities and (b) boundary tour formed by four extreme cities

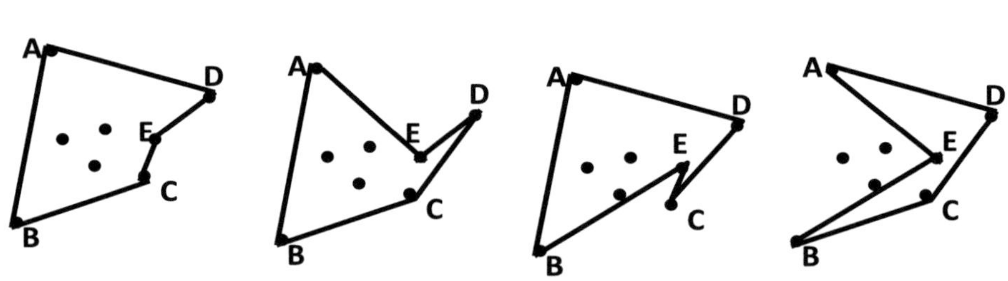

(a) (b) (c) (d)

Fig. 2. Various possible tours, which can be formed by moving city 'E' to the output list

2.1.3 The *Removesharp* heuristics

The *RemoveSharp* algorithm removes sharp increase in the tour cost due to a city, which is badly positioned. It rearranges the sequence of a tour by considering the nearest cities of a badly positioned city such that the *tour_cost* is reduced. A list containing the nearest *m* cities to a selected city is created. The selected city from the tour is removed and a tour with *N-1* cities is formed. Now the selected city is reinserted in the tour either before or after any one of the cities in the list previously constructed with m nearest cities and the cost of the new tour length is calculated for each case. The sequence, which produces the least cost, is selected. This is repeated for each city in the tour.

An example is given below:

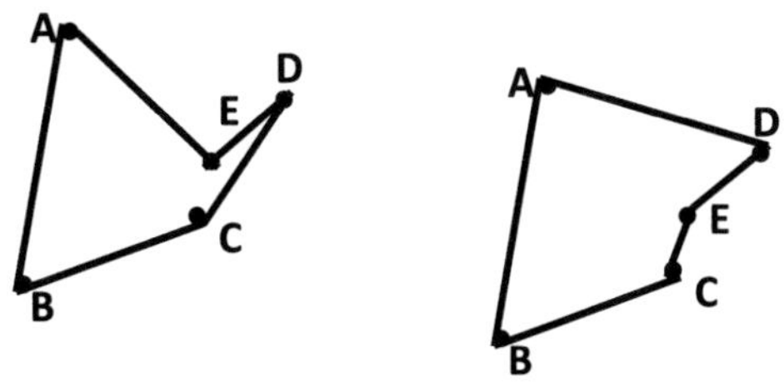

(a) (b)

Fig. 3. (a) A tour with a badly positioned city and (b) The tour after *RemoveSharp* is applied

In Figure 3(a) the city E is in between the cities A and D, while it is obvious that the nearest cities to it are city C and B. *RemoveSharp* will move city E between the cities C and D, resulting in a decrease in the tour cost as shown in Figure 3(b).

2.1.4 The local heuristics

The heuristics finds a locally optimal tour for a set of cities, by rearranging them in all possible orders. The *LocalOpt* algorithm will select q consecutive cities ($S_{p+0}, S_{p+1}, \ldots, S_{p+q-1}$) from the tour and it arranges cities $S_{p+1}, S_{p+2}, \ldots, S_{p+q-2}$ in such a way that the distance is minimum between the cities S_{p+0} and S_{p+q-1} by searching all possible arrangements. The value of p varies from 0 to $n-q$, where n is the number of cities. In Figure 4(a) it is quite clear that the distance between the cities A and G can be reduced if some rearrangements are made in the sequence of the cities between them. *LocalOpt* will make all possible rearrangements and replace them to the sequence as shown in Figure 4(b).

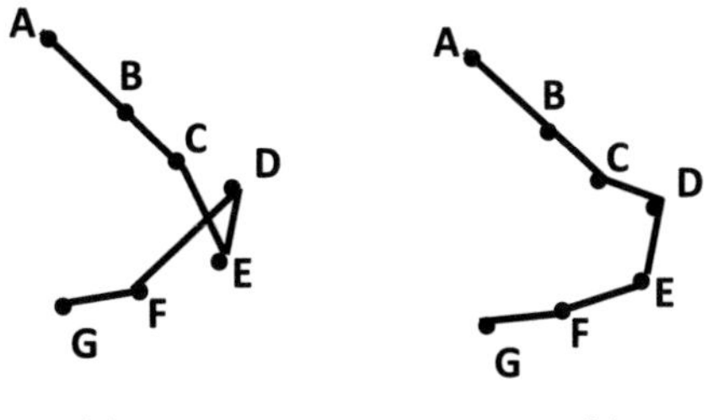

(a) (b)

Fig. 4. (a) A bad tour and (b) The tour after *LocalOpt* is applied

2.2 Results

The results for the HGA solution for 3 standard TSP problems are compared with the results for GA and SA solutions. The best integer *tour_length* and the best real *tour_length* (in parenthesis) are tabulated below in Table 2. NA represents "Not Available".

The heuristics with which Hybrid GA is compared here are GA and SA as reported by *Whitley. D* et al. and TSPLIB. The difference between integer and real tour length is that in the first case distances are measured by integer numbers, while in the second case by floating point approximations of real numbers. In TSPLIB website only Eil51 and Eil76 are available which have an additional city to Eil50 and Eil75 respectively.

Problem name	HGA	GA	SA
Eil50 **50-city problem**	426 (428.871) {for Eil51}	428 (NA)	443 (NA)
Eil75 **75-city problem**	538 (544.36) {for Eil76}	545 (NA)	580 (NA)
KroA100 **100-city problem**	21282 (21285.44)	21761 (NA)	NA (NA)

Table 2. Comparison of HGA with other heuristics on geometric instances of the symmetric TSP.

3. Design of an adaptive GA for mobile robots

Adaptive Genetic algorithms dynamically manipulate selected parameters or operators during the course of evolving a problem solution. Adaptive GAs are advantageous over SGAs in that they are more reactive to unanticipated characteristics of the problem and can dynamically acquire information about the problem characteristics and exploit them. As described by *Davis,* adaptive GAs can be categorized based on the level at which the adaptive parameters operate. *Population-level* techniques dynamically adjust parameters that are global to the entire population. *Individual-level* adaptive methods modify a particular individual within the population depending on how it is affected by the mutation operators. *Component-level* adaptive GAs dynamically alter the individual components depending on how each individual will be manipulated independently from each other. The operator probabilities play a major role in determining the solution quality and the convergence rate. Since the range of potential applications of genetic algorithms is infinite, it is difficult to measure the goodness of the parameter values. This suggested the idea of adapting the operator probabilities during the evolution of the GA.

The path planner proposed by *Jayalakshmi* et al. incorporates domain knowledge in the algorithm through domain specific operators. An initialization heuristics is used to convert infeasible paths into feasible ones. The fitness is measured based on the number of fragments, acute edges and the angle between the turns in the path. The algorithm plans the path for the current environment and the robot travels in that direction. If an obstacle is found in its path, the robot senses the presence of the obstacle before the critical time to avoid collision and calls the path planner algorithm again to find the new path from that point onwards. The following sections describe the solution.

3.1 The adaptive GA solution

The path planner algorithm design has four major phases. The first phase is the design of the *Initialization Heuristics* (IH), which includes the *Backtrack* and *Change_Y* operators. The algorithm initializes the population randomly and then repairs the population by applying these two operators. This leaves the initial population free of any infeasible solutions and reduces the search region considerably. The heuristics is discussed in detail in a later section.

The second phase is the design of domain specific genetic operators *End_New, Mid_New, Shake* and *Adjacent* to change the characteristics of the path. The operators tune the path generated by removing the sharp edges, inserting adjacent segments in the path and introducing new vertices. The operators are discussed in detail in section 3.1.4.

The third phase is the design of the objective function which is designed to include the smoothness factors of the path in calculating the fitness value. This ensures that the path is not only optimal in length but also smooth without any sharp turns. The objective function is discussed in detail in section 3.1.5. The fourth phase of the algorithm is the design of the adaptive rules to evolve the operator probabilities. The operator probabilities are adapted based on the smoothness factors. The adaptive rules are discussed in section 3.1.6. The binary tournament selection scheme is used to select a parent chromosome and the reproduction is carried out by refining the paths in the previous generation using the domain specific genetic operators described in section 3.1.4. The complete algorithm is given below:

```
Adaptive_Path_Planner()
Begin
        Initialize the population using the heuristics
        Calculate the objective function and evaluate the population
        While not convergence do
        Begin
                Repeat
                        Select the parent using binary tournament selection
                        Apply the domain specific genetic operators with the optimal
                        probability and produce offspring
                        Replace the parent with the offspring
                Until the next generation is filled
                Evaluate the population based on the objective function
                Tune the operator probability
        End
End
```

3.1.1 The chromosome structure

A chromosome in a population represents a feasible/infeasible path for the robot to reach the goal location. It consists of the starting point followed by the intersection points of the line segments of the path, and finally the goal position or the ending point. A path can have a varied number of intermediate nodes. And hence the length of the chromosomes will be a variable. An initial population of chromosomes is randomly generated such that it has a random number of intermediate edges with random coordinates. A sample path and its representation are given below:

Fig. 5. Example path and its equivalent chromosome

3.1.2 The selection scheme

The binary tournament selection scheme is used to select a parent chromosome. In binary tournament selection, pairs of individuals are picked at random from the population; whichever has the higher fitness is copied into a mating pool. This is repeated until the mating pool is full. In this, the better individual wins the tournament with probability p, where $0.5 < p < 1$. By adjusting tournament size or win probability, the selection pressure can be made arbitrarily large or small.

3.1.3 The initialization heuristics

Each chromosome shall represent a feasible or an infeasible path. A random initialization generally leads to a large number of infeasible paths and hence a heuristics is used to convert infeasible paths to feasible ones. The heuristics involve two operators: *Change_Y* and *Backtrack*. The *Change_y* operator is used to change the value of Y coordinate of the vertex which when included leads to an infeasible path. This makes the robot go up a step to take a different path so that the collision with the obstacle is avoided. The *Backtrack* operator is used to take a different path when the path already taken by the robot leads to an obstacle. It allows the robot to go back by two steps in the original path and take up a new path. It is designed to go back two steps because when the robot realizes an obstacle, it will be very nearer to the obstacle and going back one step may not lead to a feasible path.

3.1.4 The genetic operators

The traditional genetic operators Mutation and Crossover cannot be used as such here, hence they are given new forms to accommodate the requirements of the problem to be solved. The operators are designed having in mind the nature inspired actions, a person shall take to avoid collisions with obstacles. The operators are described below:

Mid_New: A vertex is chosen randomly and the edges connecting it to the previous and the next vertices are altered so that any steep increase in the path can be eliminated. The mid points of the edges are considered recursively until a feasible path is found. This helps to take up a closer but safer path around the obstacle

End_New: A vertex is chosen randomly and is removed from the path provided the resultant path is feasible. This operator helps to reduce the number of fragments in a path.

Shake: This operator chooses a vertex randomly and changes its Y coordinate by either adding or subtracting a constant value. This works like a mutation operator.

Adjacent: This operator changes the original path by interchanging a segment of a path with a parallel segment by either adding or subtracting a constant value with the Y coordinate of each vertex in that segment.

3.1.5 The evaluation function

The fitness of a chromosome is calculated based on the following factors:

- Length
- Feasibility
- Number of acute edges
- Number of bends
- Number of fragments

Two objective functions are designed; one based on length and the other on smoothness. The objective function *Simple_Obj* calculates the fitness of a path on the basis of the length of the path and its feasibility. The objective function *Smooth_Obj* calculates the fitness based on the smoothness of the path. The objective functions are described in the following sections.

3.1.5.1 The objective function based on length and feasibility

The fitness is calculated based on the length and feasibility. The feasibility is measured by checking whether the next node on the path generated so far leads to a collision with any of the obstacles. The length_constant is a large integer value which when divides the path length will give high fitness value for the path with minimal length.

$$Fitness\ F(path) = \left(\frac{Kl}{total\ length\ (path)}\right) + Kf * Feasibility(path)$$

Where K_l is the length_constant (a large integer value) and K_f is the feasibility_constant (an integer value).

$$\text{total\_length(path)} = \Sigma\ \text{length}(i, i+1),\ 1 \le i < n,\ n = \text{No. of nodes}$$

$$feasibility(path) = \begin{cases} 1, & for\ a\ feasible\ path \\ -1, & for\ an\ infeasible\ path \end{cases}$$

3.1.5.2 The objective function based on smoothness

The objective function *Simple_Obj* discussed above takes into account only the path length and the feasibility factor whereas the *Smooth_Obj* takes into account the other factors such as the number of acute increases/decreases, turns with angle 90° and the number of fragments in the path. The new objective function *Smooth_Obj* is given below:

$$Fitness\ F(path) = \left(\frac{Kl}{total\_length(path)}\right) + Ks * smoothness(path)$$

$$Smoothness(path) = \left(\frac{50 * SA}{100}\right) + \left(\frac{40 * SB}{100}\right) + \left(\frac{10 * SF}{100}\right)$$

where

$$SA = 1 - \frac{Acute\_Count * 100}{NF}$$

$$SB = 1 - \frac{Bend\_Count * 100}{NF}$$

$$SF = \frac{Ideal\ Fragements\ * 100}{NF}$$

And NF is the number of Fragments

The *Acute_Count* is the number of sharp increases / decreases in the path, the *Bend_Count* is the number of rectangular turns and the *Ideal_Fragments* is the minimum number of fragments of all the paths that occur in a generation. Since the number of acute edges affects the smoothness of the path largely, its contribution is 50% in the calculation of the fitness value. The number of bends is given 40% share and the fragments 10% share in the calculation of the fitness value.

3.2 The dynamic environment

The robot travels in an environment where the obstacles may get introduced suddenly in the path planned by the robot for travelling. In such situations the robot has to decide at every step whether to take up the already planned path or a new path. The robot is assumed to travel at a fixed speed and has a sensor in it to detect the existence of any obstacle. When an obstacle is recognized by the robot it calls the dynamic path planner algorithm and plans its path from that position onwards. The sensor is assumed to sense the existence of any obstacle before the robot reaches the region and the path planner returns the path within the critical time. Now the new path is taken up from that position and the same procedure is repeated until the robot has reached the destination.

3.3 Results

A sample output for the adaptive GA solution is given below. The Robot is assumed to travel in a dynamic environment of dimension (5,5) to (400,400) with different kinds of obstacles placed randomly. A dynamic environment is created by adding new obstacles on the path planned by the robot.

The path planning algorithms for dynamic environments are computationally intensive and hence will take longer time to converge for an environment with non-rectilinear obstacles.

Faster genetic operators and multi-threading methods might help to speed up the path planning process in these environments.

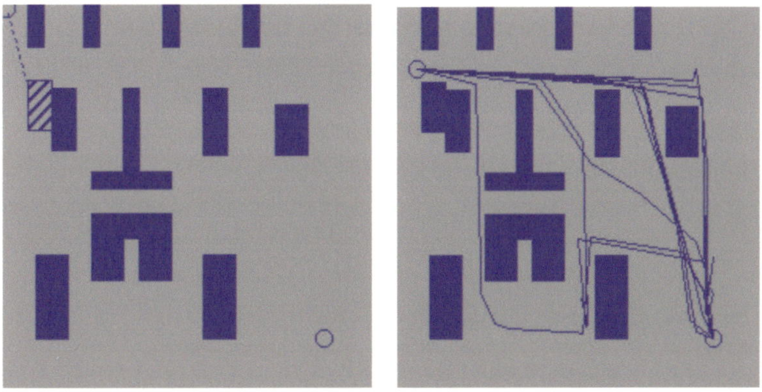

(a) (b)

Fig. 6. (a) A dynamic obstacle in the planned path and (b) the final generation of paths from the point of intervention

4. Design of a Hybrid CHC GA for VLSI routing

The Minimum Rectilinear Steiner Tree (MRST) problem arises in VLSI global routing and wiring estimation where we seek low-cost topologies to connect the pins of signal nets. The Steiner tree algorithm is the essential part of a global routing algorithm. It has been an active field of research in the recent past. This section presents a Hybrid CHC (HCHC) genetic algorithm for global routing. The *Minimum Rectilinear Steiner Tree* problem is to construct a tree that connects all the n points given in the Euclidean plane. If the edges in this tree are to be selected from all possible edges that are from the complete graph on the points, it is the familiar problem of finding a spanning tree in an undirected graph. If the edges of the tree must be horizontal and vertical, the additional points where the edges meet are called the Steiner points, and the resulting tree is a Rectilinear Steiner Tree [RST]. A shortest such tree on a set of given points is a minimum rectilinear Steiner tree.

4.1 Construction of Minimum Spanning Tree

The Spanning tree algorithm presented here is based on the shortest path heuristic as described by *Ellis Horowitz* et al. A simple genetic algorithm is used for the construction of a Minimum Spanning Tree (MST), which is then used in the generation of the Steiner minimal tree. The spanning tree is generated by initializing the population with random solutions. The random solutions are then repaired using a repair heuristics. The offspring are generated by applying one point crossover and exchange mutation which exchanges edges in an individual. The exchange of edges may lead to a totally different tree, thus justifying the purpose of mutation. The new population is evolved and the same procedure is repeated until convergence.

The algorithm for the construction of the minimum spanning tree is given below.

Minimal_Spanning_Tree()

Begin

Initialize parameters: generation count, crossover and mutation probabilities

Initialize parent population randomly

Apply repair heuristics to the parent population

While termination condition not reached

Begin

Select parents based on the total length of the Spanning tree

Apply crossover and mutation

Evolve new population

Replace previous population by new population

End

End

The repair heuristics removes cycles and repeated edges from the population and makes it a set of feasible solutions. The vertices of the graph are stored in separate sets, so that it can be later combined whenever an edge is included in the final spanning tree. The *union* algorithm unions two sets containing V_i and V_j respectively when an edge E $<V_i, V_j>$ is added to the final spanning tree. The *find* algorithm verifies whether a particular vertex belongs to a set. The *union* algorithm combines all vertices that are connected, in to a single set. When a particular edge is selected for addition into a partially constructed spanning tree, it is checked whether the vertices of that edge are already present in the same set using the *find* algorithm. If they are in the same set, then the inclusion of this edge will lead to a cycle. The repeated edges can be checked easily with the adjacency matrix.

4.2 Construction of minimum steiner tree

The Steiner tree problem can be defined as the subset of minimum spanning tree problem. In minimum spanning tree construction, a tree is constructed with vertices $V_1,V_2,...V_n$ connected without loops at the lowest cost. In the Steiner tree problem, extra vertices are added besides the existing vertices $V_1,V_2,...V_n$, to construct a lower cost tree connecting $V_1,V_2,....V_n$. The extra vertices are called the Steiner points. There are various heuristics available to construct a MRST, and most of them use MST as a starting point. The I-Steiner algorithm as discussed by *Kahng A.B* et al., constructs the MRST by evaluating all possible Steiner points for their impact on MST cost. The algorithm operates on a series of passes, in each pass the single Steiner point which provides the greatest improvement in spanning tree cost is selected and added to the set of demand points. Points are added until no further improvement can be obtained.

The heuristics used here for the construction of MRST is the "BOI" or "edge-removal technique" of *Borah, Owen* and *Irwin*. The algorithm constructs the Steiner tree through repeated modification of an initial spanning tree as discussed by Jayalakshmi et al. An edge and a vertex pair that are close to each other in an MST are determined. For each vertex V_i, edge E_i pairing of the spanning tree, an optimal Steiner point is found to merge the endpoints of the edge E_j with vertex V_i. This will create a cycle, so the longest edge on this cycle is found and a decision is made about removing this edge from the cycle based on the cost. Among all possible eliminations, whichever leads to the lowest cost is removed and the tree is modified. The edges are removed and new connections are inserted until no improvements can be obtained. The resulting tree is the minimum Steiner tree. The approach has low complexity with performance comparable to that of I-Steiner. The algorithm for the construction of the minimum Steiner tree is given below:

Steiner_Tree()

Begin

Build the routing graph G

For each net do

Begin

Initialize weights for edges

Find the minimum cost spanning tree T

For each <vertex,edge> pair of the spanning tree

Begin

Find the optimum Steiner point to connect this edge to the vertex at a suitable point

Find the longest edge on the generated cycle

Compute the cost of the modified tree and store the pair in a list if the cost is less than the MST

End

While the list is not empty do

Begin

Remove the pair from the list which results in lowest cost

Re-compute the longest edge on the cycle and the cost of the tree

If the edges to be replaced are in the tree and the cost is less then modify the tree

End

End

End

4.3 Results

The HCHC solution for four standard test problems B1, B3 B6 and B9 from the problem sets of *J.E.Beasley* are given below in Table 3. A simple GA with one point crossover and exchange mutation is compared with the HCHC solution. In HCHC, Uniform crossover and External mutation are used for reproduction.

Test problem	Optimum	Solution		Error		Generations	
		SGA	HCHC	SGA	HCHC	SGA	HCHC
B1	82	187	95	105	13	150	200
B3	138	145	140	7	2	150	200
B6	122	128	125	6	3	400	250
B9	220	241	224	21	4	150	200

Table 3. The solutions obtained by SGA and HCHC for Beasley's test problems B1, B3,B6 and B9

For HCHC algorithm the maximum error is for the test problem B1 and for the rest of the problems the error is less than 6. And SGA has performed very poorly for B1 and B9. For the other problems, SGA has performed moderately well with error less than 10.

5. Summary

With Genetic Algorithms emerging as strong alternative to traditional optimization algorithms, in a wide variety of application areas, it is important to find the factors that influence the efficiency of the genetic algorithms. The simple GAs are found to be ineffective for most of the real world problems. Hence there arises the need for the customization of the traditional GAs. This chapter explored the variants of the simple genetic algorithm and their application to solve real world problems. TSP is a problem of a specific domain and required hybridization for quicker convergence. In particular the local search algorithm chosen has a determining influence on the final performance. The heuristics used were simple and easy to implement when compared to other algorithms. The solution to the mobile robot path planning problem explored the design of different operators and showed that the adaptation of operators has a significant impact in improving the solution quality. A hybrid CHC algorithm was used to solve the VLSI global routing problem. And this example showed that the simple GA could only find a sub optimal solution and could not go beyond certain values due to the lack of techniques that avoid premature convergence.

6. References

Beasley.J.E. (1989). An SST-Based Algorithm for the Steiner Problem in Graphs, Networks, Vol.19, pp.1-16, 1989.

Borah.M, R. M. Owens & M. J. Irwin. (1994). An edge-based heuristic for Steiner routing, *IEEE Trans. Computer-Aided Design*, vol. 13, pp. 1563–1568, Dec. 1994.

Choset, H., Lynch, K. M., Hutchinson, S., Kantor, G., Burgard, W., & Kavraki, L. E. (2005). *Principles of robot motion: theory, algorithms, and implementations*. Boston, MIT Press.

Davis.L, (Editor). (1991). "*Handbook of Genetic Algorithms* ", Van Nostrand Reinhold.

Ellis Horowitz, Sartaj Sahni & Sanguthevar Rajasekaran. (2007). *Computer Algorithms*, Silicon Pr.

Eshelman,L.J. (1991). *The CHC Adaptive Search Algorithm: How to have safe search when engaging in nontraditional genetic recombination*, Rawlins G.(Editor), Foundations of Genetic Algorithms, Morgan Kaufmann, pp.265-283.

Goldberg, David E. (1989). *Genetic Algorithms in Search Optimization and Machine Learning*, Addison Wesley, ISBN 0201157675.

Goldberg, David E. (2002). *The Design of Innovation: Lessons from and for Competent Genetic Algorithms*, Addison-Wesley, Reading, MA.

Holland, John H. (1975). *Adaptation in Natural and Artificial Systems*, University of Michigan Press, Ann Arbor.

Hu.J & Sapatnekar.S.S. (2001). Performance driven global routing through gradual refinement, IEEE Int. Conf.on Comp Design, 2001, pp.481-483

Jayalakshmi.G.A, S. Sathiamoorthy & R. Rajaram. (2001). A Hybrid Genetic Algorithm - A New Approach to Solve Traveling Salesman Problem, *International Journal of Computational Engineering Science*, Volume: 2 Issue: 2 p.339 - 355.

Jayalakshmi, G.A, Prabhu. H, & Rajaram. R. (2003). An Adaptive Mobile Robot Path Planner For Dynamic Environments With Arbitrary-Shaped Obstacles, *International Journal of Computational Engineering Science*(2003),pp. 67-84

Jayalakshmi, G.A., Sowmyalakshmi.S & Rajaram.R (2003). A Hybrid CHC Genetic Algorithm For Macro Cell Global Routing, *Advances in Soft Computing* Engineering Design and Manufacturing Benitez, J.M.; Cordon, O.; Hoffmann, F.; Roy, R. (Eds.) pp. 343 - 350, Springer-Verlag, Aug 2003

Kahng.A.B and G.Robins. (1992). *A new class of iterative Steiner tree heuristics with good performance*, IEEE Trans on Computer Aided design of Integrated circuits and systems, 11(7), pp.893-902.

Li-Ying Wang, Jie Zhang & Hua Li. (2007). An improved Genetic Algorithm for TSP, *Proceedings of the Sixth International Conference on Machine Learning and Cybernetics*, Hong Kong, 19-22 August 2007.

TSPLIB: http://elib.zib.de/pub/mp-testdata/tsp/tsplib/tsplib.html retrievable as of 3rd Feb 2012

Whitley.D, Lunacek.M & Sokolov.A. (2006). Comparing the Niches of CMA-ES, CHC and Pattern Search Using Diverse Benchmarks, *Parallel Problem Solving from Nature Conference (PPSN 2006)*, Springer.

Whitley.D, T. Starkweather & D. Shaner .(1991). The Traveling Salesman and Sequence Scheduling: Quality Solutions Using Genetic Edge Recombination. *The Handbook of Genetic Algorithms*. L. Davis, ed., pp: 350-372. Van Nostrand Reinhold.

Zhou.H. (2004). Efficient Steiner Tree Construction Based on Spanning Graphs, IEEE Transactions on Computer-Aided Design of Integrated Circuits And Systems, Vol. 23, No. 5, May 2004, pp:704-710.

16

Automatic Control of the Software Development Process with Regard to Risk Analysis

Marian Jureczko
Wrocław University of Technology
Poland

1. Introduction

In this paper, we present a method for analysing risk in software development. Before describing the details, let us define the basic terms that are important for the further discussion. Among them, the most import is risk. In the context of software engineering it represents any potential situation or event that can negatively affect software project in the future and may arise from some present action. Software project risks usually affect scheduling, cost and product quality. Further details are given in Section 2.2. The risk will be discussed in the context of project requirements and releases. The requirements (or requirement model) represent requests from the customer that should be transformed into features and implemented in the software product. The releases refer to a phase in the software development life cycle. It is a term used to denote that the software product is ready for or has been delivered to the customer. This work takes into consideration agile development (i.e. processes with short iterations and many releases) therefore, it should be emphasized that some of the releases may be internal. These releases are not scheduled to be delivered to the customer; nevertheless they are 'potentially shippable'.

Considerable research has been done on controlling the software development process. The mainstream regards process improvement approaches (e.g. CMMI SEI (2002)) where time consuming activities are performed in order to monitor and control the work progress. In effect the maturity is assessed. Furthermore, mature processes indicate lower project failure risk.

The agile software development shifts the focus toward product Beck & Andres (2005). Moreover, it is recommended to ensure high level of test coverage. The tests are automated Jureczko & Młynarski (2010), and therefore it is possible to execute them frequently in a continuous integration system. Despite of the fact that the agile approach is more automated, risk assessment may be challenging. There are few risk related metrics right out of the box. Fortunately, due to the high level of automation considerable amount of data regarding quality and risk might be collected without laborious activities Stamelos & Sfetsos (2007). It is possible to build simple metrics tools (e.g. BugInfo Jureczko (Retrived on 01/11/2011a), Ckjm Jureczko & Spinellis (Retrived on 06/05/2011)) as well as a complex solution for identifying 'architectural smells' (Sotograph Hello2morrow Inc. (Retrived on 01/11/2011) into the continuous integration system. Unfortunately, it is not evident which metrics should be collected and what analysis should be performed to accomplish the risk related goals. This chapter discusses which data should be collected and how it should be done in order to obtain accurate risk estimations with smallest effort. It will be considered how to mitigate the effort by using automation. We first suggest a risk analysis method that reflects the common trends in this area (however the focus is slightly moved toward agile software development) and then we decompose the method to well–known in software engineering problems. The potential solutions to the aforementioned problems are reviewed with regard to automation in order to identify improvement possibilities. In result, a direction for further research is defined. Moreover, the main value of this chapter is the review of the status and recent development of risk related issues with regard to automation.

The remainder of the chapter is organized as follows: in the next section an approach to risk analysis is suggested and decomposed into three problems, i.e. effort estimation, defect prediction and maintainability measurement. The next three sections survey the three aforementioned issues. The sixth section discusses the automation possibilities. Finally, the seventh section presents concluding remarks and future research direction.

2. Problem formulation

There are tools that support the risk analysis. Unfortunately, such tools usually do not offer automation, but require expert assistance, e.g. RiskGuide Górski & Miler (Retrived on 01/11/2011) or are proprietary black box approaches, e.g. SoftwareCockpit Capgemini CSD Research Inc. (Retrived on 01/11/2011). The users do not know what exactly is going on inside and the software projects may differ from each other Jureczko & Madeyski (2010); Zimmermann et al. (2009). In order to make good risk related decisions, the inference mechanism should be transparent and understandable.

2.1 Requirement model

The requirement model that incorporates both functional and non-functional features might lay foundations for a method of risk assessment. Each and every requirement is characterized by its specification and so-called time to delivery. Since most of the real world projects undergo not only new feature development, but also maintenance, defect repairs are considered requisite and hence also belong to the requirement model.

In case of any functional requirement, it may or may not be met in the final product, there are only two satisfaction levels: satisfied or not satisfied. In the case of non–functional requirements the satisfaction level might be specified on the ordinal or interval scale depending upon the requirement kind.

Failing to deliver some requirement in due time usually causes severe financial losses. When a functional requirement is missing, a penalty is a function of delivery time, which in turn is a constant defined in the customer software supplier agreement. On the other hand, when a non–functional requirement is not satisfied, a penalty is a function of both: delivery time and satisfaction level. These values should be defined in the very same agreement or evaluated using empirical data from similar historical projects.

Among the non–functional requirements two are especially interesting: quality and maintainability. Both are very difficult to measure Heitlager et al. (2007); Kan (1994) and both may cause severe loss in future. Let us consider a release of a software system with low level of quality and maintainability. There are many defects since the quality is low. The development of the next release will generate extra cost because the defects should be removed. Furthermore, the removal process might be very time consuming since the maintainability is low. There is also strong possibility that developing new features will be extremely expensive due to maintainability problems. Therefore, quality as well as maintainability should be considered during risk analysis.

2.2 The risk

The risk $R(t)$ specified over a set of possible threats (t represents a threat) is defined as follows:

$$R(t) = P(t) * S(t) \tag{1}$$

where $P(t)$ stands for a probability that t actually happens and $S(t)$ represents the quantitative ramifications expressed most often in financial terms. The following threat sets a good example: a product complying with some requirement is not delivered in due time. When the requirement is delivered with delay and the ramifications depends on the value of the delay, the Equation 1 should be drafted into the following form:

$$R(t) = \int_0^{\infty} P(t,\delta) * S(t,\delta) d\delta \tag{2}$$

where δ denotes the value of delay.

When risk is estimated for a threat on the grounds of delivery time distribution, the mean loss value becomes the actual value of risk. Then, summing up risk values for all requirements belonging to the requirement model of a particular release, the mean cumulative risk for the release r is obtained:

$$RR(r) = \sum_{t \in T_r} R(t) \tag{3}$$

where T_r is the set of threats connected with release r. If all the releases are considered, the mean threat cost is obtained for the whole project p:

$$RP(p) = \sum_{r \in p} RR(r) \tag{4}$$

The risk can be used to evaluate release plan as well as the whole project. When the $RP(p)$ calculated for the whole project surpasses the potential benefits, there is no point in launching the project. Furthermore, different release plans may be considered, and the $RR(r)$ value may be used to choose the best one.

One may argue that summing risks that were estimated for single threats ($R(t)$) in order to evaluate the risk regarding the whole project can result in cumulating estimation errors. Therefore, the provided estimates of $RP(p)$ can be significantly off. Such scenario is likely when the requirements come late in the software development life cycle and are not known up front. Nevertheless, the suggested approach also has advantages. As a matter of fact, considering each threat separately can be beneficial for agile methods where plans are made for short periods of time, e.g. the Planning Game in XP Beck & Andres (2005). The agile methods usually employ high level of automation Beck & Andres (2005), Stamelos & Sfetsos (2007), and hence create a good environment for automated risk analysis. Furthermore, the mean cumulative risk for the release $RR(r)$ is designed to be the main area of application of the suggested approach and the release is usually the main concern of planning in agile methods Beck & Andres (2005), Schwaber & Beedle (2001).

The suggested risk analysis method partly corresponds with the ISO/IEC standard for risk management ISO (2004). The standard is focused on the whole software life cycle whereas our method can cover the whole life cycle but is focused on shorter terms, i.e. releases. Similar, but slightly more complex method was suggested by Penny Penny (2002), who emphasized the relevance of resources quality and environmental conditions (specifically interruptions and distractions in work). We do not question the value of those parameters. Nevertheless, they may cause difficulties in collecting the empirical data that is necessary to estimate the threat occurrence probabilities. Therefore, data regarding those parameters is not required in our risk analysis method, but when available, definitely recommended.

2.3 Threat occurrence probability

According to Equation 1 the risk consist of two elements : probability of occurrence of a threat $P(t)$ and expected loss $S(t)$. Let us assume that the threat t_i regards failing to deliver the requirement req_i in the due time. Therefore, in order to calculate $P(t_i)$ the distribution of expected effort, which is necessary to fulfill requirement req_i, must be evaluated. The methods of effort estimation are discussed in Section 3.

There is also a second factor that is relevant to the probability of occurrence. It is the availability of resources for the planned release, which means that the effort estimated for the requirements and the available manpower indicates the probability of failing to deliver a requirement in due time. Furthermore, not every requirement could be implemented with equal effort by any participant of the developers team Helming et al. (2009). Therefore, the effort estimations should be combined with a workflow model. The workflow model is indicated by the employed software development methodology, and hence should be available in every non chaotic process. Nevertheless, what derives from the methodology definition can be on a very high level of abstraction or flexibility, especially in the case of agile methods Schwaber & Beedle (2001), and detailing the release plan is not a trivial task. According to Bagnall et al. Bagnall et al. (2001) the problem is NP hard even without any dependencies between requirements. Fortunately, there are methods for semi-automated planning of release Helming et al. (2009).

Estimation of probability of occurrence of a threat $P(t)$ must take into consideration two factors: effort estimation for different requirements on the one hand and the allocation of resources on the other. Resources allocation can be done semi-automatically Helming et al. (2009) or may come from a manager decision. In both cases the resources allocation will be available and can be used for further analysis. However, in some cases (i.e. in the agile methods) it may be done on a very low level of granularity. Effort estimation is more challenging. The difficulties that may arise during automating those estimations are discussed in the next section.

2.4 Loss function

As it has been stated in Section 2.1 the value of loss function might be explicitly defined in the customer software supplier agreement. Unfortunately, such simple cases do not cover all possibilities. There are quality related aspects that might not be mentioned in the agreement, but still may cause loss. According to the ISO/IEC 9126 ISO (2001) there are six characteristics to consider: functionality, reliability, usability, efficiency, maintainability, and portability. Some of these characteristics may be covered by the customer software supplier agreement and when are not delivered on the contracted level there is a penalty. However, there are two aspects that should be explicitly considered, i.e. software defects and maintainability (the defects might regard not satisfying one of the five other ISO quality characteristics).

It is a typical case that software contains defects Kan (1994). Therefore, the customers, in order to protect their interests, put defect fixing policies in the agreement. As a result, the defects create loss. First of all they must be repaired, and sometimes there is also a financial penalty.

There are many different grounds for defects. Nevertheless, there is a general rule regarding the relation between cost, time and quality called The Iron Triangle Atkinson (1999), Oisen (1971), see Fig. 1. According to The Iron Triangle cost, time and quality depend on each other. In other words, implementation of a set of requirements generates costs and requires time in order to achieve the desired quality. When one of those two factors is not well supported, the quality goes down, e.g. skipping tests of a component may result in a greater number of defects in this component. However, it should be taken into consideration that in the case of skipping tests we have savings in current release. The issues regarding lower quality and greater number of defects will arise in the next release. Therefore, it is important to estimate the risk regarding lower quality since it sometimes could be reasonable to make savings in current release and pay the bill in the next one. Risk related to low quality regards the loss that comes from defect fixing activities in the next release. Hence, the number of defects as well as the effort that is necessary to fix them should be predicted and used to construct the loss function. The defect prediction is discussed in Section 4.

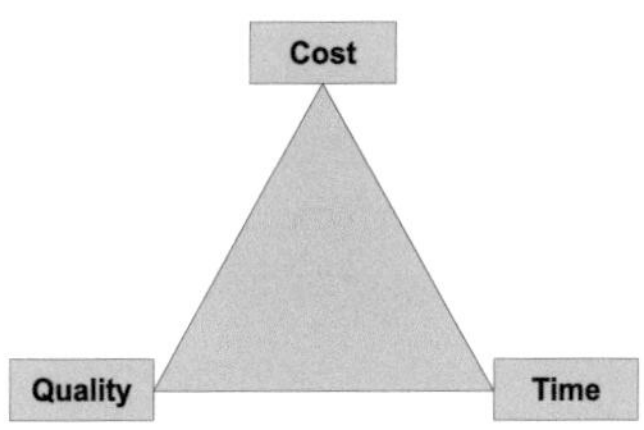

Fig. 1. The Iron Triangle

Low level of maintainability, which reflects problems with analysing, changing and testing the software increases the efforts of developing new features. Since the efforts are greater, the threat occurrence probability for forthcoming requirements grows as well. If the current release is not the last, the low maintainability related loss should be estimated and considered. Methods of estimating maintainability are discussed in Section 5.

2.5 Automation possibilities

The goal of this chapter is to present how risk can be estimated in an automated way. Risk has been decomposed into five factors: effort estimation, release planning, missing functionality penalty, defect prediction, and maintainability measuring (Fig. 2. The release planning is a manager's decision. Nevertheless, the decision may be supported by a tool. Specifically, the suggested risk evaluation approach can be used to assess a release plan. The penalty is usually defined in the customer software supplier agreement and is a function of both: delivery time and satisfaction level of the requirement implementation.

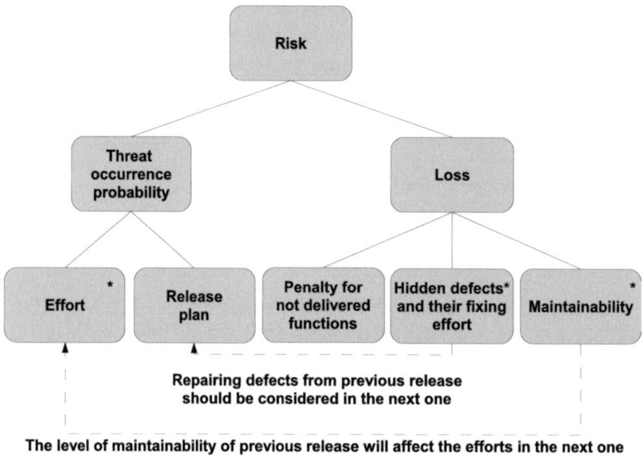

Fig. 2. Problem decomposition; elements denoted with '*' are detailed in forthcoming sections.

The three other aspects (effort, hidden defects and maintainability) are evaluated using empirical data, therefore there is room for automation. When there is no estimation method available, historical data must be collected and used to train a prediction model. If there is no historical data, cross-project prediction methods should be considered. All the aforementioned activities may be automated to some extent. We can employ tools to collect data, we can use tools to make the estimations and finally to generate the risk related reports.

Unfortunately, not all aspects of discussed above risk analysis were investigated closely enough to give conclusive remarks and not all of them are supported by tools. The following sections will discuss what has been already done and where there are gaps that must be filled by practitioners who would like to automate the risk analysis.

3. Effort estimation

Considerable research has been done on effort estimation. It regards different techniques and at least some of them may be used in risk analysis. As it was discussed in Section 2, threats like 'a product complying with some requirement is not delivered in due time' should be considered. Therefore, efforts for each requirement must be estimated to assess the risk.

In order to choose the most appropriate effort estimation approach let us review the available ones. Boehm et al. Boehm et al. (2000) summarized the leading techniques (the content of this section is mostly based on their work). Later, the estimation methods were also reviewed by Jorgensen and Shepperd Jorgensen & Shepperd (2007).

3.1 Model-based techniques

Most of the models have functional form. Metrics which describe features of the object of estimation or features of the organization, which is going to develop the object, are used as model input. After some calculations the model produces value of the expected effort. Unfortunately, many models are proprietary and hence cannot be analysed in terms of model structure. On the other hand, the proprietary solutions are usually ready to use tools with high level of automation.

3.1.1 Constructive Cost Model (COCOMO 81 and COCOMO II)

The COCOMO model was firstly suggested in Boehm (1981). It is an algorithmic software cost estimation model that uses regression formula. The intuition behind COCOMO model is that the development grows exponentially as the developed system grows in size:

$$Effort_{PM} = a * (KLOC^b) * (\prod_j EM_j) \tag{5}$$

Where, $Effort_{PM}$ is the whole effort in person months, $KLOC$ is thousands of lines of source code, EM is an effort multiplier generated via regression on historical data, a and b denote software type coefficients.

The COCOMO 81 model was based on a study of 63 projects that were created using the waterfall software development model, which reflected the practices of the 80'. Nevertheless, software development techniques changed dramatically, namely highly iterative processes were employed. Therefore, applying the COCOMO 81 model becomes problematic. The solution to the problem was to redesign the model to accommodate different approaches to software development.

The revised cost estimation model was called COCOMO II and was initially described by Boehm et al. Boehm et al. (1995). The model is focused on estimating the effort that regards the whole system. Moreover, it is based on empirical data; it is widely used and hence well validated. Web based tools that support the COCOMO II model are available on the model's web page Center for Systems and Software Engineering (Retrived on 01/11/2011). There are commercial implementations: COSTAR Softstar Systems (Retrived on 01/11/2011) and Cost Xpert Cost Xpert AG (Retrived on 01/11/2011) as well, so the automation is relatively easy. There are also derived models, e.g. COSYSMO Valerdi et al. (2003) or AFCAA REVIC Air Force Cost Analysis Agency (Retrived on 01/11/2011).

3.1.2 Putnam's Software Life-cycle Model (SLIM)

The Putnam's Software Life-cycle Model Putnam (1978) is based on analysis of software development lifecycle in terms of the well–known Rayleigh distribution of resources (project personnel) usage versus time. Quantitative Software Management offers a tools suite based on metrics collected from over 10 000 completed software projects that employ the Putnam's SLIM model Quantitative Software Management Inc. (Retrived on 01/11/2011).

The tools assure high level of automation. SLIM was originally designed to make estimations for the whole project. However, the project can be interpreted as a set of requirements, what partly corresponds with the described in Section 2 risk analysis.

3.1.3 SPR KnowledgePLAN

SPR KnowledgePLAN (following in the footsteps of previous SPR products: SPQR/20 and Checkpoint) is a knowledge-based software estimation solution Software Productivity Research (Retrived on 01/11/2009). It is based on a proprietary database of 14 531 projects (version 4.3). The estimation is calculated according to detailed project description.

The tool predicts effort at four levels of granularity: project, phase, activity and task, therefore it is suitable for risk analysis. Furthermore, since there is good support in tools the automation is relatively simple.

3.1.4 Functionality–based estimations

3.1.4.1 Function Points

The function points were suggested by Albrecht Albrecht (1979) and are a standard unit of measure for representing the functional size of a software application. The functional requirements are categorized into five types: outputs, inputs, inquiries, internal files, and external interfaces. Subsequently, the requirements are assessed for complexity and a number of function points are assigned. Since 1984 this method has been promoted by The International Function Point Users Group (IFPUG). This group produced a set of standards, known as the ISO/IEC 14143 series. Function Point Analysis was also developed by NESMA Netherlands Software Metrics Association (Retrived on 01/11/2011). It resulted in method called NESMA Functional Size Measurement that is compliant with ISO/IEC 24570. Moreover, there are good automation possibilities since the Function Point Analysis is supported by a tool called Metric Studio TSA Quality (Retrived on 01/11/2011).

3.1.4.2 MKII FPA

Mk II Function Point Analysis Symons (1991) is a software sizing method that belongs to the function point group of techniques. The most important feature of this method is the simple measurement model. There are only three components to consider: inputs (processes that enter data into software), outputs (processes that take data from software), and entity references (storage, retrieval and deletion of date from permanent storage).

3.1.4.3 Weighted Micro Function Points (WMFP)

The Weighted Micro Function Points (WMFP) is a software size estimating method developed by Logical Solutions Logical Solutions (Retrived on 01/11/2011). The method measures several different software metrics obtained from the source code that represents effort and are translated into time: flow complexity, object vocabulary, object conjuration, arithmetic intricacy, data transfer, code structure, inline data, and comments.

3.1.4.4 Common Software Measurement International Consortium (COSMIC)

The COSMIC project was launched in 1998, it is a voluntary, world-wide grouping of software metrics experts Common Software Measurement International Consortium (Retrived on 01/11/2009). The focus is on developing a method of measuring functional size of software. The method is entirely open, and all the documentation is available for free of charge download. Unfortunately, the consortium provides only the method description and hence automation may be time consuming. However, there are third party vendors that support the method[1].

3.1.4.5 Object Points

Object Points is an object oriented software size estimating method developed by Banker et al. Banker et al. (1991). The method is based on the following object types (type definitions after Banker et al. Banker et al. (1991)): rule set (a collection of instructions and routines written in a high level language of a CASE tool), 3GL module (a precompiled procedure, originally written using third-generation language), screen definition (logical representation of an on-screen image), and user report. The object points are obtained by summing the instances of objects of the aforementioned types and weighting each type by the development effort associated with it.

3.1.4.6 Use Case Points (UCP)

UCP is a measure of software size that takes into consideration use cases. It is calculated by counting the number of actors and transactions that were included in the flow of use case scenarios. Transaction is an event that occurs between the actor and the system and can be made entirely or not at all.

There are tools that support the UCP, e.g. Kusumoto et al. (2004). Nevertheless, identifying transactions in the use cases is not a trivial task. Natural language processing methods are usually employed Ochodek & Nawrocki (2007) and there is a considerable mistake possibility.

3.1.5 PRICE-S

The PRICE-S is a parametric cost model that was originally developed for internal use in NASA software projects. It is implemented in the TruePlanning PRICE Systems (Retrived on 01/11/2011), which is a tool that encapsulates the experience with 3 212 projects from over 30 years.

According to the authors, risk refers to the fact that because a cost estimate is a forecast, there is always a possibility of the actual cost being different from the estimate (such situation very well corresponds with our goal). Furthermore, using TruePlanning tool allows risk analysis that quantifies probability.

The tool makes estimations on work package level hence it is appropriate for the described in Section 2 risk analysis. Unfortunately, according to Boehm et al. (2000) the model equations were not released in public domain, still, it is a well–designed commercial solution and therefore the automation is supported.

[1] http://www.cosmicon.com/vendorsV3.asp

3.1.6 SEER for Software (SEER-SEM)

SEER-SEM is an algorithmic, parametric project management system that supports estimating, planning and tracking effort and resource consumption for software development and maintenance Galorath Inc. (Retrived on 01/11/2011). The estimations are calculated using effective size (S_e). It is a generic metric that covers development and integration of new, reused, and commercial off-the-shelf software modules. Several sizing metrics can be translated into S_e, this includes lines of code (LOC), function points, IFPUG function based sizing method, and use cases. Although the tool is focused on estimation for the whole project, there are great automation possibilities due to the support for integration with MicroSoft and IBM products.

3.1.7 PROxy-Based Estimating (PROBE)

Proxy based estimating uses previous developments in similar application domains. The method is a part of the Personal Software Process Humphrey (1995). A proxy is a unit of software size that can be identified early in a project (e.g. screen, file, object, logical entity, function point). Proxies can be translated into lines of source code according to historical sizes of similar proxies in earlier developed projects.

The concept of proxies corresponds well with the risk analysis. Nevertheless, according to author's knowledge there is no tool that supports the method. Therefore, the automation may be challenging.

3.2 Expert–based techniques

The expert–based techniques capture the knowledge of domain experts and use it to obtain reliable estimations. Those techniques are especially useful when there is not enough empirical data to construct a model. Jorgensen and Shepperd Jorgensen & Shepperd (2007) pointed out two important facts regarding those methods of estimation. Firstly, they concluded that the formal estimation techniques have no documented accuracy higher than the expert–based approach. Secondly, the expert–based technique is the most commonly used method in the industry. Unfortunately, using expert knowledge is very inconvenient with regard to automation.

Two expert–based techniques have been developed. These are the Work Breakdown Structure and the Delphi method.

3.2.1 Work Breakdown Structure (WBS)

The WBS is a method of organizing project into a hierarchical structure of components or activities (as a matter of fact, two structures can be created; one for the software product itself, and one for the activities that must be performed in order to build the software) in a way that helps describe the total work scope of the analysed project. The hierarchical structure is developed by starting with the end objective and dividing it into gradually smaller ones. A tree, which contains all actions that must be executed in order to achieve the mentioned objective is obtained as the result. The estimation is done by summing the efforts or resources assigned to single elements.

3.2.2 Delphi

The Delphi method is a structured, interactive estimation method that is based on experts judgements. A group of experts is guided to a consensus of opinion on an issue. Participants (experts) make assessments (estimations) regarding the issue without consulting each other. The assessments are collected and presented to each of the experts individually. Subsequently, the participants make second assessment using the knowledge regarding estimations provided by other experts. The second round should result in a narrowing of the range of assessments and indication of a reasonable middle value regarding the analysed issue.

The Delphi technique was developed at The Rand Corporation in the late 1940s as a way of making predictions regarding future events Boehm et al. (2000); currently a similar method, i.e. the planning poker is very popular.

3.3 Lerning-oriented techniques

Learning-oriented techniques include a wide range of different methods. There are traditional, manual approaches like case studies as well as highly automated techniques, which build models that learn from historical data, e.g. neural networks, classification trees. Unfortunately, even the most highly automated approaches require some manual work since the historical data must be collected and the estimation models must be integrated with the software development environment. The learning-oriented techniques are commonly employed in other approaches.

3.3.1 Case studies

Case studies represent an inductive process, whereby estimators and planners try to learn useful general lessons and estimation heuristics by extrapolation from specific examples Boehm et al. (2000). Case studies take into consideration the environmental conditions and constraints, the decisions that influenced software development process, and the final result. The goal is focused on identifying relations between cause and effect that can be applied in other projects by analogy. Case studies that result in a well–understood cause effect relations constitute good foundation for designing automated solutions that can be applied to similar project. Nevertheless, in order to define the boundaries of similarity further analysis is usually required.

According to Jorgensen and Shepperd Jorgensen & Shepperd (2007) there are few case studies focused on the effort estimation. In consequence, there is a possibility that we still can learn a lot from well described real–life cases.

3.3.2 Regression based techniques

Regression based techniques are the most popular method in the effort estimation studies Jorgensen & Shepperd (2007). Moreover, they were used in many of the model based approaches, e.g. COCOMO, SLIM.

The regression based methods refer to the statistical approach of general linear regression with the least square technique. Historical data is employed to train the model and then the model is used for making estimations for new projects. The regression based methods are well suited for automation since they are supported by almost all statistic tools (including spreadsheets). Unfortunately, the data which is necessary to train the model is not always available and its collection may cause automation related issues.

3.3.3 Neural networks

There are several artificial intelligence approaches to effort estimation and the neural networks are the most popular among them. Nevertheless, according to Jorgensen and Shepperd Jorgensen & Shepperd (2007) less than 5% of research papers oriented to effort estimation considered the neural networks. There is no no good reason for such low popularity since the neural networks have proven to have adequate expression power. Idri et al. Idri et al. (2002) used backpropagation threelayer perceptron on the COCOMO 81 dataset and according to the authors, the obtained accuracy was acceptable. Furthermore, the authors were able to interpret the obtained network by extracting if-then fuzzy rules from it, hence it alleviates to some extent the main neural networks drawback, namely low readability.

Other example of neural networks usefulness was provided by Park and Baek Park & Baek (2008), who investigated 148 software projects completed between 1999 and 2003 by one of the Korean IT service vendors. The authors compared accuracy of the neural networks model with human experts' judgments and two classical regression models. The neural network outperformed the other approaches by producing results with MRE (magnitude of relative error) equal to 59.4 when expert judgements and the regression models produced results with MRE equal to 76.6, 150.4 and 417.3 respectively.

3.3.4 Analogy

The analogy–based estimation methods are based on project's characteristic in terms of features, e.g. size of the requirements, size and experience of the team, software development method. The features are collected from historical projects. In order to forecast effort for a new project the most similar historical projects must be found. Similarity can be defined as Euclidean distance in n-dimensional space where n is the number of project features. The dimensions are standardized since all of them should have the same weight. The most similar projects, i.e. the nearest neighbours, are used as the basis for effort prediction for the new project. According to Jorgensen and Shepperd Jorgensen & Shepperd (2007) the popularity of research on analogy–based estimation models is increasing.

Good prediction abilities of the analogy–based methods were shown by Shepperd and Schofield in a study conducted on 9 datasets, a total of 275 software projects Shepperd & Schofield (1997). The analogy (case based reasoning) was compared upon stepwise regression analysis and produced superior predictive performance for all datasets when measured in MMRE (mean magnitude of relative error) and for 7 datasets when measured in Pred(25) (percentage of predictions that are within 25% of the actual value).

Gupta et al. Gupta et al. (2011) surveyed the recent analogy based techniques for effort estimation and identified following approaches (some of them overlap estimation methods mentioned in other subsections):

- Machine learning based approaches (Multi Layer Perceptrons, Radial Basis Functions, Support Vector Regression, Decision Trees)
- Fuzzy logic based approaches (Fuzzy analogy, Generalized Fuzzy Number Software Estimation Model)
- Grey Theory based approaches (Grey Relational Analysis based software project effort (GRACE), Improved GRA, GLASE, Weighted GRA, GRACE^{+})
- Hybrid And Other Approaches (Genetic Algorithms with GRA, Fuzzy Grey Relational Analysis, AQUA (case-based reasoning & collaborative filtering))

Gupta et al. also investigated the efficiency of the above listed methods. Moreover, the authors collected and presented information regarding the prediction accuracy of those methods. The best result was obtained in the case of Decision Trees. Nevertheless, it is not conclusive since the methods were evaluated on different data sets.

3.3.5 Analysis effort method

The analysis effort method is suited to producing the initial estimates for a whole project. The method is based on doing some preliminary detailed design and measuring the time consumed to accomplish these tasks. Then, the results are used to estimate effort for the rest of the analysis, design, coding and testing activities. For the purpose of this method, a simplified, waterfall based software development lifecycle is assumed. This method uses three key factors that apply to the estimated project in terms of its size, familiarity and complexity. Unfortunately, the Analysis Effort Method requires some manual work in the initial phase, and therefore impedes the automation. A detailed description of the Analysis Effort Method was given by Cadle and Yeates Cadle & Yeates (2007).

3.4 Dynamics–based techniques

The dynamics based techniques take into consideration the dynamics of effort over the duration of software development process. The environment conditions and constraints (e.g. available resources, deadlines) may change over time. Some of the changes are scheduled, some not, but all of them may have significant influence on productivity and, in consequence, the likelihood of project success.

3.4.1 System dynamics approach

System dynamics is a continuous simulation modeling methodology whereby model results and behavior are displayed as graphs of information that change over time. Models are represented as networks modified with positive and negative feedback loops. Elements within the models are expressed as dynamically changing levels or accumulations (the nodes), rates or flows between the levels (the lines connecting the nodes), and information relative to the system that changes over time and dynamically affects the flow rates between the levels (the feedback loops). Description after Boehm et al. Boehm et al. (2000).

There are examples of application of the system dynamics approach e.g. Madachy (1996), Ruiz et al. (2001). Detailed description of the concept was provided by Abdel-Hamid and Madnick Abdel-Hamid & Madnick (1991).

3.5 Composite techniques

Composite techniques employ two or more different effort estimation approaches (including those mentioned above) to produce the most appropriate model.

3.5.1 Bayesian approach

Bayesian approach is a method of inductive reasoning that minimizes the posterior expected value of a loss function (e.g. estimation error). Bayesian Belief Nets are an approach that incorporates the case effect reasoning. The cause–based reasoning is strongly recommended by Fenton and Neil Fenton & Neil (2000). The authors gave a very interesting example of providing wrong conclusion when ignoring the cause–effect relations. The example regards analysing car accidents. The relation between month of year and the number of car accidents shows that the safest to drive months are January and February. The conclusion, given the data involved, is sensible, but intuitively not acceptable. During January and February the weather is bad and causes treacherous road conditions and fewer people drive their cars and when they do they tend to drive more slowly. According to the authors, we experience similar misunderstanding in the effort estimation since most models use equation with following form: $effort = f(size)$. The equation ignores the fact that the size of solution cannot cause effort to be expanded.

Fenton and Neil employed Bayesian Belief Nets to represent casual relationships between variables investigated in the model Fenton & Neil (1999a). In consequence, they obtained model that has the following advantages:

- explicit modeling of ignorance and uncertainty in estimations,
- combination of diverse type of information,
- increasing of visibility and auditability in the decision process,
- readability,
- support for predicting with missing data,
- support for 'what–if' analysis and predicting effects of process change,
- rigorous, mathematical semantic.

3.5.2 COSEEKMO

COSEEKMO is an effort-modeling workbench suggested by Menzies et al. Menzies et al. (2006) that applies a set of heuristic rules to compare results from alternative models: COCOMO, Local Calibration Boehm (1981), Least Squares Regression and Quinlan's M5P algorithm Quinlan (1992). The model was trained and evaluated using data from 63 projects originally used to create the COCOMO 81 model, 161 projects originally used to create the COCOMO II model and 93 projects from six NASA centers.

The model is focused on estimating the effort that regards the whole application. According to the authors the COSEEKMO analysis is fully automated; all model procedures were published as programs written in pseudo–code or sets of heuristics rules.

3.5.3 ProjectCodeMeter

The ProjectCodeMeter Logical Solutions (Retrived on 01/11/2011) is a commercial tool that implements four different cost models: Weighted Micro Function Points (WMFP), COCOMO 81, COCOMO II, and Revic 9.2. When consider which models are supported, it is not surprising that the tool is focused on making estimations for the whole project. There is a great automation possibility since the tool integrates with many IDEs, i.e. MicroSoft Visual Studio, CodeBlock, Eclipse, Aptana Studio, Oracle JDeveloper, and JBuilder.

3.5.4 Construx Estimate

Estimate Construx (Retrived on 01/11/2011) predicts effort, budget, and schedule for software projects based on size estimates. It is calibrated with industry data. However, it is possible (and recommended) to calibrated Estimate with own data. Estimate is based on Monte Carlo simulation and two mentioned above estimation models: Putnam's SLIM and COCOMO. It is focused on estimations for the whole project.

3.5.5 CaliberRM ESTIMATE Professional

CaliberRM ESTIMATE Professional Borland Software Corporation (Retrived on 01/11/2011) uses two models in order to make initial estimate of total project size, effort, project schedule (duration) as well as staff size and cost: COCOMO II and Putnam's SLIM. Later, a Monte Carlo Simulation may be used to generate a range of estimates with different probabilities later the most appropriate one can be selected and executed.

4. Defect prediction

The described in Section 2 risk analysis considers defect repairs as a vital part of the risk assessment process. Delivering low quality software will result in a number of hidden defects, which is why the development of the next release will generate extra cost (these defects must be removed). In consequence, it is important to estimate the number of defects and the effort that is needed to fix them. Considerable research has been conducted on defect prediction. The current state of the art has been recently reported in systematic reviews Catal (2011); Kitchenham (2010).

4.1 Predicting the number of defects

Traditionally, the defect prediction studies were focused on predicting the number or density of defects Basili & Perricone (1984); Gaffney (1984); Lipow (1982). However, these approaches were extensively criticized Fenton & Neil (2000; 1999b). It was concluded that the models offer no coherent explanation of how defect introduction and detection variables affect defect count and hence there is no cause–effect relation for defect introduction. Furthermore, the employed variables are correlated with gross number of defects, but are poor in predicting. In consequence, these approaches are not adequate for risk analysis.

Recent defect prediction studies are focused on supporting resource management decisions in the testing process. The studies investigate which parts of a software system are especially defect prone and hence should be tested with special care and which could be defect free and hence may be not tested. The most popular approaches are focused on organizing modules in an order according to the expected number of defects Jureczko & Spinellis (2010); Weyuker et al. (2008) or on making boolean decision regarding whether a selected module is defect free or not Menzies et al. (2007); Zimmermann & Nagappan (2009). Unfortunately, such approaches are not effective in estimating the overall number of defects. Probably the most appropriate approach to defect prediction with regard to risk analysis is Bayesian belief nets Fenton & Neil (1999a),Fenton et al. (2007). This approach can be employed to predict the number of defects Fenton et al. (2008) and can provide the cause–effect relations. The Bayesian belief nets approach might be a source of automation related issues. Unfortunately, designing and configuring such a net without expert assistance is challenging. However, when the net is defined, it is readable and understandable Fenton & Neil (2000) and hence no further expert assistance in necessary.

There is also an alternative approach that is much easier to automate, but unfortunately does not support the cause–effect analysis, namely, reliability growth models. The reliability growth models are statistical interpolation of defect detection in time by mathematical functions, usually Weibull distribution. The functions are used to predict the number of defects in the source code. Nevertheless, this approach has also drawbacks. The reliability growth models require detailed data regarding defects occurrence and are designed for the waterfall software development process. It is not clear whether they are suitable for highly iterative processes. Detailed description of this approach was presented by Kan Kan (1994).

4.2 Predicting the defect fixing effort

The number of residual defects does not satisfy the requirements of risk analysis, when the fixing effort remains unknown. There are two types of studies focused on fixing time. First type estimates the time from bug report till bug fix Giger et al. (2010); Gokhale & Mullen (2010); Kamei et al. (2010). However, this type of estimation is not helpful in risk analysis. Fortunately, the second type of studies is more appropriate since it is focused on the fixing effort.

Weiss et al. Weiss et al. (2007) conducted an empirical study regarding predicting the defect-fixing effort in the JBoss project. The authors used text similarity techniques to execute analogy (nearest neighbor approach) based prediction. All types of issues were considered (defects, feature requests, and tasks). However, the best results were obtained in the case of defects; the average error was below 4 hours. Nevertheless, the study has low external validity and according to the authors should be considered as a proof of concept. Further research in similar direction was conducted by Hassouna and Tahvildari Hassouna & Tahvildari (2010). The authors conducted an empirical study regarding predicting the effort of defect–fixing in JBoss (the same data set as in Weiss et al. (2007)) and Codehaus projects. The suggested approach (Composite Effort Prediction Framework) enhanced the technique suggested by Weiss et al. (2007) by data enrichment and employing majority voting, adaptive threshold, and binary clustering in the similarity scoring process. From the automation point of view, this study provides us with an important finding. The authors developed PHP scripts to crawl and collect issue information from a project tracking system called JIRA. Hence, the authors showed a method of gathering empirical data with regard to defect fixing effort.

Zeng and Rine Zeng & Rine (2004) used dissimilarity matrix and self–organizing neural networks for software defects clustering and effort prediction and obtained good accuracy when applied to similar projects (average MRE from 7% to 23%). However, only 106 samples corresponding to 15 different software defect fix efforts were used to train the model, which significantly affects the external validity.

Datasets from NASA were also used by Song et al. Song et al. (2006). The study investigated 200 software projects that were developed over more than 15 years and hence the external validity was significantly better. The authors used association rule mining based method to predict defect correction effort. This approach was compared to C4.5, Naive Bayes and Program Assessment Rating Tool (PART) and showed the greatest prediction power. The accuracy was above 90%.

The studies regarding predicting defect fixing effort present various approaches. Nevertheless, the external validity is limited and they do not conclude which approaches is optimal or how it can be automated.

4.3 Cross–project prediction

In order to employ a defect prediction model, it must be trained. When analysing the first release of a project, or when no historical data is collected, the model must be trained using data from other projects. Unfortunately, the cross-project defect prediction does not give good results. Some studies have investigated this issue recently He et al. (2011); Jureczko & Madeyski (2010); Zimmermann et al. (2009). However, all of them reported low success rate.

5. Measuring maintainability

A software system that delivers all functionalities correctly may still cause problems in future. Specifically, the effort connected with implementing new features may differ depending on the character of already written source code. There are phenomena that affect the easiness of modifying and improving an application, e.g. the code smells (symptom in the source code that possibly indicates a deeper problem). Pietrzak and Walter Pietrzak & Walter (2006) investigated the automatic methods of detecting code smells. The work was later extended in Martenka & Walter (2010) by introducing a hierarchical model for software design quality. In consequence, the model employs the interpretation of software metrics as well as historical data, results of dynamic behavior, and abstract syntax trees to provide traceable information regarding detecting anomalies, especially code smells. Thanks to the traceability feature it is possible to precisely locate the anomaly and eliminate it. The model is focused on evaluating high–level quality factors that can be a selection of quality characteristic.

Unfortunately, code smells and readability do not cover all aspects of software that are related to the easiness of modifying, improving and other maintenance activities. These aspects are covered by a term called 'software maintainability', which is defined in the IEEE Std 610.12-1990 as: the ease with which a software system or component can be modified to correct faults, improve performance or other attributes, or adapt to a changed environment IEEE (1990).

Considerable research has been conducted on measuring the maintainability; therefore there is guidance regarding which approach can be used in the mentioned in Section 2 risk analysis with regard to the automation.

5.1 ISO Quality model

The ISO/IEC 9126 Quality Model ISO (2001) identifies several quality characteristics, among them maintainability. The model also provides us with sub-characteristics. In the case of maintainability they are: analyzability (the easiness of diagnosing faults), changeability (the easiness of making modifications), stability (the easiness of keeping software in consistent state during modification), testability (the easiness of testing the system), maintainability–compliance (the compliance with standards and conventions regarding maintainability). Such multidimensional characteristic of maintainability seems to be reasonable. Software that is easy to analyse, change, test, and is stable and complies with standards and conventions should be easy to maintain.

The ISO Quality Model recommends sets of internal and external metrics designed for measuring the aforementioned sub–characteristics. For instance, for the changeability there is the 'change implementation elapse time' suggested as external metric and the 'change impact' (it is based on the number of modifications and the number of problems caused by these modifications) as internal metric. Nevertheless, the ISO Quality Model was criticized by Heitlager et al. Heitlager et al. (2007). The authors noted that the internal as well as the external metrics are based on the observations of the interaction between the product and its environment, maintainers, testers and administrators; or on comparison of the product with its specification (unfortunately, the specification is not always complete and correct). Heitlager et al. Heitlager et al. (2007) argued that the metrics should be focused on direct observation of the software product.

5.2 Maintainability Index (MI)

The Maintainability Index was proposed in order to assess the software system maintainability according to the state of its source code Coleman et al. (1994). It is a composite number obtained from an equation that employs a number of different software metrics:

$$MI = 171 - 5.2 * ln(\overline{Vol} - 0.23 * \overline{V(g')} - 16.2 * ln(\overline{LOC} + 50 * sin(\sqrt{2.46 * perCM} \quad (6)$$

Where $\overline{Vol}$, $\overline{V(g')}$, and $\overline{LOC}$ are the average values of Halstead Volume, McCabe's Cyclomatic Complexity, and Lines Of Code respectively, and perCM is the percentage of commented lines in module.

The MI metric is easy to use, easy to calculate and hence to automate. Nevertheless, it is not suitable to the risk analysis since it does not support cause–effect relations. When analysing risk, it is not enough to say that there is a maintainability problem. An action that can mitigate the problem should always be considered. Unfortunately, in the case of the MI we can only try to decrease the size related characteristics (Vol, V(g'), and LOC) or increase the percentage of comments, what not necessarily will improve the maintainability. The MI is correlated with the maintainability, but there are no proofs for cause-effect relation and hence we cannot expect that changing the MI will change the maintainability.

5.3 Practical model for measuring maintainability

Heitlager et al. Heitlager et al. (2007) not only criticized other approaches, but also suggested a new one, called Practical model for measuring maintainability. The authors argued that the model was successfully used in their industrial practice and to some extend is compatible with the widely accepted ISO Quality Model ISO (2001). The model consists of five characteristics (source code properties that can be mapped onto the ISO 9126 maintainability sub–characteristics: Volume, Complexity per unit, Duplication, Unit size, and Unit testing. The authors defined metrics for each of the aforementioned properties that are measured with an ordinal scale: ++, +, o, –, – –; where ++ denotes very good and – – very bad maintainability.

5.3.1 Volume

Volume represents the total size of a software system. The main metric of volume is Lines Of Code. However, the authors noted that this metric is good only within a single programming language since different set of functionalities can be covered by the same number of lines of code written. Therefore, the Programming Languages Table Jones (Retrived on 01/11/2011) should be considered in order to express the productivity of programming languages. Heitlager et al. gave an ready to use example for three programming languages (Tab. 1).

rank	Java	Cobol	PL/SQL
++	0–66	0–131	0–46
+	66–246	131–491	46–173
o	246–665	491–1,310	173–461
–	665–1,310	1,310–2,621	461–992
– –	>1,310	>2,621	>992

Table 1. Values of the volume metric in correspondence with source code size measured in KLOC according to Heitlager et al. (2007).

5.3.2 Complexity per unit

Complex source code is difficult to understand and hence to maintain. The complexity per unit (method) can be measured using the McCabe's Cyclomatic Complexity (CC). Heitlager et al. pointed out that summing or calculating the average value of CC does not reflect the maintainability well, since the maintenance problems are usually caused by few units, which has outlying values of the complexity metric. To mitigate the problem, the authors suggested to assign risks to every unit according to the value of CC: $CC \in [1, 10] \rightarrow$ *low risk*; $CC \in [11, 20] \rightarrow$ *moderate risk*; $CC \in [21, 50] \rightarrow$ *high risk*; $CC \in [51, \infty] \rightarrow$ *very high risk*. Using the risk evaluations, the authors determine the complexity rate according to Tab. 2.

5.3.3 Duplication

Heitlager et al. suggested calculating duplication as the percentage of all code that occurs more than once in equal code blocks of at least 6 lines. Such measure is easy to automate since there are tools that support identifying code duplication (e.g. PMD). Details regarding mapping the value of duplication onto maintainability are presented in Tab. 3.

rank	Moderate risk	High risk	Very high risk
++	25%	0%	0%
+	30%	5%	0%
o	40%	10%	0%
–	50%	15%	5%
– –	–	–	–

Table 2. Values of complexity per unit according to the percentage of units that do not exceed the specified upper limit for percentage of classes with given risk according to Heitlager et al. (2007).

rank	Duplication	rank	Unit test coverage
++	0–3%	++	95–100%
+	3–5%	+	80–95%
o	5–10%	o	60–80%
–	10–20%	–	20–60%
– –	20–100%	– –	0–20%

Table 3. Rating scheme for duplication and test code coverage according to Heitlager et al. (2007).

5.3.4 Unit size

The size per unit should be measured using LOC in a similar way to cyclomatic complexity per unit. Unfortunately, the authors do not provide us with the threshold values that are necessary to map the measured values to risk categories and rank scores.

5.3.5 Unit testing

Heitlager et al. suggested that the code coverage should be used as the main measure of unit tests. Details are presented in Tab. 3.

Measuring test code coverage is easy to automate due to good tool support (e.g Clover[2], Emma[3]. Unfortunately, this measure does not reflect the quality of tests. It is quite easy to obtain high code coverage with extremely poor unit tests, e.g. by writing tests that invoke methods, but do not check their behavior. Therefore, the authors recommend supplementing the code coverage with counting the number of assert statements. Other measure of unit tests quality is becoming very popular nowadays, namely mutation score, see Madeyski & Radyk (2010) for details.

5.4 Other approaches

The presented above approaches do not cover all maintainability measuring techniques that are used or investigated. The researchers also study the usefulness of regression based models, Bayesian Networks, fuzzy logic, artificial intelligence methods and software reliability models. A systematic review of software maintainability prediction and metrics was conducted by Riaz et al. Riaz et al. (2009).

[2] http://www.atlassian.com/software/clover/
[3] http://emma.sourceforge.net/

6. Automation possibilities and limitations

Three areas of risk analysis were discussed: effort estimation, defect prediction and maintainability measurement. The most common techniques and tools were described. Unfortunately, the automation of these analyses is not as trivial as installing off–the–shelf products. There are still gaps that should be filled. This section points out what is missing in this area and what is already available.

6.1 Effort estimation

The research conducted on effort estimation as well as the set of available tools is very impressive. Unfortunately, the focus is oriented towards estimating the whole project what is not suitable for the defined in Section 2 risk analysis. There is support for estimating effort for single requirement (e.g. SLIM, SPR KnowledgePLAN, PRICE-S). However, the methods are not optimised for such task and not as well validated as in the case of larger scope estimation.

For the risk analysis it is crucial to incorporate the uncertainty parameters of the prediction. Let's assume that there are 5 requirements and each of them is estimated to 5 days. Together it makes 25 days. And on the other hand, there are 30 man–days to implement these requirements. If the standard deviation of the estimations is equal to 5 minutes, the release will presumably be ready on time. However, when the standard deviation is equal to 5 days, there will be great risk of delay. Fortunately, some of the effort estimation techniques incorporate uncertainty, e.g. the COCOMO II method was enhanced with the Bayesian approach Chulani et al. (1999); Putnam considered mapping the input uncertainty in the context of his model Putnam et al. (1999).

6.2 Defect prediction

Defect prediction is the most problematic part in risk analysis. There are methods for predicting a number of defects and some pioneer works regarding estimating the effort of the removal process. However, there is no technique of combining these two methods. Furthermore, there are issues related to model training. Cross–project defect prediction has low success rate and intra–project prediction always requires extra amount of work for collecting data that is necessary to train the model.

Full automation of defect prediction requires further research. Currently, there are methods for evaluating the software quality using defect related measures, e.g. the Bayesian belief nets approach or the reliability growth models (both described in Section 4). However, it is not clear how such measures are correlated with the defect fixing effort and hence further empirical investigation regarding incorporating these measures into risk analysis is necessary.

6.3 Measuring maintainability

There are a number of maintainability measuring methods. We recommend the Practical model for measuring maintainability suggested by Heitlager et al. Heitlager et al. (2007). This approach not only measures the maintainability, but also shows what should be done in order to improve it. The authors put the model in use in a number of projects and obtained satisfactory results. Nevertheless, no formal validation was conducted and hence it is hard to make judgments about the external validity. There is also a possibility that automation related issues will arise, since the authors omitted detailed description of some of the model dimensions. It is also not clear how to map the model output to project risk. Therefore, further research regarding incorporating the model into the risk analysis is necessary.

6.4 Data sources

Further research requires empirical data collected from real software projects. The collection process requires tools that are well adjusted to the software development environment. Such tools can be also beneficial during the risk analysis, since usually they can be used to monitor the projects and hence to assess its current state. A number of tools were listed in the Section 3. We are also working on our own solutions (Ckjm Jureczko & Spinellis (Retrived on 06/05/2011), and BugInfo Jureczko (Retrived on 01/11/2011a)). It is possible to collect data from open–source projects with such tools. Limited number of projects track the effort related data, but fortunately, there are exceptions e.g. JBoss and some of the Codehaus projects Codehaus (Retrived on 01/11/2011). There are also data repositories that can be easily employed in research. That includes public domain datasets: Promise Data Repository Boetticher et al. (Retrived on 01/11/2011), Helix Vasa et al. (Retrived on 01/11/2011), and our own Metrics Repository Jureczko (Retrived on 01/11/2011b); as well as commercial: ISBSG ISBSG (Retrived on 01/11/2011), NASA's Metrics Data Program NASA (Retrived on 01/05/2006), SLIM, SPR KnowledgePLAN, or PRICE-S.

7. Summary and future works

Method of software project risk analysis was suggested and decomposed into well known in software engineering problems: effort estimation, defect prediction and maintainability measurement. Each of these problems was investigated by describing the most common approaches, discussing the automation possibilities and recommending suitable techniques.

The potential problems regarding automation were identified. Many of them can be addressed using the available tools (see Section 3). Unfortunately, some of them are not covered and require further research or new software solutions.

Validation in the scope of single requirement is necessary in the case of effort estimation. The tools are adjusted to the scope of the whole project and hence the quality of estimations for single requirement should be proven. Such validation requires great amount of empirical data. Potential data sources were given in Section 6.4. We are developing a tool (extension of our earlier solutions: BugInfo and Ckjm; working name: Quality Spy) that will be able to collect the data and once we have finished we are going to conduct an empirical study regarding the validation.

The discussion regarding defect prediction was not conclusive due to a number of major issues related to incorporating the prediction methods into the risk analysis. It is not clear how to combine the predicted number of defects with the expected fixing effort. Furthermore, the studies regarding prediction of fixing effort have low external validity and the cross-project defect prediction has low success rate. We are going to conduct further studies regarding estimating effort of defect removal and formalizing the method of using results of defect prediction in the risk analysis by providing a mapping to the loss function.

A recommendation regarding measuring maintainability was made (the model suggested in Heitlager et al. (2007)). Unfortunately, we have no empirical data to support it and therefore, we are going to implement the model in a tool and validate it against open–source projects. We are expecting that the validation will help to solve the second maintainability related issue, namely, mapping the Heitlager's model output to values of the loss function.

8. Acknowledgement

Thanks to Z. Huzar, M. Kowalski, L. Madeyski and J. Magott for their great contribution to the design of the risk analysis method presented in this paper.

9. References

Abdel-Hamid, T. & Madnick, S. E. (1991). *Software project dynamics: an integrated approach*, Prentice-Hall, Inc., Upper Saddle River, NJ, USA.

Air Force Cost Analysis Agency (Retrived on 01/11/2011). AFCAA REVIC. URL: *https://sites.google.com/site/revic92*

Albrecht, A. J. (1979). Measuring Application Development Productivity, *Proceedings of the Joint SHARE, GUIDE, and IBM Application Development Symposium*, Monterey, California, USA, pp. 83–92.

Atkinson, R. (1999). Project management: cost, time and quality, two best guesses and a phenomenon, its time to accept other success criteria, *International Journal of Project Management* 17(6): 337 – 342.

Bagnall, A. J., Rayward-Smith, V. J. & Whittley, I. M. (2001). The next release problem, *Information and Software Technology* 43(14): 883 – 890.

Banker, R. D., Kauffman, R. J. & Kumar, R. (1991). An empirical test of object-based output measurement metrics in a computer aided software engineering (case) environment, *J. Manage. Inf. Syst.* 8: 127–150.

Basili, V. R. & Perricone, B. T. (1984). Software errors and complexity: an empirical investigation0, *Commun. ACM* 27: 42–52.

Beck, K. & Andres, C. (2005). *Extreme Programming Explained: Embrace Change; 2nd ed.*, Addison-Wesley, Boston, MA.

Boehm, B., Abts, C. & Chulani, S. (2000). Software development cost estimation approaches – a survey, *Ann. Softw. Eng.* 10: 177–205.

Boehm, B., Clark, B., Horowitz, E., Westland, C., Madachy, R. & Selby, R. (1995). Cost models for future software life cycle processes: Cocomo 2.0, *Annals of Software Engineering*, pp. 57–94.

Boehm, B. W. (1981). *Software Engineering Economics*, Prentice Hall.

Boetticher, G., Menzies, T. & Ostrand, T. (Retrived on 01/11/2011). Promise repository of empirical software engineering data. URL: *http://promisedata.org*

Borland Software Corporation (Retrived on 01/11/2011). CaliberRM ESTIMATE Professional. URL: *http://www.ktgcorp.com/borland/products/caliber/index.html#estimate_pro*

Cadle, J. & Yeates, D. (2007). *Project Management for Information Systems*, 5th edn, Prentice Hall Press, Upper Saddle River, NJ, USA.

Capgemini CSD Research Inc. (Retrived on 01/11/2011). Software Cockpit.
URL: *http://www.de.capgemini.com/capgemini/forschung/research/software/*

Catal, C. (2011). Review: Software fault prediction: A literature review and current trends, *Expert Syst. Appl.* 38: 4626–4636.

Center for Systems and Software Engineering (Retrived on 01/11/2011). COCOMO II.
URL: *http://csse.usc.edu/csse/research/COCOMOII/cocomo_main.html*

Chulani, S., Boehm, B. & Steece, B. (1999). Bayesian analysis of empirical software engineering cost models, *Software Engineering, IEEE Transactions on* 25(4): 573 –583.

Codehaus (Retrived on 01/11/2011). Codehaus.
URL: *http://jira.codehaus.org*

Coleman, D., Ash, D., Lowther, B. & Oman, P. (1994). Using metrics to evaluate software system maintainability, *Computer* 27: 44–49.

Common Software Measurement International Consortium (Retrived on 01/11/2009). Cosmic.
URL: *http://www.cosmicon.com*

Construx (Retrived on 01/11/2011). Estimate.
URL: *http://www.construx.com*

Cost Xpert AG (Retrived on 01/11/2011). Cost xpert.
URL: *http://www.costxpert.eu/en/research/COCOMOII/cocomo_comsoftware.htm*

Fenton, N. E. & Neil, M. (1999a). Software metrics and risk, *Proceedings of the 2nd European Software Measurement Conference*, FESMA '99, pp. 39–55.

Fenton, N. E. & Neil, M. (2000). Software metrics: roadmap, *Proceedings of the Conference on The Future of Software Engineering*, ICSE '00, ACM, New York, NY, USA, pp. 357–370.

Fenton, N. & Neil, M. (1999b). A critique of software defect prediction models, *Software Engineering, IEEE Transactions on* 25(5): 675 –689.

Fenton, N., Neil, M., Marsh, W., Hearty, P., Radlinski, L. & Krause, P. (2007). Project data incorporating qualitative factors for improved software defect prediction, *ICSEW '07: Proceedings of the 29th International Conference on Software Engineering Workshops*, IEEE Computer Society, Washington, DC, USA.

Fenton, N., Neil, M., Marsh, W., Hearty, P., Radliński, L. & Krause, P. (2008). On the effectiveness of early life cycle defect prediction with bayesian nets, *Emp. Softw. Engg.* 13: 499–537.

Gaffney, J. E. (1984). Estimating the number of faults in code, *Software Engineering, IEEE Transactions on* SE-10(4): 459 –464.

Galorath Inc. (Retrived on 01/11/2011). SEER for Software.
URL: *http://www.galorath.com/*

Giger, E., Pinzger, M. & Gall, H. (2010). Predicting the fix time of bugs, *Proceedings of the 2nd International Workshop on Recommendation Systems for Software Engineering*, RSSE '10, ACM, New York, NY, USA, pp. 52–56.

Gokhale, S. S. & Mullen, R. E. (2010). A multiplicative model of software defect repair times, *Empirical Softw. Engg.* 15: 296–319.

Górski, J. & Miler, J. (Retrived on 01/11/2011). RiskGuide.
URL: *http://iag.pg.gda.pl/RiskGuide/*

Gupta, S., Sikka, G. & Verma, H. (2011). Recent methods for software effort estimation by analogy, *SIGSOFT Softw. Eng. Notes* 36: 1–5.

Hassouna, A. & Tahvildari, L. (2010). An effort prediction framework for software defect correction, *Information and Software Technology* 52(2): 197 – 209.

He, Z., Shu, F., Yang, Y., Li, M. & Wang, Q. (2011). An investigation on the feasibility of cross-project defect prediction, *Automated Software Engineering* pp. 1–33.

Heitlager, I., Kuipers, T. & Visser, J. (2007). A practical model for measuring maintainability, *Proceedings of the 6th International Conference on Quality of Information and Communications Technology*, IEEE Computer Society, Washington, DC, USA, pp. 30–39.

Hello2morrow Inc. (Retrived on 01/11/2011). Sotograph.
URL: *http://www.hello2morrow.com/products/sotograph*

Helming, J., Koegel, M. & Hodaie, Z. (2009). Towards automation of iteration planning, *Proc. of the 24th ACM SIGPLAN conference companion on Object oriented programming systems languages and applications*, OOPSLA '09, ACM, New York, NY, USA, pp. 965–972.

Humphrey, W. S. (1995). *A Discipline for Software Engineering*, 1st edn, Addison-Wesley Longman Publishing Co., Inc., Boston, MA, USA.

Idri, A., Khoshgoftaar, T. & Abran, A. (2002). Can neural networks be easily interpreted in software cost estimation?, *Fuzzy Systems, 2002. FUZZ-IEEE'02. Proceedings of the 2002 IEEE International Conference on*, Vol. 2, pp. 1162 –1167.

IEEE (1990). IEEE Standard Glossary of Software Engineering Terminology, *Technical report*, IEEE.

ISBSG (Retrived on 01/11/2011). The global and independent source of data and analysis for the it industry.
URL: *http://www.isbsg.org*

ISO (2001). ISO/IEC 9126-1:2001, Software engineering – Product quality – Part 1: Quality model, *Technical report*, International Organization for Standardization.

ISO (2004). ISO/IEC 16085:2001, systems and software engineering – life cycle processes – risk management, *Technical report*, International Organization for Standardization.

Jones, C. (Retrived on 01/11/2011). Programming languages table.
URL: *http://www.cs.bsu.edu/homepages/dmz/cs697/langtbl.htm*

Jorgensen, M. & Shepperd, M. (2007). A systematic review of software development cost estimation studies, *IEEE Trans. Softw. Eng.* 33: 33–53.

Jureczko, M. (Retrived on 01/11/2011a). BugInfo.
URL: *http://kenai.com/projects/buginfo*

Jureczko, M. (Retrived on 01/11/2011b). Metrics repository.
URL: *http://purl.org/MarianJureczko/MetricsRepo*

Jureczko, M. & Madeyski, L. (2010). Towards identifying software project clusters with regard to defect prediction, *PROMISE'2010: Proceedings of the 6th International Conference on Predictor Models in Software Engineering*, ACM.

Jureczko, M. & Młynarski, M. (2010). Automated acceptance testing tools for web applications using test-driven development, *Electrotechnical Review* 86(09): 198–202.

Jureczko, M. & Spinellis, D. (2010). *Using Object-Oriented Design Metrics to Predict Software Defects*, Vol. Models and Methodology of System Dependability of *Monographs of System Dependability*, Oficyna Wyd. Politechniki Wroclawskiej, Wroclaw, Poland, pp. 69–81.

Jureczko, M. & Spinellis, D. (Retrived on 06/05/2011). Ckjm extended.
URL: *http://gromit.iiar.pwr.wroc.pl/p_inf/ckjm/*

Kamei, Y., Matsumoto, S., Monden, A., Matsumoto, K.-i., Adams, B. & Hassan, A. (2010). Revisiting common bug prediction findings using effort-aware models, *Software Maintenance (ICSM), 2010 IEEE International Conference on*, pp. 1 –10.

Kan, S. H. (1994). *Metrics and Models in Software Quality Engineering*, 1st edn, Addison-Wesley Longman Publishing Co., Inc., Boston, MA, USA.

Kitchenham, B. (2010). What's up with software metrics? - a preliminary mapping study, *J. Syst. Softw.* 83: 37–51.

Kusumoto, S., Matukawa, F., Inoue, K., Hanabusa, S. & Maegawa, Y. (2004). Estimating effort by use case points: Method, tool and case study, *Proceedings of the Software Metrics, 10th International Symposium*, IEEE Computer Society, Washington, DC, USA, pp. 292–299.

Lipow, M. (1982). Number of faults per line of code, *IEEE Trans. Softw. Eng.* 8: 437–439.

Logical Solutions (Retrived on 01/11/2011). Project Code Meter.
URL: *http://www.projectcodemeter.com*

Madachy, R. J. (1996). System dynamics modeling of an inspection-based process, *Proceedings of the 18th international conference on Software engineering*, ICSE '96, IEEE Computer Society, Washington, DC, USA, pp. 376–386.

Madeyski, L. & Radyk, N. (2010). Judy – a mutation testing tool for java, *Software, IET* 4(1): 32 – 42.

Martenka, P. & Walter, B. (2010). Hierarchical model for evaluating software design quality, *e-Informatica* 4: 21–30.

Menzies, T., Chen, Z., Hihn, J. & Lum, K. (2006). Selecting best practices for effort estimation, *IEEE Trans. Softw. Eng.* 32: 883–895.

Menzies, T., Greenwald, J. & Frank, A. (2007). Data mining static code attributes to learn defect predictors, *IEEE Transactions on Software Engineering* 33: 2–13.

NASA (Retrived on 01/05/2006). Nasa's metrics data program.
URL: *http://mdp.ivv.nasa.gov/repository.html*

Netherlands Software Metrics Association (Retrived on 01/11/2011). NESMA Function Point Analysis.
URL: *http://www.nesma.nl/section/home/*

Ochodek, M. & Nawrocki, J. (2007). Automatic transactions identification in use cases, *Proceedings of the Second IFIP TC 2 Central and East European Conference on Software Engineering Techniques*, Springer, pp. 33–46.

Oisen, R. P. (1971). Can project management be defined?, *Project Management Quarterly* 2(1): 12 – 14.

Park, H. & Baek, S. (2008). An empirical validation of a neural network model for software effort estimation, *Expert Syst. Appl.* 35: 929–937.

Penny, D. A. (2002). An estimation-based management framework for enhancive maintenance in commercial software products, *Software Maintenance, 2002. Proceedings. International Conference on*, pp. 122 – 130.

Pietrzak, B. & Walter, B. (2006). Leveraging code smell detection with inter-smell relations, Vol. 4044 of *Lecture Notes in Computer Science*, pp. 75–84.

PRICE Systems (Retrived on 01/11/2011). TruePlanning.
URL: *http://www.pricesystems.com/products/true_h_price_h.asp*

Putnam, L. H. (1978). A general empirical solution to the macro software sizing and estimating problem, *IEEE Trans. Softw. Eng.* 4: 345–361.

Putnam, L. H., Mah, M. & Myers, W. (1999). First, get the front end right, *Cutter IT Journal* 12(4): 20–28.

Quantitative Software Management Inc. (Retrived on 01/11/2011). SLIM Suite. URL: *http://www.qsm.com/tools*

Quinlan, J. R. (1992). Learning With Continuous Classes, *Proceedings of the 5th Australian Joint Conf. Artificial Inteligence*, pp. 343–348.

Riaz, M., Mendes, E. & Tempero, E. (2009). A systematic review of software maintainability prediction and metrics, *Proceedings of the 2009 3rd International Symposium on Empirical Software Engineering and Measurement*, ESEM '09, IEEE Computer Society, Washington, DC, USA, pp. 367–377.

Ruiz, M., Ramos, I. & Toro, M. (2001). A simplified model of software project dynamics, *Journal of Systems and Software* 59(3): 299 – 309.

Schneider, G. & Winters, J. P. (2001). *Applying use cases, Second Edition*, Addison-Wesley Longman Publishing Co., Inc., Boston, MA, USA.

Schwaber, K. & Beedle, M. (2001). *Agile Software Development with Scrum*, Prentice Hall PTR, Upper Saddle River, NJ, USA.

SEI (2002). Capability Maturity Model Integration (CMMI), Version 1.1 - Continuous Representation, *Technical Report CMU/SEI-2002-TR-011*, Software Engineering Institute.

Shepperd, M. & Schofield, C. (1997). Estimating software project effort using analogies, *Software Engineering, IEEE Transactions on* 23(11): 736 –743.

Softstar Systems (Retrived on 01/11/2011). Costar. URL: *http://www.softstarsystems.com/*

Software Productivity Research (Retrived on 01/11/2009). SPR KnowledgePLAN. URL: *http://www.spr.com/project-estimation.html*

Song, Q., Shepperd, M., Cartwright, M. & Mair, C. (2006). Software defect association mining and defect correction effort prediction, *IEEE Trans. Softw. Eng.* 32: 69–82.

Stamelos, I. & Sfetsos, P. (2007). *Agile Software Development Quality Assurance*, IGI Publishing, Hershey, PA, USA.

Symons, C. R. (1991). *Software sizing and estimating: Mk II FPA (Function Point Analysis)*, John Wiley & Sons, Inc., New York, NY, USA.

Thaw, T., Aung, M. P., Wah, N. L., Nyein, S. S., Phyo, Z. L. & Htun, K. Z. (2010). Comparison for the accuracy of defect fix effort estimation, *Computer Engineering and Technology (ICCET), 2010 2nd International Conference on*, Vol. 3, pp. 550 – 554.

TSA Quality (Retrived on 01/11/2011). Metric Studio 2011. URL: *http://www.tsaquality.com/pages/english/nav/produtos.html*

Valerdi, R., Boehm, B. & Reifer, D. (2003). Cosysmo: A constructive systems engineering cost model coming age, *Proceedings of the 13th Annual International INCOSE Symposium*, pp. 70–82.

Vasa, R., Lumpe, M. & Jones, A. (Retrived on 01/11/2011). Helix - Software Evolution Data Set. URL: *http://www.ict.swin.edu.au/research/projects/helix*

Weiss, C., Premraj, R., Zimmermann, T. & Zeller, A. (2007). How long will it take to fix this bug?, *Proceedings of the Fourth International Workshop on Mining Software Repositories*, MSR '07, IEEE Computer Society, Washington, DC, USA, pp. 1–.

Weyuker, E. J., Ostrand, T. J. & Bell, R. M. (2008). Do too many cooks spoil the broth? using the number of developers to enhance defect prediction models, *Empirical Softw. Engg.* 13: 539–559.

Zeng, H. & Rine, D. (2004). Estimation of software defects fix effort using neural networks, *Computer Software and Applications Conference, 2004. COMPSAC 2004. Proceedings of the 28th Annual International*, Vol. 2, pp. 20 – 21.

Zimmermann, T. & Nagappan, N. (2009). Predicting defects with program dependencies, *Proceedings of the 2009 3rd International Symposium on Empirical Software Engineering and Measurement*, ESEM '09, IEEE Computer Society, Washington, DC, USA, pp. 435–438.

Zimmermann, T., Nagappan, N., Gall, H., Giger, E. & Murphy, B. (2009). Cross-project defect prediction: a large scale experiment on data vs. domain vs. process, *Proceedings of the the 7th joint meeting of the European software engineering conference and the ACM SIGSOFT symposium on The foundations of software engineering*, ESEC/FSE '09, ACM, New York, NY, USA, pp. 91–100.

17

Automation in the IT Field

Alexander Khrushchev
Instream Ltd.
Russia

1. Introduction

To begin with, what is Information Technology (IT)? If you ask any person not related to IT you would most likely receive an answer that the term IT means any activity related to computers, software and communications. According to the definition adopted by UNESCO, the IT is "a set of interrelated scientific, technological and engineering disciplines are studying methods of effective labor organization of people employed information processing and storage, computing, and methods of organization and interaction with people and manufacturing facilities, their practical applications, as well as associated with all these social, economic and cultural issues.".

This formal definition does not give a complete picture of the IT industry as a whole, since it does not enumerate specific areas of activity usually included in the concept of IT: design and development of computer technology, software development, testing, technical support, customer care, communication services etc.

Over the last three decades IT has made a qualitative and quantitative leap. Everything started with the first attempts of informational support of production, bulky scientific computers and the first simple personal computers. Over time the IT industry has made a huge step toward with modern technologies: supercomputers, mobile computers, virtualization of calculations, the Internet and many others. Computing performance has increased by hundreds of thousands of times, while the cost of computers has decreased significantly, making information technologies available to billions of people all over the planet.

The progress of IT is impressive. Things that until recently seemed fantastic, such as a device transferring mental signals to a computer, systems of three-dimensional printing and android robots hardly distinguishable from a human, have already been created and are operating.

But still automation of daily human activities remains one of the main objectives of IT. Computer systems and robotized lines perform routine, complex and dangerous operations instead of human staff. Activity fields where it was formerly impossible to manage without human control nowadays are operated by robots and unmanned machines operating with jeweller accuracy.

Having freed millions of people from the necessity to work manually, at the same time IT has opened up a huge market for intellectual labour for millions of IT specialists. Moreover, the IT industry constantly acquires new areas and technologies, generating deeper division of labour between experts.

Year on year the IT industry faces an increased array of problems, starting with technological restrictions and finishing with solutions with moral and ethical implications. Some problems are solved, while others constantly appear. Among them there is a big and important issue: who can automate the work of the one who automates the work of others?

There is no universal solution. Ironic as it may be, many IT processes are extremely difficult to be automated. As an example, the process of software development, testing and support is quite creative work, despite the continuing tendency to automate the processes. This is what makes programmers, testers and support staff much like street artists. The initial requirements are a little like a client who wants a portrait painted, and the result, i.e., a finished portrait, depends on the client's preferences and mood, as well as on the artist's style of painting.

The issue of automation of IT processes is too complex and it is extremely difficult to cover this area completely, however, in this chapter we will try to examine the possibilities and specific features of automation in the main areas IT.

2. Features of the IT business

Let's consider a significant difference between the IT business and a classic industry, such as heavy industry. This is well illustrated by two giants: GM and IBM.

Company	Turnover	Revenue	Total number of employees
IBM	$ 99.9 billion	$ 14.8 billion	~400,000
GM	$ 137 billion	$ 6.5 billion	~202,000

Table 1. Activity indicators of IBM and GM in 2010.

Both of them have comparable turnovers ($137 billion and $100 billion in 2010, respectively), but the number of IBM employees is twice that of GM. At the same, the revenue of IBM is twice that of GM ($14.8 billion vs. $6.5 billion in 2010, respectively).

Most of IBM's revenue is provided by Global Services. This is the largest division of IBM (with over 190,000 employees) and is the world's largest business and technology services provider. Second place in IBM's revenue structure is Software. Therefore, most of the services and products offered by IBM are intellectual products created directly by the employees. There are no automated assembly lines, no conveyors and no additional costs for any resources (except for electricity maybe).

At the same time, GM's main income is still provided by car sales and sales of car parts. The production of car parts and the assembly of complete vehicles have a high degree of automation. In addition, there are high costs for raw materials and resources (such as steel, electricity, etc.).

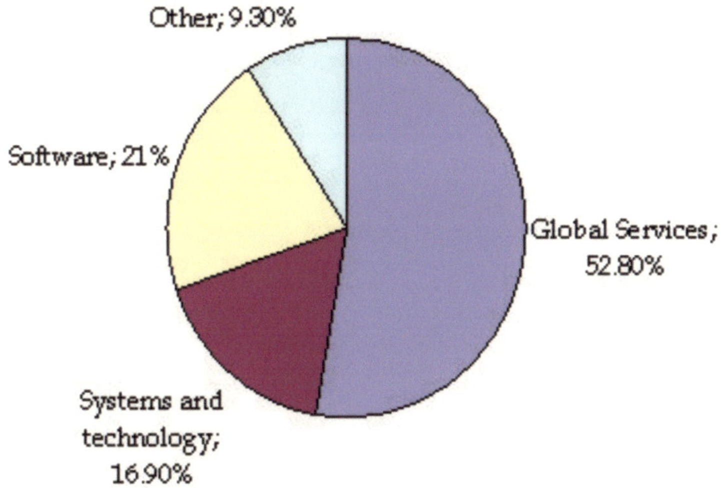

Fig. 1. Structure of IBM's income in 2010.

The main difference of the IT from industrial production is that industrial production requires complex organization of the product copies production (technological processes, components supply and quality control at the output). For software product copies production you only need to spend on the protection of intellectual property rights. In addition, there is a huge market for special products (tailor-made), where products can be used only by one or several clients without reaching the mass market. If we draw an analogy with car manufacturing, even big IT companies are more like workshops that produce hand-made automobiles rather than giant manufacturers.

At the same time the IT field is the driving force behind automation processes in any other industries. The work of IT specialists can be related to automated production processes, data collection systems, billing systems, etc. They produce products and solutions that reduce manual labour and calculations.

All this suggests a very different approach to automation in the IT business. In the IT business automation is an auxiliary tool. Many IT companies have a minimal set of automated processes (or have no automated processes), but their business is still successful. Of course IT professionals try to automate as many processes as possible.

3. Areas of the IT business

The IT business can be clearly divided into three areas:

- Software
- Hardware
- Services

The software production process includes many parts which starts from programming and finishes by software sales. Each process in the production of software looks like a brick in a wall. If one of the bricks is destroyed or removed then the entire building (production) will collapse. Each of these processes has its own characteristics and a degree of possible automation.

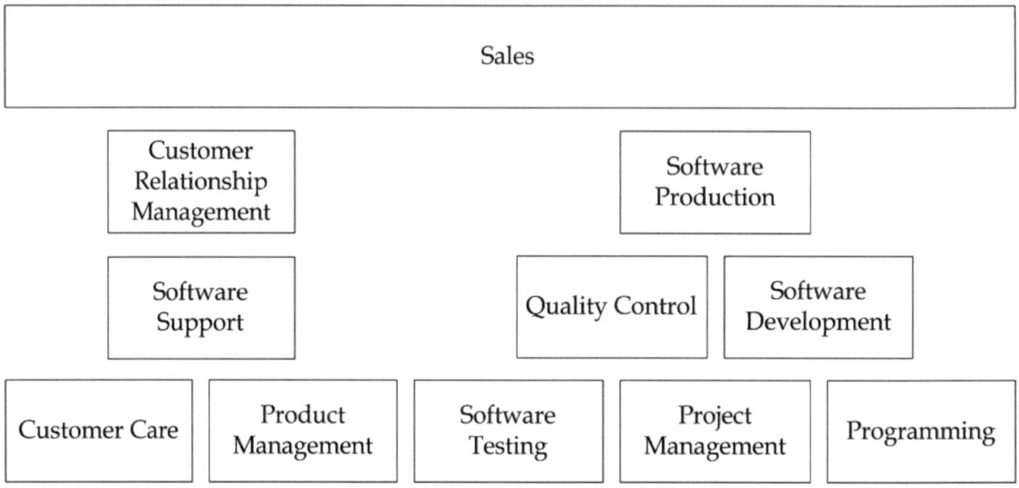

Fig. 2. Structure of software production

Hardware production has the same types of processes as software production (with minimal difference). Therefore, the approach to automate the development of hardware is not very different from the automation of software production. But it also includes direct production of components and computers (servers, network equipment, etc.). Since this process is a classic industrial production, it will not be considered in this chapter.

Providing services is an extensive part of the IT business. Some services may be provided in fully automatic mode, while others can only be provided directly by employees of the service division. Examples of services will be considered more below.

4. Software production

4.1 Software testing

Usually when people involved in IT business talk about automation in software production, they mean software testing. It is true that in software testing automation technology is most widely used. Historically, this was due to the rapid development of IT technology from the 1990s to the present.

The constant growth in the complexity of information systems on the one hand and appearance of new automation technologies on the other hand has pushed the development of testing automation over the last 20 years.

Software testing includes a set of different operations, some of which are recurrent while others are unique and product/line specific. It is quite obvious that routine recurrent processes are much easier to be automated. Testing is the easiest part of the software development for automation, since it consists of many such processes. Let's consider the most typical parts of the testing process.

4.1.1 Functional testing

Functional testing can be partly automated. It especially refers to recurrent basic functionality check-ups from one version to another, which is known as "regression" testing. Such automation is only justified when the product development (several versions production) is guaranteed and the product life cycle lasts at least one year. Besides, there are tools that permit conducting automated user interface testing. Functional testing consists of three phases: unit- testing, complex testing and regression testing.

4.1.1.1 Smoke-testing and unit-testing

As a rule, these tests are made by the developer. They consist of a set of calls of concrete modules with a defined set of parameters. Usually, developers do not have enough time to generate the necessary data sets and to make complex tests. In addition, tools which the developer has do not allow proper control of the quality of an output. There is another problem: while performing a module check, the developer is determined to prove the functionality of this module (a psychological feature of most developers). The testers have different motivation - their goal is to prove that the module does not meet the accepted requirements. The tester also looks at the working capacity of the unit from the user's point of view. Therefore, the most reasonable solution to this dilemma is to leave developers to generate and to make smoke-testing independent, assigning the making of unit-tests to testers. Meanwhile, in order to save testers' time, developers can carry out automatic unit-tests during a development cycle and correct errors at once, without activating the bug tracking process and thus keeping testers free of the tens of iterations of checks.

Meanwhile, it is very important that developed unit-tests could to be used, in the same or modified form, on complex or regression testing in the sequel.

4.1.1.2 Complex (system) testing

Complex testing tests the complete system (bundled software) to check its working capacity. If autotests were not used at earlier stages of testing, development of autotests just for complex testing does not always justify itself. A good case in point is the application of the modified autotests used on unit-testing.

4.1.1.3 Regression testing

At this stage the autotests developed earlier for other stages of testing (for current or earlier versions of the product) should be used. If the scale regression testing of system is significant, but special regression autotests are not available, their creation makes sense only if the "full regress" would be done again several times (in the case of enduring product life cycle).

4.1.2 Load testing

Load testing is a check of the system capability in the real world. Usually, the "real world" is understood like productive volume and composition of the processed data. It helps to define the limit of system performance or compliance of performance requirements.

Load testing is the easiest and the most effective process to automate. Load testing uses systems that imitate simultaneous work of thousands of users. Also to simulate the real content of the database, testers use data generators. This is quite natural and 100% justified, since it is pretty difficult to module the functioning of a popular Internet service or hundreds of user calls coming into a call centre in any other way.

Stress testing. Stress testing is a kind of load testing. It is usually applied to software and hardware complexes, and critical information systems (such as a hardware and software complexes of aircraft). This testing includes verification of complex systems as a whole under extreme conditions far from the operational environment. As part of this test, resistance to extreme environmental conditions, component redundancy, alarm systems and disaster recovery capabilities are checked. Sometimes, the conditions for a stress test can be simulated only by automation.

Integration testing. This is type of testing verifies the interaction of the object under test with adjacent systems or complexes. In practice, manual integration testing with external systems is much more effective than automatic testing. Firstly, this is because the majority of man-hours (up to 50%) within the limits of integration testing are spent on various options and interaction adjustments between adjacent systems. Practically, this work cannot be automated.

4.1.3 Advantages of automated testing

4.1.3.1 Cost

- Decrease of man-hours spent on testing;
- Reduce to a minimum time for each test-iteration after bug fixing;
- Automatic creation of test-reporting, tester does not waste time on reports;
- Automatic registration of issues in bug tracking systems;
- After hours performance of autotests (overnight or over weekend).

4.1.3.2 Quality

- Reduced human factor;
- More tests per unit of time make testing more qualitative.

4.1.3.3 Productivity

- More iterations in time unit;
- Higher productivity than in manual testing;
- Limitation on the number of tests by system performance only, not on the available float time of a tester.

4.1.3.4 Employee motivation

In many IT companies testers have a less profitable position with respect of both prestige and wages than developers. Naturally, some of them are not satisfied and view the position of tester as an intermediate step in their career. Instead of the development of testing skills, they prefer to improve their software development skills in order to switch to more profitable positions in the current company or find a job as a developer in another company. As a result, the quality of work of both testers and developers may be affected.

Participation in the automation of the testing process is an excellent chance for career growth for these workers. A good specialist in the field of automated tests is more valuable to a company than an ordinary developer or tester. In addition, they can develop their skills and build a career without leaving the testing process or without changing their place of work. Anyway, the making of automated tests is a more creative process than manual testing. It brings more satisfaction to workers, increasing their motivation.

4.1.3.5 Other

Efficiency of autotesting is expressed not only in the saving of man-hours on testing. One of the main advantages of the autotest is earlier error detection and lower total cost for their correction.

4.1.4 Automated testing is not advantageous in the following cases

- The scale of an autotest's covering of the system is incorrectly estimated. In such cases, the cost of producing autotests is many times greater than the benefit of its use;
- Employees don't have appropriate qualifications. In this case, autotest quality is low and the cost of their creation exceeds that planned;
- Autotests cannot be used in a sequel. Usually, if autotests cannot be re-used as is or with minimal modification, the cost of their creation will never be paid back (except in special cases, e.g., load testing);
- The cost of licensing the autotesting tools exceeds the economic profit engendered from using autotesting in projects in the first place. It is necessary to remember that autotesting tools often require separate servers that have to be purchased.

4.2 Software development

At first sight you might think that the term "software development automation" belies common sense. It sounds pretty much like "novel-writing automation" or "picture-painting automation". But if we look closer at the process of software development over the last two or three decades, we will see that it makes total sense. Yes, the artistic part of programming cannot be automated by the existing means of automation, but they can provide the developers with handy tools that multiply the speed and efficiency of their work. There is a reason why the development tools market is worth billions of dollars and still continues to grow ($6.9 billion in 2007, $7.3 billion in 2008 according to Gartner).

The automation of software development is generally intended for reducing the required qualification level of the workers and for speeding up the software development process. This is clearly shown by the programming languages, technologies and platforms evolution.

Many environments with possibilities in visual development permit minimizing the expenses on user interface routine operations. Frameworks serve as components and protocols connectors. Automated syntax checks, code formatting and debugging make the programmer's job significantly easier. To help developers there are tools to ensure the normal functioning of the development team, such as version control systems, tools and team coding. Modern means of communication and techniques of project management allow constructing the distributed team, staffed by specialists from different countries and areas of programming. Thus, the developers have more time for the artistic part of programming and experience exchange.

The same process of natural evolution led to the creation of business integration systems and technologies. Also these systems can hardly be called automated, but the aim of their creation was to reduce man-hours during corporate applications development and integration. Practically these systems made it possible to delegate part of the corporate applications development and integration job to analysts and integrators who do not need any special knowledge of programming languages or technologies to perform their jobs. In addition, this helps to reduce time spent on worker communication with each other in the course of software product development.

4.2.1 An approach to the development of a program code

The possibility of applying autotests depends very strongly on the methods of development of a program code. Therefore, the use of test automation can strongly influence the development. Implementation of some principles of development makes autotest using easier and reduces time for autotest development. The features of adaptation of a program code for autotests depend on technology and testing tools used. The following basic manoeuvres can be used:

- Option of quick rollback of changes for great volumes of data (this is especially important for databases);
- Functionality switches. Flags or the parameters that limit a set of operations of standard modules;
- "Hidden" parameters of a test mode, including additional logging or emulation of any operations;
- System messages (logs, warnings etc.) that should be easy for parse in the software tools;
- Option to fix a unit's state at the intermediate stages of work.

Testers can put up additional requirements to a program code, based on the selection of the available tools of testing and features of the testing process. Usually, compliance to these requirements does not consume lots of developer man-hours, but it makes autotest making easier and decreases the number of total man-hours on development and testing. The majority of tester's requirements should probably be included as a part of the process of program code development.

4.3 Automation of project management

It is often said about project management that it is 'The Art and Science of Getting Things Done'. In fact, the process of project management cannot be automated, as with any of the arts. There are only auxiliary tools which can help the project manager, like an easel or a palette for the artist.

Among the tools the project manager can identify the main ones are: a planning tool and means of gathering information.

At the moment there are lots of tools for planning projects. The criterion for selecting a planning tool is ease of use.

In general, useful information provided by the ERP system makes the work of the project manager more automated. The keyword is "useful", since just the multiplicity of information in the ERP system is not the key to success.

At present, there are comfortable online tools of project management that combine the following features:

- Ability to manage requirements;
- Possibility of automatic notification to interested parties on changes to the project;
- Ease of use;
- Assigning and tracking tasks;
- Storing, categorization and versioning of files;
- Forums to discuss tasks and projects;
- Project planning;
- Tracking the progress of the project by all participants;
- Tracking time spent;
- Report about the project state at any time;
- Implementation of communication on the project with reference to the aims and objectives (comments).

Currently, among such solutions the best known are Basecamp and TeamLab.

4.4 An adhesive force of the processes within an IT company

It makes no sense to consider further the automation of processes involved in software development without mention of the force capable to link these processes together.

Lack of integration can negate the benefits of automation of individual processes. Therefore, the task of information exchange between processes is one of the most important in the construction of software production.

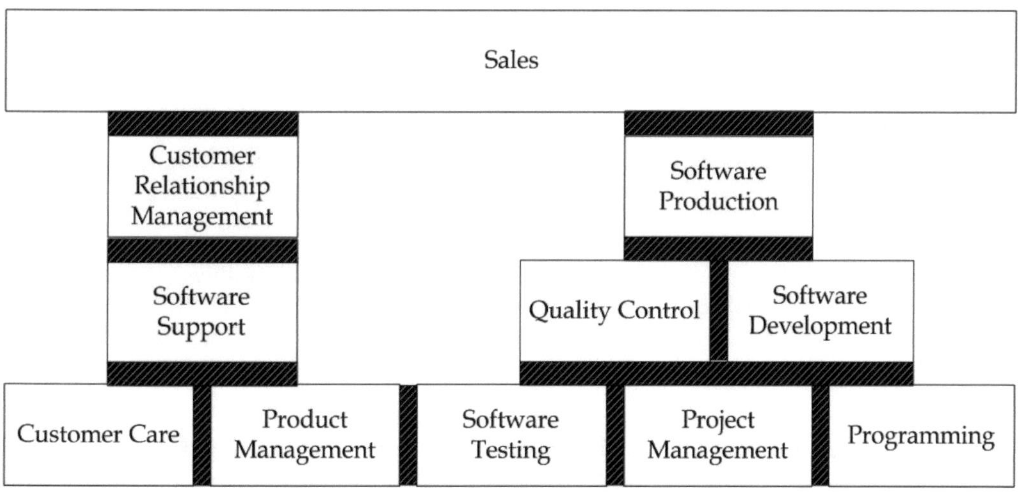

Fig. 3. Areas of IT business. Integrated processes of software production.

4.5 Enterprise resource planning

ERP it is an information system used to manage information flows in the organization and to automate the various areas of the company. In 1990 Gartner Group first employed the acronym ERP as an extension of material requirements planning (MRP), later manufacturing resource planning and computer-integrated manufacturing. Not all ERP packages were developed from a manufacturing core. Vendors of ERP solutions variously began with accounting, maintenance and human resources. By the mid–1990s ERP systems were adjusted to all core functions of enterprises. Naturally, IT companies widely use ERP systems to automate their processes. Moreover, many IT companies are trying to automate all internal and external processes in order to reduce the influence of human factors on the quality and outcome of their work.

Implementation of ERP systems allows automating the many workflows in the IT company. Conventionally, these processes can be divided into two groups: major (productive activities of the company) and minor (non-productive activities of the company).

Major workflows are:

- Manufacturing (engineering, work orders, manufacturing process, manufacturing projects, manufacturing flow, activity-based costs);
- Project management (human resource planning, costing, billing, time and expense, performance units, activity management, risk management);
- Product management (capacity planning, workflow management, quality control, cost management, product lifecycle management);
- Customer relationship management (sales and marketing, commissions, service, customer contact, call centre support, demand management, customer care);
- Data services (self-care, knowledge base).

Minor workflows are aimed at the support of normal functioning of the company and not directly aimed at the production or provision of services:

- Finance/Accounting (general ledger, payables, cash management, budgeting);
- Human resources (payroll, training, recruiting, time tracking);
- Communications and business structure (business structure, integrated channels of communication, document repository, access control, planning meetings and events);
- Supply chain management (order to cash, inventory, order entry, purchasing, supply chain planning, supplier scheduling, inspection of goods, claim processing, commissions).

Different ERP systems can include support of different sets of workflows or components, but the characteristic feature of all of these systems is the tight integration between the components. These solutions also support the common protocols, interfaces and integration tools. Thus, companies can integrate multiple solutions to meet the requirements for automation.

There are some difficulties in the implementation of generic ERP solutions around the world that relate to the peculiarities of doing business, the laws and taxation in each country. This allows local markets of similar solutions to exist. Due to this, specific solutions of local vendors can compete with solutions offered by industry giants such as SAP, Oracle or Microsoft. A good example is the company 1C (Russia), which in 2008 took second place (after SAP) in the ERP market in Russia, increasing its stake to 18.7% in monetary terms. Among them, many Russian IT companies prefer to use some components of the solutions 1C for running their businesses.

4.6 Automating the minor enterprise workflows

4.6.1 Finance

Accounting for financial operations, budgeting, accounting, financial planning and other financial operations are all critical to the business of any company. The complexity of their implementation within the information system in part relates to complex laws (in many countries the law requires a special licence for similar systems) and increased requirements for accuracy and reliability; also finance is a complex subject area and without the help of financial advisors a similar system cannot be properly developed.

For these reasons, the range of existing solutions automating finances is not so wide. The implementation and development of such systems produces qualified professionals. Even in IT, there are not many companies that have their own solutions of full automation for financial operations.

4.6.2 Human resources, communications, business structure and supply chain management

These workflows of any company (including IT) are very easily automated. To do this, there are many solutions in complex ERP systems and in single solutions (including free solutions). Here are just some processes of these workflows that can be automated by existing ERP solutions:

- Maintaining a register of positions and instructions;
- Management of the structure of the company;
- Maintaining personnel files;
- Time management (tracking and registration);
- Calculation of the efficiency of staff;
- Training of employees;
- Registration of skills, experiences and work results of staff;
- Planning of meetings;
- Control of room reservation;
- Accounting of inventory;
- Document circulation;
- Maintenance of purchasing equipment and supplies;
- Maintaining a register of vendors;
- Recruitment of staff;
- Control of authority staff (up to integration with access control system).

The main criteria for successful application of such solutions are flexibility and ease of use. These two requirements usually mean a flexible adjustment of the instrument according the company's business processes, user-friendly interface and easy integration with other solutions.

4.7 Integration of minor processes with major processes

Information obtained by an ERP system with the support of secondary processes is often used to adjust the main processes and vice versa. The relationship between internal and external processes within the company's ERP system is just as important as it is in real life. As example, information of results and quality of work of employees are aggregated and can be used for calculation of wages or benefits. Another example, the statistics obtained from the accounting systems of work are used in project planning and information about the skills and productivity of employees helps planning of the project.

4.8 Customer care (self-care systems)

Customer care organization is one of the most important processes. The customer care service quality reflects a lot on the reputation of the product and the company itself. At the same time it is quite difficult to forecast the quantity and content of the user calls which makes it necessary for IT companies to reserve a lot of support staff or resort to the services of external call centres. In addition, the higher the level of the customer care service, the more expensive it gets, which in the end reflects on the price of the software product.

Customer care automated systems can handle a great part of user calls, transferring them to operator only if necessary. The processing cost of such a self-care system, licence and technical support inclusive, is ten times lower than having it processed by operators. Besides, modern automated systems can integrate many channels of call processing like email, telephone, IP telephony, SMS, Internet portals, etc., which makes it possible to create customer care solutions with higher capacity than a standard help desk.

At the same time it is necessary to take into account that most users (45% according to Forrester) still prefer to be attended by a real person on the telephone, rather than by an automated system. Modern systems offer compromises in the form of IVR[1] services and animated digital characters integrated into customer care portals. However, these systems cannot process even half of the calls entering.

Therefore, despite the successful experience of implementation of self-care solutions in many large companies, the developers of such systems must still solve many difficult problems in the future, such as:

- Increasing intellectualization of these systems;
- Implementation of new technologies (speech recognition, the formation of visual images, etc.);
- Increase the capacity of self-care systems;
- Support for customers with disabilities, etc.

The successful solution to these problems can raise the level of user confidence in these systems, expand their range of applications and improve the quality of the automated service. That in turn may reduce costs and increase CSI[2] of customers.

4.9 Customer relationship management

Despite the fact that the term customer relationship management (CRM) has only arisen recently, it is based on systematic studies in marketing relationships that have been carried out systematically for more than 20 years.

Today it is very difficult to sustain an IT business with many thousands of customers without the help of automated customer interaction. Automated information systems CRM is now

[1] IVR (Interactive Voice Response) is system of pre-recorded voice messages, performing the function of routing calls within the call-center, using the information entered by the client using tone dialing.

[2] Customer Satisfaction Index

firmly established in the business processes of many companies. IT companies are also very widely using similar solutions for their businesses. Basically, these systems are used by companies operating in the business to customer (B2C) sector. These companies have many customers, a wide range of products and many partners around the world. Additionally it is these companies who obtain greater benefit from an automation of processes of interaction with customers. It is especially effective if such systems are integrated with the company's business processes and existing information systems, such as self-care and bug tracking.

The implementation and development of CRM systems in companies may vary. CRM systems are very complex and critical to the business, therefore, most non-IT companies prefer to use third-party solutions. Only a few companies are willing to create their own departments to develop and implement CRM solutions. However, many IT companies, with large resources of specialists and experience in developing complex solutions, decide to create their own solutions. Some companies start to develop their own CRM solutions based on experience in the use of automated CRM third-party solutions.

An example of the successful implementation of a self-made solution is the experience of Microsoft. This company has millions of customers and hundreds of thousands of partners. For such a big company automating routine tasks is critical to ensure the efficient operation of sales. For a long time Microsoft used a third-party CRM solution, but this solution was too complex and inconvenient for business users. In addition, this solution has sufficiently high TCO[3], because it requires high maintenance costs, extension of warranty, licensing and acquisition of equipment. The procedure of adding new features to the old platform was too long.

This had a negative effect on the company's flexibility, which is necessary in order for the company to be at the forefront of the software industry.

To remove these problems and improve the efficiency of sales, in 2009, Microsoft implemented Global Sales Experience (GSX) solution based on Microsoft Dynamics CRM, Microsoft SQL Server and other Microsoft products. The effect of the implementation of the solution has been very positive:

- Time spent on training sales managers was reduced;
- Total time savings amounted to about of 3 hours per week;
- More user friendly interface made users' CSI higher (CSI has grown by 25% as of 2009);
- Total cost of ownership of the CRM system was decreased.

In 2011 Microsoft was recognized by Gartner[4] as a "leader" and "visionary" in sales force automation.

4.10 Automation of product management

Product management is an extremely important component of any IT business. It includes management of the image of products, pricing and management of user satisfaction. Product management is often viewed as an integral part of CRM, but in the IT business this process is particularly important, so it should be considered separately.

[3] TCO (total cost of ownership)

[4] According Gartner's Magic Quadrant. Gartner rates vendors upon two criteria: completeness of vision and ability to execute. "Leaders" score higher on both criteria. "Visionaries" score higher on the completeness of vision only.

The range of tasks in product management also includes forecasts for capacity, markets and user expectations. A proper product management strategy helps to increase user satisfaction of the products or services of the company and helps to extend the life cycle of products. It also avoids negative effects such as "requirements death spiral"[5]. In its turn it increases the range of proposed solutions, comprehensive income and enhances the company's image.

So which tools exist for product management in IT companies? First of all, the market can offer a lot of solutions for product lifecycle, gathering of user's expectations, products for market analysis and long-term forecasting. Typically, such solutions are already integrated into the existing complex ERP systems or can be added to them by a simple customization. Some single solutions also exist that can solve a separate task of product management or a wide range of those tasks.

However, small IT companies often rely on their abilities in the automation of product management. Automation of only a part of the process of product management allows those companies to make proper product policy and constantly improve products. Automated tools are also widely used to monitor usage and diagnose problems (such as those made by a third-party and those made by the company itself). In the case of corporate clients, these tools can be integrated with customer information systems (CRM, ERP). This can be beneficial for both the vendor and customer (the vendor's reaction time is reduced and its awareness is increased).

4.11 Integration of Information systems

The effectiveness of implementing new information systems in the IT infrastructure of the company will drop sharply if they are not integrated with existing processes or solutions. IT companies as owners of a wide range of information systems must solve the problem of integration. IT companies are suitable to the tasks of integration in different ways.

4.11.1 Fragmented integration environment

The easiest way to integrate information systems is the creation of several integration applications that can convert data streams from different systems. Each such application has a simple logic. It's easy and cheap to develop and maintain, but has some significant drawbacks:

- The complexity of the configuration. If the data format of the one of information systems has been changed then integration application must also be changed;
- The lack of uniformity in the technologies and techniques of development. Development of information technologies and changes in staff skills occur in parallel with the development of the IT infrastructure of a company. It often happens that in an environment of integration there are two types of applications running simultaneously: applications that use the old technologies and the latest solutions. This phenomenon has been called the "zoo". This phenomenon is usually harmless, except in two situations:
  - The company has no experts in the old technology (specialists were fired or switched to using other technologies);
  - The solution based on the old technology can no longer cope with the increasing volumes of data.

---

[5] Customer expectations permanently warm up by advertising and promises of management. At the same time production of goods and services cannot keep up with customer expectations.

Normally, the second problem entails the first one, since until there are no problems with the old solution, usually no one thinks about the lack of the necessary specialists. Therefore, this method of integrating information systems into a single space is practiced in small IT companies (because of the economy), or giving way to a common integration environment.

4.11.2 Common integration environment

The alternative is a single integration environment that is capable of taking on the task of integrating all of the company's information systems. This environment is the complex solution and has following abilities:

- Own internal data format;
- Ability to convert the output data formats of external systems into internal data format;
- Ability to convert the internal data format to output data formats of external systems;
- Ability to route internal data streams;
- Ability to build the simplest logical processes.

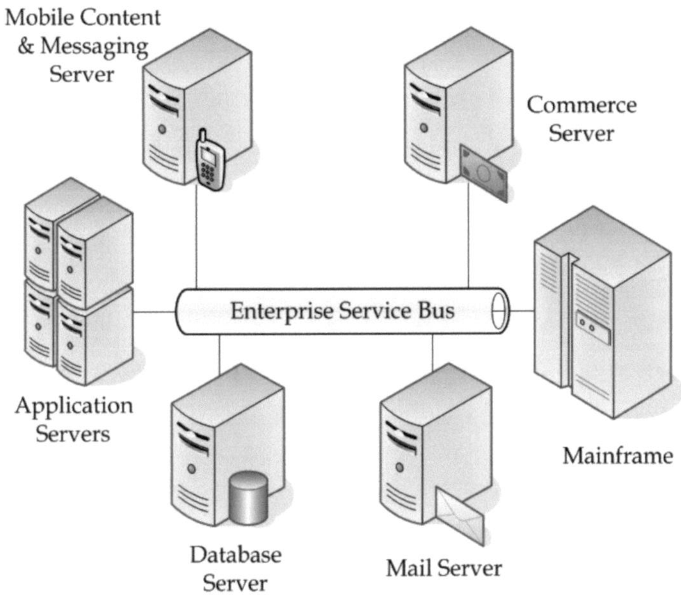

Fig. 4. Usage of enterprise service bus

This principle of interaction of information systems is called ESB (enterprise service bus). This concept has been developed in analogy to the bus concept found in computer hardware architecture combined with the modular and concurrent design of high-performance computer operating systems. Motivation was to find a standard, structured and general purpose concept for describing implementation of loosely coupled software components (called services) that are expected to be independently deployed, running, heterogeneous and disparate within a network.

Companies that have the ability to create these custom solutions typically use their own unified integration environment, but many IT companies prefer to use third-party solutions or do not use a single integration environment. The integration solutions market is constantly growing (up to the 2008 crisis, the growth was about 10% per year, according to the Forrester report, 2008), existing solutions are also continuing to develop.

Integration solutions offer a variety of flexible solutions that are suitable for large and medium-sized businesses, but existing solutions that implement the concept of ESB have the same drawbacks (in comparison with fragmented environment integration):

- Low performance. Multiple transformations of data formats and the universalization of the components lead to significant reduction in overall system performance;
- High hardware requirements. Due to the fact that the system becomes a central part of the entire IT infrastructure of the company, its low performance is unacceptable, because this reduces the productiveness of all involved processes. In addition, such systems have special requirements for equipment reliability and redundancy.

In some companies, ESB solutions coexist with specialized simple applications of integration. The companies take these measures if eliminating the disadvantages of ESB (listed above) is impossible. This situation occurs also during the gradual implementation of a common integration environment. Such a principle of building the integration environment combines the drawbacks of both approaches with all of the possible negative effects.

5. Services

Services provided by IT companies can be divided into two groups: consulting and resources providing.

It is clear that consulting services principally cannot be provided automatically and the production of such services is always based on human activity. The exception may be a variety of specialized knowledge bases or guided help systems. But such services are closer to the second group - the provision of resources.

Providing of resources is a large group, which includes the following specific services:

- Hosting;
- Rental computing resources;
- Provision of communications services;
- Internet service providing;
- Data storage;
- Creation and storage of special accounts;
- Virtual cash;
- Social networks, etc.

All of these services have one major feature - they are provided by automated systems and solutions. The basis of any of these services is the product manufactured by the IT business. Therefore, the chain of automation of these services starts from the development of relevant IT products. A review of solutions and products to automate the providing of these services is beyond the scope of this chapter.

6. Conclusion

As we can see, automation of labour is widely used in the IT field, however, due to the inherent features of this area, there are a number of typical problems faced by the majority of IT companies during automation of its operations.

6.1 Common problems of automation in IT field

The capabilities of automation tools are considerably inferior to the flexibility of the human brain, so the main drawback is a necessity to add the ability of using automation tools to specific processes. The wider use of automation in the processes of developing and supporting software makes these processes less flexible.

Application of automation technologies reduces the influence of the human factor on the results, however, the total cost of a bug or failure in the automation process can be much higher than the price of an error in manual labour. This happens for several reasons: first, the result of an automatic process is more trusted than the result of human activity; in any case, failure of an automated process can stop all related processes, either manual or automatic.

Finally, the most important factor: automation is very expensive. Automation tools, hiring experts in automation, the necessity to adjust the company's processes for automation - all of these requires a significant investment. For this reason, management often looks sceptically at the benefits of automation.

IT companies often want "quick" profits, i.e., the investment paying off as soon as possible. But investment in automation looks like an investment in the basic research:

- It is too expensive;
- It gives no obvious benefit;
- It has a very long payback period.

6.2 How to minimize the impact of these problems?

Automation is not for the sake of automation, and not a passing fad. Automation offers convenience and time saving. The implementation of automation into a company's processes must be based on a principle that can be called RGB:

- Relevant, that need to be automated only those processes that need it;
- Gradual - the simultaneous and widespread introduction of automated technologies adversely affects the stability of the workflow in the company. Automation should be implemented from simple to complex;
- Balanced - an initiative to implement various processes must come from a place where it is required. Centralized implementation of automation processes in the company does not bring real benefits.

Building automated processes must be done by the most experienced staff that well understand the subject of automation. It is desirable for complex processes or calculations at the first introduction of automation to be duplicated manually. Also, you must have a backup scheme in case of failure.

6.3 A few words about automation tools

There is no standard tool for all possible problems in the market. Many IT areas deal with highly tailored tasks, but even if such a tool did exist, sometimes a company, especially a small one, could not afford it. Licensing costs for some automation products can be compared to the monthly turnover of a small company. In such cases a company has some possible alternatives:

- Continue using manual labour;

- Buy the expensive product;
- Use freeware tools or technologies;
- Develop the necessary product themselves.

The choice of a free tool seems the easiest, however, it is very difficult to find a free tool to automate a particular process, because most of them have a limited range of options.

Developing a good tool for automating is very expensive, therefore, among freeware automation tools there are few good or universal ones available.

Development of automation tools by the company itself seems quite reasonable, but in many cases it turns out to be a trap. The sum invested into the development of the tool can exceed many times the cost of the ones that already exist. This is quite logical, as it is impossible to create in just one week what has been developed by another company over many years. However, sometimes the risk pays off.

But what can we do if there is no tool on the market that could serve for a specific problem? Should the process be automated anyway? In any case, automation always pays off over the long-term. So the company management has to decide on whether the company is ready to start implementing automation right now to realize the profits in about a year's time or whether it is better to leave everything as it is. With the constantly changing market conditions, in a year a small company could change its specialization or reconstruct the whole development process, and therefore all the efforts would be wasted.

7. References

Definition of Information Technology. 2011, Available from: http://definitionofinformationtechnology.com. Retrieved on 02/10/2011.

Vinnichenko, I. (2005) *Automation of process of testing*, Piter, *ISBN*, St.-Petersburg, Russian Federation.

Enterprise Resource Planning. (October 16, 2011), In: *Wikipedia The Free Encyclopedia*. 2011, Available from: http://en.wikipedia.org/wiki/Enterprise_resource_planning. Retrieved on 12/10/2011.

Genina, N. Enterprise Management: "1C" surpassed Oracle in Russia. (September 18, 2009), In: *C-News, Russian IT-Review*. 2009, Available from: http://www.cnews.ru/news/top/index.shtml?2009/09/18/362481. Retrieved on 20/10/2011.

Smulders, C. Magic Quadrants and Market Scopes: How Gartner Evaluates Vendors Within a Market (April 26, 2011), In: *Gartner*, 2011, Available from: http://www.gartner.com/resources/154700/154752/magic_quadrants_and_marketsc_154752.pdf . Retrieved on 20/10/2011.

Payne, I. (June 6, 2005) *Handbook of CRM: Achieving Excellence in Customer Management.* Butterworth–Heinemann, 0750664370

Lawley, B.; Cohen, G. & Lowell L. (October 14, 2010) *42 Rules of Product Management: Learn the Rules of Product Management from Leading Experts "from" Around the World,* Super Star Press, 1607730863

18

VHDL Design Automation Using Evolutionary Computation

Kazuyuki Kojima
Saitama University
Japan

1. Introduction

This chapter describes the automatic generation method of VHDL which lays out the control logic of a control system. This framework releases a designer from the work of describing the VHDL directly. Instead, the designer inputs the equation of motion of a system and target operation.

In this chapter, first, FPGA, CPLD, VHDL and evolutionary computation are outlined. This is basic knowledge required for an understanding of this chapter. Next, the framework of automatic generation of FPGA using evolutionary computation is described. VHDL description is expressed by several kinds of data structures called a chromosome. VHDL expressed in a chromosome is changed using evolutionary computation and changes to a more suitable code for controller purposes. Finally two example applications are shown. The first one is the controller for a simple inverted pendulum. After that, the framework is applied to a more complicated system, an air-conditioning system. The simulation results show that the controller automatically generated using this framework can control the system appropriately.

2. Computer-aided controller design using evolutionary computation

2.1 FPGA/CPLD/ASIC and VHDL

CPLDs and FPGAs are both sorts of programmable LSIs. The internal logic of both can be designed using HDL. The ASIC is one example of a device that can be designed using HDL in the same way as programmable LSIs. CPLDs and FPGAs can be immediately evaluated on the system for the designed logical circuit. In addition, they are flexible for the rearrangement of a specification. These merits make them suitable for the intended use in the case of a rapid prototyping. For this reason, a CPLD is used as a controller. However, the proposed framework is applicable to all devices that can be designed by HDL. VHDL is one of the most popular HDLs and is therefore used in this paper.

The logic described by VHDL is verified and synthesized using a simulator or a logic synthesis tool so that it can be written into a device. When CPLD or FPGA serves as target devices, the programming code which determines the function of the target device can be, through a download cable, written into it in order to obtain the target LSI easily. The VHDL for a simple logical circuit is shown in Fig. 1.

```
library IEEE;
use IEEE.std_logic_1164.all;

entity HALF_ADDER is
  port(
    A,B : in std_logic;
    S,CO : out std_logic);
end HALF_ADDER;

architecture DATAFLOW of HALF_ADDER is
  signal C, D : std_logic;
begin
  C  <= A or B;
  D  <= A nand B;
  CO <= not D;
  S  <= C and D;
end DATAFLOW;
```

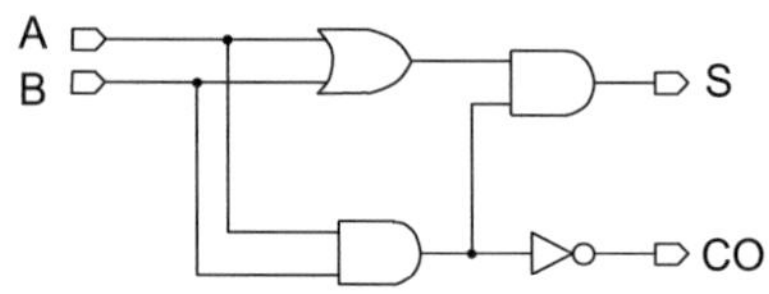

Fig. 1. VHDL for simple logical circuit

2.2 Genetic algorithm

The genetic algorithm used as a basis of this framework is outlined below (Fig. 2). The decision-variable vector x of an optimization problem is expressed with the sequence of N notations $s_j(j = 1, \cdots, N)$ as follows:

$$x : s = s_1 s_2 s_3 \cdots s_N \tag{1}$$

It is assumed that symbol string is a chromosome consisting of N loci. s_j is a gene in the j-th locus and value s_j is an allelomorph. The value is assumed to be a real number, a mere notation, and so on of the group of a certain integer or a certain range of observations as an allelomorph. The population consists of K individuals expressed with Eq. (1). Individual population $p(n)$ in generation n changes to individual population $p(n+1)$ in next generation $n+1$ through the reproduction of a gene. If reproduction in a generation is repeated and if the individual who expresses solution x nearer to an optimum value is chosen with high probability, then the value increases and an optimum solution is obtained (Goldberg, 1989; Koza, 1994).

2.3 Evolvable hardware

Higuchi et al. proposed evolvable hardware that regards the architecture bit of CPLD as a chromosome of a genetic algorithm to find a better hardware structure by genetic algorithm (Higuchi et al., 1992). They applied this to myoelectric controllers for electrical prosthetic hands, image data compression and so on (Kajitani et al., 2005; Sakanasi et al., 2004). This approach features coding of an architecture bit directly for a chromosome. The designer must determine which CPLD is used beforehand, implicitly determining the meaning of the architecture bit. If hardware is changed for the problem of circuit scale or structure, it is necessary to recalculate evolutionary computation.

Henmi et al. evolved a hardware description language (HDL) called structured function description language (SFL) and applied it to robot walking acquisition (Hemmi et al., 1997).

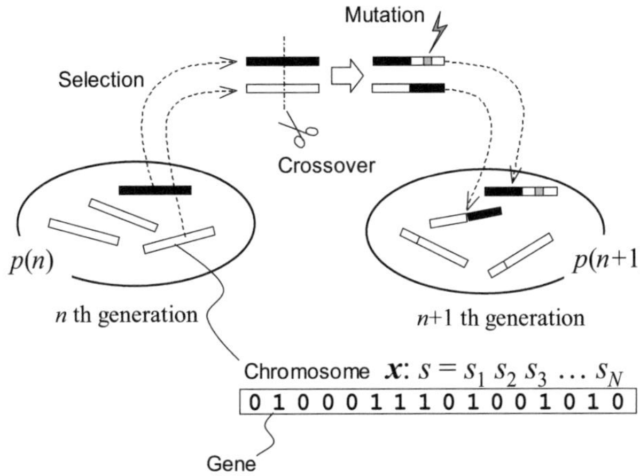

Fig. 2. Genetic algorithm overview

The basic motion, called an "action primitive", must be designed in a binary string. In our approach, the designer needs only to define I/O pins for CPLD. VHDL rather than an architecture bit is coded directly onto the chromosome, so the chromosome structure does not depend on CPLD scale or type, and after VHDL is generated automatically, CPLD is selected appropriately to the VHDL scale. Pin assignment is set after VHDL generation.

2.4 Controller design framework using evolutionary computation

The study of optimizing rewritable logical-circuit ICs, such as CPLD, using a genetic algorithm has increased. The framework that changes an internal logic circuit IC configuration to attain evolution is called evolvable hardware (EHW). With this framework, the designer needs only to define the criteria used to evaluate a controller. We explain the controller design framework using evolutionary computation with XC95144 (Xilinx, 1998) as the test device. Internal blocks of XC95144 are shown in Fig. 3. XC95144 is a small CPLD that has 144 pins (117 user input-outputs), 144 macro cells and 3200 usable gates. A designer chooses input and output signals from 117 user I/Os and assigns these pins. Each signal is defined for each I/O. If CPLD is used in control, sensors and actuators can be associated with CPLD I/O pins (Fig. 4). Here, I/O pins are associated with two sensors and one actuator. The sensor values are inputs to CPLD as two 16-bit digital signals and a 10-bit digital signal is output as a reference signal to the actuator.

VHDL, which describes internal CPLD logic, is encoded on a chromosome. An example of the VHDL generated is shown in Fig. 5, corresponding to Fig. 4. This VHDL consists of three declarations — (a) entity declaration, (b) signal declaration, and (c) architecture body. CPLD I/O signals are defined in (a) and internal CPLD signals are in (b). Signals mainly used in VHDL are a `std_logic` type and a `std_logic_vector` type. The `std_logic` type is used when dealing with a signal alone and the `std_logic_vector` type is used when dealing with signals collectively. The `std_logic` and the `std_logic_vector` types should be combined to optimize maintenance and readability. A description of VHDL can be restored if all of the input output signals and internal signals are used as the same `std_logic` type and only the number is encoded on a chromosome. The number of input signals, output

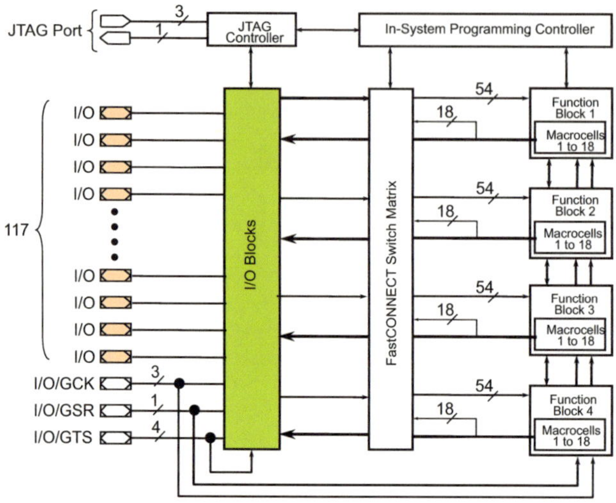

Fig. 3. XC95144 architecture

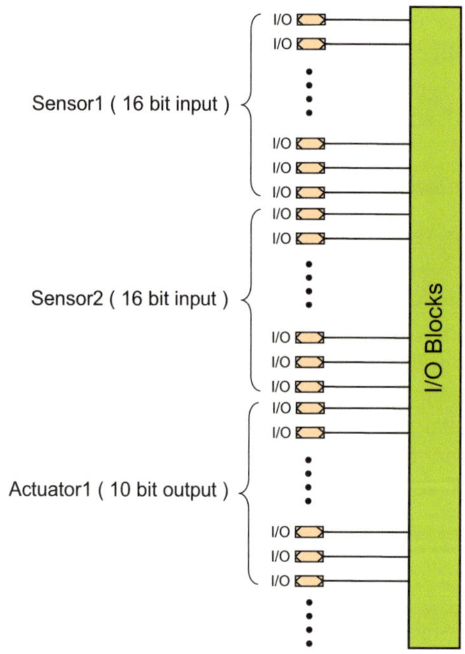

Fig. 4. Example of CPLD application

signals and internal signals are encoded on the head of a chromosome (Fig. 6). In Fig. 4, two input signals are set up with 16 bits and an output signal with 10 bits.

A chromosome that represents a VHDL assignment statement is shown in Fig. 7. A chromosome structure corresponding to a process statement is shown in Fig. 8. The value currently described is equivalent to the process statement in which "`DI009`" and "`DI014`" are enumerated in the sensitivity list.

```
----------------------------------------------------------------------------
-- VHDL for I/P controller generated by _hiGA_AC with _hiGA.dll
--     (c) 2011, Saitama Univ., Human Interface Lab., programmed by K.Kojima
--     Chromo#:E/50 Length:13 Inputs:32 Outputs:10 Signals:2
--     Simulation: Initial condition: 180.0, dt=0.0100
----------------------------------------------------------------------------
library IEEE;
use IEEE.std_logic_1164.all;
use IEEE.std_logic_arith.all;
use IEEE.std_logic_unsigned.all;

entity GA_VHDL is
    port(
        DI000 : in std_logic;
        DI001 : in std_logic;
        DI002 : in std_logic;
        DI003 : in std_logic;
        DI004 : in std_logic;
        DI005 : in std_logic;
        DI006 : in std_logic;
        DI007 : in std_logic;
        DI008 : in std_logic;
        DI009 : in std_logic;
        DI010 : in std_logic;
        DI011 : in std_logic;
        DI012 : in std_logic;
        DI013 : in std_logic;
        DI014 : in std_logic;
        DI015 : in std_logic;
        DI016 : in std_logic;
        DI017 : in std_logic;
        DI018 : in std_logic;
        DI019 : in std_logic;
        DI020 : in std_logic;
        DI021 : in std_logic;
        DI022 : in std_logic;
        DI023 : in std_logic;
        DI024 : in std_logic;
        DI025 : in std_logic;
        DI026 : in std_logic;
        DI027 : in std_logic;
        DI028 : in std_logic;
        DI029 : in std_logic;
        DI030 : in std_logic;
        DI031 : in std_logic;
        DO000 : out std_logic;
        DO001 : out std_logic;
        DO002 : out std_logic;
        DO003 : out std_logic;
        DO004 : out std_logic;
        DO005 : out std_logic;
        DO006 : out std_logic;
        DO007 : out std_logic;
        DO008 : out std_logic;
        DO009 : out std_logic
    );
end GA_VHDL;

```

(a) Entity declaration

Fig. 5. (a) Automatically generated VHDL (List-1)

The VHDL description has an if-statement inside of a process statement and the description has two nesting levels. The hierarchy of the list structure is deep compared to the substitution statement. When the gene of such a multilist structure is prepared, it is possible to represent various VHDL expressions.

2.5 Variable length chromosome and genetic operations

The structure of a chromosome changes with the control design specification. The number of internal signals is set arbitrarily and different descriptions in VHDL are expressed with different locus lengths. The length of a chromosome is determined by the VHDL line count. The length determines the number of internal signals enumerated on the sensitivity list or

```
architecture RTL of GA_VHDL is

    signal S000 : std_logic;
    signal S001 : std_logic;

begin


    process(DI013) begin
        S000 <= not DI013;
    end process;


    process(DI002, DI009) begin
        S001 <= DI009;
    end process;

    DO000 <= (((DI002 or not DI011) nor DI002) nand DI003);
    DO001 <= (((((((((((((((((((((((((((((DI028 nor not DI011) or DI001) or
      DI008) nor DI017) and not DI008) nor DI016) or not DI030) or DI025) or not
      DI015) or not DI010) and DI010) nand not DI028) nand not DI010) and not DI012)
      and DI008) nand DI002) or not DI009) nor not DI031) nor DI007) nor not DI009)
      nand DI023) or DI014) nor DI019) and not DI025) nand DI024) nand not DI010) or
      not DI031) nor not DI019) nand DI000);

    process(DI006, DI029, DI014) begin
        DO002 <= ((not DI014 or DI029) or DI006);
    end process;


    process(DI029) begin
        DO003 <= DI029;
    end process;


    process(DI009, DI014) begin
        if(DI009'event and DI009='0')then
            DO004<=DI009 nand not DI007;
        end if;
    end process;

    DO005 <= ((((((((((((((((((((((((((((DI014 and DI016) nand not DI007) and
      not DI022) nor not DI005) nand DI018) nor DI021) and not DI000) nand not DI013) or
      not DI022) or DI030) or not DI027) and DI031) and DI020) or DI023) nor not DI025)
      or not DI000) nor DI002) nand DI002) and not DI027) nand not DI017) nand not
      DI025) nand not DI007) nor DI031) and DI012) nand not DI012) or not DI030) nand
      not DI026) and not DI011);
    DO006 <= ((((((((((((((((((((((((DI010 and not DI028) nand DI009) nor DI011) nand
      not DI016) or DI029) nand not DI001) and DI011) or DI017) or DI011) nand not
      DI030) nor DI022) or not DI005) or DI001) and DI019) nor DI027) nand DI010) or not
      DI025) or not DI023) and DI019) or not DI019) and DI023) nor not DI008) or not
      DI005) nand not DI021);
      DO007 <= DI009;

    process(DI025, DI013, DI023) begin
        DO008 <= not DI023;
    end process;


    process(DI014, DI024) begin
        DO009 <= DI014;
    end process;


end RTL;
```

(b) Signal declaration

(c) Architecture body

(d) Assignment statement

(e) Process statement

Fig. 5. (b) Automatically generated VHDL (List-2)

the length of the right-hand side of an assignment statement. When dealing with such a variable length chromosome, the problem is that genetic operations will generate conflict on a chromosome. To avoid this problem, we prepared the following restrictions:

1. With a top layer, the length of a chromosome is equal to the number that added one to the summary of the number of internal signals and the number of output signals.
2. All signals are encoded on a chromosome using a reference number.

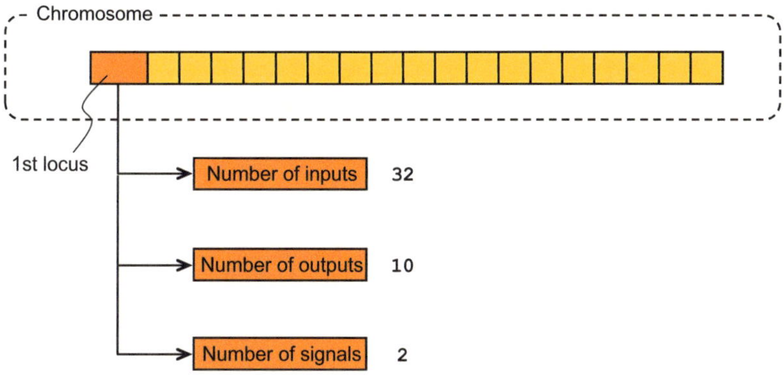

Fig. 6. Signal definition on the first locus

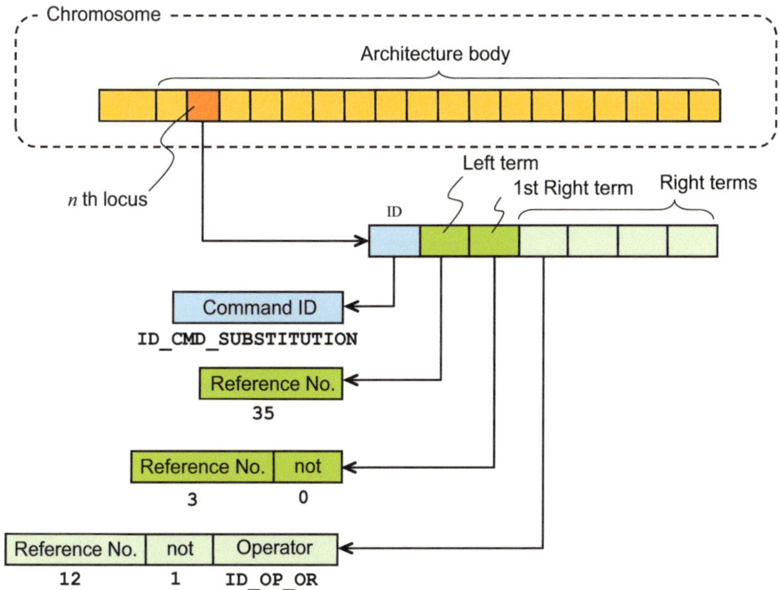

Fig. 7. Gene structure for assignment statement

3. The signal with a large reference number is described by only the signal whose reference number is smaller than the signal.
4. The top layer of a chromosome describes the entity declaration using all internal signals and output signals in order with a low reference number. Each signal can be used only once.
5. Crossover operates on the top layer of a chromosome.

These restrictions avoid the conflict caused by genetic operations.

Figure 9 shows an example of crossover operation. The back of the 6th gene is chosen in this example. Chromosome (A) and chromosome (B) cross and change to chromosome (A′) and chromosome (B′). Only the gene before and behind the crossover point of each chromosome

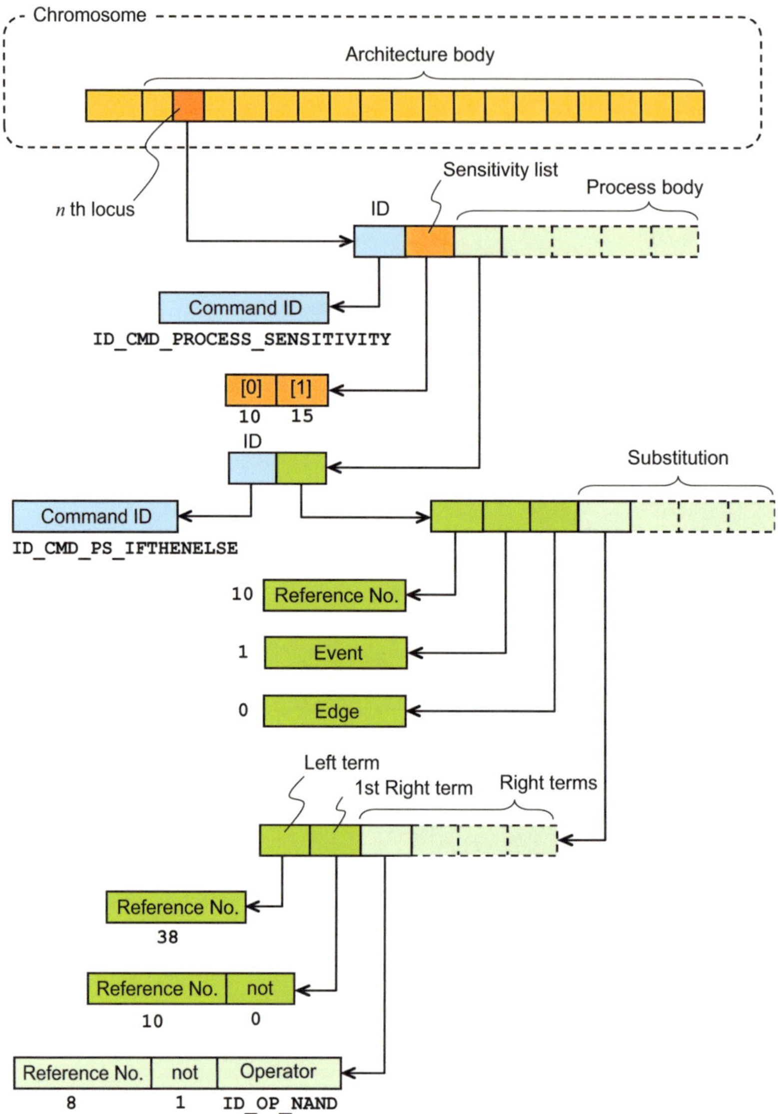

Fig. 8. Gene structure for if-then-else statement in process body

show the gene of a lower layer. In the figure, chromosome (A) has two sensitivity lists and chromosome (B) has two assignment statements. The structure of a chromosome changes by replacing the gene from the back of a top gene to before a crossover point. Both chromosome (A′) and chromosome (B) came to have an assignment statement and a sensitivity list.

3. Application to HDL-based controller of inverted pendulum

In this section an application to an HDL-based controller for the inverted pendulum is described (Kojima, 2011).

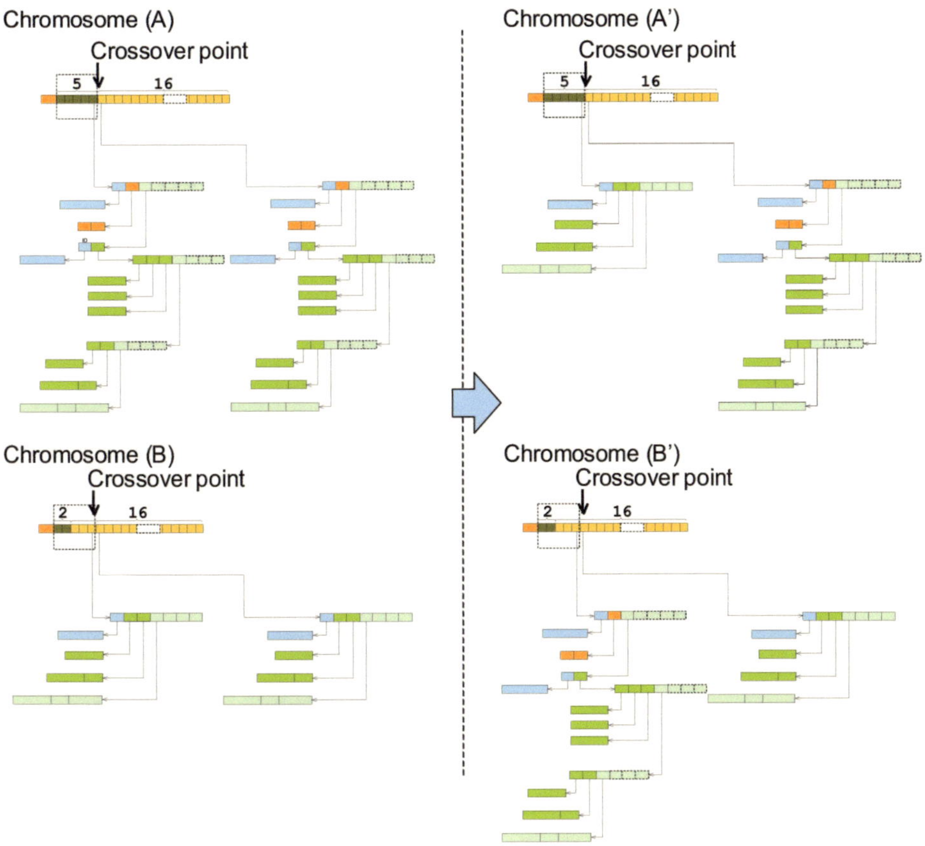

Fig. 9. Crossover operation

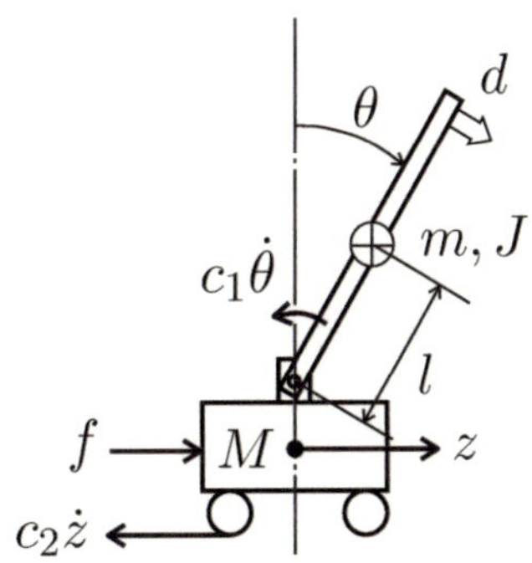

Fig. 10. Inverted pendulum

3.1 Equations of motion

Figure 10 shows the model of the inverted pendulum. The equations of motion are given by

$$(ml^2 + J)\ddot{\theta} = mgl \sin\theta - ml\ddot{z}\cos\theta - c_1\dot{\theta} + d \tag{2}$$

$$(M + m)\ddot{z} = f - c_2\dot{z} \tag{3}$$

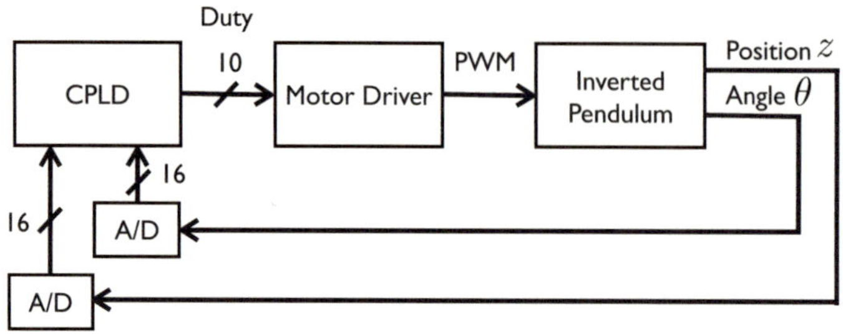

Fig. 11. Control system block diagram

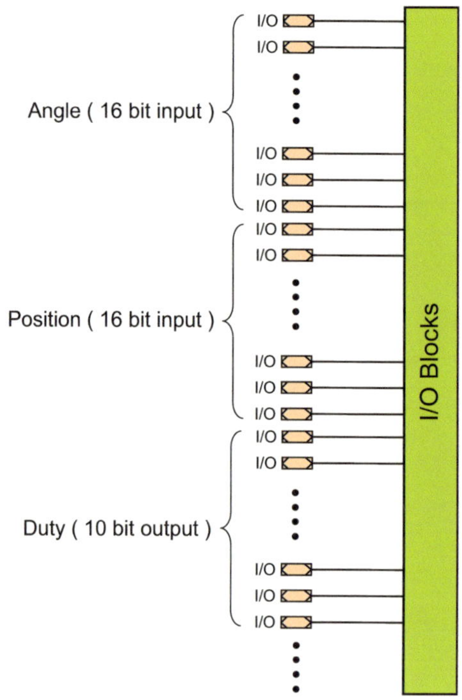

Fig. 12. CPLD pin assignment

$$f = Kv \tag{4}$$

where θ [rad] is the angle of the pendulum, m [kg] is the mass of the pendulum, M [kg] is the mass of the cart, J [kgm^2] is the inertia of the rod, l [m] is the length between the gravity point and fulcrum, z [m] is the position of the cart, d [Nm] is the disturbance torque, c_1 [Ns/rad] is the viscous resistance of the pendulum, c_2 [Ns/m] is the viscous resistance of the cart, f [N] is the controlling force, v is the parameter for the control input and K is the gain. $\bar{J}$, a and b are defined as:

$$\bar{J} = J + ml^2 \tag{5}$$

$$a = \frac{c_2}{M + m} \tag{6}$$

$$b = \frac{K}{M + m} \tag{7}$$

then, equations (2) and (3) become:

$$\ddot{\theta} = \frac{mgl}{\bar{J}}\sin\theta - \frac{ml\ddot{z}}{\bar{J}}\cos\theta - \frac{c_1}{\bar{J}}\dot{\theta} + \frac{d}{\bar{J}} \tag{8}$$

$$\ddot{z} = -a\dot{z} + bv. \tag{9}$$

Each parameter can be determined by measurement and experiment as $l = 0.15$, $m = 46.53 \times 10^{-3}$, $\bar{J} = 1.58 \times 10^{-3}$, $c_1 = 2.05 \times 10^{-2}$, $a = 4.44$ and $b = 2.46 \times 10^{-1}$. Using these parameters, evolutionary simulations are conducted.

3.2 Control system and CPLD pin assignment

Figure 11 shows the control system block diagram. The CPLD is used as the controller of the inverted pendulum. The position of the cart and the angle of the pendulum are converted 16 bit digital signals respectively and input to the 32 CPLD pins. The control logic in the CPLD, which is formed using the framework previously mentioned, determines the 10 bit control signal driving the motor of the inverted pendulum. Figure 12 shows the CPLD pin assignment for this application. In this case, 32 bit parallel inputs and 10 bit parallel outputs are adopted. Instead of using them, we can use serial I/Os connected with A/D converters and D/A converters. In this case, serial to parallel logics should be formed in the CPLD and even though serial inputs are used, automatically generated VHDL can be used as a VHDL component.

3.3 Fitness

The fitness value is calculated as a penalty to the differences in the rod angle and the cart position.

$$\text{fitness} = -\frac{1}{\pi}\int_0^{30} |\theta| dt - \int_0^{30} |z| dt \tag{10}$$

Disturbances are given as a random torque during the control simulation. When calculating fitness value, disturbance torque is always initialized. Therefore, all individuals are given different disturbance at each evaluation. This kind of disturbance makes the controller robust to various disturbances.

3.4 Simulation

Simulations are conducted under two conditions – (1) $\theta_0 = 1^\circ, z_0 = 0$, (2) $\theta_0 = 180^\circ, z_0 = 0$. Population size is 50, mutation rate is 0.5, crossover rate is 1.0, tournament strategy, tournament size is 10 and the elite strategy is adopted.

Figures 13 and 14 show the simulation results. Figure 13 shows the result of condition (1), Figure 14 shows the result of condition (2). (a) the result at 0 generation and (b) the result at 1000 generation are represented respectively. The angle of the rod, the position of the cart, disturbance and the control signal are shown at each generation. At zero generation, in both conditions (1) and (2), the obtained controller cannot control the inverted pendulum adequately. The rod moves in a vibrating manner at around 180°. At 1000 generation, the controller controls the inverted pendulum successfully in both conditions. Further, in condition (2), swing up motion can be observed (Fig.14(b)). The control signal at 1000 generation has more various patterns than the signal at 0 generation.

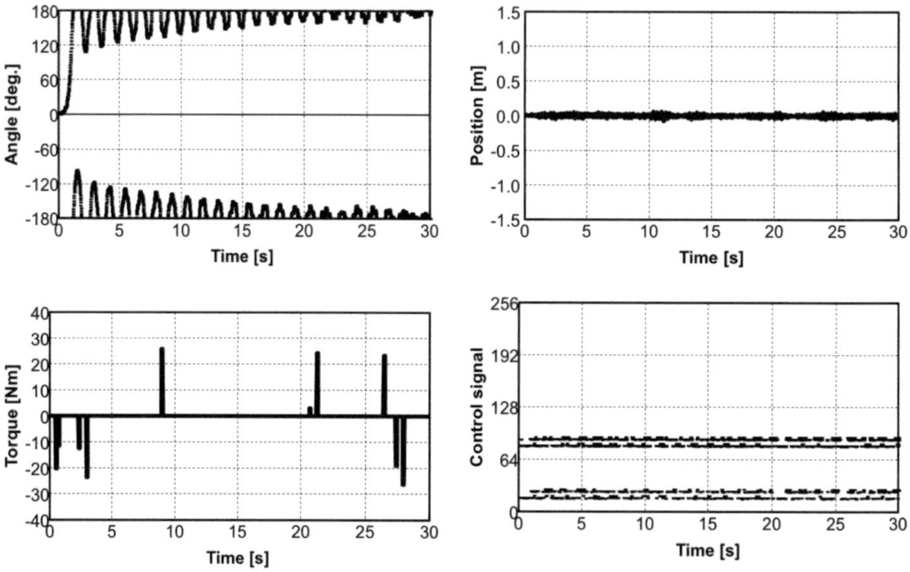

(a) 0th generation

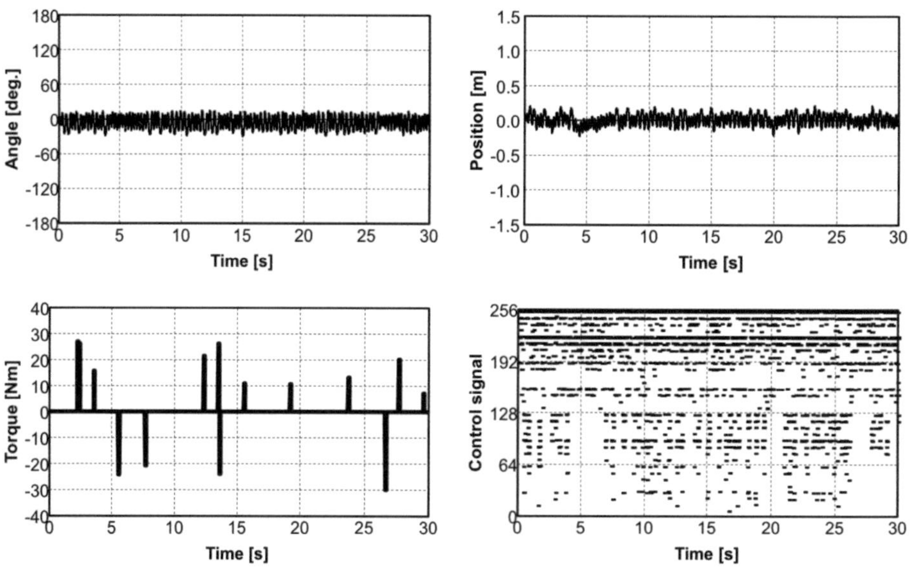

(b) 1000th generation

Fig. 13. Simulation results ($\theta_0 = 1^\circ, z_0 = 0$)

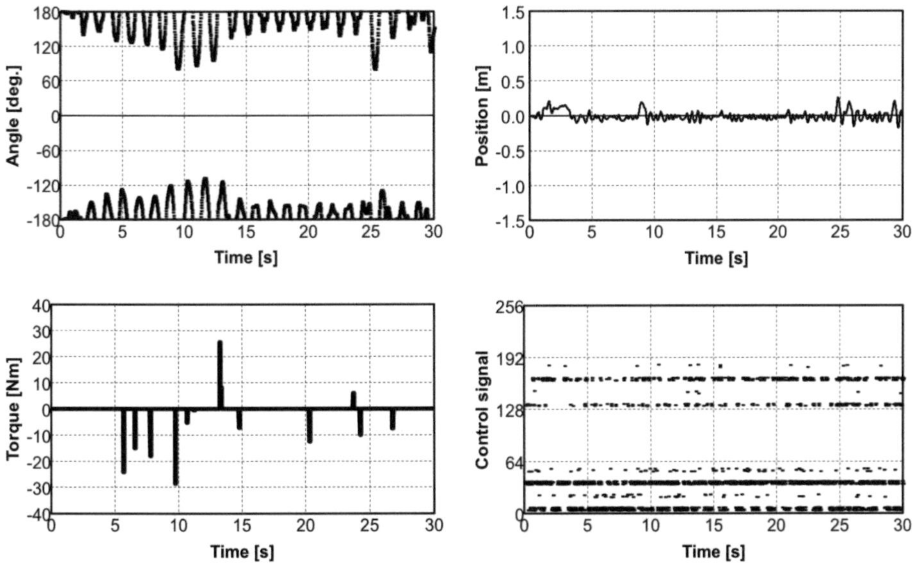

(a) 0th generation

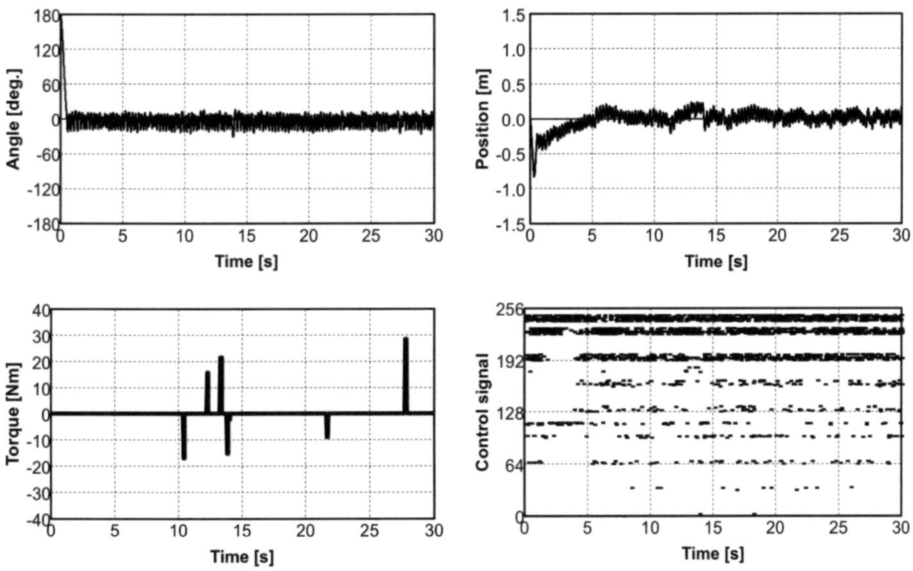

(b) 1000th generation

Fig. 14. Simulation results ($\theta_0 = 180°, z_0 = 0$)

4. Application to an air-conditioning controller

Next, the framework is applied to an air-conditioning controller. Simulation model, task definition and fitness function are described in this section. Simulation results will be shown at the end of the section (Kojima et al., 2007; Kojima, 2009).

4.1 Simulation model

In the targeted air-conditioning system(Fig. 15), the outside air imported is cooled or warmed and sent to the console to adjust temperature. Air entering from the inlet (a) is cooled by the refrigerator (c). Part of the cooled air is warmed with a heater (e), then mixed with cool air to adjust the temperature. The angle of the mixture door (d) controls the mixing ratio of warm and cold air. Mixed air is sent to console (f), changing the indoor temperature. The system controller controls blower motor rotation and the angle of the air mixture door.

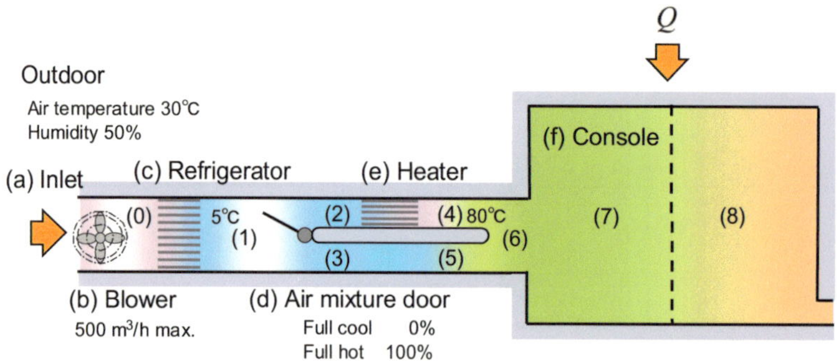

Fig. 15. Air-conditioning system

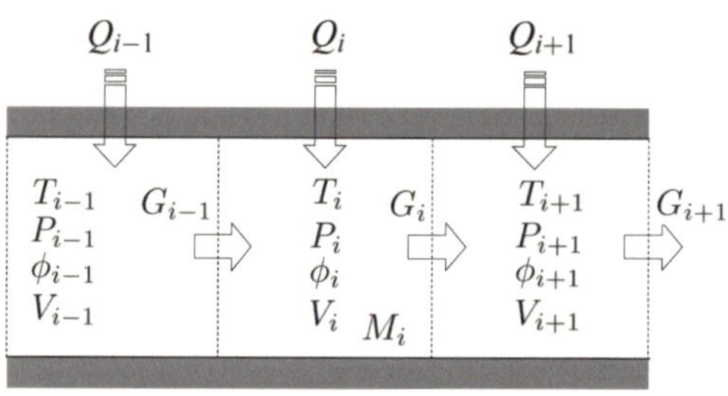

Fig. 16. Control volumes

To evaluate the controller performance, an air-conditioning simulation model is required in evolutionary computation. We consider an air-conditioning model combining the console and duct using nine control volumes (Fig. 16). Temperature and humidity, but not compressibility, are considered in this model. To calculate the predicted mean vote (PMV), which indicates thermal comfort, we require globe temperature, metabolization, flow rate, insulation of clothes and external work.

Figure 16 shows three control volumes in this system. We assume that air flows toward the control volume indexed with $(i+1)$ from the control volume indexed with (i). Air temperature T_{i+1}[K] and relative humidity ϕ_{i+1} vary with temperature T_i[K], relative humidity ϕ_i, mass flow G_i[kg/s] and heat transfer Q[kJ/s]. In air-conditioning simulation, mass flow is proportional to the opening of blower motor α.

$$G_0 = G_i = \alpha G_{max} \ (0 \leq \alpha \leq 1) \tag{11}$$

where G_{max} is maximum flow at full blower opening. Flow rate G_0 is the sum of air flow G_{a0} and steam flow G_{w0}.

$$G_0 = G_{a0} + G_{w0} \tag{12}$$

Water vapour pressure P_{w0} is as follows:

$$P_{w0} = \phi_0 \cdot P_{s0} \tag{13}$$

where P_{s0} saturation water vapour pressure from Tetens' formula (Tetens, 1930):

$$P_{s0} = 610.78 \times e^{\frac{17.2694(T_0 - 273.15)}{(T_0 - 273.15) + 238.3}} \tag{14}$$

Specific humidity x_0 is given by water vapour pressure P_{w0}.

$$x_0 = 0.622 \frac{P_{w0}}{P_0 - P_{w0}} \tag{15}$$

Air mass flow G_{a0} is calculated from the following gas equation:

$$G_{a0} = P_{a0} \cdot \frac{V}{R_a T_0} \tag{16}$$

where R_a is a gas constant. Water vapour mass flow G_{w0} is given by:

$$G_{w0} = x_0 \cdot G_{a0} \tag{17}$$

Air flow rate G_a is constant. Steam flow G_w is constant except during dehumidification. Considering the air flow into control volume (i) from control volume $(i-1)$ in unit time dt[s], temperature T_i', humidity x_i' and mass of control volume M_i' after dt is given as follows:

$$T_i' = \frac{G_{i-1}C_{i-1}T_{i-1}dt + (M_i - G_i dt)C_i T_i}{G_{i-1}C_{i-1}dt + (M_i - G_i dt)C_i} + \frac{Q_i dt}{M_i C_i} \tag{18}$$

$$x_i' = \frac{M_i x_i + (1 + x_i)(G_{wi-1} - G_{wi})dt}{M_i + (1 + x_i)(G_{ai-1} - G_{ai})dt} \tag{19}$$

$$M_i' = M_i + (G_{i-1} - G_i)dt \tag{20}$$

where specific heat C_i[kJ/kg $\cdot$ K] is given by

$$C_i = \frac{1.005 + x_i\{(2501.6/T_i) + 1.859\}}{1 + x_i} \tag{21}$$

At the mixture door, air is divided into two flows. The ratio of the mass of divided air depends on the mixture door angle. Here G_1 is the mass flow at location (1) in Fig. 15, G_2 that at location (2) and G_3 that at location (3),

$$G_2 = \beta G_1 \tag{22}$$

$$G_3 = (1 - \beta)G_1 \tag{23}$$

where β $(0 \leq \beta \leq 1)$ is the opening ratio of the mixture door. Adding two mass flow rates enables us to calculate the downstream mass at a juncture. Here G_4 is the mass flow at location (4) in Fig.15, G_5 that at location (5) and G_6 that at location (6),

$$G_6 = G_4 + G_5 \tag{24}$$

Temperature and humidity at a juncture are given in the same way as for when three control volumes are considered.

4.2 Predicted mean vote (PMV)

PMV is the predicted mean vote of a large population of people exposed to a certain environment. PMV represents the thermal comfort condition on a scale from -3 to 3, derived from the physics of heat transfer combined with an empirical fit to sensation. Thermal sensation is matched as follows: "+3" is "hot." "+2" is "warm." "+1" is slightly warm." "0" is "neutral." "-1" is slightly cool." "-2" is "cool." "-3" is "cold." Fanger derived his comfort equation from an extensive survey of the literature on experiments on thermal comfort (Fanger, 1970). This equation contains terms that relate to clothing insulation I_{cl}[clo], metabolic heat production M[W/m^2], external work W[W/m^2], air temperature T_a[°C], mean radiant temperature T_r[°C], relative air speed v[m/s] and vapour pressure of water vapour P[hPa].

$$\begin{aligned}
\text{PMV} = {} & \{0.33\exp(-0.036M) + 0.028\}\Big[(M - W) \\
& -3.05\{5.73 - 0.007(M - W) - P\} \\
& -0.42\{(M - W) - 58.1\} \\
& -0.0173M(5.87 - P) \\
& -3.96 \times 10^{-8} f_{cl}\{(T_{cl} + 273.15)^4 \\
& -(T_{mrt} + 273.15)^4\} \\
& -f_{cl}h_c(T_{cl} - T_a)\Big]
\end{aligned} \tag{25}$$

f_{cl} is the ratio of clothed and nude surface areas given by:

$$\begin{aligned}
f_{cl} &= 1.0 + 0.2I_{cl}(I_{cl} \leq 0.5) \\
f_{cl} &= 1.05 + 0.1I_{cl}(I_{cl} > 0.5)
\end{aligned} \tag{26}$$

where T_{cl} is the clothing surface temperature given by repeated calculation of:

$$\begin{aligned}
T_{cl} = {} & 35.7 - 0.028(M - W) \\
& -0.155I_{cl}\Big[3.96 \times 10^{-8} f_{cl}\{(T_{cl} + 273.15)^4 \\
& -(T_{mrt} + 273.15)^4\} + f_{cl}h_c(T_{cl} - T_a)\Big]
\end{aligned} \tag{27}$$

where h_c is the heat transfer coefficient,

$$h_c = \max\{2.38(T_{cl} - T_a)^{0.25}, 0.0121\sqrt{v}\} \tag{28}$$

and T_{mrt} is mean radiant temperature. PMV is detailed in (Fanger, 1970).

4.3 Task definition

The task is to adjust PMV in the console despite heat transfer from outside changes. The air-conditioning controller controls the opening ratio of blower n and the opening ration of mixture door m appropriately.

4.4 Control system and CPLD pin assignment

Figure 17 shows the control system block diagram. PMV is input to the CPLD as a 8 bit signal. The control logic in the CPLD determines the 8 bit blower opening control signal and the 8 bit mixture door control signal. Figure 18 shows the CPLD pin assignment for this application.

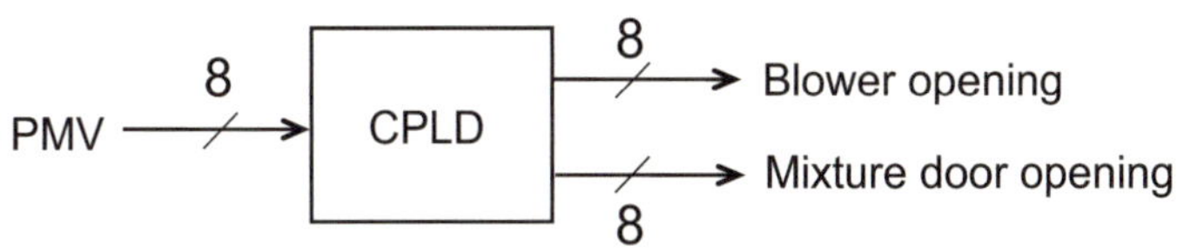

Fig. 17. Control system block diagram

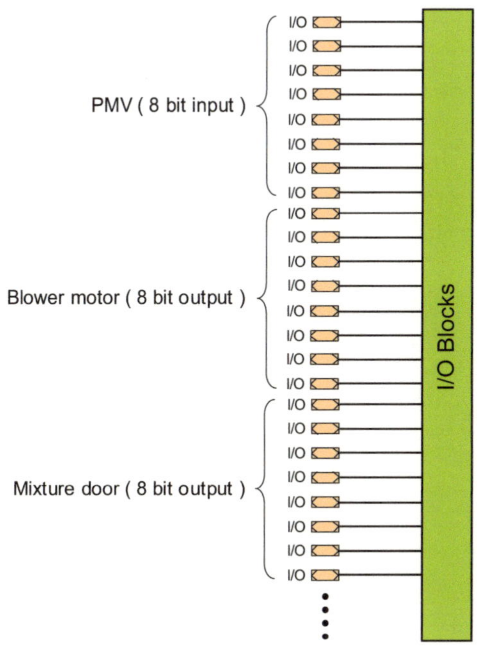

Fig. 18. Control system block diagram

4.5 Fitness

The fitness function is as follows:

$$\text{fitness} = -\int_0^{t_{end}} |\text{PMV}_{\text{ctrl}} - \text{PMV}_{\text{target}}|dt \tag{29}$$

where t_{end} is the end of simulation time. The difference between target and controlled PMV is integrated as a penalty in the controller simulation.

4.6 Simulation results

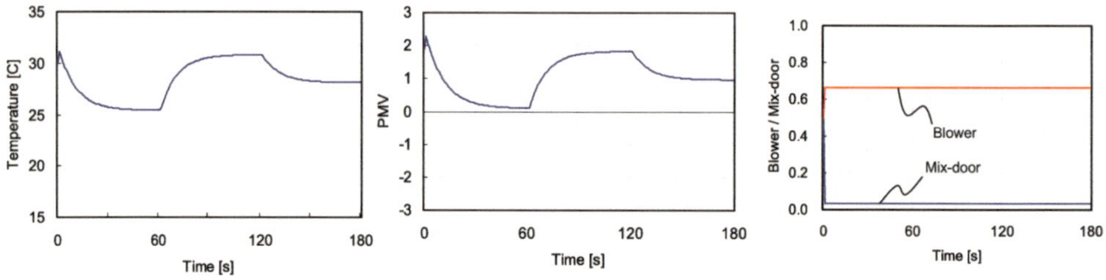

(a) 0th generation

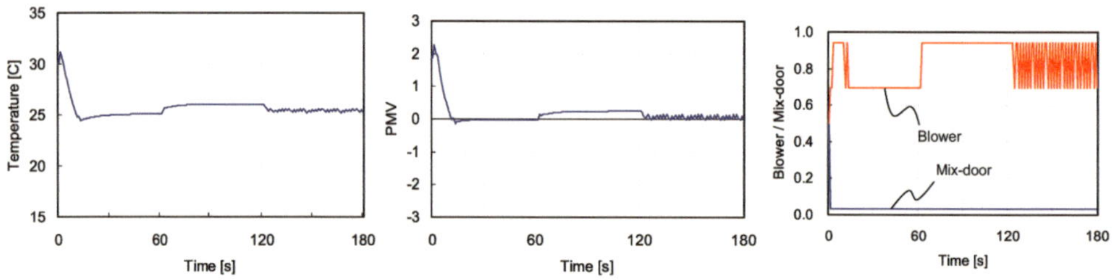

(b) 100th generation

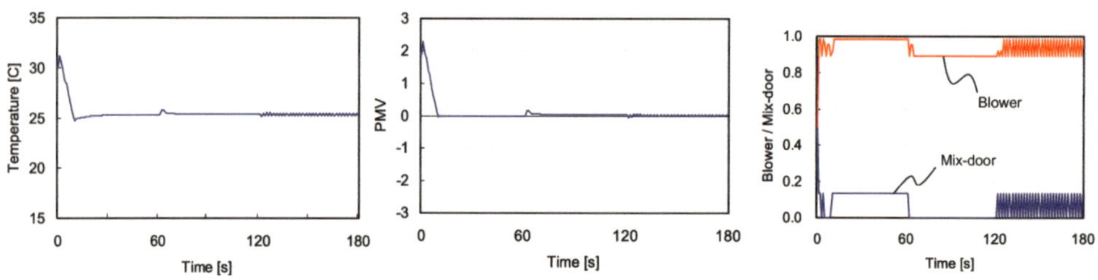

(c) 10000th generation

Fig. 19. Simulation results

Figure 19 shows the simulation results. PMV in a console is fed back to the controller. In the graphs, trends change every 60 seconds. Variations are based on the load effect change each 60 seconds. At zero generation (Fig. 19 (a)), temperature rises or falls with the change in heat load. PMV also changes simultaneously. This means that simply optimizing a controller is not enough. After 100 generations of calculation, the difference between the target and the estimated value decreases (Fig. 19 (b)). In the 10,000th generation (Fig. 19 (c)), tolerance decreases further. These results show that hardware corresponding to the purpose is obtained automatically using this framework.

Figure 20 shows simulation results under other conditions. Nine graphs result for three thermal loads. Three graphs — temperature, PMV and control — are shown for each thermal condition. The controller used under three conditions was obtained after 10,000 generations of calculation. Although thermal load and load change timing were random, the blower and mix door were controlled so that PMV is set to zero.

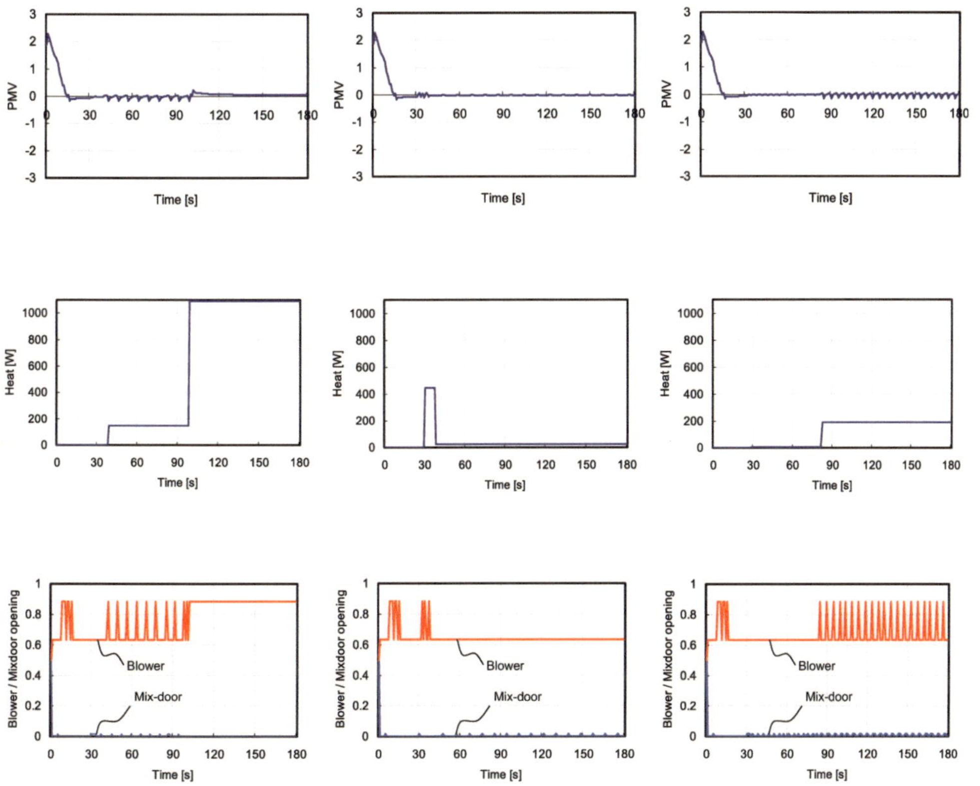

Fig. 20. Simulation results

5. Conclusion

In this chapter, to automate controller design, CPLD was used for controller data-processing and VHDL to describe the logical circuit was optimized using the genetic algorithm. Two example cases (an inverted pendulum and an air-conditioner) were shown and we confirmed this framework was able to be applied to both systems. Since this framework is a generalized framework, so in these kinds of systems which process data from some sensors and drive some actuators, this framework will work functionally.

6. References

Balderdash, G., Nolfi, S., & Parisi, D. (2003). Evolution of Collective Behavior in a Team of Physically Linked Robots, *Proceedings of the Applications of Evolutionary Computing, EvoWorkshops 2003*, pp.581-592, Ti£¡bingen, April 2009.

Barate, R. & Manzanera, A. (2008). Evolving Vision Controllers with a Two-Phase Genetic Programming System Using Imitation, *Proceedings of the 10th international conference on Simulation of Adaptive Behavior: From Animals to Animats*, pp.73-82, ISBN: 978-3-540-69133-4, Osaka, July 2008.

Fanger, P. O. (1970). *Thermal Comfort*, McGraw-Hill.

Goldberg, D. E. (1989). *Genetic Algorithm in Search, Optimization and Machine Learning*, Addison-Wesley.

Hemmi, H., et al., (1997). AdAM: A Hardware Evolutionary System, *Proc. 1997 IEEE Conf. Evolutionary Computat.(ICEC'97)*, pp.193-196.

Higuchi, T. et al. (1992). Evolvable Hardware with Genetic Learning: A First Step Towards Building a Darwin Machine, *Proceedings of the 2nd International Conference on the Simulation of Adaptive Behavior*, MIT Press, pp.417-424.

Kajitani,I. & Higuchi, T. (2005). Developments of Myoelectric Controllers for Hand Prostheses, *Proceedings of the Myoelectric Controls Symposium 2005*, pp.107-111, Fredericton, August 2005.

Kojima, K. et al. (2007). Automatic Generation of VHDL for Control Logic of Air-Conditioning Using Evolutionary Computation, *Journal of Advanced Computational Intelligence and Intelligent Informatics*, Vol.11, No.7, pp.1-8.

Kojima, K. (2009). VHDL Design Automation Using Evolutionary Computation, *Proceedings of 2009 International Symposium on Industrial Electronics (IEEE ISIE 2009)*, pp.353-358, Seoul, July 2009.

Kojima, K. (2011). Emergent Functions of HDL-based Controller of Inverted Pendulum in Consideration for Disturbance, *Proceedings of 2011 IEEE/SICE International Symposium on System Integration*, Kyoto, December 2011.

Koza, J. (1994). *Genetic Programming*, A Bradford Book, ISBN0262111705.

Sakanashi, H. et al., (2004). Evolvable Hardware for Lossless Compression of Very High Resolution Bi-level Images, *IEEE Proceedings-Computers and Digital Techniques*, Vol.151, No.4, pp.277-286, Ibaraki, August 2004.

Tetens, O. (1930). *Uber einige meteorologische Begriffe, Zeitschrift fur Geophysik*, Vol.6, pp.297-309. (in German)

wyns10] Wyns, B., et al., (2010). Evolving Robust Controllers for Corridor Following Using Genetic Programming, *Proceedings of the International Conference on Agents and Artificial Intelligence*, volume 1 : artificial intelligence; pp.443-446, Valencia, January 2010.

Xilinx. (1998). *DS056(v.2.0) XC95144 High Performance CPLD Product Specification*, Xilinx; 2.

19

Applicability of GMDH-Based Abductive Network for Predicting Pile Bearing Capacity

Isah A. Lawal[1] and Tijjani A. Auta[2]
[1]*University of Genoa*
[2]*ATBU, Bauchi*
[1]*Italy*
[2]*Nigeria*

1. Introduction

A pile is a type of foundation commonly used in civil construction. They are made using reinforced concrete and pre tensioned concrete to provide a firmer base where the earth around a structure is not strong enough to support a conventional foundation [Pile, 2011]. Accurate prediction of the ultimate bearing capacity of a structural foundation is very important in civil and construction engineering. Conventional method of estimating the pile bearing capacity has been through pile load test and other in situ test such as standard penetration test and cone penetration test [Bustamante & Gianeselli, 1982]. Though these tests may give useful information about ground conditions, however the soil strength parameters which can be inferred are approximate.

In recent time, advances in geotechnical and soil engineering research have presented more factors that can affect pile ultimate bearing capacity. However, due to nonlinearity of these factors, the use of statistical model analysis and design has proved difficult and impractical [Lee & Lee, 1996]. So there is need to provide civil and structural engineers with intelligent assistance in the decision making process. Soft computing and intelligent data analysis techniques offer a new approach to handle these data overload. They automatically discover patterns in data to provide support for the decision-making process. Tools used for performing such functions include: Artificial Neural Networks (ANN) [Abu- keifa, 1998; Chow et al, 1995; Teh et al, 1997], Support Vector Machine (SVM) [Samui, 2011],Genetic programming [Adarsh, in-press] and Gaussian process regression [Pal & Deswal, 2008, 2010]. The results from using these tools suggest improved performances for various datasets. However, neural networks like other tool suffer from a number of limitations, e.g. long training times, difficulties in determining optimum network topology, and the black box nature with poor explanation facilities which do not appeal to structural engineers [Shahin et al, 2001]. This research work proposes the use of self-organising Group Method of Data Handling (GMDH) [Mehra, 1977] based abductive networks machine learning techniques that has proved effective in a number of similar applications for performing modelling of pile bearing capacity.

Recently, abductive networks have emerged as a powerful tool in pattern recognition, decision support [El-Sayed & Abdel-Aal, 2008], classification and forecasting in many areas [Abdel-Aal, 2005, 2004]. Inspired by promising results obtained in other fields, we explore the use of this approach for the prediction of pile bearing capacity.

2. Related work

In recent years some researchers have developed computational intelligence models for the accurate prediction of pile bearing capacity. In [Lee and Lee, 1996], the authors used back-propagation neural networks to predict the ultimate bearing capacity of piles. A maximum error of prediction not exceeding 20% was obtained with the neural network model developed by using data set generated from calibration chamber. Also, in [Pal,2008], the author investigates the potential of support vector machines based regression approach to model the static pile capacity from dynamic stress-wave data set. The experiments shows excellent correlation coefficient between the predicted and measured values of the static pile capacity investigated. Similarly, in [Samui, 2011], the author studied the potential of Support Vector Machine (SVM) in prediction of bearing capacity of pile from pile load data set. In the study the author introduces ε-insensitive loss function and the sensitivity analysis of the model developed shows that the penetration depth ratio has much effect on the bearing capacity of the pile.

However, in [Pal, 2010], the author took a different approach and investigated the potential of a Gaussian process (GP) regression techniques to predict the load-bearing capacity of piles. The results from the study indicated improved performance by GP regression in comparison to SVM and empirical relations. However, the author reported that despite the encouraging performance of the GP regression approach with the datasets used, it will be difficult to conclude if the method can be used as a sole alternative to the design methods proposed in the literature. The reason been that soft computing based modelling techniques are data-dependent. Their results may change depending on the dataset, the scale at which the experiments are conducted or the number of data available for training.

The potential for GMDH-based abductive network in pile bearing capacity prediction has not been explored before in the literature. However, compared to neural networks and other learning tools, the method offers the advantages of faster model development requiring little user intervention, faster convergence during model synthesis without the problems of getting stuck in local minima, automatic selection of relevant input variables, and automatic configuration of the model structure.

3. GMDH and AIM abductive networks

Abductory Inductive Mechanism (AIM) is a powerful supervised inductive learning tool for automatically synthesizing network models from a database of input and output values [AbTech, 1990]. The model emerging from the AIM synthesis process is a robust and compact transformation implemented as a layered abductive network of feed-forward functional elements as shown in Figure 1. An abductive network model numerical input output relationships through abductive reasoning. As a result, the abductive network can be used effectively as a predictor for estimating the outputs of complex systems [Lee, 1999], as a classifier for handling difficult pattern recognition problems [Lawal et al, 2010] or as a system identifier for determining which inputs are important to the modelling system[Agarwal. 1999]. With the model represented as a hierarchy of polynomial expressions, resulting analytical model relationships can provide insight into the modelled phenomena, highlight contributions of various inputs, and allow comparison with previously used empirical or statistical models.

3.1 Abductive machine learning

The abductive machine learning approach is based on the self-organizing group method of data handling (GMDH) [Fallow, 1984]. The GMDH approach is a proven concept for iterated polynomial regression that can generate polynomial models in effective predictors. The iterative process involves using initially simple regression relationships to derive more accurate representations in the next iteration in an evolutionary manner.

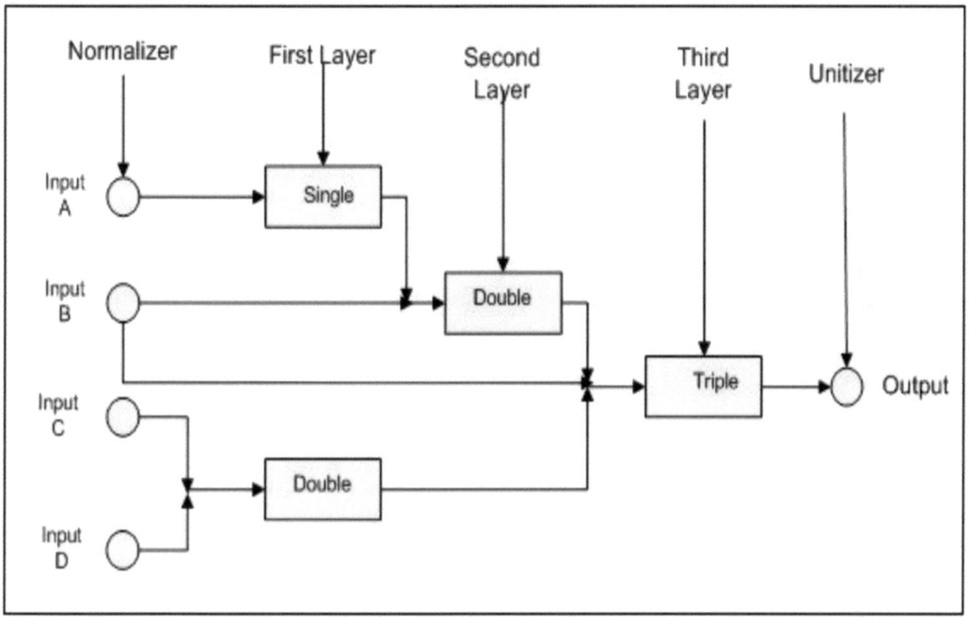

Fig. 1. Abductive network showing various types of functional elements

The algorithm selects the polynomial relationships and the input combinations that minimize the prediction error in each phase. This prevents exponential growth in the polynomial model generated. Iteration is stopped automatically at a point in time that strikes a balance between model complexity for accurate fitting of the training data and model simplicity that enables it to generalise well with new data. In the classical GMDH-based approach abductive network models are constructed by the following 6 steps [Fallow, 1984]:

i. Separating the original data into training data and testing data.

The available dataset are split into training dataset and testing dataset. The training dataset is used for estimating the optimum network model and the testing dataset is used for evaluating the network model obtained on the new data. Usually a 70-30 splitting rule is employed on the original data, but in this work a pre-determined split used by earlier published work was adopted to allow direct comparison of results.

ii. Generating the combinations of the input variables in each layer.

Many combinations of r input variables are generated in each layer. The number of combinations is $p!/((p-r)!\,r!)$. Here, p is the number of input variables and the value of r is usually taking as 2.

iii. Calculating the partial descriptors

For each input combination, a partial descriptor which describes the partial characteristics of the model is calculated by applying regression analysis on the training data. The following second order polynomial regression relationship is usually used

$$y_k = b_0 + b_1 x_i + b_2 x_j + b_3 x_j + b_4 x_i^2 \tag{1}$$

The output variables y_k in Eq. (l) are called intermediate variables.

iv. Selecting optimum descriptors.

The classical GMDH algorithm employs an additional and independent selection data for selection purposes. To prevent exponential growth and limit model complexity, the algorithm selects only relationships having good predicting powers within each phase. The selection criterion is based on root mean squared (RMS) error over the selection data. The intermediate variables which give the smallest root mean squared errors among the generated intermediate variables (y_k) are selected.

v. Iteration

Steps III and IV are iterated where optimum predictors from a model layer are used as inputs to the next layer. At every iteration, the root mean squared error obtained is compared with that of the previous value and the process is continued until the error starts to increase or a prescribed complexity is achieved. An increasing root mean squared error is an indication of the model becoming overly complex, thus over-fitting the training data and will more likely perform poorly in predicting the selection data.

vi. Stopping the multi-layered iterative computation

Iteration is stopped when the new generation regression equations start to have poorer prediction performance than those of the previous generation, at which point the model starts to become overspecialized and, therefore, unlikely to perform well with new data.

Computationally, the resulting GMDH model can be seen as a layered network of partial descriptor polynomials, each layer representing the results of iteration. Therefore, the algorithm has three main elements: representation, selection, and stopping. Figure 2 shows the flow chart of the classical GMDH-based training.

Abductory Inductive Mechanism (AIM) is a later development of the classical GMDH that uses a better stopping criterion that discourages model complexity without requiring a separate subset of selection data. AIM adopts a well-defined automatic stopping criterion that minimizes the predicted square error (PSE) and penalises model complexity to keep the model as simple as possible for best generalization. Thus, the most accurate model that does not overfit the training data is selected and hence a balance is reached between accuracy of the model in representing the training data and its generality which allows it to fit yet unseen new evaluation data.

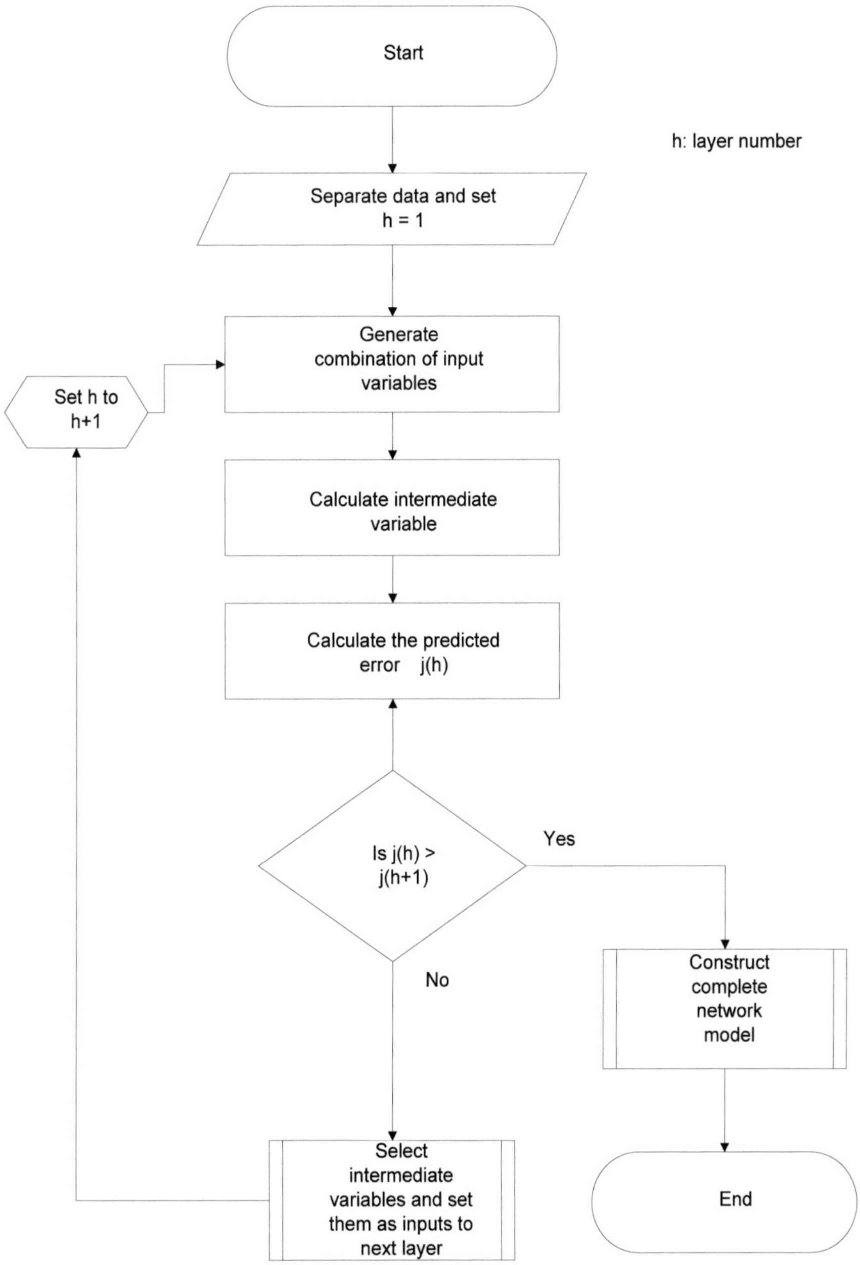

Fig. 2. The flow chart of the classical GMDH-based training.

The PSE consists of two terms [AbTech, 1990]:

$$PSE = FSE + KP \tag{2}$$

Where FSE is the average fitting squared error of the network for fitting the training data and KP is the complexity penalty for the network, expressed as [AbTech, 1990].

$$KP = CPM * (2\sigma^2/N) * K \quad (3)$$

Where CPM is the complexity penalty multiplier, K is the number of coefficients in the network, and σ^2 is a prior estimate of the model error variance. Usually, a complex network has a high fitting accuracy but may not generalize well on new evaluation data unseen previously during training. Training is automatically stopped to ensure a minimum value of the PSE for the CPM parameter used, which has a default value of 1. The user can also control the trade-off between accuracy and generality using the CPM parameter. CPM values greater than 1 will result in less complex models that are more likely to generalise well with unseen data while values less than the default value will result in a more complex models that are likely to over fit training data and produce poor prediction performance. Figure 3 shows the relationship between PSE, FSE and KP, [AbTech, 1990].

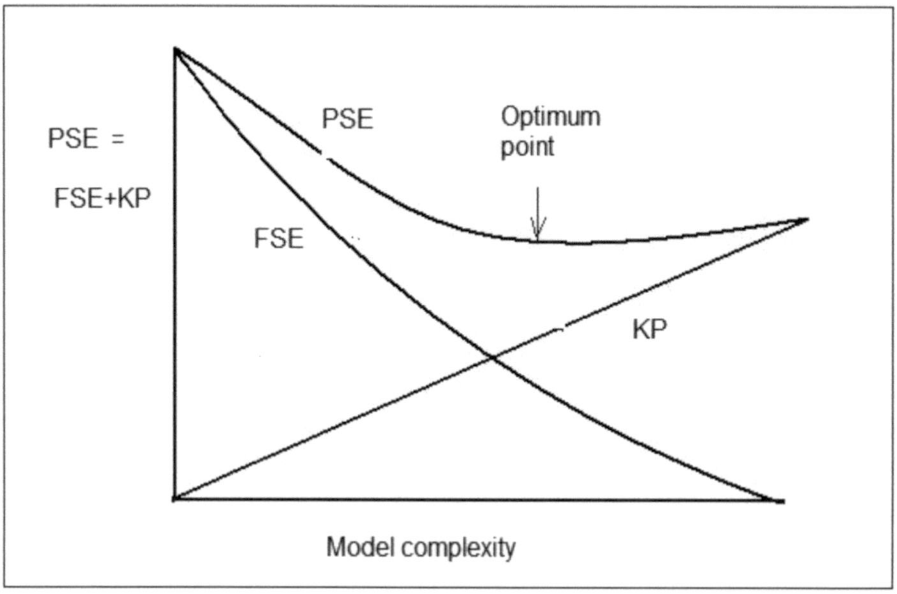

Fig. 3. The predicted square error

3.2 AIM functional elements

The used version of AIM supports several functional elements [AbTech, 1990], see Figure 1, including:

Normaliser: Transforms the original input into a normalized variable having a mean of zero and a variance of unity.

$$y = z_0 + z_1 x \quad (4)$$

Where x is the original input, y is the normalized input, z_0 and z_1 are the coefficients of the normaliser

Unitizer: Converts the range of the network outputs to a range with the mean and variance of the output values used to train the network.

Single Node: The single node only has one input and the polynomial equation is limited to the third degree, i.e.

$$y = z_0 + z_1 + z_2x^2 + z_3x^3 \quad (5)$$

Where x is the input to the node, y is the output of the node and z_0, z_1, z_2 and z_3 are the node coefficients.

Double Node: The double node takes two inputs and the third-degree polynomial equation includes cross term so as to consider the interaction between the two inputs, i.e.

$$y = z_0 + z_1x_i + z_2x_j + z_3x_i^2 + z_4x_j^2 + z_5x_ix_j + z_6x_i^3 + z_7x_j^3 \quad (6)$$

Where x_i, x_j are the inputs to the node, y is the output of the node and z_0, z_1, z_2 ...and z_7 are the node coefficients

Triple Node: Similar to the single and double nodes, the triple node with three inputs has a more complicated polynomial equation allowing the interaction among these inputs.

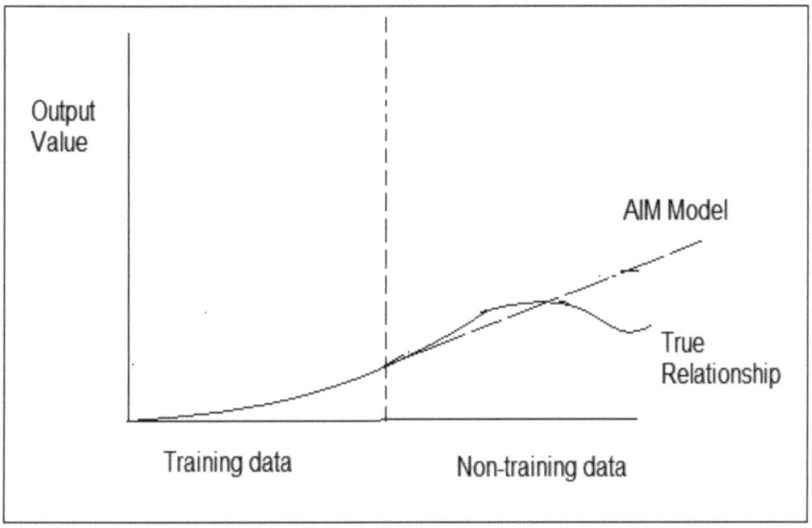

Fig. 4. Non-representative training data

However, not all terms of an element's equation will necessarily appear in a node since AIM will throw away or carve terms that do not contribute significantly to the solution. The eligible inputs for each layer and the network synthesis strategy are defined as a set of rules and heuristics that form an integral part of the model synthesis algorithm as described earlier.

On a final note, any abductive network model is only as good as the training data used to construct it. To build a good model it is important that the training database be representative of the problem space. Figure 4 [AbTech, 1990], illustrates a scenario where the training database used to create the AIM model does not cover an important portion of the problem. Training AIM using only the data to the left of the dotted line will result in a model that generalizes well within the training data range but will be inaccurate in the other region.

4. The dataset and feature discussion

To evaluate and compare the performance of the proposed approach we used the experimental dataset developed in [Lee & Lee, 1996]. The dataset were generated from a calibration chamber, in which field stress conditions were simulated with poorly graded, clean, fine and uniformly layered sand that was dried in air below 2% of water content. The Sand was deposited in the calibration chamber using a method that allows a uniform sand deposit of known relative density to be obtained. The setup was allowed to settle for 24 hours before the model pile was driven into it using a guided steel rod (hammer). The ultimate bearing capacities of the model pile were assumed to be affected by the penetration depth ratio of the model pile, the mean normal stress and the number of blows. So the dataset consist of the following features: penetration depth ratio (i.e. penetration depth of pile/pile diameter), the mean normal stress of the calibration chamber and the number of blows as input and the ultimate capacity (kN) as the output. More detailed description of the dataset generation experiment can be found in [Lee & Lee, 1996].

5. Experiments and results

This section describes the development of abductive networks model for predicting the ultimate bearing capacity using the experimental dataset described in section 3. To allow direct comparison of results, the same splitting used by earlier published work using the dataset was adopted. Two experiments were conducted. In the first experiment, the 28 instances in the dataset were split into a training set of 21 instances and an evaluation of 7 instances. While in the second experiment, 14 instances were selected for training purposes and 14 instances for evaluation. The full training set was used to synthesize an abductive network model with all the 3 features present in the dataset enabled as network inputs. The best model was obtained by adjusting the CPM value. The effect of the CPM values on the models' performance is shown in Table 1. The numbers (e.g. Var_4) indicated at the model input represent the feature selected as input to the model during training, while Var_6 represent the network output corresponding to ultimate pile bearing capacity. It is noted that the model uses 2 inputs out of the 3 inputs which indicates that almost all the features are relevant, with only little redundancy in the feature set

To measure the performance of the trained model after evaluation, two statistical measures namely Root Mean Squared Error (RMSE) and Correlation Coefficient (R^2) were used. A brief description and mathematical formulae are shown below:

5.1 Root Mean-Squared Error

The root mean square error (RMSE)) is a measure of the differences between values predicted by a model and the values actually observed from the phenomenon being modeled or estimated. Since the RMSE is a good measure of accuracy, it is ideal if it is small. This value is computed by taking the square root of the average of the squared differences between each predicted value and its corresponding actual value.

The formula is:

$$RMSE = \sqrt{\frac{[(x_1 - y_1)^2 + (x_2 - y_2)^2 + \cdots + (x_i - y_i)^2]}{n}} \tag{7}$$

Where x_i and y_i are the predicted and actual values respectively while n is the size of the data used.

Model	CPM Values	Number of input features selected during model synthesis	Model Performance with training dataset in experiment 1
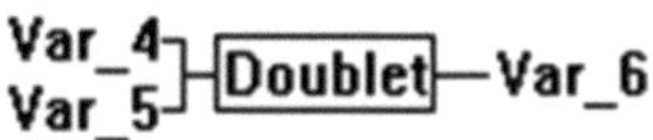	0.5	2 out of the 3 features	RMSE = 34.11 kN Correlation = 0.98
	1	2 out of the 3 features	RMSE = 70.22 kN Correlation = 0.91
	2.5	2 out of the 3 features	RMSE = 70.22 kN Correlation = 0.91

Table 1. The effect of the CPM values on the synthesized model's performance

5.2 Correlation coefficient

A correlation coefficient is measure that determines the degree to which two variable's movements are associated. It gives statistical correlation between predicted and actual values. This coefficient is unique in model evaluations. A higher number means a better model, with a value of one (1) indicating a perfect statistical correlation and a value of zero (0) indicating there is no correlation.

$$Correlation\ Coefficient = \frac{\Sigma(y_a - y_a^1)(y_p - y_p^1)}{\sqrt{[\Sigma(y_a - y_a^1)^2]\,\Sigma(y_p - y_p^1)^2}} \quad (8)$$

Where y_a and y_p are the actual and predicted values while y_a^1 and y_p^1 are the mean of the actual and predicted values.

In experiment 1, the best model with CPM = 0.5 was evaluated using the evaluation sets. A satisfactory agreement between the predicted and measured values of the ultimate pile bearing capacity was obtained, which is shown by the cross plots in Figure 5. The maximum error of prediction was 18%. A Root Mean Squared Error (RMSE) and a correlation coefficient (R^2) of 59.22kN and 0.82 were obtained respectively in the first experiment as shown in Table 2.

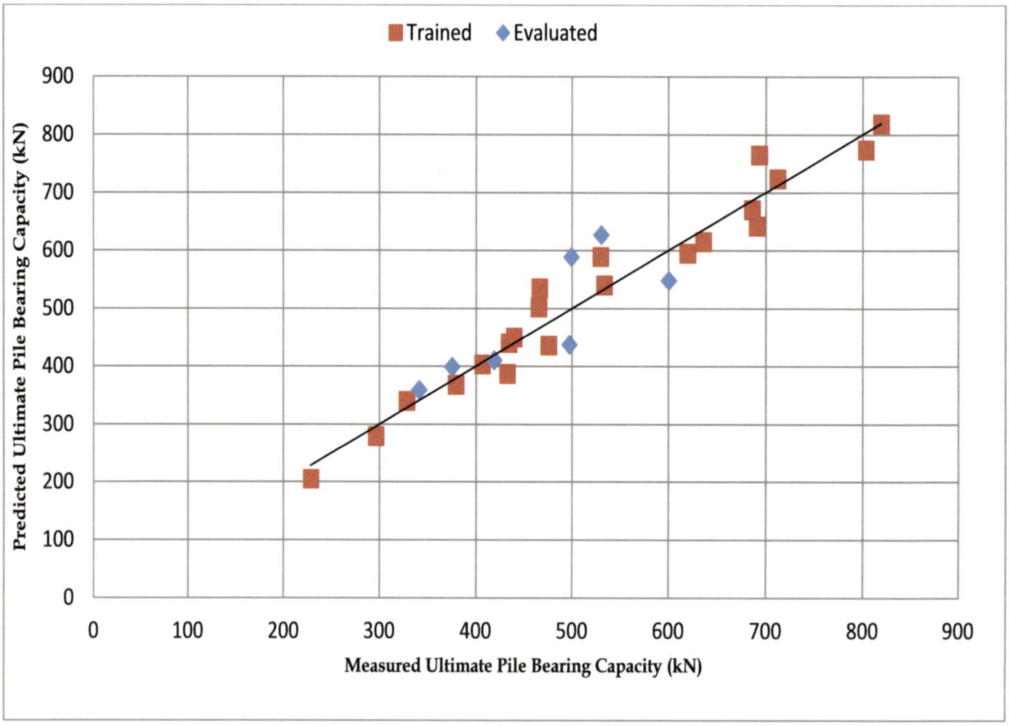

Fig. 5. The cross plot of predicted and measured values of the pile bearing capacity for experiment 1

In the second experiment, after training the abductive network with just 14 instances, the best model was evaluated with the evaluation set of 14 instances. The cross plots in Figure 6 shows the trained and the predicted values of the ultimate bearing capacity. The result showed widely scattered plots, with a RMSE of 92.5kN and Correlation Coefficient of 0.83. The reason for the poor result in this case can be attributed to the small number of the training data which was not enough for the model to learn the entire pattern in the data set. Therefore, it could be concluded that a certain number of training data sets was needed to obtain reasonable predictions.

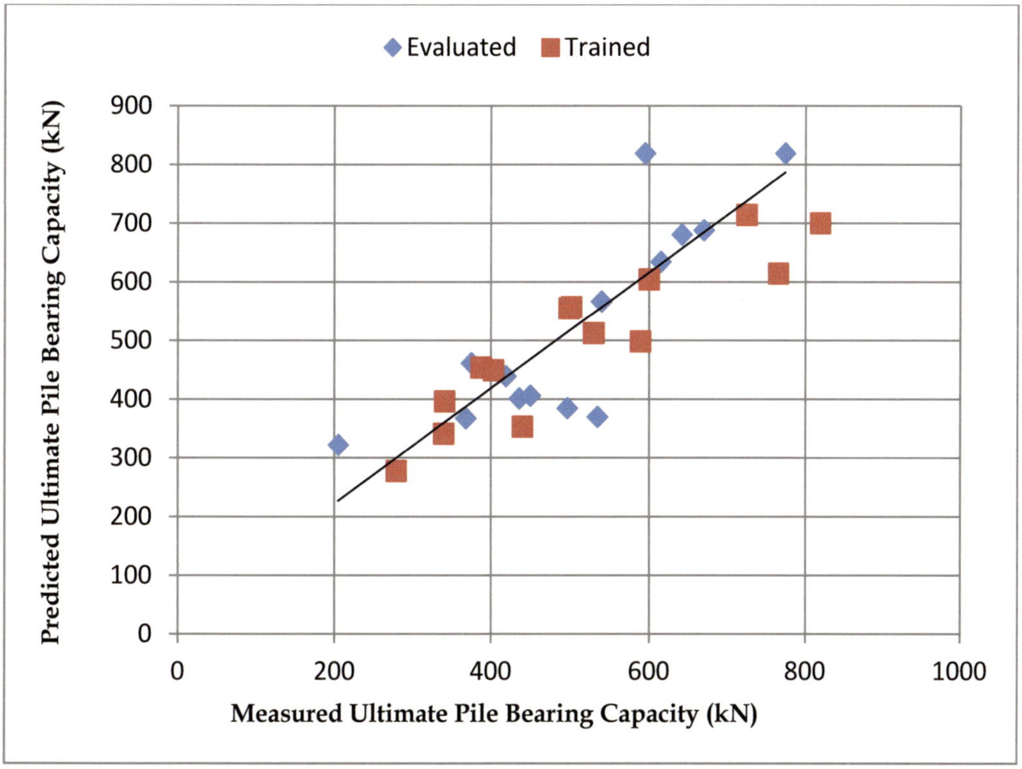

Fig. 6. The cross plot of predicted and measured values of the pile bearing capacity for experiment 2

Finally, the maximum prediction error of other modeling tool used previously on this dataset was compared with that of abductive model synthesized in this work as shown in Figure 5. The comparison indicates that the abductive network model performed much better in terms of prediction error, with almost 9% improvement compared to that obtained with Neural Network reported in [Lee & Lee 1996].

	Experiment 1		Experiment 2	
	Training set (21 instances)	Evaluation set (7 instances)	Training set (14 instances)	Evaluation set (14 instances)
RMSE (kN)	34.11	59.22	70.2	92.5
Correlation(R^2)	0.98	0.82	0.91	0.83

Table 2. Performance measure of the two experiments on the training and test data set

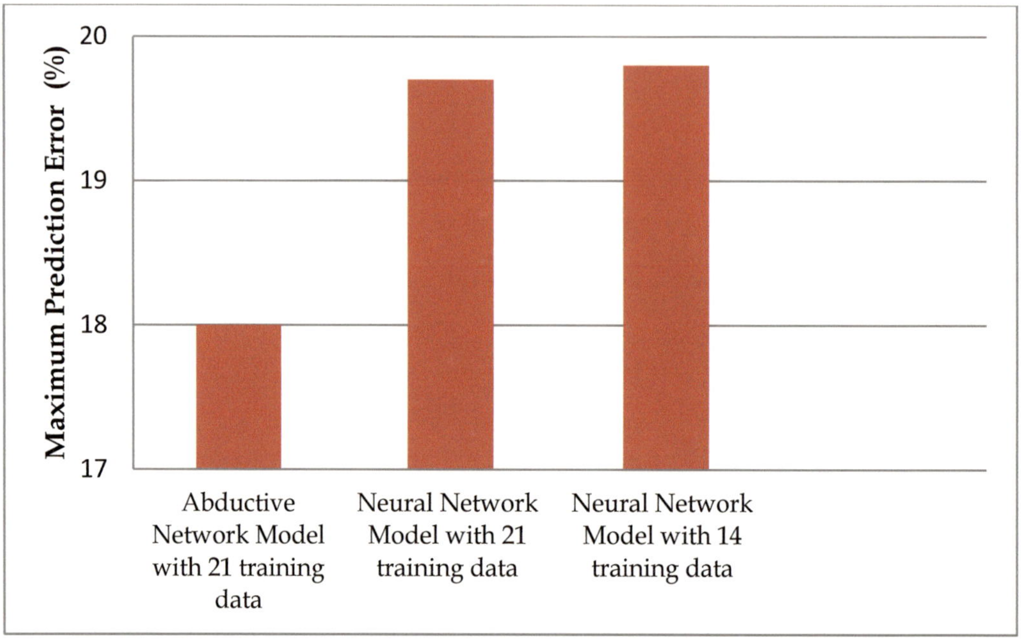

Fig. 7. Comparison of GMDH-based abductive network model with other previous model used on this dataset in term of Maximum Prediction Error (%)

6. Conclusion

This work demonstrates the use of abductive machine learning techniques for the prediction of pile bearing capacity. A RMSE value of 59.22kN and 92.5kN and a correlation coefficient of 0.82 and 0.83 were obtained with respect to the pile bearing capacity values predicted in two separate experiment conducted respectively. An improvement of almost 9% in terms of prediction error was recorded. This result indicated that the proposed abductive network approach yields a better performance compared to the other already implemented technique using the same dataset mentioned in the introduction section. However, the experiments conducted revealed that for a good prediction, a large number of training set is required to train the model before evaluation. So, to validate the performance of the abductive network approach it is recommended that data set obtained from the fields are use in further studies. This will help realize the full potential of abductive network approach in pile bearing capacity prediction.

Meanwhile, the work has outlined the advantages of abductive networks and has placed it in the perspective of geotechnical engineering problem computing point of view. Thus, researchers are encouraged to consider them as valuable alternative modeling tool. Hopefully, future work will consider the possibility of extending the approach to modelling of soil behaviour and site characterization.

7. Acknowledgement

The authors will like to acknowledge Dr. R. E. Abdel-Aal of the Department of Computer Engineering, King Fahd University of Petroleum & Minerals for providing the tool used in this study.

8. References

Abdel-Aal, R.E. (2004). Short-Term Hourly Load Forecasting Using Abductive Networks, *IEEE Transaction on Power Systems*, vol. 19, No.1, pp. 164 -173

Abdel-Aal, R.E. (2005). Improved classification of medical data using Abductive network committees trained on different feature subsets, *Computer Methods and Programs in Biomedicine*, vol. 80, no.2, pp. 141-153

AbTech, Corporation (1990). *AIM user's manual*, Charlottesville, VA

Abu-Kiefa, M. A. (1998). General regression neural networks for driven piles in cohesionless soils, *Journal of Geotechnical & Geoenviromental Engineering*, Vol. 124, No.12,

Adarsh, S.; Dhanya,R. Krishna,G. Merlin,R. & Tina,J. (in press). Prediction of Ultimate Bearing Capacity of Cohesionless soils using Soft Computing Techniques, *Journal of ISRN Artificial Intelligence*.

Agarwal, A. (1999). Abductive networks for two-group classification: A comparison with neural networks. *The Journal of Applied Business Research*, Vol. 15, No. 2, pp. 1–12.

Bustamante, M. & Gianeselli, L. (1982). Pile bearing capacity prediction by means of static penetrometer CPT. *Proceeding of the second European symposium on penetration testing*, Amsterdam, May

Chow, Y.K; Chan, W.T. Liu, I.F. & Lee, S.L. (1995). Predication of pile capacity from stress wave measurements: a neural network approach, *International Journal of Numerical & Analytical Methods in Geomechanics*, Vol. 19, pp. 107–126

El-Sayed, E.M., Abdel-Aal, R.E. (2008). Spam Filtering with Abductive Networks, *International Joint Conference on Neural Networks*, Hong Kong

Farlow, S. J. (1984). The GMDH algorithm. In *Self-organizing methods in modeling: GMDH type algorithms*, S. J. Farlow (Ed.), pp. 1–24, New York: Marcel-Dekker

Lawal, I. A.; Abdel-Aal, R. E. & Mahmoud, S. A. (2010). Recognition of Handwritten Arabic (Indian) Numerals Using Freeman's Chain Codes and Abductive Network Classifiers, *IEEE 20th International Conference on Pattern Recognition*, Istanbul, pp.1884-1887

Lee, B. Y; Liu, H. S. & Tarng, Y. S. (1999). An Abductive Network for Predicting Tool Life in Drilling, *IEEE Transactions on Industry Applications*, Vol. 35, No. 1, January

Lee, I. & Lee, J. (1996). Prediction of Pile Bearing Capacity Using Artificial Neural Networks, *Journal of Computers and Geotechnics*, Vol. 18, No. 3, pp. 189-200,

Mehra, R.K. (1977). Group Method for Data handling (GMDH): Review and Experience, *Proceedings of the IEEE Conference on Decision and Control*, New Orleans, pp. 29-34

Pal, M. & Deswal, S. (2008). Modeling Pile Capacity Using Support Vector Machines and Generalized Regression Neural Network, *Journal of Geotechnical and Geoenviromental Engineering*, Vol.134, No. 1021 pp. 134-7

Pal, M. & Deswal, S. (2010). Modeling pile capacity using Gaussian process regression, *Journal of Computers and Geotechnics*, Vol. 37, pp. 942–947,

Pile, (2011).In *Encyclopeadia Britannica*, Retrieved from http://www.britannica.com/EBchecked/topic/460377/pile pp. 1177-1185

Samui, P. (2011). Prediction of pile bearing capacity using support vector machine, *International Journal of Geotechnical Engineering*, pp. 95-102,

Shahin, M. A.; Jaksa, M. B. & Maier, H. R. (2001) Artificial Neural Network Applications In Geotechnical Engineering, *Australian Geomechanics*, Vol. 36, No.1, pp. 49-62

Teh, C.I; Wong, K.S. Goh, A.T.C. & Jaritngam, S. (1997). Prediction of pile capacity using neural networks, *Journal of Computing Civil Engineering* Vol. 11, No.2, pp. 129-38.

20

An End-to-End Framework for Designing Networked Control Systems

Alie El-Din Mady and Gregory Provan
Cork Complex Systems Lab (CCSL), Computer Science Department
University College Cork (UCC), Cork
Ireland

1. Introduction

Designing a control system over Wireless Sensor/Actuator Network (WSAN) devices increases the coupling of many aspects, and the need for a sound discipline for writing/designing embedded software becomes more apparent. Such a WSAN-based control architecture is called a Networked Control System (NCS). At present, many frameworks support some steps of the NCS design flow, however there is no end-to-end solution that considers the tight integration of hardware, software and physical environment. This chapter aims to develop a fully integrated end-to-end framework for designing an NCS, from system modelling to embedded control-code generation. This framework aims to generate embedded control code that preserves the modelled system properties, and observes the hardware/software constraints of the targeted platform.

Existing approaches for control code design typically ignore the embedded system constraints, leading to a number of potential problems. For example, network delays and packet losses can compromise the quality of control that is achievable Mady & Provan (2011). Designing embedded control that accounts for embedded system constraints requires dealing with heterogeneous components that contain hardware, software and physical environments. These components are so tightly integrated that it is impossible to identify whether behavioural attributes are the result of computations, physical laws, or both working together.

Contemporary embedded control systems are modelled using hybrid system Henzinger (1996) that captures continuous aspects (e.g. physical environment) and discrete-event behaviour (e.g. control decision). Even though many tools support model-based code generation (e.g. Simulink), the emphasis has been performance-related optimizations, and many issues relevant to correctness are not satisfactorily addressed, including: (a) the precise relationship between the model and the generated code is rarely specified or formalized; (b) the generated code targets a specific embedded platform and cannot be generalized to multi-targeted platforms, moreover there is no generator considers the embedded WSAN; (c) the generated code does not respect the targeted platform hardware/software constraints; (d) the continuous blocks are either ignored, or discredited before code generation. Therefore, the correspondence between the model and the embedded code is lost.

This chapter proposes a framework for designing embedded control that explicitly accounts for embedded system constraints. We develop an end-to-end framework for designing an NCS where we model the system using a hybrid systems language. We focus on adopting a distributed control strategy that explicitly considers hardware/software constraints. Our approach enables us to generate code for multiple embedded platforms.

Fig. 1 shows an overview of the framework. The framework consists of four design stages, outlining the role for each stage as follows: (i) the reference model captures the control/diagnosis[1] strategies and the physical plant using a hybrid system; (ii) the control model is a projection of the source model, in which the physical/continuous aspects are abstracted/transformed to a generic control specification; (iii) the embedded control model combines the control model and the hardware/software constraints to define an embedded control model satisfying the platform constraints; (iv) the target embedded model, that captures the embedded platform code, is generated from the control model considering the hardware/software constraints.

Consequently, the reference and control models abstract the hardware/software constrains effect on the control algorithm. These constrains can be classified to: (a) processing resources constrains; and (b) memory space constrains. The processing resources constrains consider the hardware/software factors that affect the algorithm execution time (e.g., CPU speed), which can lead to incompatibility between the processing capacity and the algorithm execution. Whereas, the memory space constrains check if there is enough memory space for executing the control algorithm.

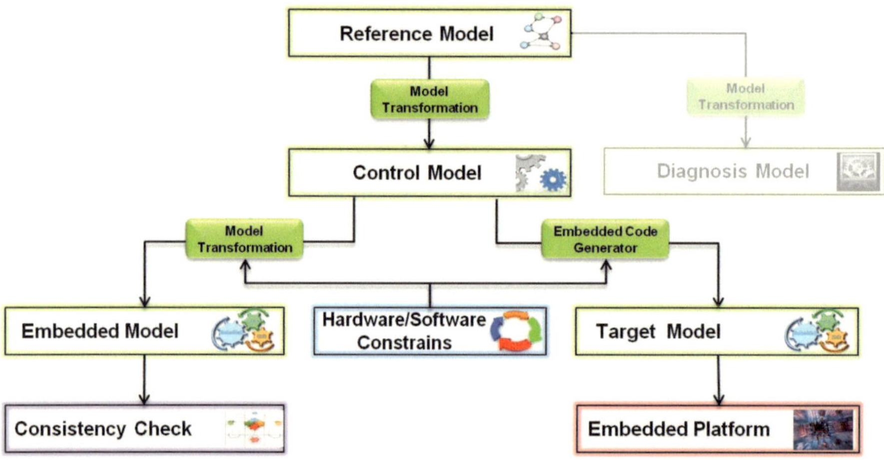

Fig. 1. Framework Overview

An NCS is widely used in many applications, such as habitat monitoring, object tracking, fire detection and modern building control systems. In particular, Building Automation Systems (BAS) often uses a large wireless/wired sensor network. BAS is selected as our application domain as it is considered as a cornerstone application for decreasing energy consumption; around 40% of total energy use in the West is consumed in the industrial building sector, which accounts for nearly one-third of greenhouse gas emissions. Of this figure, approximately

[1] In this chapter, we consider only the control aspects in the reference model.

one-third can be attributed to Heating, Ventilation and Air-Conditioning (HVAC) systems present in buildings. In this chapter, we consider an Air Handling Unit (AHU) control system as a case study for the framework development. We apply the framework design stages on this case starting from the reference model to the embedded control code.

We assume that we are given as input a set $\mathfrak{R}$ of top-level system requirements, for example user comfort requirements that any set-point will differ from operating value by less or equal to 10%. We provide empirical guarantees that the requirements $\mathfrak{R}$ are met by the generated models and the corresponding transformation rules, by empirically evaluating the system property for each generated model (i.e. reference, control, target model). If the system property is respected for each design stage model, then the models are transformed correctly.

Our contributions in this chapter will be as follows:

- we formulate an end-to-end framework for designing a network control system. This framework preserves the modelled system's properties under hardware/software constraints;
- we identify the transformation rules between the framework design stages;
- we formulate a typical AHU control system as a case-study of the framework;
- we empirically check that our framework preserves the system's requirements $\mathfrak{R}$ by applying it to an Air Handling Unit (AHU) model as a case-study.

The remainder of the chapter is organized as follows: Section 2 provides a survey covering the related work and discusses our contribution comparing to the state-of-the-art. The framework architecture and the description for each design stage modelling are discussed in Section 3. The model transformation rules between the design stages are explained in Section 4. The application domain for the case-study are highlighted in Section 5, and its experiments design is shown in Section 6. We end in Section 7 by giving a discussion of our work and outlining future perspectives.

2. Related work

Modelling Frameworks: Modeling languages define a representation method for expressing system design. Given the heterogeneity of engineering design tasks, modelling languages consequently cover a wide range of approaches, from informal graphical notations (e.g., the object modelling technique (OMT) Rumbaugh et al. (1991)), to formal textual languages (e.g., Alloy for software modelling Jackson (2002)).

Semantic meta-modelling is a way to uniformly abstract away model specificities while consolidating model commonalities in the semantics meta-model. This meta-modelling results in a mechanism to analyze and design complex systems without renouncing the properties of the system components. Meta-modelling enables the comparison of different models, provides the mathematical machinery to prove design properties, and supports platform-based design.

An abstract semantics provides an abstraction of the system model that can be refined into any model of interest Lee & Sangiovanni-Vincentelli (1998). One important semantic meta-model framework is the *tagged signal model* (TSM) Lee & Sangiovanni-Vincentelli (1998), which can compare system models and derive new ones.

There are several prior studies on the translational semantics approach. For example, Chen et al. (2005) use the approach to define the semantic anchoring to well-established formal models (such as finite state machines, data flow, and discrete event systems) built upon AsmL Gurevich et al. (2005). Further, they use the transformation language GME/GReAT (Graph Rewriting And Transformation language) Balasubramanian et al. (2006). This work, through its well-defined sets of semantic units, provides a basis for similar work in semantic anchoring that enables for future (conventional) anchoring efforts.

Tools Several embedded systems design tools that use a component-based approach have been developed, e.g., MetaH Vestal (1996), ModelHx Hardebolle & Boulanger (2008), Model-Integrated Computing (MIC) Sztipanovits & Karsai (1997), Ptolemy Lee et al. (2003), and Metropolis Balarin et al. (2003). These tools provide functionality analogous to the well-known engineering tool MATLAB/Simulink. In particular, Metropolis and Ptolemy II are based on semantic metamodelling, and hence obtain the entailed abstract semantics (and related abstract metamodels) of the approach. In these tools, all models conforming to the operational versions of the TSM's abstract semantics also conform to the TSM's abstract semantics. One drawback of these tools is that "components" can be assembled only in the supporting tool. As a consequence, different systems and components must all be developed in the same environment (tool) to stay compatible. However, the most recent version of Metropolis, Metropolis II, can integrate foreign tools and heterogeneous descriptions.

One tool that is closely related to our approach is the Behaviour Interaction Priority (BIP) tool Basu et al. (2006). BIP can combine model components displaying heterogeneous interactions for generating code for robotics embedded applications. BIP components are described using three layers, denoting behaviour, component connections and interaction priorities. Our approach focuses more on the higher-level aspects, in that it uses two levels of meta-model (i.e., meta-model and meta-meta-model) to define all underlying specifications. Moreover, our approach considers the consistency check between the system model and the hardware/software constraints for the embedded platform.

To our knowledge, our approach is unique in its use of two levels of meta-models, a single centralized reference model with a hybrid systems semantics, and its generation of embeddable code directly from the centralized meta-model considering hardware/software constrains.

3. Modelling framework architecture

In this section, we provide a global description for modelling framework. In addition, each design stage of our framework is formulated.

3.1 Modeling objectives

Our objective is to abstract the essential properties of the control generation process so that we can automate the process. It is clear that the process has two quite different types of inputs:

Control constraints The control-theoretic aspects concern sequences of actions and their effects on the plant. These constraints cover order of execution, times for actions to be executed, and notions of forbidden states, etc. Note that these constraints assume infinite computational power to actually compute and effect to stated control actions.

Embedded System constraints These aspects concern the capabilities of the hardware and software platforms, and are independent of the applications being executed on the platform.

It is clear that both property types are needed. Because of the significant different between the two types, we must use different constraint representations for each type. As a consequence, we then must use a two-step process to enforce each constraint type. This is reflected in the fact that we have a two-step model-generation process. Step 1 maps from a reference model ϕ_R to a control model ϕ_C using mapping rules $\mathcal{R}_C$; the second step maps ϕ_C to ϕ_E using mapping rules $\mathcal{R}_E$. If we represent this transformation process such that $\phi_C = f(\phi_R, \mathcal{R}_C)$, and $\phi_E = f(\phi_C, \mathcal{R}_E)$, then we must have the full process represented by $\phi_E = f(f(\phi_R, \mathcal{R}_C), \mathcal{R}_E)$.

3.2 Meta-meta-model formal definition

The meta-meta-model is formulated using a typical hierarchal component-based modelling Denckla & Mosterman (2005), as shown in Fig. 2. We can define a meta-meta-model Γ as $\Gamma = \langle C, Y \rangle$. C represents a set model components, where control components C_C, plant components C_P and building-use components C_B are C instances, i.e., $C \in \{C_C, C_P, C_B\}$. The connection between the components C are described by Y.

Each component $c \in C$ is represented as $c = \langle P_{io}, \phi, C \rangle$, where P_{io}[2] is a set of component input/output ports and ϕ describes the relation between the input and output ports. Moreover, c can contain a set of components C to represent hierarchical component levels. Relation ϕ can be expressed based on the use (application domain) of the framework. In this article, we consider four meta-model instances of ϕ: $\phi \in \{\phi_R, \phi_C, \phi_E, \phi_T\}$, where ϕ_R is the reference meta-model, ϕ_C is the control meta-model, ϕ_E is the embedded meta-model and ϕ_T is the target embedded meta-model. Each of these meta-models captures the DSML abstraction for the framework design stage. In this case, for each design stage we can use any Domain-Specific Modelling Language (DSML) that can be represented using the corresponding meta-model. In this article we have selected the DMSLs that support designing NCS for BAS system. However, this design methodology can be adapted to match any other application domain.

3.3 Reference meta-model formal definition

In the context of BAS modelling, Hybrid Systems (HS) are used to create our reference model ϕ_R. We have used HS to present BAS models as it captures both discrete (e.g. presence detection) and continuous (e.g. heat dissipation) dynamics, where the continuous dynamic is represented using algebraic/differential equation. Hence, Linear Hybrid Automata (LHA) become a suitable HS candidate for this model.

[2] In our framework, we assume that the used ports are unidirectional ports.

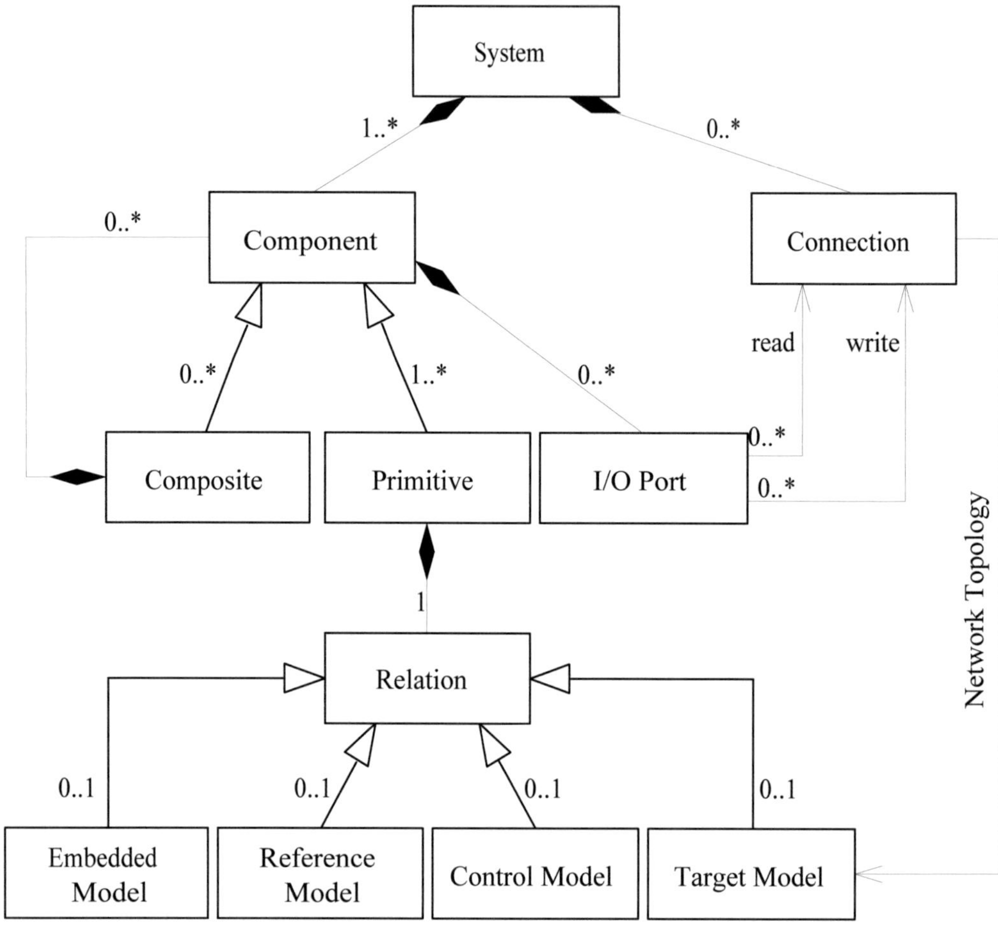

Fig. 2. Framework Meta-Meta-Model

The reference meta-model is considered as one instance of the component relation $\phi = \phi_R$. As shown in Fig. 3, we represent ϕ_R as a standard LHA ($\phi_R = H$) representation, as described in Def. 3.1.

Definition 3.1. ***[Linear Hybrid Automata]*** *A linear hybrid automaton H is a 3-tuple, i.e. $H = \langle V, M_s, E \rangle$, with the following structural extensions:*

- *V is a finite set of component binding variables. It is used to bind/connect the component ports P_{io} with ϕ. V is represented as following: $\bar{V} = \{\bar{v}_1, ..., \bar{v}_n\}$ for real-valued variables, where n is the dimension of H. $\dot{V} = \{\dot{v}_1, ..., \dot{v}_n\}$ represents the first derivatives during continuous change. $\acute{V} = \{\acute{v}_1, ..., \acute{v}_n\}$ represents values at the conclusion of discrete change.*
- *M_s is a set of hierarchical system-level modes that describe the system statuses. $m_s \in M_s$ captures the system-level status using either H or single mode m, i.e., $m_s \in \{H, m\}$. m can be diagnosis mode m_d or control mode m_c, i.e. $m \in \{m_d, m_c\}$. We assume that m_c controls nominal system behaviour and does not consider fault modes (e.g., fault actuation). Two vertex functions assigned to each mode $m \in M$ or $m_s \in M_s$. Invariant ($inv(m)$) condition is a predicate whose free variables*

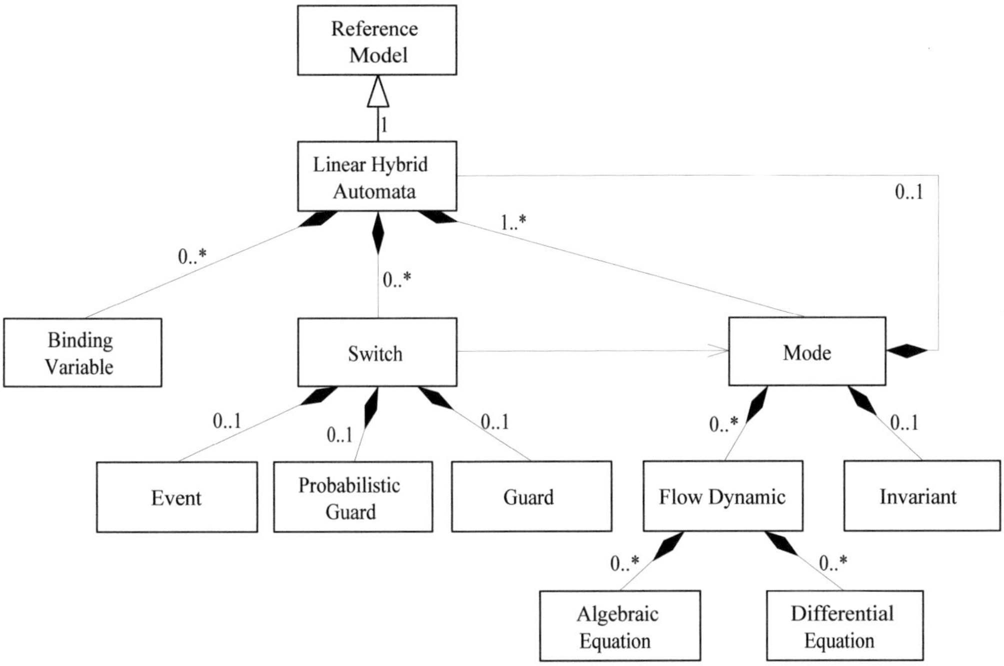

Fig. 3. Meta-Model for the Framework Reference Model

are from V and flow dynamic $flow(m)$ *is described using algebraic equation* $flow_{alg}(m)$ *and/or differential equation* $flow_{dif}(m)$ *for* $\bar{V} \cup \dot{V}$ *binding variables.*

- *E is set of switches (edges). An edge labelling function j that assigns to each switch* $e \in E$ *a predicate. E is also assigned to a set* Σ *of events, where event* $\sigma \in \Sigma$ *is executed if the corresponding predicate j is true, where j can be presented using probabilistic or deterministic predicate. Each jump/guard condition* $j \in J$ *is a predicate whose free variables are from* $\bar{V} \cup \acute{V}$*. For example,* $e(m_i, m_l)$ *is a switch that moves from mode* m_i *to* m_l *under a guard j then executes* σ*, i.e.,* $e(m_i, m_l) : m_i \xrightarrow{j/\sigma} m_l$.

3.4 Control meta-model formal definition

The control meta-model ϕ_C aims to capture the control components C_C in a discrete behaviour. Therefore, Fig. 4 shows a Finite-State Machine (FSM) F that used to describe the components relation $\phi = \phi_C$. In Def. 3.2, we formally define the control meta-model using FSM, i.e., $\phi_C = F$.

Definition 3.2. ***[Finite-State Machine]*** *A finite-state machine F is a 3-tuple, i.e.* $H = \langle V, S, T \rangle$*, with the following structural extensions:*

- *V is a finite set of component binding variables. It is used to bind/connect the component ports* P_{io} *with* ϕ*. V is represented as discrete change only.*
- *S is a set of states that used to identify the execution position.*
- *T is set of transitions to move from one state s to another. Similar to E in LHA, a set* Σ *of events (actions) and jump/guard condition J are assigned to T. For example, a transition* $t(s_i, s_l) \in T$ *is*

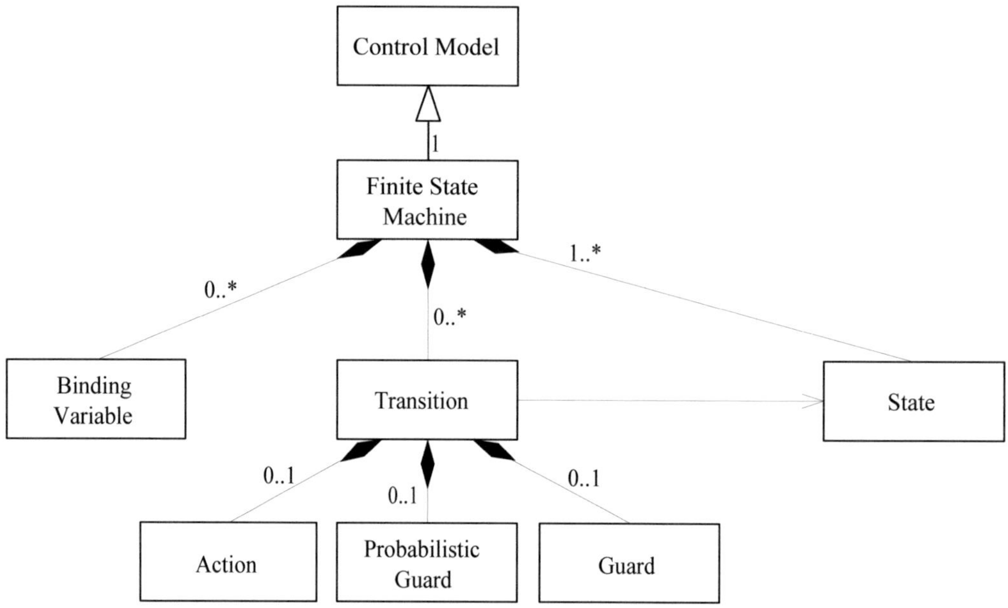

Fig. 4. Meta-Model for the Framework Control Model

used to reflect the move from state s_i to s_l under a guard j (i.e., deterministic or probabilistic) then executing σ, i.e., $t(s_i, s_l) : s_i \xrightarrow{j/\sigma} s_l$.

3.5 Embedded meta-model formal definition

In order to check the consistency between our control model and the hardware/software constrains, we identify the embedded model following its corresponding meta-model ϕ_E shown in Fig. 5. One standard modelling language that can present our embedded model is Analysis and Design Language (AADL)[3] (i.e., SAE Standard). However, any other language/tool that can capture ϕ_E elements can be used in the hardware/software consistency check.

Definition 3.3. ***[Embedded Model]*** *An embedded model ϕ_E is a 5-tuple, i.e. $\phi_E = \langle Mem, Pro, Bus, Thr, Dev \rangle$, with the following structural extensions:*

- *Mem is a RAM memory identification used in the embedded platform during the control algorithm execution. This memory contains 3-tuple used to identify the memory specification, i.e., $Mem = \langle Spc, Wrd, Prt \rangle$, where Spc identifies the memory space size, Wrd identifies the memory word size, and Prt identifies the memory communication protocol (i.e., read, write, read/write).*
- *Pro is a processor identification used in the embedded platform to execute the control algorithm. In ϕ_E, the Pro specification is identified using the processor clock period Clk, i.e., $Pro = \langle Clk \rangle$.*
- *Bus is a bus identification used to communicate between different hardware components, such as Mem and Pro. This Bus contains 2-tuple to identify its specification such as the bus latency Lcy and massage size Msg, i.e., $Bus = \langle Lcy, Msg \rangle$.*

[3] http://www.aadl.info/

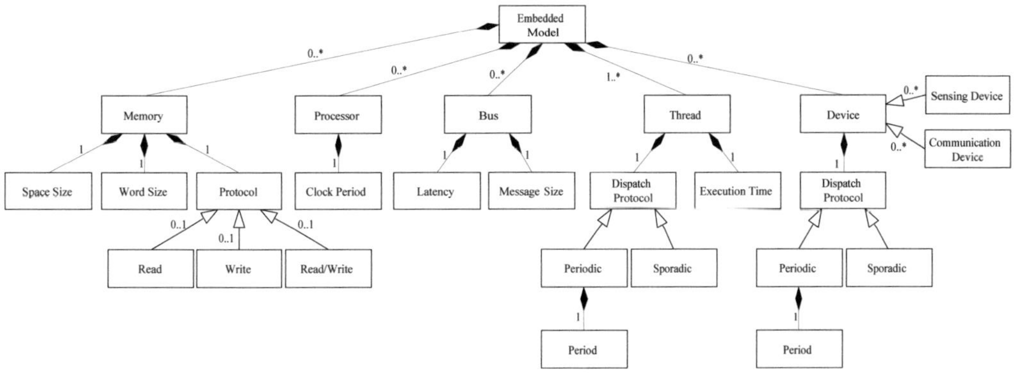

Fig. 5. Meta-Model for the Framework Embedded Model

- *Thr is a thread identification used to run/execute the control algorithm over Pro. This Thr contains a 2-tuple to identify its specification, such as the dispatch protocol Dprt and execution time Tex, i.e., Thr* $= \langle Dprt, Tex \rangle$. *The Dprt can be either sporadic or periodic. In case of periodic dispatch protocol, we need to identify the period for the thread switching.*
- *Dev is a device identification used to sense/actuate the physical plant, or activate (sending or receiving) the wireless communication devices. In Dev, we identify dispatch protocol Dprt to trigger Dev, i.e., Dev* $= \langle Dprt \rangle$. *The Dprt can be either sporadic or periodic. In case of periodic dispatch protocol, we need to identify the Dev activating period.*

We can deduce that the hardware/software constraints needed for the consistency check are:

1. Memory space size.
2. Memory word size.
3. Memory communication protocol.
4. Processor clock period (CPU).
5. Bus latency.
6. Bus massage size.
7. Each thread execution time.
8. Each thread switching protocol.
9. Device activation/fire protocol.

The embedded model evaluates:

1. The binding consistency between the hardware and software description, considering the hardware/software constrains.
2. The processing capacity, i.e., how much the described processor is loaded with the execution for the described software. This evaluation factor must be less than 100%, otherwise the designer has to increase the processing resources (e.g., number of the processors).

3.6 Target embedded meta-model formal definition

The target embedded model presents the embedded code that will be deployed on the embedded platform. Fig. 6 depicts the meta-model ϕ_T for the target embedded model. This ϕ_T captures the primitive building elements that needed to describe an embedded code. Consequently, we can use any embedded language by identifying the corresponding semantic for each ϕ_T element. In this article, we focus on embedded Java for Sun-SPOT[4] sensors and base-station.

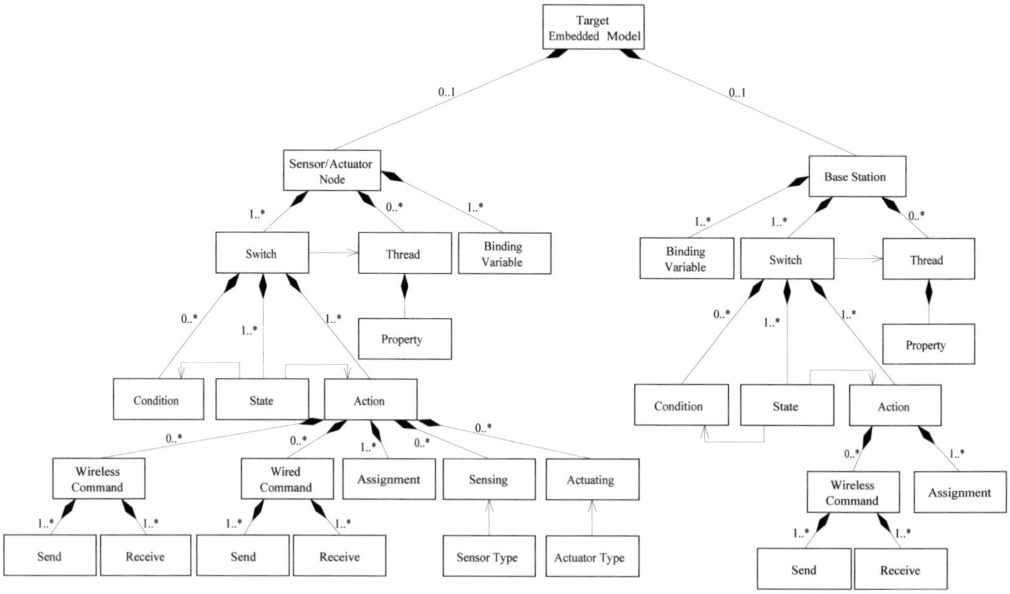

Fig. 6. Meta-Model for the Framework Targeted Model

The target embedded meta-model can formally be described as in Def. 3.4.

Definition 3.4. ***[Target Embedded Model]*** *An embedded model ϕ_T is a 2-tuple, i.e. $\phi_T = \langle N_S, N_B \rangle$, where $N_S = \langle T, Thr, V \rangle$ and $N_B = \langle T, Thr, V \rangle$ are the semantic definitions needed in case of sensor node and base-station deployment, respectively, with the following structural extensions:*

- *T is a set of switch semantic definition. Considering a switch $t \in T$, $t = \langle j, s, \sigma \rangle$, where j is the predicate that allows executing the action σ, and s the a Boolean variable to identify the execution position. State s is included in the semantic of j and σ, as following:*

```
if (state && condition){
    action;
    assignment(s);
}
```

The action σ semantic can contain wireless/wired commands, typical assignment, and/or sensing/actuation commands. However, at this stage some hardware/software constrains are needed

[4] http://www.sunspotworld.com/

in order to identify the sensor/actuator type and the sensors ID, which identified from the network topology. For example, in case of sending a wireless data from sensor node, the semantic is as follows:

```
try {
        SendDg.reset();
        SendDg.writeInt(NodeID);
        SendDg.writeDouble(DataToBeSent);
        SendConn.send(SendDg);
}
catch (Exception e) {
                        throw e;
}
```

- *The thread Thr is used to identify the semantics of a thread as follows.*[5]

```
public class Thread_name extends Thread{
      public void run(){
            while(true/SynchCond){
                  Switches semantics;
            }
      }
}

Thread_name.sleep(period);
```

- *V is a finite set of binding variables that can be identified as Integer, Double, Boolean.*

Regarding the base-station node N_B, it has the same aforementioned description with less actions. Typically, N_B is used to communicate wirelessly between the sensors and the actuator, therefore its action contains the wireless command and assignment semantics only.

4. Model transformation

In this section, we provide the transformation rules R used to transform an instance ϕ of the source relation to the corresponding instance of the target representation. As mentioned before, the developed framework aims to generate control code, therefore R considers the control components C_C (i.e., sensor, actuator, controller, base-station). The connections Y and Ports P_{io} that identified in the meta-meta-model remain the same. However, each model has a corresponding representation for Y and P_{io}. For example, in the reference model the components can be connected through a variable, whereas in the target model the connection can be a wireless channel.

[5] If there is only one thread under execution, the thread doesn't have a critical role in this case. Moreover, *Thr* has a period property to be identified.

4.1 Reference model to control model transformation

The transformation rules R_C transforms a ϕ_R model to a ϕ_C model. For simplifying the transformation rules R_C, we have used the same symbols for the unchanged terms between ϕ_R and ϕ_C. Using R_C, the control model can be generated from the reference model as $\phi_C = f(\phi_R, R_C)$, where R_C abstracts the flow dynamics, invariants, and hierarchy in ϕ_R. In this section, we have highlighted the transformation rules as follows:

- Binding variables V in H are transformed to an equivalent format in F, i.e., $V^H \xrightarrow{R_C} V^F$. In this case, the continuous change variables in H will be updated discretely in F.
- The control modes M_c in H are transformed to states S in F, i.e., $M_c^H \xrightarrow{R_C} S^F$. In this case, the states S capture the positions of M_c without its operations (e.g. invariant, flow dynamics). This transformation assumes that there is no fault control modes, therefore the diagnosis modes M_d (presenting faulty modes) is not transformed.
- The switches E in H (E^H) are transformed to transitions T in F, i.e., $E^H \xrightarrow{R_C} T^F$.
- The invariant $inv(m)$ and flow dynamics equations $flow(m)$ for each mode in H are transformed in F to a transition from/to the same state, i.e., $t(s,s) : s \xrightarrow{j/\sigma} s$. The transition guard j is triggered if the mode invariant is true. In order to avoid activating multiple transitions at the same time, we add (logical and) to j the set $\bar{j}$ of the inverse guards for the transitions moving out from state s, i.e., $j = inv(m) \wedge \bar{j}$, where $\bar{j}$ is the guards of the transitions leaving s, i.e., $t(s,*)$. This multiple transitions activation can happen if the invariant and any other transition guard are true, which is acceptable in H but not in F. The flow dynamics equations $flow(m)$ are transformed to the transition action σ, i.e., $flow(m) \xrightarrow{R_C} \sigma$. However, differential equations $flow_{dif}(m)$ in the form of $\dot{x(t)} = f$ are approximated to discrete behaviuor using Euler's method, i.e., $x(k+h) = x(k) + fh$, where k is the step number and h is the model resolution that reflects the time step (as much h is small, as the approximation is accurate).

  These rules can be summarised as: $flow(m), inv(m) \xrightarrow{R_C} t(s,s) : s \xrightarrow{inv(m) \wedge \bar{j} / flow(m)} s$.

The aforementioned rules do not consider the hierarchical feature in the system-level modes M_s. Given a system-level mode $m_s \in M_s$ that contains a set of inherited control modes M_c, i.e., $m_s \preceq M_c$. An initial control mode m_{init} is identified in M_c, m_{init} is the first mode under execution whenever m_s is triggered. The transformation of these hierarchical modes is as follows:

- The set of control modes M_c in m_s is transformed to a set of states S_c. Consequently, m_{init} is transformed to s_{init}.
- The union of: (a) the set of switches E_c^H that enters m_{init}, i.e., $E_c^H(*, m_{init})$; and (b) the set of the switches E_s^H that enters m_s, i.e., $E_s^H(*, m_s)$ is transformed to a set of transitions T_c in F that enters the initial state s_{init}, called $T_c^F(*, s_{init})$. Therefore, $E_c^H(*, m_{init}) \cup E_s^H(*, m_s) \xrightarrow{R_C} T_c^F(*, s_{init})$.

- The set of switches E_c^H that enters $m_c \in M_c$, i.e., $E_c^H(*, m_c)$ is transformed to a set of transitions T_c in F that enters $s_c \in S_c \setminus s_{init}$[6], i.e., $T_c^F(*, s_c)$. Therefore, $E_c^H(*, m_c) \xrightarrow{R_C} T_c^F(*, s_c)$.
- The set of switches E_c^H that exits m_c, i.e., $E_c^H(m_c, *)$, is transformed to a set of transitions T_c in F that exits $s_c \in S_c$, called $T_c^F(s_c, *)$. Therefore, $E_c^H(m_c, *) \xrightarrow{R_C} T_c^F(s_c, *)$. However, in order to give a higher priority to the system transitions in case of activating control and system transitions at the same time, we introduce the inverse of the condition for the system transition to J of $T_c^F(s_c, *)$, i.e., $J(T_c^F(s_c, *)) = J(E_c^H(m_c, *)) \wedge \overline{J(E_s^H(m_c, *))}$.
- The set of the switches E_s^H that exits m_s, i.e., $E_s^H(m_s, *)$ is transformed to be added to the set of transitions T_c in F that exits $s_c \in S_c$. Therefore, $E_s^H(m_s, *) \xrightarrow{R_C} T_c^F(s_c, *)$.
- The invariants $inv(m_c)$ and $inv(m_s)$, and flow dynamics equations $flow(m_c)$ and $flow(m_s)$ for m_c and m_s, respectively, are transformed in F to a transition from/to the same state, i.e., $t(s_c, s_c) : s_c \xrightarrow{j/\sigma} s_c$. The transition guard j is triggered if the mode invariant for m_c or m_s, and $\bar{J}$ (explained earlier) are true, i.e., $j = (inv(m_s) \vee inv(m_c)) \wedge \bar{J}$. The flow dynamics equations $flow(m_c)$ and $flow(m_s)$ are assigned to $t(s_c, s_c)$ action σ, i.e., $flow(m_c) \uplus flow(m_s) \xrightarrow{R_C} \sigma$.

4.2 Control model to embedded model transformation

The embedded model is performed from the control model as $\phi_E = \phi_C \bigotimes R_E$ and can be performed from the reference model as $\phi_E = \phi_R \bigotimes (R_C \bigoplus R_E)$, where the transformation rules R_E considers the hardware/software constrains and the hardware/software architecture.

As mentioned before, the control model ϕ_C focuses on the control components C_C. In a BAS system, C_C can be classified to a set of sensor component instances C_C^S, actuator component instances C_C^A and processing/controller component instances C_C^P (which execute the control algorithm), i.e., $C_C \in \{C_C^S, C_C^A, C_C^P\}$. Based on the type of the C_C instance, ϕ_E creates the corresponding hardware architecture, e.g., Fig. 7, Fig. 8, and Fig. 9 show a sensor component architecture, an actuator component architecture, and a processing/controller component architecture, where each architecture contains the appropriate *Pro*, *Mem*, *Bus*, *Dev*. As shown in Fig. 5, each of these hardware components has its own hardware/software constrains.

In order to add the software impact to this architecture, we consider a transformation rule R_E from control model ϕ_C. Typically, each control model ϕ_C contains only one F, therefore F transformed to a thread in ϕ_E, i.e., $F \xrightarrow{R_E} Thr^E$, where Thr^E captures the algorithm execution constrains.

The multi-threading appears in case if a control component C_C contains several inherited components. In this case, each component is transformed to a thread and all the transformed threads will be executed under the same processor.

[6] This rule does not consider s_{init}

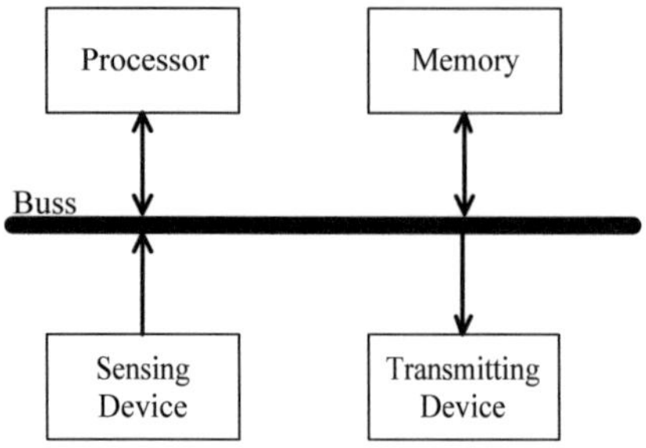

Fig. 7. Sensor Hardware Architecture

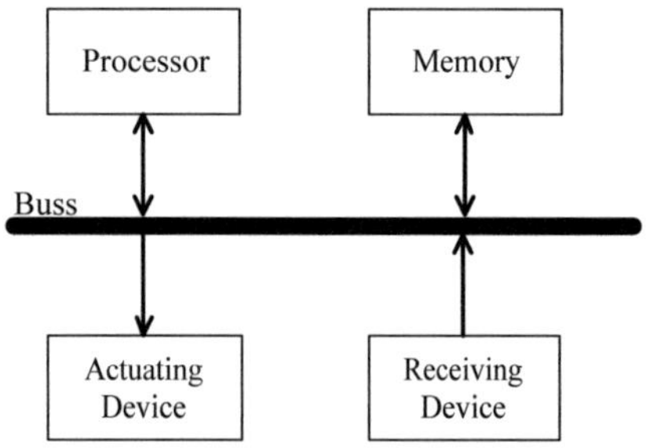

Fig. 8. Actuator Hardware Architecture

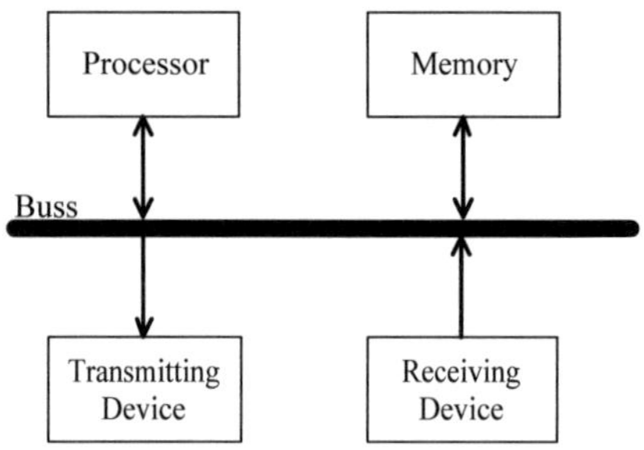

Fig. 9. Controller Hardware Architecture

4.3 Control model to target embedded model transformation

The target embedded model is performed from the control model as $\phi_T = \phi_C \otimes R_T$ and can be performed from the reference model as $\phi_T = \phi_R \otimes (R_C \oplus R_T)$. The transformation rules R_T considers only the hardware/software constrain, whereas R_E considers both hardware/software constrains and architecture. The R_T considers only constrains as ϕ_T directly uses the hardware architecture on the targeted platform.

In this section, we show the transformation rules R_T between the control model ϕ_C and the target embedded model ϕ_T, as follows:

- The set of binding variables V of ϕ_C (called V^C) is transformed to an equivalent set of variables of corresponding type (e.g., Double, Integer, Boolean) in the binding variables V^T of ϕ_T, i.e., $V^C \xrightarrow{R_T} V^T$.

- The set of transitions T in ϕ_C (called T^C) is transformed to a set of switches T^T of ϕ_T, i.e., $T^C \xrightarrow{R_T} T^T$.
- The set of guards J, including probabilistic and deterministic, of T^C is transformed to a set of conditions J of T^T, i.e., $J(T^C) \xrightarrow{R_T} J(T^T)$.
- The set of actions Σ of ϕ_C is transformed to a set of actions Σ of ϕ_T, i.e., $\Sigma(T^C) \xrightarrow{R_T} \Sigma(T^T)$. However, additional P_{io} constrains of N_B and N_S are needed to R_T for identifying the P_{io} specification. For example, for a sensor node N_B one port should read from a sensor, this port name should be identified and the corresponding sensor type. Therefore, whenever an action reads from this port, a sensor reading action is added. The same is done for the ports read/write from/to a wireless network, and these ports are identified with the connected ID node, where this process identifies the network topology.
- The set of states S of ϕ_C is transformed to a set of states (i.e., Boolean variables) in ϕ_T, i.e., $S^C \xrightarrow{R_T} S^T$. In addition, the set of transitions $T^T(s^T, *)$ that exiting $s^T \in S^T$ includes s^T in its conditions and $s^T = false$ in its actions. The set of transitions $T^T(*, s^T)$ that enter s^T includes $s^T = true$ in its actions. This transformation rule should be followed in the specified order, which means starting from the transitions exiting the states and then applying the ones entering the states.

5. Application domain

Heating, Ventilating, and Air-Conditioning (HVAC) systems provide a specified ambient environment for occupants with comfortable temperature, humidity, etc. One way to regulate the temperature in a room is Air Handling Unit (AHU), it is a set of devices used to condition and circulate air as part of an HVAC system. Several control strategies have been introduced to control the temperature regulation, where an operating scheduling is pre-defined. In this context, standard PI control algorithms are adequate for the control of HVAC processes Dounis & Caraiscos (2009). Therefore, we consider an AHU model used to regulate the temperature for a single room, as out framework case-study. The AHU heating/cooling coils are controlled using a PI algorithm, where the switch between cooling and heating coil is performed using a system level decision.

We describe in this section the system specification for the used model and evaluation metrics for the system property. Moreover, we apply the end-to-end transformation process for the system reference model.

5.1 System evaluation metric

In this case-study, we regulate the temperature (for maintaining thermal comfort) with respect to user discomfort. The expected discomfort metric penalizes the difference between the measured indoor air temperature $y(k)$ at time-step k (where k varies from k_s to k_f), and the reference temperature $r(k)$. We use the Root Mean Square Error (RMSE) to reflect the indoor temperature variation around the reference temperature in ^oC, during only the scheduling period $p(k) = 1$:

$$DI = \sqrt{\sum_{k=k_s}^{k_f} \frac{(y(k) - r(k))^2}{k_f} p(k)} \quad (1)$$

In order to validate the transformation rules between the models, each generated model[7] has to respect the system property $DI \leq 2\ ^oC$.

5.2 System specification

As shown in Fig. 10, an AHU uses coils to heat or cool the indoor temperature. The heated/cooled air is pumped to the room using a supply-air fan. In order to use the heated/cooled indoor air more efficiently, the AHU recycles some of the return air via an air loop, which has a return-air fan to mix the indoor air with the outdoor air, as controlled by three dampers. In our case-study, we assume fixed settings for the fan speeds and the damper settings, and we control only the valve settings for the coils.

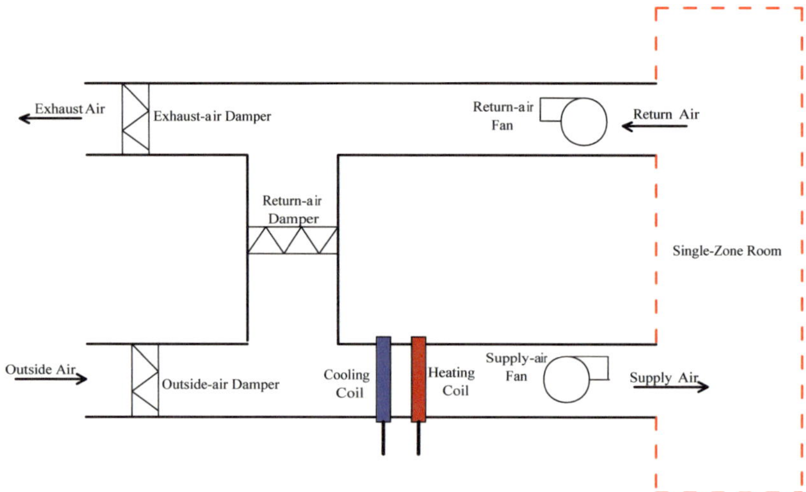

Fig. 10. AHU Structure

Fig. 11 shows the reference model that used as our framework case-study. This model contains two main groups of components: (a) environment/plant components reflect the plant physics (e.g., walls, coils); (b) control components show the control/sensing algorithm that used to modify/monitor the environment.

5.2.1 Environment components

All models described below are lumped-parameter models. Two variables are identified to evaluate the model behaviour: external temperature T_{ext} and set-point temperature T_{sp}. Moreover, five environment/plant components have been used: Wall, Window, Heating/Cooling Coil, Return Air and Indoor Air models, as follows:

[7] This rule will not be applied to embedded model ϕ_E, as this model used to validate the hardware/software consistency

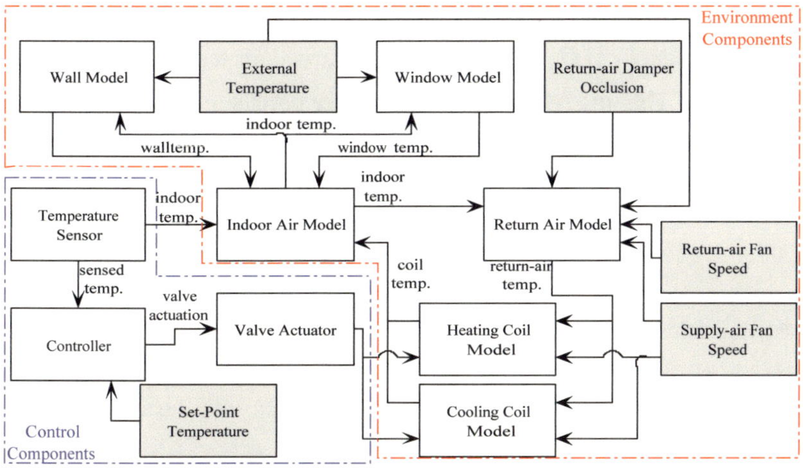

Fig. 11. AHU Component-Based Model

1. *Wall Model*: One of the room walls is facing the building façade, which implies heat exchanges between the outdoor and indoor environments. In general, a wall can be modelled using several layers, where greater fidelity is obtained with increased layers in the wall model. In our case, four layers have been considered to reflect sufficient fidelity Yu & van Paassen (2004), using the following differential equation; Eq. 2:

$$\rho_{wall} V_{wall} c_{wall} \frac{dT_{wall}}{dt} = \alpha_{wall} A_{wall} (T_{ext} - T_{wall}) \tag{2}$$

ρ_{wall} is the wall density $[kg/m^3]$, V_{wall} is the wall geometric volume $[m^3]$, c_{wall} is the wall specific heat capacity $[J/kg.K]$, T_{wall} is the wall temperature $[^oC]$, α_{wall} is the wall thermal conductance $[W/(m^2.K)]$, A_{wall} is the wall geometric area $[m^2]$ and T_{ext} is the outdoor temperature $[^oC]$.

2. *Heating/Cooling Coil Model*: The coil model uses the temperature difference between the water-in and water-out in order to heat/cool the room. In this case, the temperature is controlled through the radiator water flow using the valve occlusion (called actuation variable u). Moreover the coil exchanges temperature with its environment, such as the indoor air temperature and returned air temperature (considering the supply fan speed) as shown in equations: Eq. 3 and Eq. 4.

$$M_{wtr} c_{wtr} \frac{dT_{coil}}{dt} = m^{\cdot}_{wtr} c_{wtr} (T_{wtrin} - T_{wtrout}) - Q \tag{3}$$

$$Q = \alpha_{air} A_{coil} (T_{coil} - T_{air}) + m^{\cdot}_{supfan} c_{air} (T_{coil} - T_{retair}) \tag{4}$$

Where, M_{wtr} is water mass $[kg]$, c_{wtr} is water specific heat capacity $[J/kg.K]$, T_{coil} is coil temperature $[^oC]$, $m^{\cdot}_{wtr}$ is water mass flow rate throw the coil valve $[kg/s]$, T_{wtrin} is water temperature going to the coil $[^oC]$, T_{wtrout} is water temperature leaving from the coil $[^oC]$, α_{air} is air thermal conductance $[W/(m^2.K)]$, A_{coil} is coil geometric area $[m^2]$, T_{air} is indoor air temperature $[^oC]$, $m^{\cdot}_{supfan}$ is air mass flow rate throw the supply-air fan $[kg/s]$, c_{air} is air specific heat capacity $[J/kg.K]$, and T_{reair} is returned air temperature $[^oC]$.

3. *Return Air Model*: This model acts as an air mixer between the indoor air flow and the external air based on return-air damper occlusion ε, as shown in equations: Eq. 5.

$$T_{retair} = T_{air} + \varepsilon\left(\frac{m^{\cdot}_{retfan}}{m^{\cdot}_{supfan}}\right)(T_{air} - T_{ext}) \tag{5}$$

Where, $m^{\cdot}_{retfan}$ is air mass flow rate throw the return-air fan $[kg/s]$.

4. *Indoor Air Model*: In order to model the indoor temperature T_{air} propagation, all HVAC components have to be considered as they exchange heat with the air inside the controlled room following equation 6 .

$$\rho_{air} V_{air} c_{air} \frac{dT_{air}}{dt} = Q_{air} + Q_{wall} + Q_{window} \tag{6}$$

$$Q_{wall} = \alpha_{air} A_{wall}(T_{wall} - T_{air}) \tag{7}$$

$$Q_{air} = m^{\cdot}_{supfan} c_{air}(T_{coil} - T_{air}) \tag{8}$$

$$Q_{window} = \alpha_{air} A_{window}(T_{window} - T_{air}) \tag{9}$$

Where, ρ_{air} is air density $[kg/m^3]$, V_{air} is air geometric volume $[m^3]$, A_{window} is window geometric area $[m^2]$, and T_{window} is window temperature $[^oC]$.

5. *Window Model*: A window has been modelled to calculate the effects of glass on the indoor environment. Since the glass capacity is very small, the window has been modelled as an algebraic equation, Eq. 10, that calculates the heat transfer at the window node.

$$\alpha_{air}(T_{ext} - T_{window}) + \alpha_{air}(T_{air} - T_{window}) = 0 \tag{10}$$

5.2.2 Control components

A temperature sensor, PI-controller, and valve actuator are used to monitor, control and actuate the environment model, respectively. The temperature sensor samples the indoor temperature (T_s=1 min) and sends the sampled value to the controller. In case the sampled value within the operating time slot (identified by the operating schedule), then the PI-controller calculates the next actuation value and sends it to the valve actuator in order to adjust the coil valve occlusion.

The reference model for the control algorithm has two levels of hierarchy, as shown in Fig. 12. The high level represents the system modes of the system decisions, where the heating mode is activated if T_{ext} is greater than T_{sp}, and the cooling mode is activated if T_{sp} is greater than T_{ext}. In these system modes, a PI-control algorithm is used to adjust the valve occlusion for the corresponding coil and deactivate the other coil. Consequently, the controller updates the actuation value each sampling period implemented using a continuous variable to present the controller timer.

In this document, we consider the controller component as an example to apply the end-to-end transformation rules. Fig. 12 shows the reference model for the controller component with its transitions, actions, invariants and flow dynamics. However, we have used the same design approach to design the reference model for the sensor and actuator components.

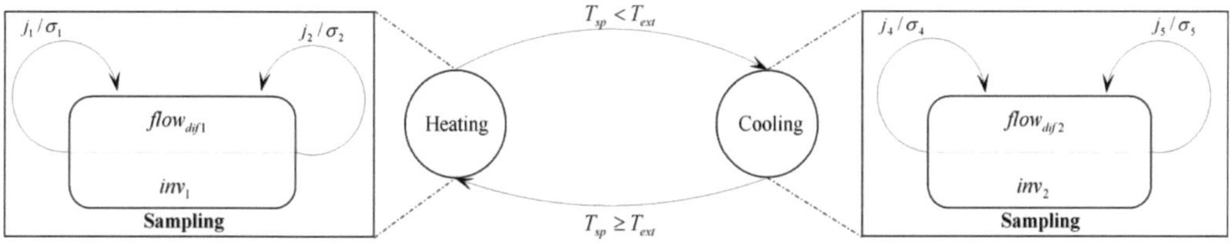

$$
\begin{aligned}
j_1 &: ControlTime > SamplingPeriod \wedge Schedule \\
\sigma_1 &: ControlTime = 0 \\
&: u_{heating} = f_{PI}(T_{sp}, T_{air}) \\
&: u_{cooling} = 0 \\
j_2 &: ControlTime > SamplingPeriod \wedge \overline{Schedule} \\
\sigma_2 &: ControlTime = 0 \\
&: u_{heating} = 0 \\
&: u_{cooling} = 0 \\
flow_{dif1} &: \frac{dControlTime}{dt} = 1 \\
inv_1 &: ControlTime \leq SamplingPeriod
\end{aligned}
$$

Fig. 12. Reference model for the controller component, where the high level automata describes the system level modes M_C and the inherited level represents the control modes M_C.

Where, $ControlTime$ is a continuous variable used to reflect the controller time, $SamplingPeriod$ is the sampling period needed to receive the sensor sample and then update the actuator, $Schedule$ is a Boolean variable to indentify if the operation schedule is triggered or not, $u_{heating}$ is the valve occlusion value for the heating-coil, f_{PI} is the heating PI optimization function, and $u_{cooling}$ is the valve occlusion value for the cooling-coil.

The cooling system mode uses similar LHA as in heating. However, it actuates on $u_{cooling}$ using $\grave{f}_{PI}$ instead of $u_{heating}$, where $\grave{f}_{PI}$ is the cooling PI optimization function.

5.3 Case-study control model

We have followed the transformation rules R_C in Sec. 4.1 to transform the controller reference model in Fig. 12 to the controller control model in Fig. 13 with the corresponding transition conditions J and actions Σ.

We can deduce that system modes hierarchy has been removed by adding the complement of its transition conditions to the conditions in the control model transitions. Moreover, the mode dynamic flow and invariant are transformed to $t_3(j_3, \sigma_3)$.

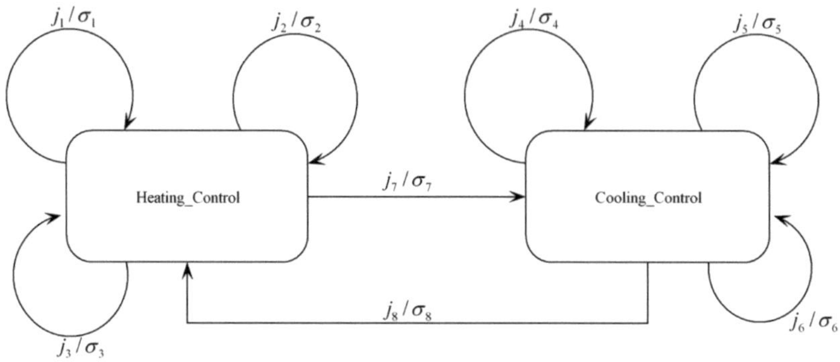

$$
\begin{aligned}
j_1 &: ControlTime > SamplingPeriod \wedge Schedule \wedge \overline{T_{ext} > T_{sp}} \\
\sigma_1 &: ControlTime = 0 \\
&: u_{heating} = f_{PI}(T_{sp}, T_{air}) \\
&: u_{cooling} = 0 \\
j_2 &: ControlTime > SamplingPeriod \wedge \overline{Schedule} \wedge \overline{T_{ext} > T_{sp}} \\
\sigma_2 &: ControlTime = 0 \\
&: u_{heating} = 0 \\
&: u_{cooling} = 0 \\
j_3 &: ControlTime \leq SamplingPeriod \wedge \overline{T_{ext} > T_{sp}} \\
\sigma_3 &: ControlTime = ControlTime + h \\
j_7 &: T_{ext} > T_{sp} \\
\sigma_7 &: ControlTime = 0
\end{aligned}
$$

Fig. 13. Control model for the controller component, where the system mode hierarchy, continuous variables are abstracted.

The rest of the control model transitions follow the same conditions and actions as the aforementioned transitions, but for the cooling-coil.

5.4 Case-study embedded model

In our case-study, we create the controller component hardware architecture for the embedded model as in Fig. 9 following R_E in Sec. 4.2. We have emulated the embedded model using Sun-SPOTs with the following hardware specifications: (a) CPU=180MHz, 32 bit ARM 920T Processor, and (b) RAM=512 Kbyte. Therefore, hardware/software constraints captured by the embedded model as follows:

1. Memory space size: $Spc(Mem) = 512kbyte$.
2. Memory word size: $Spc(Wrd) = 32bit$.

3. Memory communication protocol: $Spc(Prt) = read/write$.
4. Processor clock period (CPU): $Clk(Pro) = 1/180M$.
5. Bus latency: $Lcy(Bus) = 11ns$.
6. Bus massage size: $Msg(Bus) = 32bit$.
7. Thread execution time: $Tex(Thr) = 5 - 10ms$. In this example, we have considered only one thread as the controller component is a primitive component. However, in case of a composite component, it is translated to multi-thread.
8. Thread switching protocol: $Dprt(Thr) = periodic$ with switch period $20ms$. This constrain is not critical for our case-study as the processor execute only one thread.
9. Sending/Receiving Communication Device activation/fire protocol: $Dprt(Dev) = periodic$ with a period equals to the network sampling rate T_s.

5.5 Case-study target embedded model

In the target embedded model, we follow R_T in Sec. 4.3 to generate the embedded language semantic. In our case-study we have used embedded Java for Sun-SPOT devices. In the controller component, $u_{heating}$, $u_{cooling}$ and T_{air} are identified as the component wireless P_{io}, which are corresponding to u_heating, u_cooling, and IndoorTemp variables in Fig. 14, respectively. These P_{io} identifications are transformed to wireless communication commands in case of reading/writing from/to P_{io}. Moreover, the sending/receiving IDs are identified for each port in order to structure the network topology, and consequently these IDs are included in the wireless commands. For example, Fig. 14 shows the first transition transformation from the control model, where $sensorID$ is the sensor ID that sends T_{air} ($IndoorTemp$) to the controller and the controller ID is identified as $hostID$.

6. Experimental design

In order to check if the developed framework preserves the systems properties, we run a set of experiments for the AHU system as a case-study at each model (i.e., reference, control, and target). Then, we evaluate if each model respects the identified system property $\mathfrak{R} = DI \leq 2^oC$.

In our case-study, we vary external temperature and the set point temperature (T_{ext}, T_{sp}) and then evaluate the corresponding DI. For example, if we have an input 2-tuple $(15, 23)$ to present (T_{ext}, T_{sp}), we will measure outputs $DI = 0.73^oC$.

6.1 Dependent/independent variables

In our case-study, we evaluate each model over the domains of the independent variables as follows:

1. External Temperature $T_{ext} \in \{5, 10, 15, 20\}$;
2. Set-point Temperature $T_{sp} \in \{18, 20, 23\}$;

where some constraints are identified for the variables' search space to eliminate the physically unrealizable solutions as $\mid T_{ext} - T_{sp} \mid \leq T_{max}$, $T_{max} = 13^oC$ is the maximum allowable temperature difference and $T_{ext} \neq T_{sp}$.

```
if (Heating_Control && ControlTime > SamplingPeriod && Schedule
      && !(ExternalTemp > OptimalTemp)){

   ControlTime = 0.0;
   //Receiving indoor temperature from sensor with ID: sensorID.
   try {
      RecConn.receive(RecDg);
      recAddr = RecDg.getAddress();
      sensorID=RecDg.readInt();
      if (ID=sensorID) IndoorTemp = RecDg.readDouble();
   } catch (Exception e) {
      throw e;
   }
   //Updating the heating actuation value using PI algorithm.
   u_heating=heatingPIOptimization(OptimalTemp,IndoorTemp);
   //Sending the heating actuation value packet to the actuator
   //with the base station ID (hostID).
   try {
      recAddr = RecDg.getAddress();
      SendDg.reset();
      SendDg.writeInt(hostID);
      SendDg.writeDouble(u_heating);
      SendConn.send(SendDg);
   } catch (Exception e) {
      throw e;
   }
   //Deactivating the cooling-coil.
   u_cooling=0.0;
   //Sending the actuation value packet to the actuator
   //with the base station ID (hostID).
   try {
      recAddr = RecDg.getAddress();
      SendDg.reset();
      SendDg.writeInt(hostID);
      SendDg.writeDouble(u_cooling);
      SendConn.send(SendDg);
   } catch (Exception e) {
      throw e;
   }
   //Update the execution state.
   Heating_Control=true;
   Cooling_Control=false;
}
```

Fig. 14. Target embedded model of the first transition t_1 of the controller component.

The dependent variable for each model is the average of DI over the independent variable cross-product space given by $T_{ext} \times T_{sp}$.

6.2 Empirical results

Fig. 15, 16 and 17 show the DI experiments for ϕ_R, ϕ_C, and ϕ_T. The reference and control models are empirically evaluated using the CHARON tool [8], whereas the target embedded model is emulated using Sun-SPOT nodes, with the following hardware specifications: (a) CPU=180MHz, 32 bit ARM 920T Processor, and (b) RAM=512 Kbyte. The embedded model is captured using the AADL language and then evaluated using the OSATE tool[9].

Figs. 15, 16 and 17 show that the DI decreases when the difference between T_{ext} and T_{sp} decreases, as the initial overshoot the settling time decreases. Moreover, we see that control model gives the worst DI values, the reference model improves the DI and the target model gives the best DI values. Because of the model resolution h for updating the dynamic flow $(flow_{dif}, flow_{alg})$ decreases in case of the target model (in range of msec) in comparison to the reference mode (in range of sec), and it increases for the control model (in range of 3 sec) over the reference model. However, the average of DI for all models respects the system property $DI \leq 2^oC$, where $DI = 0.95, 2, 0.44^oC$ for ϕ_R, ϕ_C, and ϕ_T, respectively.

We have evaluated the binding consistency and the processing capacity for the embedded model ϕ_E to show a consistency binding and 50% processing capacity.

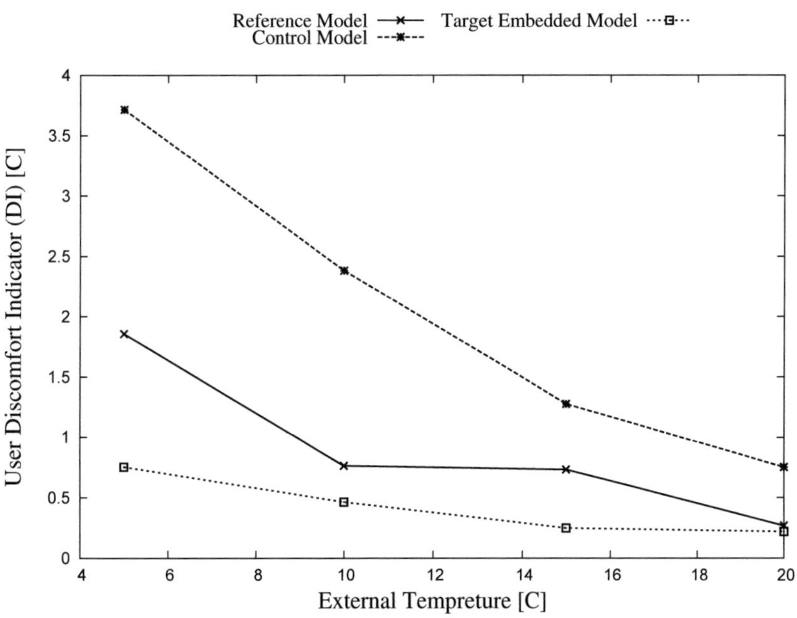

Fig. 15. User Discomfort Evaluation of Framework Models for $T_{sp} = 18$

[8] http://rtg.cis.upenn.edu/mobies/charon/index.html
[9] http://www.aadl.info/aadl/currentsite/tool/index.html

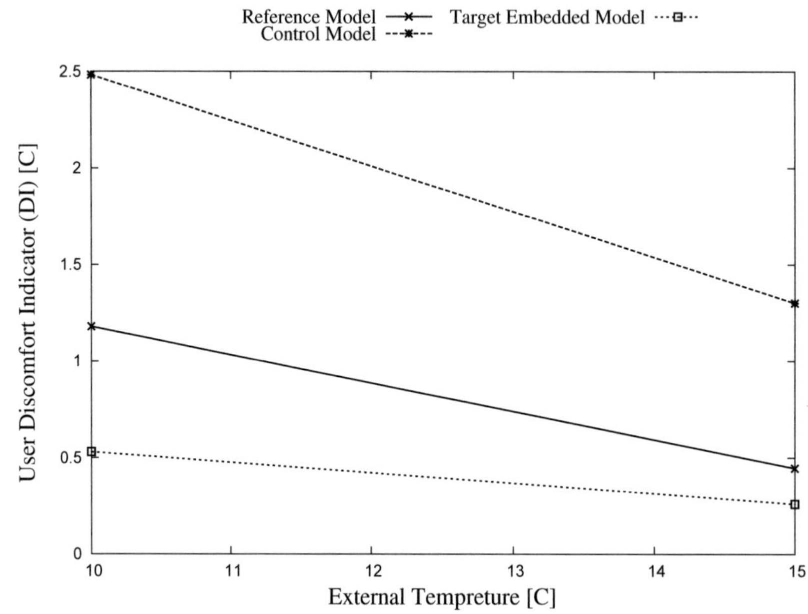

Fig. 16. User Discomfort Evaluation of Framework Models for $T_{sp} = 20$

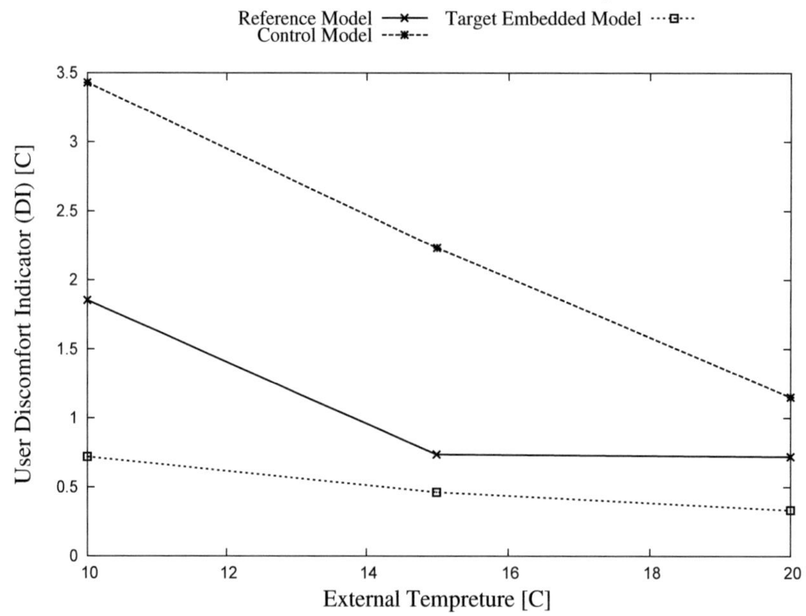

Fig. 17. User Discomfort Evaluation of Framework Models for $T_{sp} = 23$

7. Conclusion

In this article, we have proposed a novel end-to-end framework for designing an NCS for BAS system embedded over a distributed WSAN. This framework considers the design flow stages along from modelling to embedded control-code generation. The developed framework is empirically validated using AHU system as a case-study.

A future trend of this framework is integrating a *de facto* standard tool, such as Simulink, to the reference model. Moreover, we need to consider the timing and delay for the generated embedded code.

8. References

Balarin, F., Watanabe, Y., Hsieh, H., Lavagno, L., Passerone, C. & Sangiovanni-Vincentelli, A. (2003). Metropolis: An integrated electronic system design environment, *Computer* pp. 45–52.

Balasubramanian, D., Narayanan, A., van Buskirk, C. & Karsai, G. (2006). The graph rewriting and transformation language: Great, *Electronic Communications of the EASST* 1.

Basu, A., Bozga, M. & Sifakis, J. (2006). Modeling heterogeneous real-time components in BIP, *SEFM*, Vol. 6, Citeseer, pp. 3–12.

Chen, K., Sztipanovits, J. & Neema, S. (2005). Toward a semantic anchoring infrastructure for domain-specific modeling languages, *Proceedings of the 5th ACM international conference on Embedded software*, ACM New York, NY, USA, pp. 35–43.

Denckla, B. & Mosterman, P. (2005). Formalizing causal block diagrams for modeling a class of hybrid dynamic systems, *Proc. 44th IEEE Conference on Decision and Control,and European Control Conference (CDC-ECC'05)* pp. 4193–4198.

Dounis, A. & Caraiscos, C. (2009). Advanced control systems engineering for energy and comfort management in a building environment-a review, *Proc. of Renewable and Sustainable Energy Reviews* pp. 1246–1261.

Gurevich, Y., Rossman, B. & Schulte, W. (2005). Semantic essence of AsmL, *Theoretical Computer Science* 343(3): 370–412.

Hardebolle, C. & Boulanger, F. (2008). ModHel'X: A component-oriented approach to multi-formalism modeling, *Lecture Notes In Computer Science* 5002: 247–258.

Henzinger, T. (1996). The theory of hybrid automata, *Proc. 11th Annual IEEE Symposium on Logic in Computer Science (LICS 96)* pp. 278–292.

Jackson, D. (2002). Alloy: a lightweight object modelling notation, *ACM Transactions on Software Engineering and Methodology (TOSEM)* 11(2): 256–290.

Lee, E., Neuendorffer, S. & Wirthlin, M. (2003). Actor-oriented design of embedded hardware and software systems, *Journal Of Circuits Systems And Computers* 12(3): 231–260.

Lee, E. & Sangiovanni-Vincentelli, A. (1998). A framework for comparing models of computation, *IEEE Transactions on computer-aided design of integrated circuits and systems* 17(12): 1217–1229.

Mady, A. & Provan, G. (2011). Co-design of wireless sensor-actuator networks for building controls, *the 50th IEEE Conference on Decision and Control and European Control Conference (IEEE CDC-ECC)*.

Rumbaugh, J., Blaha, M., Premerlani, W., Eddy, F. & Lorensen, W. (1991). *Object oriented modeling and design*, Prentice Hall, Book Distribution Center, 110 Brookhill Drive, West Nyack, NY 10995-9901(USA).

Sztipanovits, J. & Karsai, G. (1997). Model-integrated computing, *IEEE computer* 30(4): 110–111.

Vestal, S. (1996). MetaH programmerŠs manual, *Technical report*, Version 1.09. Technical Report, Honeywell Technology Center.

Yu, B. & van Paassen, A. (2004). Simulink and bond graph modeling of an air-conditioned room, *Simulation Modelling Practice and Theory, Elsevier* .

21

The Role of Automation in the Identification of New Bioactive Compounds

Pasqualina Liana Scognamiglio[1], Giuseppe Perretta[2] and Daniela Marasco[1]
[1]*Department of Biological Sciences, University "Federico II" of Naples, Naples,*
[2]*Institute of Motors, CNR, Naples,*
Italy

1. Introduction

Automation is nowadays implemented in many areas of the drug discovery process, from sample preparation through process development. High-Throughput Screening (HTS) is a well-established process for lead discovery that includes the synthesis and the activity screening of large chemical libraries against biological targets via the use of automated and miniaturized assays and large-scale data analysis. In recent years, high-throughput technologies for combinatorial and multiparallel chemical synthesis and automation technologies for isolation of natural products have tremendously increased the size and diversity of compound collections. The HTS process consists of multiple automated steps involving compound handling, liquid transfers and assay signal capture. Library screening has become an important source of hits for drug discovery programmes. Three main complementary methodologies are actually used: 1) *in silico* virtual screening of libraries to select small sets of compounds for biochemical assays, 2) fragment-based screening using high-throughput X-ray crystallography or NMR methods to discover relatively small related compounds able to bind the target with high efficiency and 3) HTS of either diverse chemical libraries or focused libraries tailored for specific gene families. Furthermore a variety of assay technologies continues to be developed for high-throughput screening; these include cell-based assays, surrogate systems using microbial cells and systems to measure nucleic acid-protein and receptor-ligand interactions. Modifications have been developed for in vitro homogeneous assays, such as time-resolved fluorescence, fluorescence polarization and the scintillation proximity. Innovations in engineering and chemistry have led to delivery systems and sensitive biosensors for Ultrahigh-Throughout Screening working in nanoliter and picoliter volumes. Spectroscopic methods are now sensitive to single molecule fluorescence. Technologies are being developed to identify new targets from genomic information in order to design the next generation of screenings. As HTS assay technologies, screening systems, and analytical instrumentation the interfacing of large compound libraries with sophisticated assay and detection platforms will greatly expand the capability to identify chemical probes for the vast untapped biology encoded by genomes.

As a consequence the growing demand for new, highly effective drugs is driven by the identification of novel targets derived from the human genome project and from the understanding of complex protein-protein interactions that contribute to the onset and maintaining of pathological conditions. To illustrate the dynamics of quantitative and qualitative process approaches to accelerated drug development, a model pipeline is depicted in Fig. 1.

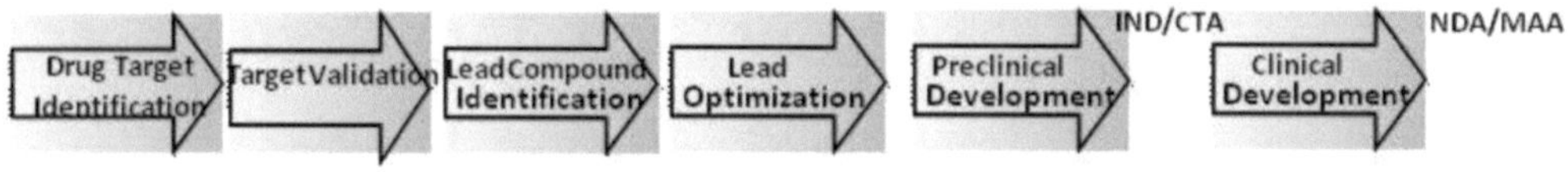

Fig. 1. Drug Discovery pipelines: **IND**: Investigational New Drug, **CTA**: Clinical Trial Application, **NDA**: New Drug Application, **MAA**: Marketing Authorization.

At the basis of HTS is the simultaneous employment of different sets of compounds that are rapidly screened for the identification of active components. This approach is nowadays regarded as a powerful tool for the discovery of new drug candidates, catalysts and materials. It is also largely utilized to improve the potency and/or the selectivity of existing active leads by producing analogues derived by systematic substitution or introduction of functional groups or by merging active scaffolds. In Combinatorial Chemistry (CC) the higher is the number of building blocks and the number of transforming synthetic steps, the more attainable molecules can be had. However, the higher is the number of attainable molecules, the higher should be the handling capacity of so many reagents and products; thus automation in CC plays a very critical role. Automation in chemical synthesis has been essentially developed around the solid phase method introduced by R. B. Merrifield for the synthesis of peptides, which, after its introduction, has been continuously improved in terms of solid supports, linkers, coupling chemistry, protecting groups and automation procedures, making it nowadays one of the most robust and well-established synthetic methods (Shin et al., 2005). For these reasons, the basic concepts of CC, such as compound libraries, molecular repertoires, chemical diversity and library complexity have been developed using peptides and later transferred to the preparation of libraries of small molecules and other oligomeric biomolecules (Houghten et al., 2000). The easiness of preparation, characterization and the robustness of the available chemistry provide high purity levels and the built-in code represented by their own sequences have promoted the employment of large but rationally encoded mixtures instead of single compounds, leading to the generation and manipulation of libraries composed of hundreds of thousands and even millions of different sequences. The broad complexity of mixture libraries has also led to the development of several screening procedures based on iterative or positional scanning deconvolution approaches for soluble libraries.

2. HTS

High-throughput screening (HTS) has achieved a dominant role in drug discovery over the past two decades. Its aim is to identify active compounds (hits) by screening large numbers of diverse chemical compounds against selected targets and/or cellular phenotypes. The HTS process consists of multiple automated steps involving compound handling, liquid transfers, and assay signal capture, all of which unavoidably contribute to systematic variation in the screening data. It represents the process of testing a large number of diverse chemical structures against disease targets to identify 'hits'.

Compared to traditional drug screening methods, HTS is characterized by its simplicity, rapidness, low cost, and high efficiency, taking the ligand-target interactions as the principle, as well as leading to a higher information harvest.

Independent of the precise nature of the applied screening technology, lead discovery efforts can always be analyzed and optimized along the same fundamental principles of performance management: time, costs, and quality of the process (Fig. 2).

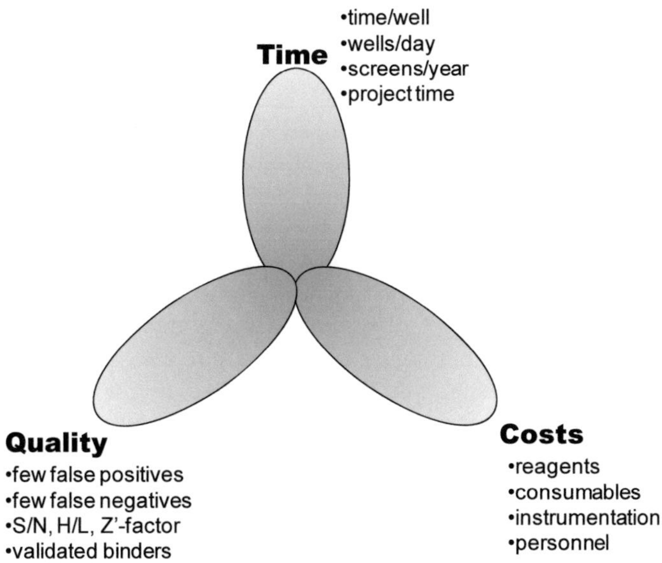

Fig. 2. HTS "magic triangle".

As a multidisciplinary field, HTS involves an automated operation-platform, highly sensitive testing system, specific screening model (in vitro), abundant component libraries, and a data acquisition and processing system. Several technologies such as fluorescence, nuclear-magnetic resonance, affinity chromatography, surface plasmon resonance and DNA microarray, are actually available, and the screening of more than 100 000 samples per day is already possible.

The data analysis challenge is to detect biologically active compounds from assay variability. Traditional plates, controls-based and non-controls-based statistical methods have been widely used for HTS data processing and active identification by both the pharmaceutical industry and academic sectors. Recently, the introduction of improved robust statistical methods has reduced the impact of systematic row/column effects in HTS data. In practice, no single method is the best hit detection method for every HTS data set. Nevertheless, to help in the selection of the most appropriate HTS data-processing and active identification methods a 3-step statistical decision methodology has been developed: Step 1) to determine the most appropriate HTS data-processing method and establish criteria for quality control review and active identification from the 3-day assay signal window and validation tests. Step 2) to perform a multilevel statistical and graphical review of the screening data to exclude data that fall outside the quality control criteria. Step 3) to apply the established active criterion to the quality-assured data to identify the active compounds.

The principles and methods of HTS find their application for screening of combinatorial chemistry, genomics, protein, and peptide libraries. For the success of any HTS assay or screening several steps -like target identification, reagent preparation, compound management, assay development and high-throughput library screening- should be carried out with utmost care and precision. Historically, the majority of all targets in HTS-based lead discovery fall into a rather small set of just a few target families (Table 1). Enzymes such as kinases, proteases, phosphatases, oxidoreductases, phosphodiesterases, and transferases comprise the majority of biochemical targets in today's lead discovery efforts. Among cell-based targets, many GPCRs (G-protein-coupled receptors, 7-transmembrane receptors), nuclear hormone receptors, and some types of voltage- and ligand-gated ion channels (e.g., Ca2+-channels) are very well suited for screening large compound collections. Despite the large number of human genes (>25,000) and the even larger number of gene variants and proteins (>100,000), the number of molecular targets with drugs approved against the target is still fairly limited (~350 targets). The explanations for this discrepancy can be due to some targets might not be feasible at all for modulation via low molecular weight compounds. Others, however, might simply not be approachable by current technologies and therefore constitute not only a great challenge but also a tremendous potential for future lead discovery. Among those, a large number of ion channels, transporters, and transmembrane receptors but also protein-protein, protein-DNA, and protein-RNA interactions, even RNA/DNA itself, might form innovative targets for modulation via low molecular weight compounds. With better tools with sufficient predictivity, one might be able to expand the current HTS portfolio into novel classes of pharmaceutical targets. Besides the obvious classes with targets on the cell surface, such as ion channels, transporters and receptors, we predict that modulation of intracellular pathways, particularly via protein-protein interactions, might have a great potential for pharmaceutical intervention. In this regard, some modern technologies such as subcellular imaging (High Content Screening [HCS]) will certainly enable novel and powerful approaches for innovative lead discovery.

Established Target Classes	**Novel Target Classes**
• Kinases/Phosphatases • Oxidoreductases • Transferases • Proteases • Nuclear Hormone Receptors • G-protein-coupled receptors, 7-transmembrane receptors (GPCRs) • Ion Channels (some) • Signaling Pathways (some)	• Ion Channels • Transporters • Transmembrane Receptors • Signaling Pathways • Protein-Protein Interactions • Protein-DNA Interactions • Protein-RNA Interactions • DNA/RNA

Table 1. Established and potential future target classes for HTS

HTS not only helps in drug discovery but it is also important in improving present drug moieties to optimize their activity. In past years many advances in science and technology and economic pressures have kept every researcher to develop speedy and precise drug discovery and screening technologies to tackle the ever increasing diseases and the many pathogens acquiring resistance to currently available drugs. This also applies to screening the ever increasing compound libraries waiting to be screened due to increase in the parallel and combinatorial chemical synthesis. Research is also carried out so to cut the drug development costs, so that industries keep abreast with ever increasing competition.

2.1 Screenings of libraries

The greatest potential of combinatorial chemistry is represented by the number and variety of screenable compounds. Major efforts of researchers in the last decades have been focused on the development of methodologies to further increase molecular diversity. An interesting approach is based on general reversible reactions that produce "dynamic mixtures": in these, reactants and products are present in thermodynamic equilibrium. The assayed biological system selects the best binding structures among the different mixture components, thus producing a shifts of the reaction equilibrium by subtracting the product. Importantly, dynamic libraries do not need a deconvolution step and ligands are selected directly in the reaction mixture. Although innovative and promising, dynamic combinatorial libraries have been so far limited only to a small number of reversible reactions and libraries of moderate size. Similarly, the methodology named "libraries from libraries" has represented an innovation compared to the traditional concepts of combinatorial chemistry. By this approach, combinatorial libraries of peptides are built on a solid phase and are subsequently modified in order to maximize the chemical diversity. Oxidations, reductions, alkylations and acylations can be performed, exponentially increasing the number of new compounds. Once synthesized, libraries are employed in screening processes to determine active components for a given target. The choice of the assay is of utmost importance to succeed with a screening program. Indeed, different assays can be chosen depending on whether binders for an unspecified site are searched for, or ligands with predetermined properties are needed. Competition assays ensure the selection of active molecules with a specificity for the interacting interface of the proteins employed as targets in the screening. Binding assays can also be performed "on bead", in this case the libraries have been prepared with the Mix and Split method, the use of bead bound molecules has the advantage of having a local high concentration of ligands (several picomoles on a very limited surface area). The assay principle is simple and is based on the interaction between molecules and a labelled target-any natural or artificial receptor, enzymes, antibodies, nucleic acids or even small molecules in solution. When the labelled target binds the bound molecule, the bead will also be labelled and therefore can be visualized/detected by chromogenic methods and using micromanipulators. The labeled bead can be isolated and the molecule/peptide microsequenced. Though the on-bead assay is faster and easier to perform homogeneous phase assays in solution it can provide more specific functional evaluations. However highly charged or very hydrophobic molecules, when locally highly concentrated, can result in a high number of false-positive hits. Partially releasable libraries, based on light sensitive chemical linkers have also been described. These linkers, being very resistant to cleavage under acidic conditions, allow the complete removal of amino acid side chain protections, while they can be easily cleaved by irradiation at a defined wavelength. By tuning light intensity and duration, small amounts of peptides can be released in solutions, while the molecules bound to the bead can be microsequenced to reveal the molecule identity. The deconvolution of libraries bound to solid supports can be performed following different approaches. Peptide arrays can be distinguished on the basis of their preparations: some ones involve the immobilization of pre-synthesized peptide derivatives and others the *in situ* synthesis directly on the array surface. The pre-synthesis approach requires a chemoselective immobilization on solid supports that can provide a useful method for controlling the orientation and the density of the immobilized peptides.

A critical factor for the screening of compound arrays is the accessibility to the immobilized molecules by the target proteins. Several spacers between peptides and chip surface have been proposed, such as 11 mercaptoundecanoic acid, hydrophilic polyethylene glycol chain, dextran, bovine serum albumin or human leptin and water-compatible supramolecular hydrogels.

In situ parallel synthesis methods can provide cheaper and faster miniaturized spatially addressed peptide arrays. Two general approaches have been introduced for the preparation of peptide arrays on solid supports: the photolithographic and the SPOT synthesis. The photolithographic approach consisted in the translation of the solid phase method for peptide synthesis into the preparation of supported peptide arrays by mean of photolabile protecting groups that allow the synthesis to proceed only on defined surface spots that are illuminated by light at a given wavelength. In the SPOT synthesis, the peptides arrays are synthesized in a stepwise manner on a flat solid support, such as functionalized cellulose membrane, polypropylene and glass, following the standard Fmoc-based peptide chemistry. Each spot is thus considered as an independent microreactor: thereby the selection of the solid support is a key factor. The support has to meet the chemical and the biological requirements of the target and it determines the synthesis and screening methods. It also dictates the functionalization type and the insertion of spacers and/or linkers. Described supports include ester-derivatized planar supports, CAPE [celluloseamino-hydroxypropyl ether] membranes, amino-functionalized polypropylene membranes and glass surface. Libraries can thus be seen as a source of bioactive molecules that are selected on the basis of the biochemical properties pre-defined by appropriate assay settings. The more diverse will be the library, the higher will be the probability to select good "hits". The concept of diversity is therefore of utmost importance in choosing a library for a particular screening and good libraries must first fulfill the requirement of highest diversity instead of the highest complexity. Molecules with overlapping structures will contribute little or nothing at all to the overall probability to find out a positive hit. The diversity of a library is generally associated to its complexity, that in turn depends on the number of different components. However this is not always true: small but smart libraries can display a higher diversity than libraries with a huge number of components. Indeed the synthesis of random combinatorial libraries of peptides generates a large number of "quasi-duplicates" deriving from the strong similarity between several side chains. Using common amino acids, in L- or D- configuration, sequences where Glu is replaced by Asp, Leu by Ile or Val, Gln by Asn and so on, can display very similar properties. Such residues, although being different in their propensity to adopt secondary structures, can be considered almost equivalent in terms of intrinsic physico-chemical properties, as for example the capacity to establish external interactions or to fit in a given recognition site. In large libraries, the need to manage large arrays of tubes and codes can puzzle the way to the identification of lead compounds and slows down the synthetic and deconvolution steps. To simplify the synthesis and deconvolution procedures without significantly affecting the probability to find out active peptides, we have reasoned on the general properties of L and D-amino acids, reaching a compromise between the need to maintain the highest possible diversity and that of reducing the number of building blocks. We have called these new libraries "Simplified Libraries", intending with this any new ensemble of possible sequences achievable with a reduced and non redundant set of amino acids. The set of "non redundant" different residues (12 instead of 20) is chosen following several simple rules. Firstly "quasi-identical" groups of amino acids are selected: Asp and Glu; Asn and Gln; Ile and Val; Met and Leu; Arg and Lys; Gly and Ala; Tyr and Trp; Ser and Thr. Then, from each group, one is selected trying to keep a good distribution of properties, including hydropathic properties, charges, pKa, aromaticity. The cysteine is converted to the stable acetamidomethyl-derivative to prevent polymerization and to increase the number of polar residues within the final set. The distribution of residue molecular weights is also an important parameter in cases of deconvolution of the library by mass spectrometry approaches. The molecular weights do not overlap, allowing unambiguous assignment of sequences by tandem mass spectrometry. Using this set of amino acids in the L- or D-configuration, several libraries have been designed and synthesized (Marasco et al, 2008).

2.2 Automated libraries characterization

A key element for a successful screening of peptide libraries is the preparation of good quality libraries. Unclear screening results are often ascribed to the random generation of large populations of impurities, therefore even for very large mixture libraries, gross analytical characterizations must be performed to assess the relative amounts and the distribution of experimental molecular weight of library components. While classical analytical methods, as for example LC-MS, can be in most cases enough for the characterization of single compounds, more sophisticated techniques are required for the analysis of complex mixtures. The recent progresses in parallel separation methods, such as orthogonal chromatography, associated to synergic combinations of different detectors of complementary selectivities, have allowed the development of high-throughput analytical technologies based on traditional methods as HPLC, capillary electrophoresis (CE), NMR, FTIR, LC-MS, evaporative light scattering (ELSD), chemiluminescent nitrogen (CLND). A list of useful techniques is reported in **Table 2**.

Mass spectrometry (MS) is the best method to assess compound identity and purity due to its sensitivity, speed and specificity. Electrospray ionization (ESI) coupled to quadrupole analyzers or Matrix Assisted Laser Desorption Ionization Time-Of-Flight (MALDI-TOF) are useful for peptide library analysis. LC-MS systems using ESI and different analyzers (single or triple quadrupoles, ion traps, TOF), offer the advantage of chromatographic separation prior to on-line mass analysis, thus enormously simplifying data interpretation. Single quadrupole instruments typically provide unit mass resolution and therefore do not resolve isobaric ions (i.e. two compounds with the same nominal mass but different exact mass), which are often encountered in combinatorial libraries. LC-MS systems equipped with ESI and TOF analyzers (ESI-TOF) or with combined quadrupole-TOF (Q-TOF) analyzers provide increased mass accuracies up to 5 ppm and great versatility for tandem mass analysis. Limitations of MS include the inability to distinguish isomers and, with the increase of potential molecular formulae, differences in masses are too close to measurement error.

Detection mode	Advantages	Disadvantages
UV-absorbance spectroscopy	• High sensibility and response linearity • Easy on-line integration	• Relative purity measurements • Compounds have to bear chromophores • Responses depende on extinction coefficients
Evaporative light scattering	• Response factors of diverse compounds • Easy on-line integration	• Relative purity measurements • Limited for compounds with MW>300 amu • Right response depends on geometry molecule
Chemiluminescent nitrogen	• Universal response to number of nitrogen structure • Quantitative purity (if used with an internal standards) • Easy on-line integration	• Compounds must contain nitrogen
MS	• Excellent sensibility • Exact mass identification • Easy on-line integration	• Response factors depend on compound ionizability and matrix effects • Limited resolution with compounds with a high MW
NMR	• Universal response indipendent from structure • Quantitative purity (if used with an internal standards) • Detailed structure information	• Complex data interpretation • UV-absorbance Peak assignement difficult in presence of impurities • Difficult on-line integration

Table 2. Rivelator Method for HTS Characterization Of Compound Libraries

In addition to standard HPLC columns systems for interfacing to MS, several alternative separation techniques or platforms have been investigated for high-throughput purity analysis. An interesting application regards imaging time-of-flight secondary ion mass spectrometry (TOF-SIMS) to perform high-throughput analysis of solid phase synthesized combinatorial libraries. Also NMR techniques have been extensively utilized for the identification and quantification of combinatorial compounds. NMR has the capability of elucidating compound structure and it is a quantitative detector whose response factor is directly proportional to the number of nuclei for a given signal, independent of molecular structure. Recently advances in NMR probe design have enabled sampling from smaller sample volumes and have significantly improved sample throughput overcoming some limitations of traditional NMR for combinatorial analysis included its low sensitivity, low sample throughput and complexity of data interpretation. The characterization of large mixture libraries has also been performed by pool amino acid analysis and by pool N-terminal sequencing techniques. Such techniques, though not providing single compounds purity and identity, can offer a rough but useful indication of equimolarity and representativeness of components.

2.3 Assays preparation

Automation is an important element in HTS's usefulness. Typically, an integrated robot system consisting of one or more robots transports assay-microplates from one station to another for sample and reagent addition, mixing, incubation and finally readout or detection. A HTS system can usually prepare, incubate, and analyze many plates simultaneously, further speeding the data-collection process. Currently HTS robots are able to test up to 100,000 compounds per day. The key labware or testing vessel of HTS is the microtiter plate: a small container, usually disposable and made of plastic, that features a grid of small, open divots called wells. Modern microplates for HTS generally have either 384, 1536, or 3456 wells. These are all multiples of 96, reflecting the original 96 well microplate with 8 x 12 9mm spaced wells. Most of the wells contain experimentally useful matter, often an aqueous solution of dimethyl sulfoxide (DMSO) and some other chemical compounds, the latter of which is different for each well across the plate. The other wells may be empty, intended for use as optional experimental controls. A screening laboratory typically holds a library of stock plates, whose contents are carefully catalogued, and each of which may have been created by the same lab or obtained from a commercial source. These stock plates themselves are not directly used in experiments; instead, separate assay plates are created as needed. An assay plate is simply a copy of a stock plate, created by pipetteing a small amount of liquid (often measured in nanoliters) from the wells of a stock plate to the corresponding wells of a completely empty plate. To prepare an assay each well of the plate is filled with some quantity of a protein, or an animal embryo. After some incubation time has passed to allow the biological matter to absorb, bind to, or otherwise react (or fail to react) with the compounds in the wells, measurements are taken across all the plate's wells, either manually or by a machine. Manual measurements are often necessary when the researcher is using microscopy to (for example) seek changes or defects in embryonic development caused by the wells compounds, looking for effects that a computer could not easily determine by itself. Otherwise, a specialized automated analysis machine can run a number of experiments on the wells. In this case, the machine outputs the result of each experiment as a grid of numeric values, with each number mapping to the value obtained from a single well. A high-capacity analysis machine can measure dozens of plates in the space of a few minutes like this, generating thousands of experimental datapoints very quickly. Depending on the results of this first assay, the researcher can perform follow up assays within the same screen by "cherrypicking" liquid from the source wells that gave interesting results into new assay plates, and then re-running the experiment to collect further data on this narrowed set, confirming and refining observations. Problems associated with screening robotics have included long design and implementation time, long manual to automated method transfer time, non-stable robotic operation and limited error recovery abilities. These problems can be attributed to robot integration architectures, poor software design and robot–workstation compatibility issues (e.g., microplate readers and liquid handlers). Traditionally, these integrated robot architectures have involved multiple layered computers, different operating systems, a single central robot servicing all peripheral devices, and the necessity of complex scheduling software to coordinate all of the above. Usually robot-centric HTS systems have a central robot with a gripper that can pick and place microplates around a platform. They typically process between 40 and 100 microplates in a single run (the duration of the run depends on the assay type). The screener loads the robotic platform with microplates and reagents at the beginning of the experiment and the assay is then processed unattended. Robotic HTS systems often have humidified CO_2 incubators and are enclosed for tissue culture work. Like in assembly-line manufacturing, microplates are passed down a line in serial fashion to consecutive processing modules. Each module has its own simple pick and place robotic arm (to pass plates to the next module) and microplate processing device. The trend towards assay miniaturization arose simultaneously with move towards automation from the direct need to reduce development cost. Although at present most HTS is still carried out in 96-well plate format, the move towards 384-well and higher density plate formats is well taking place. Instrumentation for accurate, low-volume dispensing into 384-well plates is commercially available, so they are sensitive plate-readers that accommodate this format. The combination of liquid handling automation and information management processes supports the entire compound management cycle from compound submission to delivery of assay ready compound plates. The compound management often consists in the organization of hundreds of thousands of compounds (small molecules, natural products and peptide libraries in different formats). Compounds are stored offline in an automated compound store in 96-well tube format and online in 384-well format for hitpicking operations, 1536-well for HTS operations. Often there are dedicated LC-MS systems with multimode mass spectrometry (ESI, APCI, ELSD) for quality control of incoming HTS compound libraries and hit confirmation efforts. This platform is designed for automated analysis and generates database-ready reports.

2.4 Screening assays

Assays are either heterogeneous or homogeneous. Heterogeneous assays are a bit complex, requiring additional steps like filtration, centrifugation etc. beyond the usual steps like fluid addition, incubation and reading. Homogeneous assays are simpler, consisting of the latter three usual steps - this may also be called a true homogeneous assay. However at times homogeneous assays could be complex due to the need for multiple addition and different incubation times. Though homogeneous assays are advantageous, many companies prefer to continue to use heterogeneous assay, eyeing their better precision over its counterpart, though it is true only in few number of cases. The driving force for use of homogeneous assays is the lower number of steps, which will help reduce assay cost. This simplicity may also reduce the robotic complexity requirement for automation. When performing HTS of free compounds in solution, automation, miniaturization and very sensitive detection methodologies are required. Different approaches such as ELISA, cell-based cytotoxic, antimicrobial, radiometric and fluorescence-based assays, affinity chromatography methodologies can be used. Fluorescence-based techniques are likely to be among the most important detection approaches used for HTS due to their high sensitivity and amenability to automation, given the industry-wide drive to simplify, miniaturize, and speed up assays. Fluorescence resonance energy transfer (FRET), fluorescence polarization (FP), and fluorescence correlation spectroscopy (FCS) are indeed already broadly utilized for screenings of large collections of compounds and many optical readers capable of handling multi-well plates are commercially available. FRET is based on energy transfer between appropriate energy donor and acceptor molecules. It is typically used in protein-protein interaction studies where one protein partner is labelled with a fluorescence donor and the other one is labelled with an appropriate fluorescence acceptor molecule. The donor has the specific property of being excitable, emitting a fluorescence photon at a wavelength that is able to in turn excite the acceptor. When the two proteins interact and the system is excited at the donor excitation wavelength, a specific fluorescence emission at the acceptor emission wavelength is recorded. The energy transfer occurs only when the donor-acceptor pair is within a minimum distance, the Förster distance (the distance at which energy transfer efficiency is half-maximal), which is around 50 Å. When performing a screening assay, if a library component is able to disrupt the protein-protein interaction, the effect can be quantitatively measured by a reduction of the FRET effect. Similar assays can be performed using the Time Resolved FRET (TR-FRET) whereby the fluorescence of the acceptor molecule has a duration on the milliseconds time scale, allowing fluorescence measurement after a short time delay to remove interference by the excitation energy or by inhibitors. TR-FRET can be integrated on large time intervals to increase sensitivity. Typical donor-acceptor molecules are the Allophycocyanin-Europium chelates or the fluorescein - Terbium chelates. FRET can also be used to determine enzyme activity using internally quenched probes. In these systems short peptides corresponding to the sequence for a natural cleavage site of the enzyme is synthesized and labelled at opposite ends with appropriate donor and quencher pairs. Before cleavage, donor and quencher are very close and the effective fluorescence emitted is low; once the two parts drift apart, the fluorescent signal increases. This approach has been successfully utilized to screen the substrate specificity of an alkaline serine proteinase. FP experiments allow measurements of changes in the emitted light intensity of small labelled probes on binding to larger molecules. The sample is excited with polarized light and, when a binding equilibrium is established, the observed polarization of the emitted light increases. In FCS the main detected parameter is the spontaneous intensity fluctuations caused by the minute deviations of the small system from the thermal equilibrium. FCS is an emerging technique for HTS. In this case, measurements are carried out using confocal optics to provide the highly focused excitation light. It is used to monitor binding interactions as well as other molecular events. At times it was reported that assays for biological targets cannot be conveniently designed to fit with standard cellular or biochemical assay formats. For example, in the search for new antibacterial agents, genomic experiments have indicated a large number of proteins that are essential for the survival of the bacterium but whose function in the cell is unknown. In this case there is no known biological function that will allow the design a biochemical or cellular screen. To screen these types of target, an alternative to conventional chemical or cellular screenings may be used. One alternative screening approach that does not require knowledge or analysis of the biological function of the target of choice is direct measurement of compound interaction with protein. A range of techniques are available to measure the direct binding events such as NMR, SPR and calorimetry. One advantage afforded by NMR is that it can provide direct information on the affinity of the screening compounds and the binding location of protein. The structure-activity relationship acquired from NMR analysis can sharpen the library design, which will be very important for the design of HTS experiments with well-defined drug candidates. Affinity chromatography used for library screening will provide information on the fundamental processes of drug action, such as absorption, distribution, excretion, and receptor activation; also the eluting curve can give directly the possibility of candidate drug. SPR can measure the quantity of a complex formed between two molecules in real-time without the need for fluorescent or radioisotopic labels. SPR is capable of characterizing unmodified biopharmaceuticals, studying the interaction of drug candidates with macromolecular targets, and identifying binding partners during ligand fishing experiments. They indeed allow real-time detection of soluble peptide binding to different biomolecules, like proteins, nucleic acids or sugars and also to cell membranes or whole cells. In addition, they allow the one-step measurement of kinetic rates and affinity of binding, thus providing an affinity ranking of molecules peptides.

2.5 Data analysis

In validating a typical HTS assay, unknown samples are assayed with reference controls. The sample signal refers to the measured signal for a given test compound. The negative control (usually referred to as background) refers to set of individual assays from control wells that give minimum signals. The positive control refers to the set of individual assay from control wells that give maximum signals. In validating the assay, it is critical to run several assay plates containing positive and negative control in order to assess reproducibility and signal variation at two extremes of the activity range. The positive and negative control data can then be used to calculate their means and standard deviations (SD). The difference between the mean of the positive controls and the mean of the negative controls defines the dynamic range of the assay signal. The variation in signal measurement for samples, positive control, and negative controls (i.e., SDs) may be different. The mean and SD of all the test samples are largely governed by the assay method and also by intrinsic properties of the compound library. Because the vast majority of compounds from an unbiased library have very low or no biological activity, the mean and SD of all the sample signals should be close to those of the positive controls for inhibition/antagonist type assays and near to those of the negative controls for activation/agonist types assays.

The Z-factor is a measure of statistical effect. It was proposed for use in HTS to judge whether the response in a particular assay is large enough to warrant further attention. The Z-factor is defined by four parameters: the means (μ) and standard deviations (σ) of both the positive (p) and negative (n) controls (μp, σp, and μn, σn). Given these values, the Z-factor is defined as:

$$\text{Z-factor}=1-\frac{3\times\left(\sigma_p+\sigma_n\right)}{\left|\mu_p-\mu_n\right|}$$

Generally HTS suffers from two types of errors false positives and negatives. A poor candidate or an artifact gives an anomalously high signal, exceeding an established threshold. While a perfectly good candidate compound is not flagged as a hit, because it gives an anomalously low signal. Moreover, a low degree of relevance of the test may induce a high failure rate of type. Much more attention is given to false-positive results than to false-negative results. Some of the false positives are promiscuous compounds that act non competitively and show little relationship between structure and function.

In HTS each biochemical experiment in a single well is analyzed by an automated device, typically a plate reader or other kind of detectors. The output of these instruments comes in different formats depending on the type of reader. Sometimes multiple readings are necessary, and the instrument itself may perform some initial calculation. These heterogeneous types of raw data are automatically fed into the data management software. In the next step raw data are translated in contextual information by calculating results. Data on percentage inhibition or percentage of control are normalized with values obtained from the high and low controls present in each plate. The values obtained depend on the method used for the normalization step (e.g. fitting algorithms used for dose-response curve) and have to be standardized for screens. All the plates that fail against one or more quality criteria are flagged and discarded. A final step in the process requires the experimenter to monitor visually the data that have been flagged, as a final quality check. This is a fundamental step to ensure the system has performed correctly. In addition to registering the test data, all relevant information about the assay has to be logged, e.g. the supplier of reagents, storage conditions, a detailed protocol, plate layout, and algorithms for the calculation of results. Each assay run is registered and its performance documented. HTS will initially deliver hits in targeted assays. Retrieval of these data has to be simple and the data must be exchangeable between different project teams to generate knowledge from the total mass of data.

3. Virtual screening

Even with HTS, the discovery of new lead compounds largely remains a matter of trial and error. Although the number of compounds that can be evaluated by HTS methods is apparently large, these numbers are small in comparison to the astronomical number of possible molecular structures that might represent potential drug-like molecules. Often, far more compounds exist or can be synthesized by combinatorial methods than can be reasonably and affordably evaluated by HTS. As the costs of computing decreases and as computational speeds increase, many researchers have directed efforts to develop computational methods to perform "virtual screens" of compounds. Thus since performing screens *in silico* can be faster and less expensive than HTS methods, virtual screening methods may provide the key to limit the number of compounds to be evaluated by HTS to a subset of molecules that are more likely to yield "hits" when screened. For the practical advantages of virtual screening to be realized, computational methods must excel in speed, cheapness, and accuracy. Striking the right balance of these criteria with existing tools presents a formidable challenge. An inspiring example study is related to the structure-based virtual screening applied to the selection of new thyroid hormone receptor antagonists when only a related receptor structure is available. Receptor-based virtual screening uses knowledge of the target protein structure to select candidate compounds with which it is likely to favorably interact (Schapira et al., 2003). Even when the structure of the target molecule is known, the ability to design a molecule to bind, inhibit, or activate a biomolecular target remains a daunting challenge. Although the fundamental goals of screening methods are to identify those molecules with the proper complement of shape, hydrogen bonding, electrostatic and hydrophobic interactions for the target receptor, the complexity of the problem is in reality far greater. For example, the ligand and the receptor may exist in a different set of conformations when in free solution than when bound. The entropy of the unassociated ligand and receptor is generally higher than that of the complexes, and favorable interactions with water are lost upon binding. These energetic costs of association must be offset by the gain of favorable intermolecular protein–ligand interactions. The magnitude of the energetic costs and gains is typically much larger than their difference, and, therefore, potency is extremely difficult to predict even when relative errors are small. While several methods have been developed to more accurately predict the strength of molecular association events by accounting for entropic and solvation effects, these methods are costly in terms of computational time and are inappropriate for the virtual screening of large compound databases. The challenge in developing practical virtual screening methods is to develop an algorithm that is fast enough to rapidly evaluate potentially millions of compounds while maintaining sufficient accuracy to successfully identify a subset of compounds that is significantly enriched in hits. Accordingly, structure-based screening methods typically use a minimalist "grid" representation of the receptor properties and an empirical or semiempirically derived scoring function to estimate the potency of the bound complex (Schulz-Gasch & Stahl, 2003). Several programs now employ a range of scoring functions, but it is often difficult to assess their effectiveness on difficult "real-world" problems. Virtual screening based on receptor structure therefore has the distinct advantage of aiding the discovery of new antagonist structural classes or pharmacophores. First, and most importantly, from the many case studies published over the past decade, it has become evident that the applicability and thus the usefulness of a particular virtual screening method for a given drug discovery project depends on the macromolecular target being investigated. Thus it seems more appropriate to consider virtual screening from a problem-centric rather than a method centric perspective. Depending on what is already known about a target and its ligands, different approaches to virtual screening – and consequently different sets of methods – are preferred. In addition, some virtual screening methods that have been reported might be premature or simply not sufficiently accurate. Second, the perceived success or failure of virtual screening in a particular organization depends on the depth and mode of integration of virtual screening in the organization's hit identification process, and whether expectations are realistic. For example, an important factor for the chances of success of virtual screening could be whether hits with interesting characteristics, such as structural novelty and/or patentability, can be sufficiently nurtured by medicinal chemists to produce leads that can compete with those arising from HTS. Indeed, this is typically the scenario for hits arising from alternative lead discovery approaches such as fragment-based screening. Third, a major achievement of virtual screening so far has been to help eliminate the bulk of inactive compounds (negative design), rather than to actually select bioactive molecules for a given target (positive design). Although this statement is a simplification it highlights the extent of the challenges for developing improved virtual screening methods.

It will be important to systematically determine which ligand–receptor interactions are amenable to such an approach and which require other or additional features to be considered indeed, dynamic descriptions of molecules will have to replace our predominantly static view of both targets and ligands. Molecular dynamics simulations can sample conformational ensembles of targets and ligands. However, some of the popular force-field approaches used to describe the energetics of molecular systems might be inadequate for drug design. Furthermore, although in general it might be more valuable to identify ligand chemotypes for which receptor–ligand complex formation is dominated by enthalpy changes rather than entropy changes, improvements are required to allow for a more accurate estimation of both enthalpic and entropic contributions (Freire, 2008). The thermodynamics of ligand–receptor interactions are commonly treated in a similar way as molecular reactions, and this may not always be appropriate. How can we reliably and efficiently predict that some protein–ligand interactions become stronger with increasing temperature, or identify the role of buried water on ligand binding? Questions like these must be answered by computational chemistry as the forces that govern ligand–receptor interactions are only understood at a rudimentary level: flexible fit phenomena, the role of water molecules, protonation states in proteinaceous environments, and the entropic and enthalpic contributions and compensations upon complex formation are not satisfactorily addressed by the existing virtual screening methods. In this respect, advanced and specialized computer hardware might enable extended (>100 ns) dynamics simulations of macromolecular targets and receptor–ligand complexes on a routine basis, which might help our understanding of allosteric effects and flexible fit phenomena of druglike ligands and effector molecules. As a consequence, modelling of dynamic molecular features (in contrast to static properties such as molecular mass, logP and other time-invariant molecular properties), which cannot be accurately achieved at present, could improve the accuracy of future predictions of novel bioactive compounds.

4. Fragment screening

Fragment-based drug discovery has proved too to be a very useful approach particularly in the hit-to-lead process, acting as a complementary tool to traditional HTS. Over the last ten years, fragment-based drug discovery has provided in excess of 50 examples of small molecule hits that have been successfully advanced to leads and therefore resulted in useful substrate for drug discovery programs. The unique feature of fragment-based drug discovery is the low molecular weight of the hit. It has the potential to supersede traditional HTS based drug discovery for molecular targets amenable to structure determination. This is because the chemical diversity coverage is better accomplished by a fragment collection of reasonable size than by larger HTS collections. Fragments represent smaller, less complex, molecules than either drug compounds or typical lead series compounds. It is now widely acknowledged within the pharmaceutical and biotech industries that weakly active fragment hit molecules can be efficiently optimised into lead compound series if structural insight is obtained at the outset for the binding interaction between each fragment hit and the target protein of interest. This is supported by recent reports of the progression into human clinical trials of drug molecules developed from weakly active fragment starting points (Jhoti et al., 2007). Fragment-based drug discovery can explore the drug-like chemical diversity space in an efficient and effective manner. Two key factors govern this approach, firstly the coverage of fragment chemical diversity space during the screening stage and then, that drug chemical diversity space is explored in an efficient iterative fashion during the optimisation stage as fewer combinations need to be evaluated than through a purely random screening and undirected optimisation approach. For example, a fragment collection of 10 000 molecules may virtually represent the diversity of one billion molecules if one considers the combinatorial power of fragment merging or linking (e.g. by assuming two adjacent binding sites to which fragments bind and 10 different possibilities of fragment linking) but only a small part of the larger chemical space defined by fragment merging and linking needs to be explored in the structure- directed elaboration of fragments into leads. Employing fragment-based drug discovery a relatively small number of low molecular weight fragment molecules can provide a higher degree of sampling of the chemical diversity space for fragments than a very large number of higher molecular weight compounds is able to sample the respective chemical diversity space for drug-like compounds. Furthermore lower molecular weight molecules exhibit reduced complexity than the larger molecules in drug-like collections and it can be hypothesized a model to rationalise ligand–receptor interactions in the molecular recognition process. Accordingly the theoretical probability of a useful interaction falls dramatically with increasing molecular complexity of the ligand. The selection of the fragment screening method is of key importance since there are two general factors that have an impact for all methods. The first is sensitivity of the screening method and the second is throughput. Sensitive screening methods enable weakly active fragment molecules of lower molecular weights to be identified as hit compounds and so fragment libraries with a lower molecular weight range can be used. On the contrary the use of a low throughput screening method necessitates the use of a smaller fragment library with a concomitant sparser coverage of fragment chemical diversity space. Perhaps the most elegant method of fragment screening is by X-ray crystallography, in that it provides directly structural information on the interaction between fragment ligands and the protein target. However, owing to the method's low throughput, even when fully automated, the technique can only be effectively applied to targets for which a robust crystallographic system is available that allows soaking of preformed crystals with fragment cocktail mixtures of up to 10 compounds at high concentrations. This requirement imposes two key limitations: in the number of evaluable fragment compounds, typically limited to no more than 1000 fragments, and for this, there is a significant possibility of missing active fragments owing to the protein being locked into a conformation, in the crystals used for the soaking studies, that does not allow the interaction of fragments that require induced fit to bind. Although, no data are available on the false negative rate for fragment screening by X-ray crystallography it may be significant for certain targets. Each fragment screening technique has its advantages; X-ray crystallography provides immediate structural information, NMR provides binding site and affinity information of a very high quality while bioassays provide functionally relevant activity data for larger collections of fragments. The best approach is to combine the methods in order to maximize their value to fragment-based drug discovery. NMR and biochemical screening of fragments are complementary orthogonal methods that can be used individually or in concert to provide the most effective way of addressing each new biological target of interest. The strength of biochemical screening is that its throughput allows large fragment collections to be screened in a short length of time. This ensures that the most ligand efficient diverse starting points are available for medicinal chemists to select for subsequent optimisation. A further advantage is that screening related targets using generic biochemical assay formats enables insights into target selectivity from the outset. The large number of fragment hits that are obtained through use of biochemical screening of large fragment libraries can be effectively triaged ahead of crystallography by the use of protein NMR. Thus, the most effective way to perform fragment screening is not to rely on a single method but to use orthogonal methods in concert.

5. Conclusion

HTS is the most widely applicable technology delivering chemistry entry points for drug discovery programs, however it is well recognized that even when compounds are identified from HTS they are not always suitable for further medicinal chemistry exploration. It is evident that in the future the overwhelming number of emerging target will dramatically increase the demand put on HTS and that this will call for new hit and lead generation strategies to curb costs and enhance efficiency. The collections of large pharmaceutical companies are approaching approximately one million entities, which represents historical collections (intermediates and precursors from earlier medicinal or agrochemical research programs), natural products and combinatorial chemistry libraries. This about one order of magnitude higher than ten years ago when HTS and combinatorial chemistry first emerged. However today purchasing efforts in many pharmaceutical companies are directed towards constantly improving and diversifying the compounds collections and making them globally available for random HTS campaigns. The combinatorial explosion- meaning the virtually number of compounds that are synthetically tractable-has fascinated and challenged chemists ever since the inception of the concept. Independent of the library designs, the question of which compounds should be made from the huge pool of possibilities always emerges immediately, once the chemistry is established and the relevant building blocks are identified. The original concept of "synthesize and test", without considering the targets being screened, was frequently questioned by the medicinal chemistry community and is nowadays considered of lower interest due to the unsatisfactory hit rates obtained so far. As a consequence there is now a clear trend to move away from huge and diverse "random" combinatorial libraries towards smaller and focused drug-like subsets. Hit and lead generation are key processes involved in the creation of successful new medicinal entities and it is the quality of information content imparted through their exploration and refinement that largely determines their fate in the later stages of clinical development. The combination of virtual screening and parallel and medicinal chemistry, in conjuction with multi-dimensional compound-property optimization, will generate a much-improved basis for proper and timely decisions about which lead series to pursue further.

6. References

Freire, E. (2008). Do enthalpy and entropy distinguish first in class from best in class?. Drug discovery today 13(19-20): 869-874.

Houghten, R.A., Wilson, D. B, & Pinilla, C. (2000). Drug discovery and vaccine development using mixture-based synthetic combinatorial libraries. Drug discovery today 5(7): 276-285.

Jhoti, H., Cleasby, A., Verdonk. M., & Williams, G. (2007). Fragment-based screening using X-ray crystallography and NMR spectroscopy. Current opinion in chemical biology 11(5): 485-493.

Marasco, D., Perretta, G., Sabatella, M., & Ruvo, M. (2008). Past and future perspectives of synthetic peptide libraries. Curr Protein Pept Sci. 9(5):447-67.

Schapira, M., Raaka, B.M., Das, S., Fan, L., Totrov, M., Zhou, Z., Wilson, S.R., Abagyan, R., & Samuels H.H. (2003). Discovery of diverse thyroid hormone receptor antagonists by high-throughput docking. Proceedings of the National Academy of Sciences of the United States of America 100(12): 7354-7359.

Schulz-Gasch, T. & Stahl, M.. (2003). Binding site characteristics in structure-based virtual screening: evaluation of current docking tools. Journal of molecular modeling 9(1): 47-57.

Shin, D.S., Kim, D.H., Chung, W.J., & Lee, Y.S. (2005). Combinatorial solid phase peptide synthesis and bioassays. Journal of biochemistry and molecular biology 38(5): 517-525.

22

Automatic Stabilization of Infrared Images Using Frequency Domain Methods

J. R. Martínez de Dios and A. Ollero
Robotics, Vision and Control Research Group, University of Seville
Spain

1. Introduction

In the last decade the decrease in the cost of infrared camera technology has boosted the use of infrared images in a growing number of applications. Traditional uses of infrared images, such as thermal analyses (Carvajal et al., 2011), nondestructive testing and predictive maintenance (Maldague, 2001), have been extended to new fields in which infrared cameras are mounted on vehicles and mobile robots and are used in applications such as border surveillance, building inspection, infrastructure maintenance, wildlife monitoring, search and rescue and surveillance, among many others. Image vibrations are harmful disturbances that perturb the performance of image-processing algorithms. Many of the aforementioned applications require having stabilized sequences of images before applying automatic image-processing techniques. Also, in cases where humans visualize the images, image vibrations induce significant stress and decrease the attention capacity of the operator.

Two main approaches for image stabilization have been developed. The first one aims at stabilizing the camera vibrations using different devices ranging from simple mechanical systems for handheld camcorders to high-performance inertial gyro-stabilized platforms. The first ones usually have low accuracy and perform "vibrations reduction" instead of "vibrations cancellation". The high size, weight and cost of gyro-stabilized platforms constrain their use in a good number of applications.

The second main approach aims at correcting the images by applying image-processing techniques. Classical image-processing methods for image stabilization are based on detecting a set of local features -e.g. corners, lines and high-gradient points- in one image and tracking them along the images of the sequence. The relative motion of features from one image to another is used to model the motion between both images. Once the motion between images has been estimated, the second image is compensated such that no vibration can be perceived between both images. These methods have demonstrated very good performance with images with high contrast and low noise level, where a number of local features can be robustly detected. However, they do not perform well in images where these features cannot be robustly identified. This is the case of infrared images, which often have low contrast and resolution and high noise levels.

This chapter describes an automatic image-processing technique for the stabilization of sequences of images using frequency domain image representations obtained by means of the Fourier transform. In particular, the method described in this chapter uses Fourier-Mellin transforms (FMT) and the Symmetric Phase Only Matched Filtering (SPOMF). This frequency domain representation provides advantages when stabilizing images with low contrast and high noise levels: the described method does not rely on local features but on the global structure of the image. The proposed method can correct translations, rotations and scalings between images, which is sufficient for a high number of stabilization problems. This chapter presents the main principles of the method, gives implementation details and describes its adaptation to different image stabilization applications.

The method was implemented efficiently for real-time execution with low computing resources. Several versions were tested and validated in different applications. The proposed method has also been validated with visual images but it is with infrared images where the advantages with respect to feature-based methods are more evident.

The main strengths of the proposed technique are the following:

1. it relies on the global structure of the image and thus, it is suitable for images with low contrast and resolution and high noise levels;
2. the method can be tuned balancing the stabilization accuracy and the computational burden, allowing adaptation to specific image stabilization needs.

This chapter is structured as follows:

- Section 2 introduces the problem of stabilization of sequences of infrared images.
- Section 3 presents the principles of image matching using Fourier-Mellin transform and Symmetric Phase Only Matched Filtering.
- Section 4 shows examples that illustrate the operation of the proposed method.
- Section 5 proposes some practical aspects to increase the robustness, accuracy and efficiency of the method.
- Section 6 presents various implementations in different applications.

Finally, Section 7 is devoted to the final discussions and conclusions.

2. Stabilization of infrared images

2.1 Infrared images

Infrared cameras generate images of the scene that contain the radiation intensity field within the infrared band. Infrared cameras can "see" in pitch black conditions and through smoke. They can transform radiation measurements in temperature estimations and can also generate thermograms containing temperature of the objects in the scene (Hudson, 1969). In the last decade infrared camera technology has evolved significantly. While in the decade of the 90's most infrared cameras weighted several kilograms, consumed hundreds of watts and their cryogenic cooling systems required frequent maintenance, now it is possible to find radiometric infrared cameras that weight less than 150 gr, consume less than 1,5 W and require no maintenance since they have no cooling system. These advances together with a remarkable cost decrease have motivated their use in a growing number of applications. Two infrared cameras are shown in Fig. 1, one *Mitsubishi IR-M300* (left) and one *Indigo Omega* micro-camera (right).

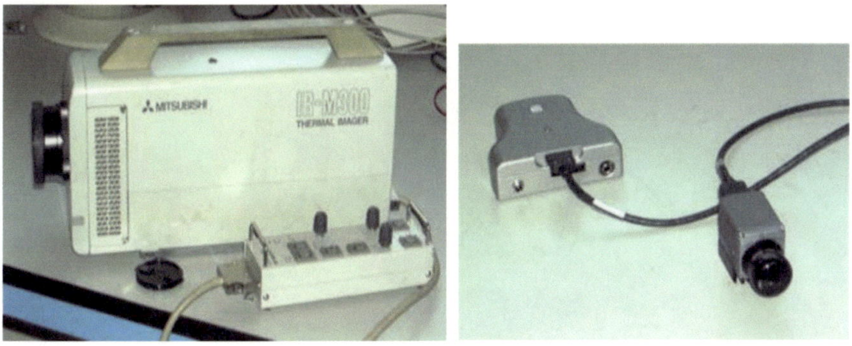

Fig. 1. Left) *Mitsubishi IR-M300* infrared camera. Right) *Indigo Omega* infrared micro-camera

However, despite these advances the quality of the images from infrared cameras is significantly lower than that of visual images due to several physical and technological reasons. Infrared camera detectors are highly affected by different types of noise including thermal noise, shot noise and flicker noise (Hudson, 1969). In fact, many cameras, particularly those operating in the mid-infrared spectral window [3-5] µm, use cooling systems -often cryogenic- to ensure that the detector operates at low and constant temperatures. Noise is crucial in infrared technology. In fact, infrared cameras usually use several characteristics to measure the influence of noise in the images, such as the *Noise Equivalent Temperature* (NET) or the *Noise Equivalent Temperature Difference* (NETD), among others (Maldague, 2001).

Also, infrared detectors usually have significantly lower sensitivity than CMOS and CCD detectors commonly used in visual cameras. This lack of sensitivity originates images with low contrast levels. Many infrared cameras compensate this lack of sensitivity by increasing camera exposure times, which originate image blurring if the camera is under vibrations. Also, infrared detectors usually have lower resolution than visual cameras. Thus, infrared images usually have lower resolutions -and details- than visual images.

Two images of the same scene are shown in Fig. 2: one taken with a visual camera (left) and one taken with a *FLIR ThermaCam P20* infrared camera (right). The differences in contrast, noise levels and detail levels are noticeable.

2.2 Brief description of image stabilization methods

Assume that we have a sequence of images taken by one camera under vibrations. Each image is represented by $Im_t(x,y)$, where t is the time when the image was captured. The camera vibrations cause relative motions between the images in the sequence. Assume image $Im_0(x,y)$ is considered the reference image. Stabilization of image $Im_t(x,y)$ consists of detecting and cancelling the motion between $Im_t(x,y)$ and $Im_0(x,y)$, such that there is no apparent motion between $Im_0(x,y)$ and the stabilized version of $Im_t(x,y)$. The process of detecting the relative motion between two images is called image matching.

There are two main groups of image-matching methods: spatial domain and frequency domain methods. Other techniques such as invariant moments (Abu-Mostafa & Psaltis, 1984) have poorer matching performance. A detailed survey can be found in (Zitová & Flusser, 2003).

Fig. 2. Two images of the same scene but taken with different cameras one visual camera (left) and one *FLIR ThermaCam P20* infrared camera (right)

Spatial domain matching methods are based on local features that can be detected in both images $Im_0(x,y)$ and $Im_t(x,y)$. Assume that both images contain enough and sufficiently distributed features perceptible in both images. These feature-based methods usually have four steps. The first step is to detect local features in both images. These features are typically selected as corners (Tomasi, 1991), high-contrast points or local patterns with invariant properties, (Bay et al., 2008). The feature detector is applied to both images. The second step is to associate features detected in $Im_0(x,y)$ with features detected in $Im_t(x,y)$. Maximum likelihood criteria are often used for feature association. This step provides associations between features in both images, which can be used to estimate the relative motion between them.

The third step is to determine the motion between $Im_0(x,y)$ and $Im_t(x,y)$ using the aforementioned associations. The homography matrix is often used to model the motion between two images since it allows describing the transformations originated by changes on the location and orientation of the camera when the scene can be approximated by a plane (Hartley & Zisserman, 2004). Methods such as Least Median of Squares and *RANSAC* (Fischler & Bolles, 1981) are used to increase the robustness of the motion estimation. Once m_{0i}, the motion from $Im_0(x,y)$ to $Im_t(x,y)$, has been estimated, the fourth step is to apply image-processing methods to induce in $Im_t(x,y)$ a motion inverse to m_{0i}.

Spatial domain techniques have good performance if both images contain clear and robust local features but are not suitable for infrared images since a low number of local features can be robustly detected in infrared images.

On the other hand, frequency domain image-matching methods exploit the properties of images in the frequency domain. Normalized cross-correlation, see e.g. (Barnea & Silverman, 1972) and (Segeman, 1992), has been a common approach to match images with relative translations. In this case, the location of the maximum peak of the cross-correlation function between $Im_0(x,y)$ and $Im_t(x,y)$ corresponds to the relative translation between $Im_0(x,y)$ and $Im_t(x,y)$. Cross-correlation is commonly applied using the Fourier transform due to significant computational savings. This method does not have good performance if $Im_0(x,y)$ and $Im_t(x,y)$ have high noise levels: it produces broad peaks, which originate inaccuracies when determining the location of the maximum peak. Some alternatives, such as the Symmetric Phase-Only Matched Filtering (SPOMF) (Ersoy & Zeng, 1989) produce significantly sharper peaks and are more resilient to noise, partial occlusions in the images and other defects.

Another disadvantage of Fourier transform methods is that they can only match translated images. The Fourier-Mellin transform (FMT) improves also this disadvantage since FMT methods can match translated, rotated and scaled images, (Chen et al., 1994).

2.3 Proposed frequency domain image stabilization method

The method proposed in this chapter uses image-matching methods based on frequency domain transformations. The method relies on the global structure of the image and not on local features. This fact provides significant robustness to noise and to the lack of contrast and resolution. Moreover, the proposed method is capable of stabilizing sequences which images are related through translations, rotations and scalings.

The proposed stabilization method adopts an incremental approach: the first image of the sequence, $Im_0(x,y)$, is considered reference and each of the remaining images in the sequence, $Im_t(x,y)$, is matched and corrected with respect to the stabilized version of the previous image, $Im^S_{t-1}(x,y)$. Section 3 briefly describes the principles of the proposed method. Section 4 illustrates the method with some examples.

Image matching can originate small -sub-pixel- errors. The accumulation of these errors in an incremental stabilization approach can originate drifts. To avoid drifts we developed another absolute stabilization approach in which all the images of the sequence are matched and corrected with respect to the same reference image $Im_0(x,y)$. This and other practical issues are discussed in Section 5.

The proposed method was customized to several problems with different stabilization requirements. In all cases the method was implemented and validated. This is described in Section 6.

3. Image matching using the Fourier-Mellin transform and Symmetric Phase-Only Matched Filtering

Assume that $Im^S_{t-1}(x,y)$ is the stabilized version of the image gathered at time t-1. In the incremental stabilization approach adopted $Im^S_{t-1}(x,y)$ is considered the reference image for stabilization of $Im_t(x,y)$ at time t. Assume that $Im_t(x,y)$ is a rotated, scaled and translated replica of $Im^S_{t-1}(x,y)$. The proposed stabilization method consists of two steps. The first one detects and corrects the rotation and scaling between $Im^S_{t-1}(x,y)$ and $Im_t(x,y)$. $Im^R_t(x,y)$ is the rotation and scaling corrected version of $Im_t(x,y)$. The second step detects and corrects the translation in axes x and y between $Im^S_{t-1}(x,y)$ and $Im^R_t(x,y)$. The stabilized version of $Im_t(x,y)$, i.e. the translation-corrected version of $Im^R_t(x,y)$, is denoted $Im^S_t(x,y)$.

Assume that $s(x,y)$ and $r(x,y)$ are the central rectangular regions of $Im_t(x,y)$ and $Im^S_{t-1}(x,y)$, respectively. Thus, $s(x,y)$ is a replica of $r(x,y)$ rotated with an angle α, scaled with a factor σ and translated with translational offsets (x_0, y_0):

$$s(x,y) = r\big(\sigma(x\cos\alpha + y\sin\alpha) - x_0,\ \ \sigma(-x\sin\alpha + y\cos\alpha) - y_0\big) \qquad (1)$$

The proposed method uses Symmetric Phase-Only Matched Filtering to match translated images. Below the Symmetric Phase-Only Matched Filtering and the two aforementioned steps in the algorithm are summarized.

3.1 Symmetric Phase-Only Matched Filtering

Assume that $b(x,y)$ is replica of $c(x,y)$ translated (x_0,y_0) in a noisy scene, $b(x,y)=c(x-x_0,y-y_0)+n(x,y)$, where $n(x,y)$ represents a white zero mean random noise. A traditional matched filter for $b(x,y)$ has the following transfer function:

$$H_{MF}(u,v)=\frac{1}{|n_w|^2}C^*(u,v), \tag{2}$$

where $C^*(u,v)$ is the complex conjugate of the Fourier transform of $c(x,y)$ and $|n_w|$ is the noise intensity. The output of the matched filter in (2) has a maximum peak at (x_0,y_0). The location of this peak determines the translational offset (x_0,y_0).

One of the main limitations of traditional matched filters is that the output is affected by the energy of the image while translation between images only influences the spectral phase. To minimize this dependence the alternative adopted in our method is to use Symmetric Phase-Only Matched Filtering (SPOMF) to determine translations between images. The SPOMF between $b(x,y)$ and $c(x,y)$ is as follows:

$$Q_{SPOMF}(u,v)=\frac{B(u,v)}{|B(u,v)|}\frac{C^*(u,v)}{|C(u,v)|}=\exp[j\varphi_B(u,v)-j\varphi_C(u,v)], \tag{3}$$

where $\varphi_B(u,v)$ and $\varphi_C(u,v)$ are the spectral phases of $b(x,y)$ and $c(x,y)$, respectively. In the absence of noise $Q_{SPOMF}(u,v)=\exp[-j2\pi(ux_0+vy_0)]$ and the inverse Fourier transform of $Q_{SPOMF}(u,v)$ is a Dirac delta located at (x_0,y_0). SPOMF yields to a sharp peak, significantly sharper than in case the matched filters in (2). The location of the maximum is easier to estimate and more tolerant to noise, which, as previously mentioned, is a significant advantage when dealing with infrared images.

3.2 Rotation and scaling correction step

This step aims to detect and correct the rotation and scaling between $s(x,y)$ and $r(x,y)$. The corrected version will be called $s^R(x,y)$. From (1) the Fourier transform of $s(x,y)$, $S(u,v)=Fourier\{s(x,y)\}$, can be expressed by the following expression:

$$S(u,v)=\sigma^{-2}\left|R\left(\sigma^{-1}(u\cos\alpha+v\sin\alpha),\ \sigma^{-1}(-u\sin\alpha+v\cos\alpha)\right)\right|\ \exp[-j\varphi_S(u,v)], \tag{4}$$

where $R(u,v)=Fourier\{r(x,y)\}$ and $\varphi_S(u,v)$ is the phase of $S(u,v)$. From (4) it is easy to notice that $|S(u,v)|$ is affected by rotations and scalings but is invariant to translations:

$$|S(u,v)|=\sigma^{-2}\left|R\left(\sigma^{-1}(u\cos\alpha+v\sin\alpha),\ \sigma^{-1}(-u\sin\alpha+v\cos\alpha)\right)\right| \tag{5}$$

The rotation and the scaling can be decoupled by converting $|R(u,v)|$ and $|S(u,v)|$ to polar coordinates. Let $r_p(\theta,\rho)$ and $s_p(\theta,\rho)$ be $|R(u,v)|$ and $|S(u,v)|$ in polar co-ordinates (θ,ρ):

$$r_p(\theta,\rho)=|R(\rho\cos\theta,\rho\sin\theta)| \tag{6}$$

$$s_p(\theta,\rho)=|S(\rho\cos\theta,\rho\sin\theta)| \tag{7}$$

Thus, it is easy to check that:

$$\sigma^{-1}(u\cos\alpha+\upsilon\sin\alpha)=\frac{\rho}{\sigma}\cos(\theta-\alpha) \tag{8}$$

$$\sigma^{-1}(-u\sin\alpha+\upsilon\cos\alpha)=\frac{\rho}{\sigma}\sin(\theta-\alpha) \tag{9}$$

Hence, from (5), $s_p(\theta,\rho)$ can be expressed as:

$$s_p(\theta,\rho)=\sigma^{-2}\ r_p(\theta-\alpha,\rho/\sigma) \tag{10}$$

Thus, the rotation of an angle α has been transformed into a translation in the θ axis. Scaling can be also transformed into a translation by applying a logarithimc scale on ρ axis, i.e. $\lambda=\log(\rho)$. In log-polar coordinates (5) can be expressed as:

$$s_{pl}(\theta,\lambda)=s_{pl}(\theta,\log(\rho))=\sigma^{-2}\ r_{pl}(\theta-\alpha,\lambda-\log(\sigma)), \tag{11}$$

where $s_{pl}(\theta,\lambda)$ and $r_{pl}(\theta,\lambda)$ are the log-polar versions of $|S(u,v)|$ and $|R(u,v)|$. Thus, in (11) the rotation and scaling have been transformed into translations. Let $S_{pl}(\upsilon,\omega)$ and $R_{pl}(\upsilon,\omega)$ be the Fourier transforms of $s_{pl}(\theta,\lambda)$ and $r_{pl}(\theta,\lambda)$, respectively. Using shift properties of the Fourier transform, (11) can be rewritten as:

$$|S_{pl}(\upsilon,\omega)|\exp\{-j\varphi_{Spl}(\upsilon,\omega)\}=|R_{pl}(\upsilon,\omega)|\exp\left[-j\varphi_{Spl}(\upsilon,\omega)-2\pi(\upsilon\log(\sigma)+\omega\alpha)\right] \tag{12}$$

Thus, $|S_{pl}(\upsilon,\omega)|=|R_{pl}(\upsilon,\omega)|$. In (12) the rotation and the scaling appear as shifts between $\varphi_{Spl}(\upsilon,\omega)$ and $\varphi_{Rpl}(\upsilon,\omega)$, the phases of $S_{pl}(\upsilon,\omega)$ and $R_{pl}(\upsilon,\omega)$, respectively. The phase shift is:

$$\delta=\varphi_{Spl}(\upsilon,\omega)-\varphi_{Rpl}(\upsilon,\omega)=2\pi(\upsilon\log(\sigma)+\omega\alpha) \tag{13}$$

Then, if we apply a SPOMF, the location of the peak of $q_r(\theta,\lambda)$, $(\theta_{max},\lambda_{max})$, represents the rotation angle $\alpha=\theta_{max}$ and the scaling factor $\lambda_{max}=\log(\sigma)$ between $s(x,y)$ and $r(x,y)$:

$$q_r(\theta,\lambda)=Fourier^{-1}\{\exp[-j\delta]\}, \tag{14}$$

where $Fourier^{-1}$ stands for the inverse Fourier transform. The corrected image, $Im^R_t(x,y)$, is obtained by rotating $Im_t(x,y)$ an angle $-\theta_{max}$ and scaling by factor $\exp(\lambda_{max})$. The central rectangular part of $Im^R_t(x,y)$, $s^R(x,y)$, is used for the translation correction step of the algorithm.

3.3 Translation correction step

Once the rotation and scaling have been corrected, $s^R(x,y)$ and $r(x,y)$ are related only by translations in axes x and y. This step detects and corrects the translational offsets between $s^R(x,y)$ and $r(x,y)$. The SPOMF can be used to determine the translations between them:

$$Q_t(u,v) = \exp[-j(\varphi_{SR}(u,v) - \varphi_R(u,v))] \quad , \tag{15}$$

where $\varphi_{SR}(u,v)$ and $\varphi_R(u,v)$ are respectively the phases of $S^R(u,v)$ and $R(u,v)$, the Fourier transforms of $s^R(x,y)$ and $r(x,y)$. The translational offset between $s^R(x,y)$ and $r(x,y)$ can be obtained by computing:

$$q_t(x,y) = Fourier^{-1}\{ Q_t(u,v)\} \tag{16}$$

The peak of $q_t(x,y)$ is located at $x=x_{max}$ and $y=y_{max}$. Thus, the translations between $s^R(x,y)$ and $r(x,y)$ in axes x and y are $x_0=x_{max}$ and $y_0=y_{max}$. The stabilized image, $Im^S{}_t(x,y)$, is computed by shifting $Im^R{}_t(x,y)$ by $-x_0$ in axis x and by $-y_0$ in axis y, respectively.

4. Experiments

This section presents some experiments that illustrate the proposed method. Two infrared images gathered from an infrared camera under vibrations are shown in Fig.3. The image at the left will be consider stabilized, $Im^S{}_{t-1}(x,y)$, and will be used as reference to stabilize the image at the right, $Im_t(x,y)$. The method determined and corrected the relative rotation angle, scaling factor and translations of $Im_t(x,y)$ with respect to $Im^S{}_{t-1}(x,y)$.

Fig. 3. Two infrared images from a sequence of images under vibrations. The image at the left is taken as reference

First, the central parts of both images, $r(x,y)$ and $s(x,y)$, are selected, see Fig. 4. Their size is chosen to be power of two in order to optimize the computational burden of the Fourier Transform FFT algorithm. The original images are 640x480 and the central parts are 256x256.

The first step of the algorithm described in Section 3 starts. First, $R(u,v)$ and $S(u,v)$, the Fourier transforms of $r(x,y)$ and $s(x,y)$ are computed. $|R(u,v)|$ and $|S(u,v)|$ are shown in Fig. 5. It can be noticed that rotation between $r(x,y)$ and $s(x,y)$ results in a rotation between $|R(u,v)|$ and $|S(u,v)|$ as predicted by (5).

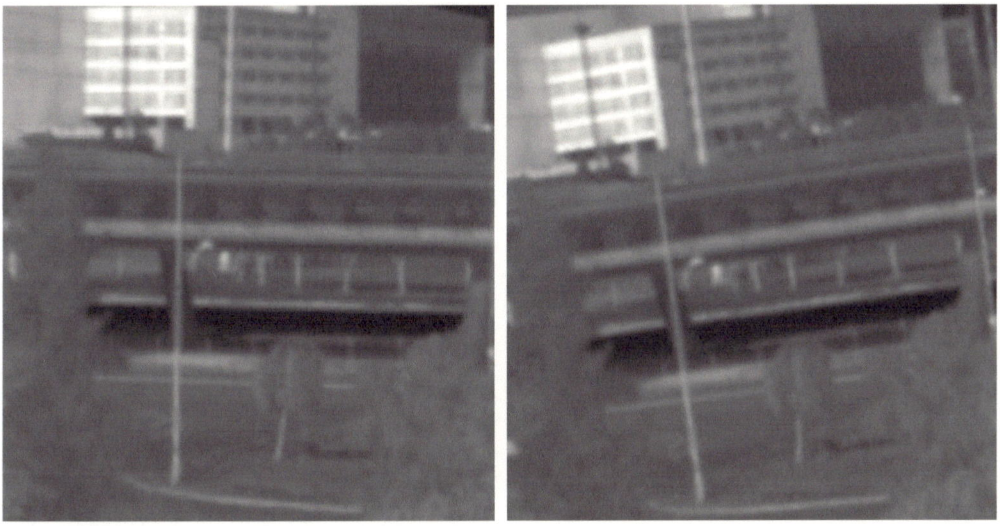

Fig. 4. $r(x,y)$ and $s(x,y)$ for images in Fig. 3

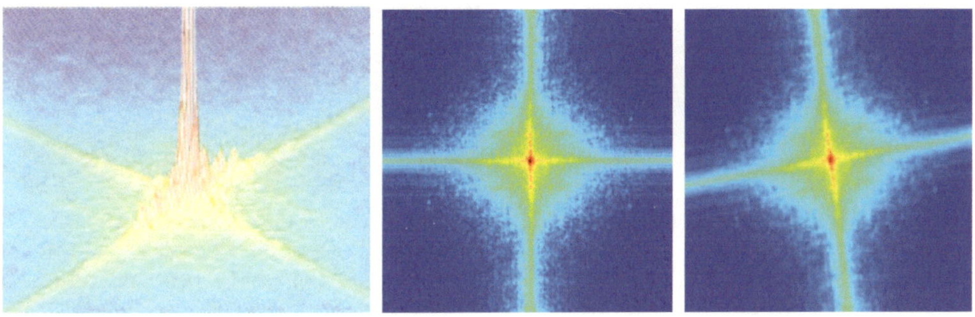

Fig. 5. Left) 3D view of $|R(u,v)|$; Center) and Right) 2D views $|R(u,v)|$ and $|S(u,v)|$, respectively

Then, $r_p(\theta,\lambda)$ and $s_p(\theta,\lambda)$, the versions of $|R(u,v)|$ and $|S(u,v)|$ in log-polar co-ordinates are computed. $r_p(\theta,\lambda)$ and $s_p(\theta,\lambda)$ are shown in Fig. 6. The colour represents the magnitude. As predicted by (11) the rotation between $|R(u,v)|$ and $|S(u,v)|$ has been transformed into a shift in the θ axis between $r_p(\theta,\lambda)$ and $s_p(\theta,\lambda)$. The value of the shift corresponds to the rotation angle between $Im^S_{t-1}(x,y)$ and $Im_t(x,y)$.

Next, we apply the SPOMF between $r_p(\theta,\lambda)$ and $s_p(\theta,\lambda)$ using (13) and (14). The resulting $q_r(\theta,\lambda)$ is shown in Fig. 7left. The maximum peak in $q_r(\theta,\lambda)$ is sharp. It is located at θ_{max}=-9.14°, λ_{max}=0. The second peak is 50% lower in magnitude and is located at high distance from the maximum peak, which facilitates the identification of the maximum peak. Notice that $\lambda_{max}=\log(\sigma_{max})=0$ involves $\sigma_{max}=1$, i.e. there is no scaling between both images. The rotation between both images is θ_{max}=-9.14°. Next, $Im_t(x,y)$ is corrected: $Im_t(x,y)$ is rotated an angle - α =9.14°. $Im^R_t(x,y)$, the rotated-corrected version of $Im_t(x,y)$, is shown in Fig. 7right. The central rectangular part of $Im^R_t(x,y)$, $s^R(x,y)$, is selected for the second part of the method.

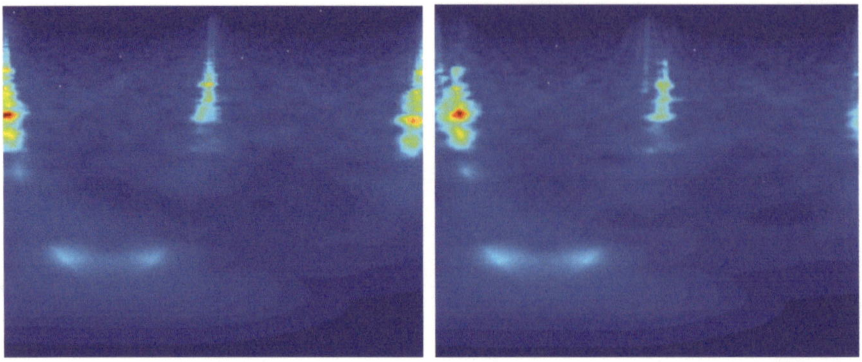

Fig. 6. Resulting $r_p(\theta,\lambda)$ (left) and $s_p(\theta,\lambda)$ (right)

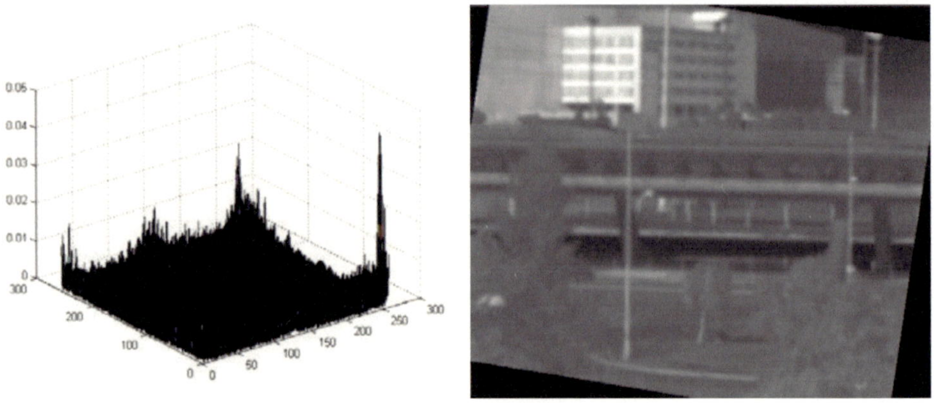

Fig. 7. Left) Resulting $q_r(\theta,\lambda)$. Right) $Im^R{}_t(x,y)$, i.e. $Im_t(x,y)$ after rotation correction

Once the rotation and scaling between $Im^S{}_{t-1}(x,y)$ and $Im_t(x,y)$ have been cancelled, the following step is to compute the translations between $s^R(x,y)$ and $r(x,y)$. The SPOMF is computed using (15) and (16). The resulting $q_t(x,y)$ is in Fig. 8left. The peak was very sharp. The maximum peak was located at x_{max}=-4 and y_{max}=7. Thus, $s^R(x,y)$ and $r(x,y)$ are translated x_0=-4 in axis x and y_0=7 in axis y. Next, $Im^R{}_t(x,y)$ is shifted by $-x_0$ and $-y_0$ in axes x and y respectively in order to compensate the vibration. Fig. 8right shows $Im^S{}_t(x,y)$, the stabilized version of $Im_t(x,y)$, i.e. the translation-corrected version of $Im^R{}_t(x,y)$.

5. Practical issues

This section presents practical aspects that have been developed to increase the robustness, accuracy and efficiency of the proposed method.

5.1 Image enhance

Preliminary experiments revealed that applying preprocessing image enhance methods improved significantly the performance of the proposed method. An efficient histogram stretching method was used. The objective is to transform the image levels such that the enhanced image has desired mean intensity -MI_{ref}- and contrast -C_{ref}- values. The histogram stretching implements the following linear transformation function:

$$Im_E(x,y) = \frac{C_{ref}}{C}\left(Im(x,y)\text{-}MI\right) + MI_{ref}\,, \tag{17}$$

where C and MI are the contrast and mean values before the histogram stretching. $Im(x,y)$ and $Im_E(x,y)$ are respectively the images before and after applying image enhance methods. Image enhance increases the level of noise in the images. Although higher noise levels have harmful effects in feature-based matching methods, the high robustness to noise of SPOMF avoids the degradation in our case. In images with very low contrast levels it is interesting to apply edge detectors. Fig. 9 shows the resulting images after applying Sobel edge detector to $r(x,y)$ and $s(x,y)$ in Fig. 4.

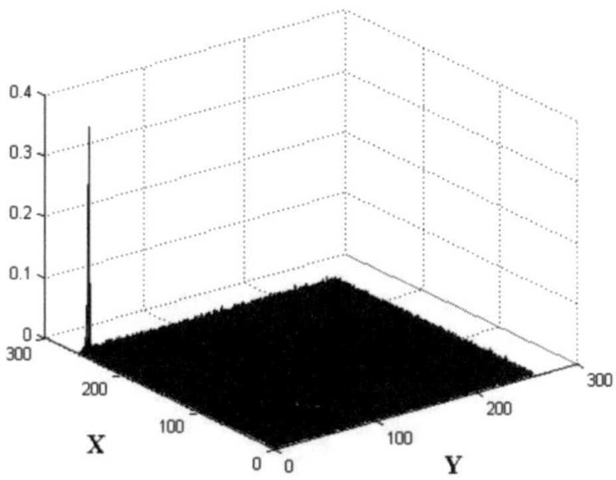

Fig. 8. Left) Resulting $q_t(x,y)$. Right) $Im^S_t(x,y)$, stabilized version of $Im_t(x,y)$

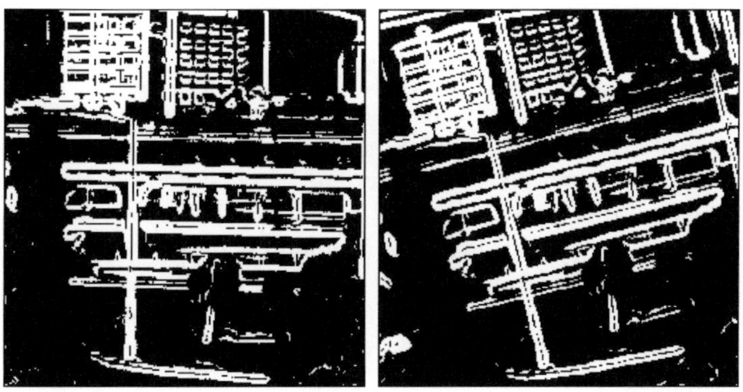

Fig. 9. Resulting images after applying Sobel edge detector to $r(x,y)$ and $s(x,y)$ in Fig. 4

5.2 Drift correction

Constrained resolutions of digital images generate small sub-pixel errors in image matching. These errors have no perceptible consequences unless they accumulate originating drifts in incremental stabilization schemes. The simplest way to avoid drifts is to adopt an absolute stabilization approach, in which images are stabilized with respect to the same reference image $Im_0(x,y)$. Different schemes were analysed to update the reference images depending on the application. The reference image is considered valid if the overlapping between $Im_t(x,y)$ and $Im_0(x,y)$ is above a certain threshold. Otherwise, a new reference image is needed.

5.3 Operating modes

The resolution in the computation of σ, α, x_0 and y_0 is highly dependent on the size of the images. Although it is not easy to establish mathematically this dependence in real noisy images, it is possible to analytically obtain some qualitative conclusions.

Assume that the size of $r(x,y)$ and $s(x,y)$ is MxM and that the size of $s_p(\theta,\lambda)$ and $r_p(\theta,\lambda)$ is WxW. The resolution of the rotation angle depends on the size of $S_p(\upsilon,\omega)$ and $R_p(\upsilon,\omega)$, i.e. WxW. The minimum detectable rotation angle is $a_{min} \approx 2/W$. The value of W also has influence on the errors in the computation of the rotation angle. The higher W, the higher accuracy in the computation of the rotation angle. SPOMF produces broader peaks in $q_r(\theta,\lambda)$ in case of using smaller W. The value of W depends on the number of different radius values considered in the polar conversion, which is constrained by the size of $S(u,v)$ and $R(\upsilon,\omega)$, which size is MxM.

The size of $r(x,y)$ and $s(x,y)$, i.e. M, has influence on the resolution and errors in the computation of the translations. Low values of M involve broader peaks in $q_t(x,y)$, which involves poorer peak detection.

To deal with different stabilization accuracies and computer burden, two operation modes have been selected: *Mode1* and *Mode2*. *Mode1* uses low values of M and W. *Mode1* has moderate stabilization capability. *Mode2* uses high values of M and W. *Mode2* stabilizes more accurately at the expense of higher computer burden. *Mode2* is applied when high accuracy is required or in case of vibrations of high magnitude.

5.4 Increase accuracy through sub-pixel resolution

As described in the above sub-section the resolutions in the computation of σ, α, x_0 and y_0 are limited by sizes of $q_r(\theta,\lambda)$ and $q_t(x,y)$ and thus, by the sizes of the images. Noise broadens the peaks of the SPOMF. A sub-pixel estimation method is used to determine with accuracy and robustness the location of the peak in $q_r(\theta,\lambda)$ or $q_t(x,y)$. This sub-pixel method detects the maximum of $q_r(\theta,\lambda)$ or $q_t(x,y)$ and defines a neighborhood around it. The centroid in the neighborhood around the maximum peak is considered as the location of the peak. This simple method allows incrementing the accuracy in peak localization. It also increases the robustness of the maximum peak detection in case $q_r(\theta,\lambda)$ and $q_t(x,y)$ have secondary peaks near the maximum peak.

6. Implementation

The proposed method has been implemented in various problems illustrating its flexibility.

6.1 Stabilization of images from a hovering UAV

The proposed technique has been tested with the HERO3 helicopter, see Fig. 10 developed by the Group of Robotics, Vision and Control (GRVC) at the University of Seville (Spain).

The application is to perform building inspection and particularly the detection of thermal heat losses on the building envelope. The main perception sensors on board HERO3 are an *Indigo Omega* infrared micro-camera and a visual camera. The *Indigo Omega* use infrared uncooled detectors, with low sensitivity and high noise levels. The helicopter is equipped with an onboard PC-104 computer for stabilizing the infrared images and applying heat losses detection.

Fig. 10. HERO3 during building inspection experiments carried out at the School of Engineering of Seville (Spain)

The vibrations in the images have diverse origins. When the UAV is hovering, the images can be affected by low-frequency oscillations originated by the compensations of the UAV control systems. Also, images are affected by high-frequency vibrations induced by the UAV engine. When the helicopter is hovering at certain position and orientation, the low-magnitude distortions between consecutive images can be approximated by translations and rotations.

The method was customized for the application. It was assumed that there is no change in the scaling factor between images: it only corrected rotations and translations. The onboard infrared micro-camera weights 120 gr. and has a resolution of 160x120 pixels. $r(x,y)$ and $s(x,y)$ were selected of size 64x64. During the inspection the objective is to stabilize images when regions of interest are present in the images. An automatic function was developed to detect when an object of interest -windows in this problem- is present. At that moment the image stabilization method is triggered and this image is used to initialize the stabilization method.

Fig. 11 shows consecutive infrared images taken from an infrared camera onboard HERO3 in an experiment carried out in December 2005. The image noise level is rather high. The translations between the images can be observed. The stabilized images are in Fig. 12. The image stabilization time was 12,6 ms in a PC-104 with computing power similar to a Pentium III 800 MHz.

6.2 Stabilization of infrared images from a ship

The proposed method has been used for stabilization of images from an infrared camera on a ship. The objective was to detect other ships and obstacles using automatic computer vision methods. The stabilization method should keep the images stable despite the sea condition and ship motion. This application required a high-performance infrared camera and optical system with high resolution and low noise level. Infrared images have some difficulties in sea scenes. Water absorbs infrared radiation: water appears in images with low radiation levels. Sky also appears in infrared images with low radiation levels. The horizon line can be well perceived in infrared images.

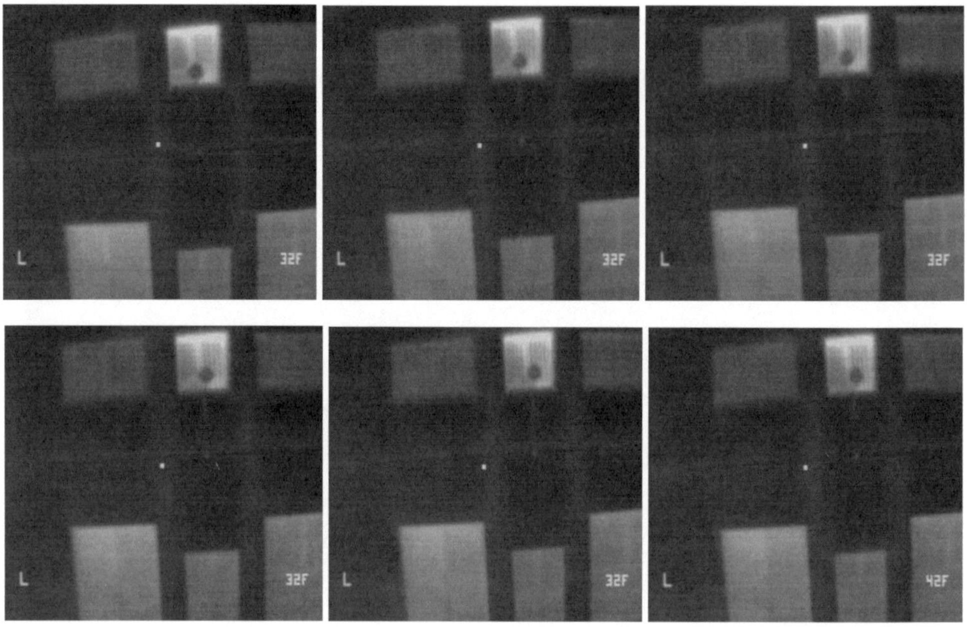

Fig. 11. Sequence of consecutive images from an *Indigo Omega* onboard HERO3

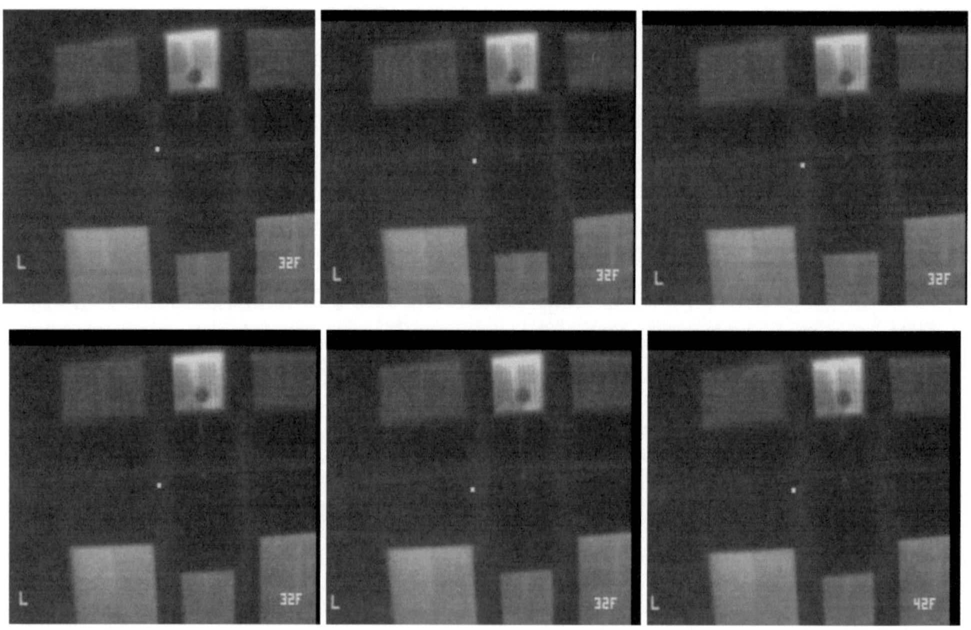

Fig. 12. Stabilized sequence of images

In images with no ships or objects, the stabilization method should greatly rely on the horizon line, what is useful to compensate rotations and translations in the vertical axis but is not enough to correct translations in the horizontal axis. If the image contains objects the method can correct also translations in the horizontal axis. Fortunately, in this application the method had to operate only if some threats for the ship navigation were found. An automatic tool was used to detect the presence of objects and to trigger the stabilization method.

The implemented version included the image enhance methods to deal with the lack of contrast. The resolution of the cameras used was 512x512. Hence, $r(x,y)$ and $s(x,y)$, input of the algorithm, were selected of size 256x256 in *Mode1* and 512x512 in *Mode2*. A short video showing the performance of the method can be found in (web1). In this video the ship was in the port and the camera was pointing at nearby forested area. Only very low magnitude vibrations can be observed. The method corrects the vibrations except the sub-pixel errors described in Section 5. The translational offsets x_0 (left) and y_0 (right) computed along one sequence of images are shown in Fig. 13. The time to stabilize one image was 102,1 ms in *Mode1* and 346,6 ms in *Mode2* in a computer with processing power similar to a Pentium III t 800 MHz.

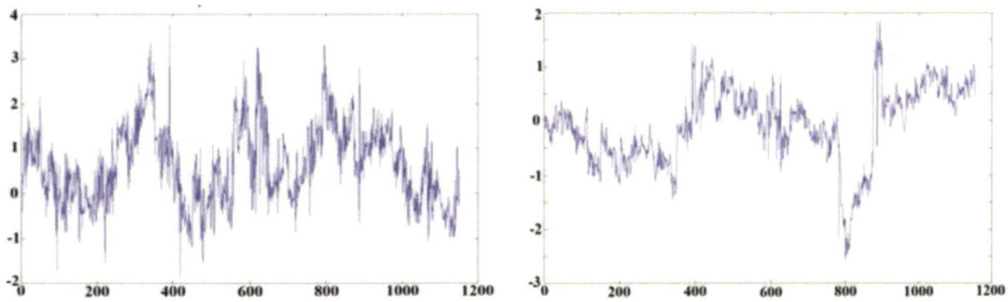

Fig. 13. Values of x_0 (left) and y_0 (right) obtained in the stabilization of a sequence of 1000 images from an infrared camera mounted on a ship

6.3 Stabilization of images from a ground vehicle in motion

The infrared camera is mounted on a ground vehicle. The camera is located on a pole at certain height and pointing such that it has a general view of an area. The vehicle can be static or moving on the ground. If the vehicle is static the method employed is similar to that used in Section 6.2. If the vehicle is moving, the stabilization method compensates rotations and vertical translations but do not cancel horizontal translations. When the vehicle is moving on the horizontal plane, even in the absence of vibrations, the images have horizontal translations. Without using any other sensors but the camera itself, it is not possible to differentiate if a horizontal translation is originated by the vehicle motion or by vibrations.

Fig. 14 shows 6 consecutive images taken from one camera under vibrations. The camera moves on the horizontal plane along the image sequence. The images resulting from the complete stabilization version of the method -taking the first image as reference- are in Fig. 15. The objects appear static along the sequence. The reduction in the overlapping between the images generates black patches of increasing size on the corrected images. The resulting images after partial stabilization, i.e. without correcting the horizontal translations, are shown in Fig. 16. The scene appears static except for the horizontal axis.

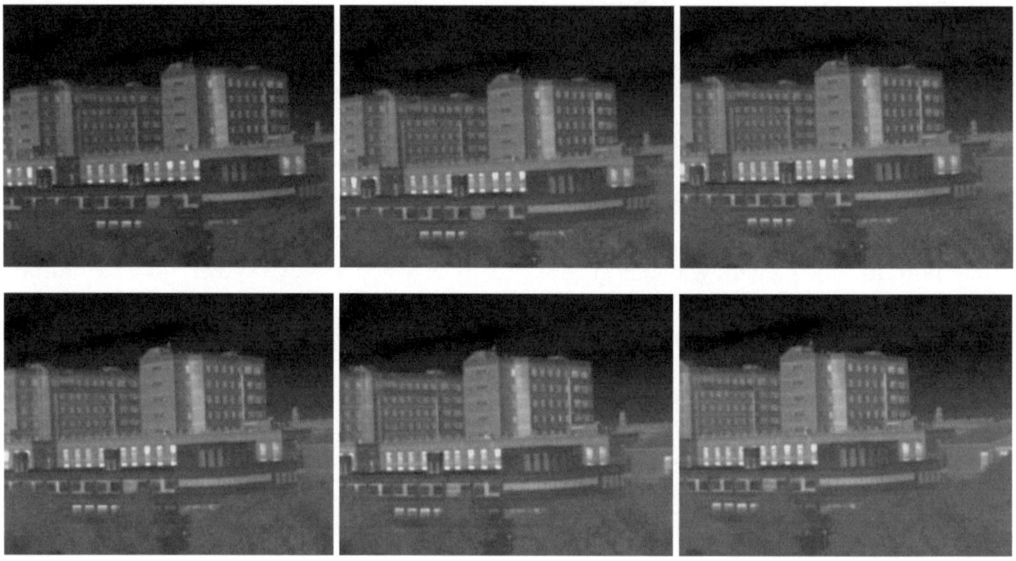

Fig. 14. Consecutive images taken from a camera under vibrations and moving on the horizontal plane

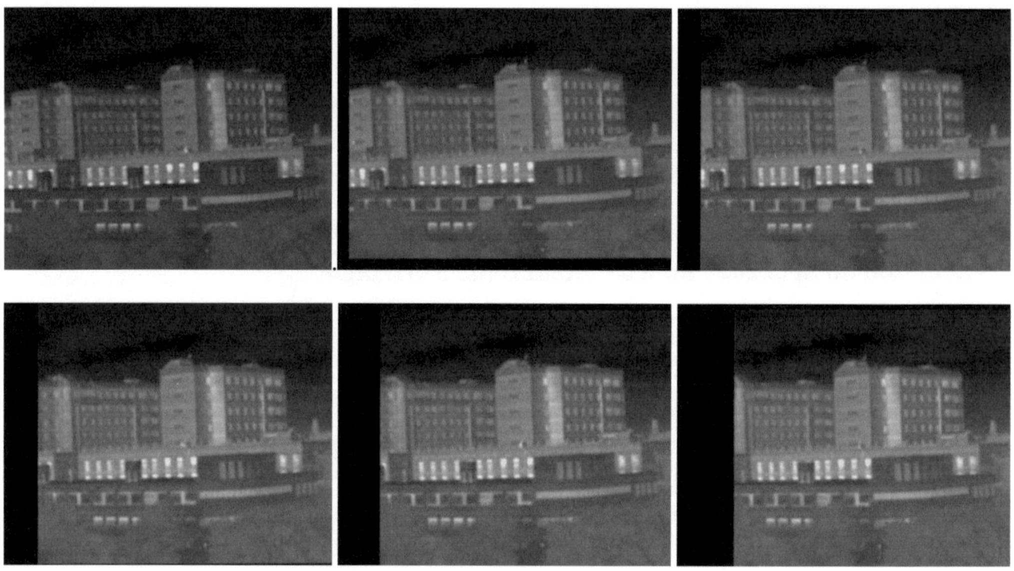

Fig. 15. Images resulting from the complete stabilization version of the method

Both versions, complete and partial stabilization, were developed, each of them with two modes: *Mode1* -image size is 256x256- and *Mode2* -image size is 512x512. An operator can change the option (complete or partial stabilization) and mode during operation. In both cases the stabilization times were similar to those in Section 6.2.

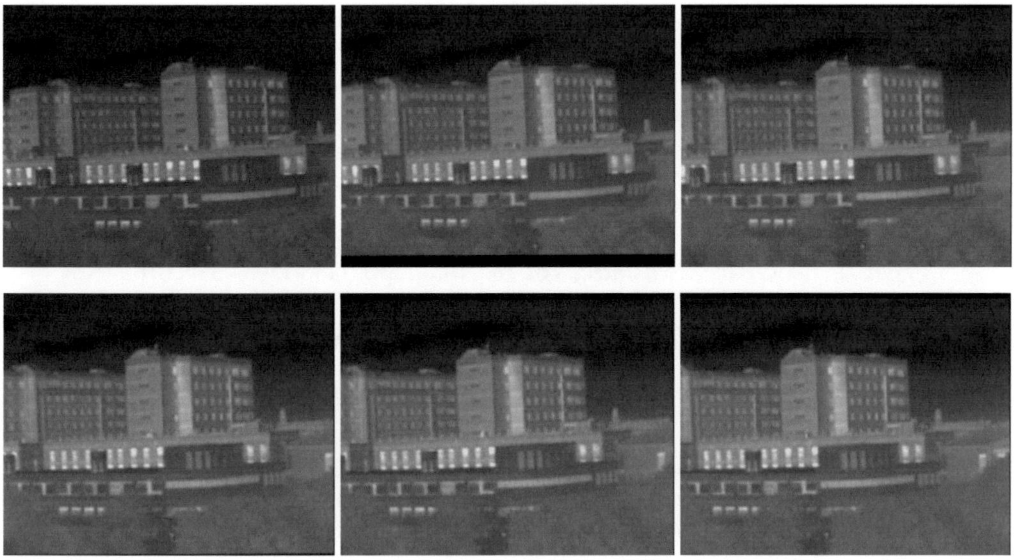

Fig. 16. Images resulting from partial stabilization. The scene appears static except for the horizontal axis

7. Conclusions

This chapter presents a technique for stabilization of sequences of infrared images using frequency domain methods. The work is motivated by the lack of local features and high noise levels commonly present in infrared images, which hampers stabilization methods based on feature matching. Instead, the proposed method relies on the global structure of the image. The described method makes extensive use of Fourier-Mellin transforms and Symmetric Phase Only Matched Filtering. The Fourier-Mellin transform allows determining the rotation, scaling and rotation between two images by converting the images to log-polar coordinates. Symmetric Phase Only Matched Filtering enhances the performance of image matching in the presence of noise.

The main strength of the proposed method is the capability to deal with images in which it is difficult to find clear and repeatable features. Its flexibility and capability for balancing between stabilization accuracy and computational cost is also remarkable and allows its customization to applications with different requirements.

The chapter also concerns several practical aspects that have been considered to increase its robustness and accuracy including image contrast enhance and the definition of modes with different performance and computer burden. The method was implemented for real-time execution with low computing resources. Different versions were implemented and validated in several applications. Three of them are briefly summarized in the chapter.

Software implementation and real-time execution have been two main requirements in the design of the methods. They originated the development of stabilization modes to balance between accuracy and computer cost. The implementation of the stabilization method in FPGA is interesting to reduce the computer burden of the main processor. In this case, *Mode2*, with better stabilization performance, could be used in all conditions.

The combination of visual and infrared images provides interesting synergies in a growing number of problems. The differences between images from cameras in different spectral bands would also hamper the application of feature-based methods. The use of the proposed method for matching images in different spectral bands is object of current work.

8. References

Abu-Mostafa, Y.S. & Psaltis, D. (1984). Recognition aspects of moment invariants. *IEEE Trans. Pattern Anal. Mach. Intel.*, Vol. 16, No. 12, pp. 1156-1168, ISSN 0162-8828

Barnea, D. I. & Silverman, H. F. (1972). A class of algorithms for fast image registration. *IEEE Trans. Computers* C-21, Vol. C-21, No. 6, pp. 179-186, ISSN 0018-9340

Bay, H; Ess, A.; Tuytelaars, T; Van Gool, L. (2008). SURF: Speeded up Robust Features. *Computer Vision and Image Understanding*, Vol. 110, No. 3, pp. 346-359, ISSN 1077-3142

Carvajal, E; Jimenez-Espadafor, F; Becerra, J.A.; Torres, M. (2011). Methodology for the Estimation of Cylinder Inner Surface Temperature in an Air-Cooled Engine. *Applied Thermal Engineering*, Vol. 31, No. 8-9, pp. 1474-1481, ISSN 1359-4311

Chen, Q.-S.; Defrise, M.; Deconinck, F. (1994). Symmetric phase-only matched filtering of Fourier-Mellin transforms for image registration and recognition. *IEEE Trans. on Pattern Analysis and Machine Intel.*, Vol. 16, No. 12, pp. 1156-1168, ISSN 0162-8828

Ersoy, O.K. & Zeng, M. (1989). Nonlinear matched filtering. *J. Opt. Soc. Am. A.*, Vol. 6, No. 5, pp. 636-648, ISSN 1084-7529

Fischler, M.A. & Bolles, R.C. (1981). Random Sample Consensus: A Paradigm for Model Fitting with Applications to Image Analysis and Automated Cartography. *Communications of the ACM*, Vol. 24, No. 6, pp. 381–395, ISSN 0001-0782

Hartley, R. & Zisserman, A. (2004). Multiple View Geometry in Computer Vision. *Cambridge University Press*, ISBN 0521540518, Cambridge (United Kingdom)

Hudson, R.D. (1969). Infrared System Engineering. *John Wiley & Sons*, ISBN 0471418501, New York (USA)

Maldague, X.P.V. (2001). Theory and Practice of Infrared Technology for Nondestructive Testing. *John Wiley & Sons*, ISBN 0471181900, New York (USA)

Segeman, J. (1992). Fourier cross correlation and invariant transformations for an optimal recogntion of functions deformed by affine groups. *J. Opt. Soc. Am. A*, Vol. 9, No. 6, pp. 895-902, ISSN 1084-7529

Tomasi, C. (1991). Shape and motion from image streams: A factorization method, PhD Thesis Carnegie Mellon University.

Zitová, B. & Flusser, J. (2003). Image registration methods: a survey. *Image and Vision Computing*, Vol. 21, No. 11, pp. 977-1000, ISSN 0262-8856

http://www.youtube.com/watch?v=inPUcDIHC4s, Retrieved on 14/03/2012

23

SITAF: Simulation-Based Interface Testing Automation Framework for Robot Software Component

Hong Seong Park and Jeong Seok Kang
Kangwon National University
South Korea

1. Introduction

Many researchers in robotics have proposed a Component-based Software Engineering (CBSE) approach to tackle problems in robot software development (Jawawi et al., 2007). Especially in the component-based robot system, the system quality depends on the quality of each component because any defective components will have bad effects on the system built with them. Thus, component interface test is critical for checking the correctness of the component's functionality. It is especially difficult to test robot software components because of the following two main problems.

First, the preparation of all hardware modules related to robot software and the configuration of a test environment is labor-intensive. Second, it is difficult to define or generate test cases for testing robot software components.

The simulation plays an important role in the process of robotic software development. The simulation allows testing of robot software components and experimentation with different configurations before they are deployed in real robots. Traditional simulation-based approaches (Hu, 2005, Martin & Emami, 2006, Michel, 2004) focus on architectures or methods (e.g., computer-based simulation, hardware-in-the-loop-simulation, and robot-in-the-loop- simulation), rather than testing. Many software engineering researchers (Buy et al., 1999, Bundell et al., 2000, Zamli et al., 2007, Momotko & Zalewska, 2004, Edwards, 2001) have investigated software component testing, but they have not considered simulation environments. Simulations can be used within a specification-based testing regime, which helps robot software developers define and apply effective test case. Note that the generation of test case is an important approach in the field of automated testing.

In this paper, we propose a Simulation-based Interface Testing Automation Framework (SITAF) for robot software components. SITAF automatically generates test cases by applying specification-based test techniques and considering simulation-dependent parameters. SITAF also performs the interface testing in distributed test environments by interacting with a simulation. SITAF controls test parameters during testing, which affect the behavior of a component under testing (CUT); examples of such parameters are simulation-dependent parameters, input/output parameters of provided/required CUT interface. The main advantage of this technique is that it identifies errors caused by interactions between the CUT and the external environment.

The primary contribution of this paper is the integration of specification-based test into the simulation for automatic interface testing of robot software components.

The rest of this paper is organized as follows. Section 2 presents the SITAF architecture. The two main functions of SITAF are presented in Section 3. Section 4 discusses the evaluation of SITAF. Finally, we have some conclusions in Section 5.

2. SITAF architecture

The main aim of SITAF is to automate as much of test process for robot software component as is possible. To achieve this aim, the architecture of proposed framework consists of a Web-based Interface Testing Automation Engine Server (ITAES), a Test Build Agent (TBA), and a robot simulator and is shown in Fig. 1.

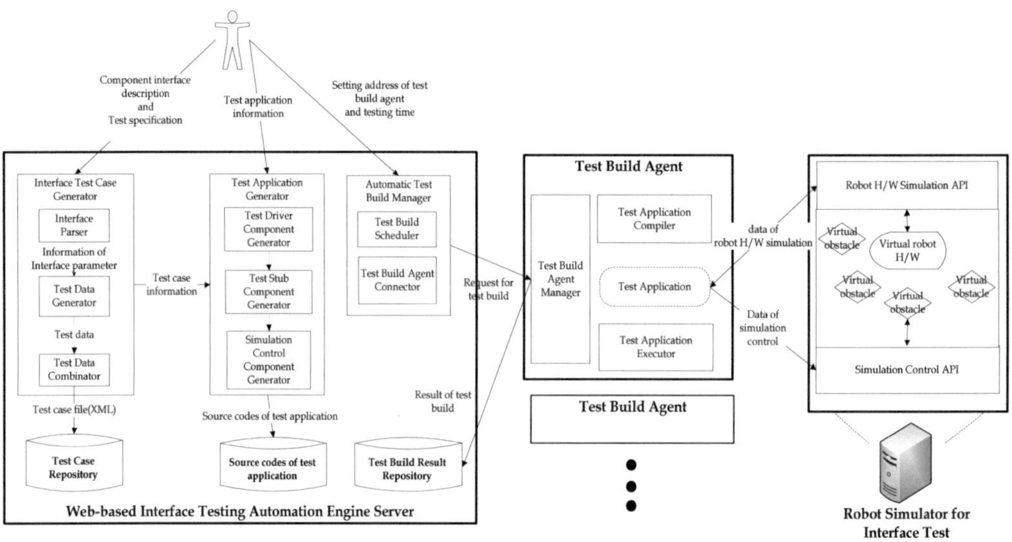

Fig. 1. Simulation-based Interface Testing Automation Framework Architecture

ITAES is the core of the framework to which the user is accessible via a Web service, and generates test cases for interface test of robot software components. And it also generates a test driver component, a test stub component, and a simulation control component, which are required for the testing, and links the generated components to each other. Furthermore ITAES manages test resources such as test cases, test applications, test results, and test logs in a unit of test activity which is a workspace for execution of test operations. The ITAES consists of three main modules of Interface Test Case Generator (ITCG), Interface Test Application Generator (ITAG), and Automatic Test Build Manager (ATBM).

The ITCG module automatically generates test cases by extended test schemes based on specification-based testing techniques such as equivalence partition (Ostrand & Balcer,

1998), boundary value analysis (Hoffman et al., 1999), random test (Ntafos, 1997, Zhu, 1997), and pairwise test (Williams, 2000). It receives the interface representation information in the form of Interface Definition Language (IDL) or eXtensible Markup Language (XML) and verifies the specification information for a CUT before test cases are automatically generated. The test cases are stored as XML files in a database. The user accesses a Web interface to modify test cases in the database and inputs the expected result values for each test case. The ITAG module generates the source code for the test application. The test application is composed of a test driver component, a test stub component, and a simulation control component. All source codes are shared with TBAs. The ATBM module connected with the distributed TBAs manages a test build which means compilation and execution of a test application. And it provides three types of the test time for the test build: immediate, reserved, and periodic.

An individual TBA can exist in different test environments, and communicates with the ATBM in ITAES. TBAs are in charge of automatic building of test application. The TBA contains three modules of Test Build Agent Manager (TBAM), Test Application Compiler (TAC), and Test Application Executor (TAE). The TBAM module manages a TBA and receives a test build request from the ATBM in ITAES, and then downloads the test application and test case files. The TAC module and the TAE module automatically compile and execute a test application. These modules upload the logs and the test results to ITAES after the completion of compilation and execution.

The robot simulator for interface test provides a simulation control API and a virtual robot hardware API. The simulation control API is used to control the virtual test environment in the simulation. The simulation control component in the test application dynamically modifies or controls a virtual test environment for each test case by using the simulation control API. The virtual robot hardware API is used to control virtual robot hardware or to receive data. If the test component is a hardware-related component, the component controls the virtual robot hardware or receives data using the API for simulation. The simulation control data is used to generate effective test cases.

3. Automatic interface test operations for robot software components

In this section, we describe two main functions of SITAF, which are the automatic generation of interface test case and the automatic execution of interface test by simulation.

3.1 Automatic generation of interface test cases

Specification-based test techniques are applied to the generation of test cases for robot software components by simulation. A test case for a robot software component consists of an input vector that requires Test Data of Input (TDI), Test Data of Simulation Dependency (TDSD), and Test Data of Test Stub (TDTS) because the behavior of CUT is affected by these test data. TDI is an input parameter of the interface under testing. Because TDSD refers the simulation control data for interface testing, it is the data affecting the CUT through the simulation. TDTS is an output data from a required interface of the CUT.

The process for the automatic interface test case generation has two steps, which is shown in Fig. 2.

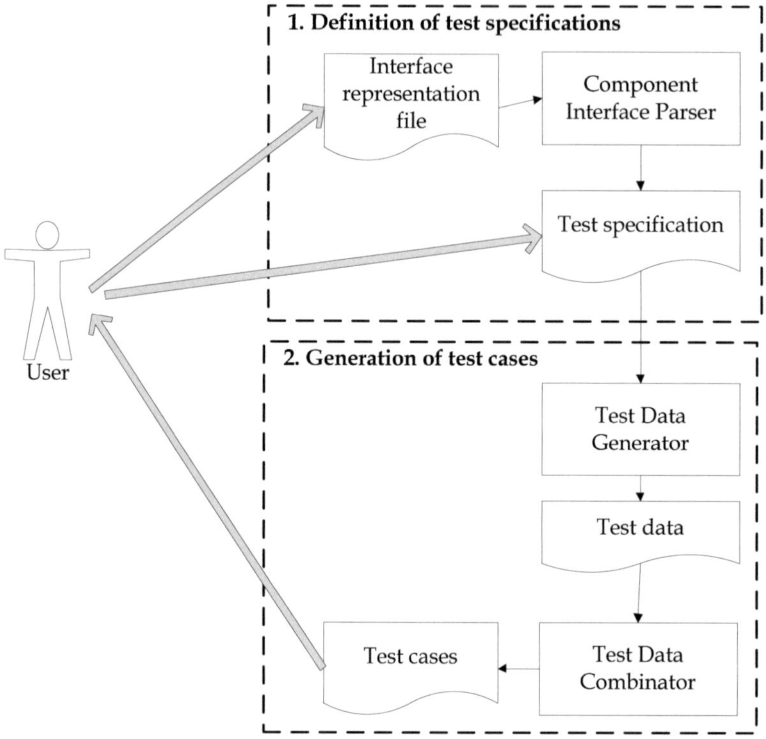

Fig. 2. Process of the automatic interface test case generation

3.1.1 Definition of test specifications

The Interface Parser module in the Interface Test Case Generator parses the interface representation of a CUT and the information on input test parameters of CUT is extracted. Fig. 3 shows a simplified XML schema of the interface representation. The main elements of the interface representation are *type_name, method,* and *param*. The *type_name* represents type of the interface and the *method* and *param* describe prototype of methods in the interface.

The test specification includes valid range values, specific candidate values, pre/post-conditions, and other values, for each test parameter. And a test model is a set of the test specifications, as shown in Fig. 4. The essential elements of the test model are *name, input_spec_list*, and *order_of_combinations*. The *name* is a name of the test model, as identifier. The *input_spec_list* includes test specifications of TDI, TDSD, and TDTS. The *order_of_combinations* is a number of interaction strength for pairwising the each test parameters.

In Fig. 4, the plus sign (+) indicates that the element further consists of *test_element* which is used to generate the test data values. The *test_element* is a key element in the test specification, which contains information on a test parameter such as type of the parameter, specific candidate value, and the method for test data generation. The *type* in *test_element* presents the type of a test parameter such as TDI, TDSD, and TDTS. The *test_spec* describes a method and additional information for generating test data, and has different XML schema for each type of *test_spec*, as shown in Fig. 5, Fig. 7, and Fig. 8.

```
<xs:schema>
  <xs:element name="service_port_type">
    <xs:complexType>
        <xs:sequence>
        <xs:element name="type_name" type="NCName"/>
        <xs:element ref="method" maxOccurs="unbounded" />
      </xs:sequence>
    </xs:complexType>
  </xs:element>
  <xs:element name="method">
    <xs:complexType>
        <xs:attribute name="name" type="xs:NCName" use="required"/>
        <xs:attribute name="return_type" type="xs:NCName" use="required"/>
        <xs:attribute name="call_type" type="xs:NCName" default="blocking"/>
      <xs:sequence>
        <xs:element maxOccurs="unbounded" ref="param"/>
      </xs:sequence>
    </xs:complexType>
  </xs:element>
  <xs:element name="param">
    <xs:complexType>
        <xs:sequence>
        <xs:element name="type" type="xs:NMTOKEN"/>
            <xs:element name="name" type="xs:NCName"/>
      </xs:sequence>
        <xs:attribute name="index" type="xs:integer" use="required"/>
    </xs:complexType>
  </xs:element>
</xs:schema>
```

Fig. 3. Simplified XML Schema of an interface representation

```
<xs:schema>
  <xs:element name="test_model">
   <xs:complexType>
    <xs:sequence>
     <xs:element name="name" type="xs:string" use="required" />
        <xs:element name="description" type="xs:string/>
     + <xs:element ref="pre_condition_spec_list" />
     + <xs:element ref="input_spec_list" use="required"/>
     + <xs:element ref="output_spec_list" />
     + <xs:element ref="post_condition_spec_list" />
        <xs:element name="order_of_combinations" type="xs:integer"
          use="required"/>
     </xs:sequence>
   </xs:complexType>
  </xs:element>
  <xs:element name="test_element">
   <xs:complexType>
    <xs:sequence name="name" type="xs:string" use="required"/>
     <xs:sequence name="type" type="xs:string" use="required"/>
     <xs:sequence name="description" type="xs:string"/>
     <xs:sequence ref="test_spec" use="required" />
     <xs:sequence ref="user_value_list" />
     <xs:attribute name="abstract_type" type="xs:string" use="required" />
     <xs:attribute name="real_type" type="xs:string" use="required"/>
   </xs:complexType>
  </xs:element>
</xs:schema>
```

Fig. 4. Simplified XML Schema of a test model

3.1.2 Generation of test cases

The Test Data Generator (TDG) module in the ITCG generates the test data satisfying the test specification for each test parameter. The TDG automatically generates the numeric test data by applying an equivalence partitioning scheme (ECP), a boundary value analysis scheme (BVA), and a random testing scheme. Furthermore this paper generates the test data of string type using BVA and random testing scheme.

The ECP scheme (Ostrand & Balcer, 1998) is a software testing technique that divides input data for a software unit into partitions of data from which test cases cane be derived. In principle, test cases are designed to cover each partition at least once. This technique aims to define test cases and uncover classes of errors, thereby reducing the total number of test cases that must be developed. Additionally this paper defines types of equivalence class, listed in Table 1. The TDG automatically generates test data by each type of equivalence class. Fig. 5 shows simplified XML schema of *test_spec* element for ECP.

Type of equivalence class	Description
NEC_NUMERIC_ONE_BOUDARY	The type includes just one boundary. If x < 10, there are two equivalence classes, x<10 and x>10.
NEC_NUMERIC_TWO_BOUDARY	The type includes two boundaries. If -1 < x < 10, there are three equivalence classes, x<-1, -1<x<10, x>10.
NEC_BOOLEAN	If x=true, there are two equivalence classes, x=true, x=false.
NEC_NUMERIC_CONSTANT	If x=3, there are two equivalence classes, x=3, x!=3
NEC_NUMERIC_SET	If x={-1,0,1}, there are two equivalence classes, x={-1,0,1}, x!= {-1,0,1}.

Table 1. Types of equivalence class for ECP scheme

```
<test_spec type="ECP" >
   <xs:complexType>
    <xs:element ref=" equiv_class" />
   </xs:complexType>
   <xs:element name="equiv_class">
     <xs:attribute name="type" use="required">
       <xs:simpleType>
         <xs:restriction base="xs:string">
           <xs:enumeration value=" NEC_NUMERIC_ONE_BOUDARY"/>
           <xs:enumeration value=" NEC_NUMERIC_TWO_BOUDARY"/>
           <xs:enumeration value=" NEC_BOOLEAN"/>
           <xs:enumeration value=" NEC_NUMERIC_CONSTANT"/>
           <xs:enumeration value=" NEC_NUMERIC_SET"/>
         </xs:restriction>
       </xs:simpleType>
   </xs:attribute>
   .......
   </xs:element>
</test_spec>
```

Fig. 5. Simplified XML schema of *test_spec* element for ECP scheme

The BVA scheme (Hoffman et al., 1999) is a software testing technique that designs tests including representatives of boundary values. Values on the minimum and maximum edges of an equivalence partition are tested. The values could be input or output ranges of a software component. Boundaries are common locations for errors that result in software faults, so they are frequently explored in test cases. Furthermore this paper defines the offset value of boundary for generation of elaborate test data, as shown in Fig. 6. This paper automatically generates the test data by the BVA such as the values of min_low_off_set, minimum boundary, min_high_off_set, max_low_off_set, maximum boundary, max_high_off_set, and additionally a middle value.

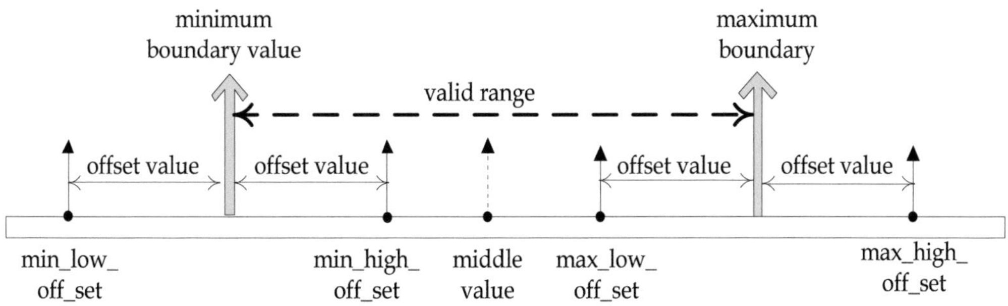

Fig. 6. Offset and test data of BVA scheme

In the BVA, there are just two types of equivalence class because the scheme needs range values, as shown in Fig. 7.

```
<test_spec type="BVA" offset="" >
   <xs:complexType>
    <xs:element ref=" equiv_class" />
   </xs:complexType>
   <xs:element name="equiv_class">
     <xs:attribute name="type" use="required">
        <xs:simpleType>
           <xs:restriction base="xs:string">
              <xs:enumeration value=" NEC_NUMERIC_ONE_BOUDARY"/>
              <xs:enumeration value=" NEC_NUMERIC_TWO_BOUDARY"/>
           </xs:restriction>
        </xs:simpleType>
   </xs:attribute>
   .......
   </xs:element>
</test_spec>
```

Fig. 7. Simplified XML schema of *test_spec* element for BVA scheme

A random testing scheme (Ntafos, 1997, Zhu, 1997) is a strategy that requires the "random" selection of test cases from the entire input domain. For random testing, values of each test case are generated randomly, but very often the overall distribution of the test cases has to conform to the distribution of the input domain, or an operational profile. In this paper, the scheme is used to generate test data of numeric and string types. In particular, we combine the random testing scheme and the BVA scheme for generation of test data of string type. This paper analyses the boundary value of minimum and maximum length of string, and then randomly generates test data, which is satisfied with the options such as alphabet, number, special character, space, and negative character, as shown in Fig. 8.

```
<test_spec type="RANDOM_STRING">
  <xs:complexType>
   <xs:sequence>
    <xs:element name="min_length" type="xs:integer" use="required" />
     <xs:element name=" max_length" type="xs:integer" use="required" />
     <xs:element name=" alphabet" type="xs:boolean" use="required" />
     <xs:element name=" number" type="xs: boolean" use="required" />
     <xs:element name=" special_char" type=" xs:boolean" use="required" />
     <xs:element name=" space" type="xs:boolean" use="required" />
     <xs:element name=" negative_char_list" type="xs:string" />
   </xs:sequence>
  </xs:complexType>
</test_spec>
```

Fig. 8. Simplified XML schema of *test_spec* element for random string scheme

The Test Data Combinator (TDC) module in ITCG combines the test data using a pairwise scheme (Williams, 2000) for reducing the number of test cases. The pairwise scheme is an effective test case generation technique, which is based on the observation that most faults are caused by interactions among input vectors. The TDC enables two-way combination, three-way combination, and all possible combinations of the test data, which allows the user to remove overlapping test cases from pairs of parameter combinations. The combined test cases are stored in an XML file, as shown in Fig. 9. The *case_param_info* describes name and type of a test parameter and the *case_list* consists of values of the test case.

```
<xs:schema>
   <xs:element name="test_suite_data">
    <xs:complexType>
     <xs:sequence>
      <xs:element name="name" type="xs:string" use="required" />
        <xs:element name="description" type="xs:string/>
        <xs:element name="test_ model_name" type="xs:string" />
      + <xs:element ref="case_param_info" />
      + <xs:element ref="case_list" />
      </xs:sequence>
    </xs:complexType>
</xs:schema>
```

Fig. 9. Simplified XML schema of the test cases

3.2 Automatic test execution by simulation

The test application performs testing by interacting with the robot simulator, as shown in Fig. 10. This paper automatically generates skeleton source codes for test applications and links the components to each other for simplifying testing.

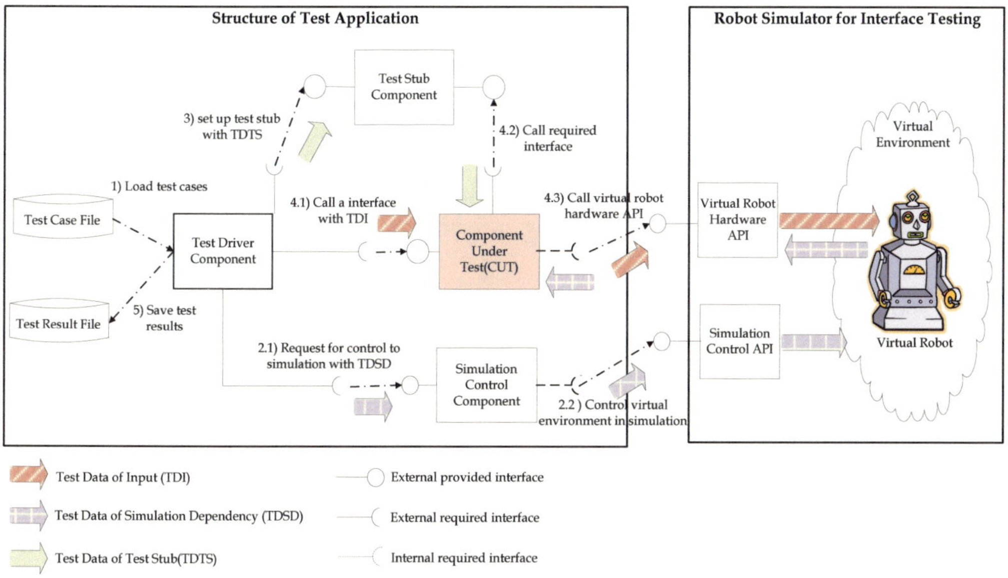

Fig. 10. Structure and operation sequence of a test application

The test application consists of a Test Driver (TD) component, a Simulation Control (SC) component, a Test Stub (TS) component, and a CUT, as shown in Fig. 10. The TD component controls the overall operation of test. During the testing, the TD component reads test cases, and sets the simulation environment and the required interface of the CUT. After the end of testing, the component stores the test results in a file. The SC component sets the simulation environment through the simulation control API in the robot simulator. TS component provides virtual interfaces of the same type as the required interface of the CUT. The TS component simulates the behavior of CUT-dependent software components. Thus, the component is used instead of an actual software component which is needed for execution of the CUT.

The test application and the robot simulator are connected to each other and the following operations shown in Fig. 10 are performed to test the CUT: 1) Read test case file, 2) Call the interface of SC component for control to the simulation environment, 3) Set up the TS component using the TDTS values, 4) Call the interface of the CUT, 5) Save the test results in a file.

The TD component reads a test case file and divides it into the TDI, the TDSD, and the TDTS. The TD component calls the interface of SC component using the TDSD values for setting the simulation environment. The SC component changes the virtual test environment through the simulation control API using TDSD values. After the virtual test environment setup is completed, the TD component set up the output of the required interface of the CUT via the TS component interface using the TDTS. After the configuration of the simulation environment and the TS component are completed, the TD component calls interface of the CUT using the TDI as the input parameters. The CUT calls the interface of the TS component and requests or receives data via the robot hardware API during the simulation. When the operation is completed, the value resulting from the operation may be returned to the TD component. The TD component compares the actual resulting value with the expected resulting value and saves the test result in a file. After all of the testing, the TBA uploads the test result file and the test log file to ITAES. The test log file contains log information on compilation and execution of the test application.

Fig. 11 shows simplified XML schema of the test result descriptor. The *summary* consists of the number of pass and fail, and information on processing times of the interface of the CUT such as a minimum time, a maximum time, an average time, and a standard deviation time. The *test_result_list* contains detailed information on the test result, such as expected and actual test results and a processing time of the interface of the CUT, for each test case. The ITAES reads the test result file and log files, and then shows the information through web interfaces which are table-based view and graphic-based view, for easily analysing the test result of the CUT.

```
<xs:schema>
  <xs:element name="test_result_descriptor">
   <xs:complexType>
    <xs:sequence>
        <xs:element name="name" type="xs:string" />
     + <xs:element ref="summary" />
     + <xs:element ref="test_result_list" />
     </xs:sequence>
   </xs:complexType>
  </xs:element>
</xs:schema>
```

Fig. 11. Simplified XML schema of a test result descriptor

4. Evaluation

In this section, the proposed framework is evaluated using an example of the test of the Infrared Ray (IR) sensor component interface.

This paper implements the Interface Testing Automation Engine Server (ITAES) and Test Build Agent (TBA) in Java and Flex. And the robot simulator used in this paper is OPRoS simulator (http://www.opros.or.kr/). This paper develops the test simulation environment for testing shown in Fig. 12. The environment consists of an IR sensor robot which has some virtual IR sensors linked to the IR sensor simulation API and an obstacle which can move by the obstacle distance control API.

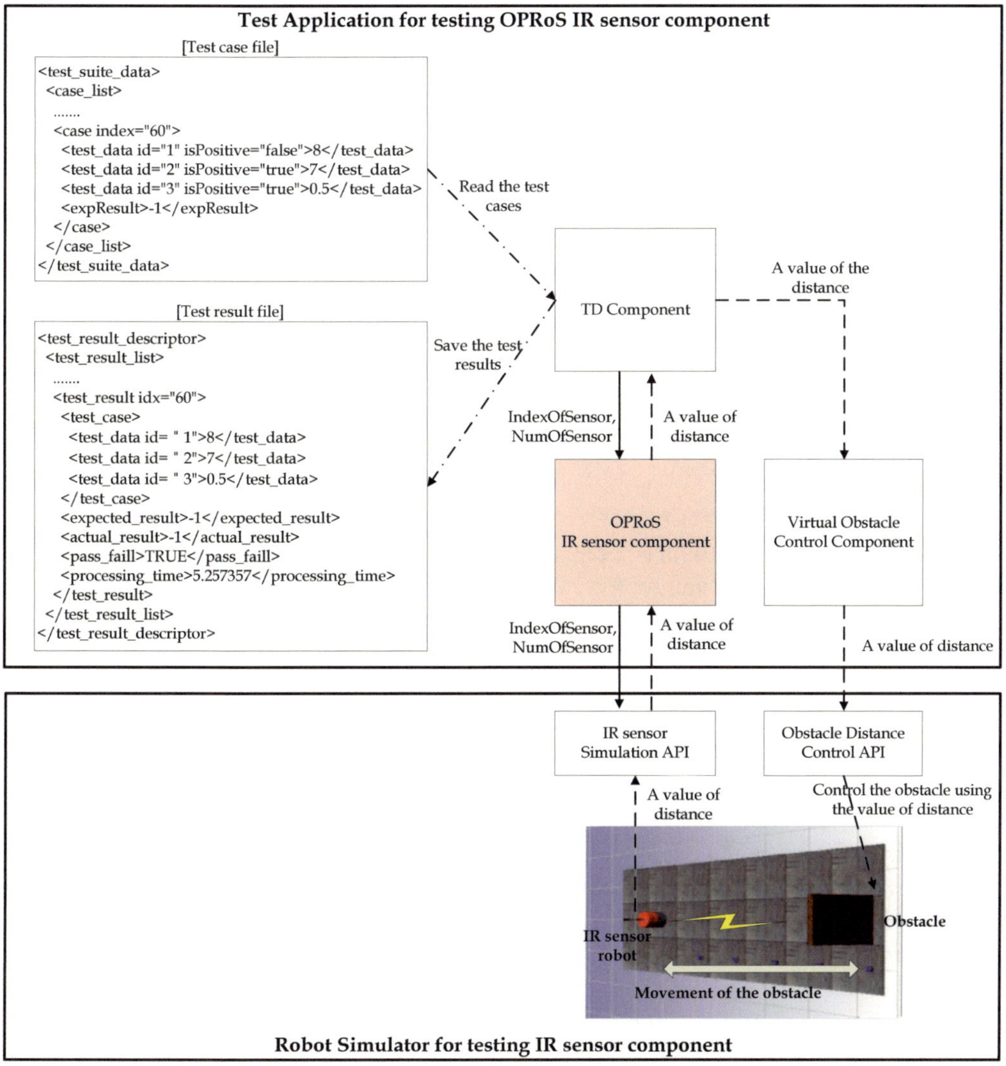

Fig. 12. Application for testing the IR component interface

We tested an Open Platform for Robotic Service (http://www.opros.or.kr/) with the Infrared Ray (IR) sensor component interface, named *GetInfraredData* interface, which had two input parameters of *IndexOfSensor* and *NumOfSensor*. The function of the interface is to get a distance value using IR sensors. This paper defines a new test parameter for TDSD, named "Distance", to control the virtual obstacle in the test simulation environment. Thus, there are two input test parameters and one simulation-dependent parameter, which are shown in Table 2.

Name	Type	Test Specification	Description
IndexOfSensor	TDI	1 <= IndexOfSensor <= 7, The offset value is 1.	It is index of IR sensor.
NumOfSensor	TDI	1 <= NumOfSensor <= 7, The offset value is 1.	It is the number of IR sensor.
Distance	TDSD	0.0 <= #Distance <= 10.0, The offset value is 0.5.	It is a distance between an IR sensor and an obstacle in the test simulation environment. The obstacle is moved using the value of the distance test parameter.
Expected Return Value		If all test parameter are valid values, the value is same of the distance value. If the test parameters are not valid values, the value is -1.	-1 means the operation is failed.

Table 2. Test Specifications and an expected test result of *GetInfraredData* Interface

The test application contains the Test Driver (TD) component, OPRoS IR sensor component, and the virtual obstacle control component. The test application does not have a test stub component because the OPRoS IR sensor component does not have a required interface. First the TD component moves the virtual obstacle in the simulation via the virtual obstacle control component using the value of *Distance* test parameter. Then the TD component calls the interface of the IR sensor component using values of the *IndexOfSensor* and the *NumOfSensor* test parameter. If the return value of the interface of the IR sensor component is same of *Distance* value, the test case is a success.

We validate three functions of proposed framework, which are the creation of test activity, the automated test case generation, and the automatic test execution by simulation.

The process of the creation of test activity has four steps, which are shown in Fig. 13(a) – Fig. 13(d). In the creation of the test activity, the information on the IR sensor component such as component profile, dll file, and interface profile, external library (optional), and the type of the skeleton test code are used.

Fig. 13(e) shows the generated directory of the test activity after completion of the process. The directory includes the test profile, the test driver component, the test stub component, and the concrete test driver.

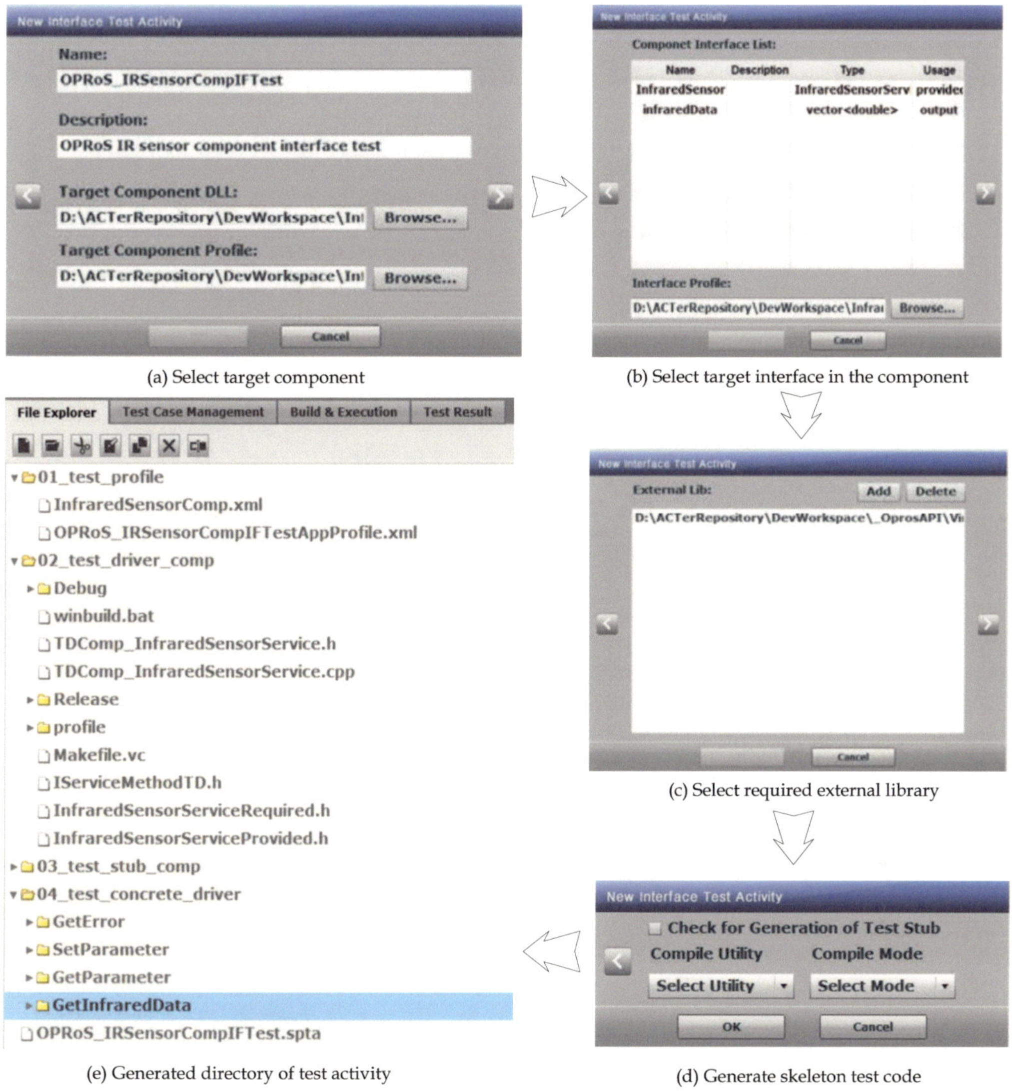

(a) Select target component

(b) Select target interface in the component

(c) Select required external library

(d) Generate skeleton test code

(e) Generated directory of test activity

Fig. 13. Process of creating the test activity for IR sensor component

The process of automatic test case generation has four steps, which are shown in Fig. 14(a) - Fig. 14(d). We input the test specifications through the web user interface, which are the *IndexOfSensor*, the *NumOfSensor*, and the *Distance*. We analyse valid range value of each test parameter using BVA scheme. And then we select "2-way", as order of combination for pairwise. After completion of the process, 60 test cases are generated as shown in Fig. 15. Then we input the expected test result into each test case.

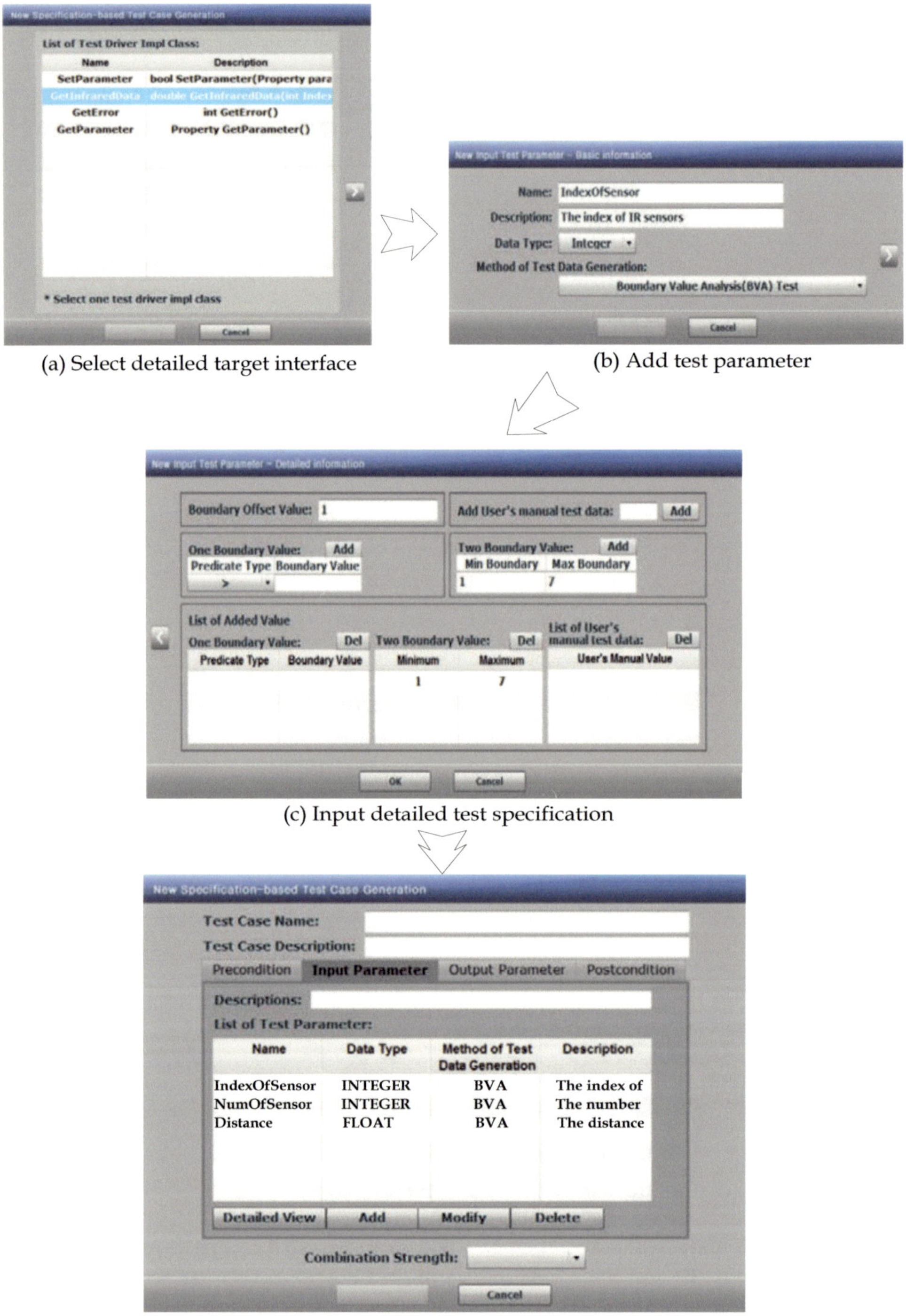

(a) Select detailed target interface

(b) Add test parameter

(c) Input detailed test specification

(d) List of the added test parameters and select the order of combinations for pairwise

Fig. 14. Process of automatic test case generation of *GetInfraredData* interface

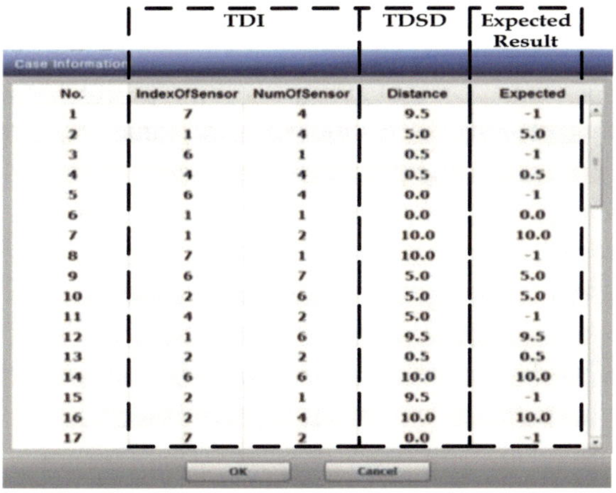

No.	IndexOfSensor	NumOfSensor	Distance	Expected
1	7	4	9.5	-1
2	1	4	5.0	5.0
3	6	1	0.5	-1
4	4	4	0.5	0.5
5	6	4	0.0	-1
6	1	1	0.0	0.0
7	1	2	10.0	10.0
8	7	1	10.0	-1
9	6	7	5.0	5.0
10	2	6	5.0	5.0
11	4	2	5.0	-1
12	1	6	9.5	9.5
13	2	2	0.5	0.5
14	6	6	10.0	10.0
15	2	1	9.5	-1
16	2	4	10.0	10.0
17	7	2	0.0	-1

Fig. 15. Test Case list of *GetInfraredData* interface

We perform and evaluate the example of test application for the IR sensor component in the test simulation environment, as shown in Fig. 12. Finally we compare the return value of the interface of the IR sensor component with the value of expected test result. Fig. 16 shows C++ code of test driver component used in this evaluation.

```
void IRConcreteTestDriver::runTest(int testCaseID){

   // Read test case
   MethodTestCase* testCase = m_methodTestCaseList->getMethodTestCase(testCaseID);

   // Type casting of each test case data
   int tcIndexOfSensor = StringToInteger(testCase->getTestDataString(0));
   int tcNumOfSensor   = StringToInteger(testCase->getTestDataString(1));
   double tcDistance   = StringToDouble(testCase->getTestDataString(2));

   // variable for measurement of interface processing time
   double elapsedTime = 0.0;

   // Moving an obstacle in the test environment
   RobotSimulator->moveWall(tcDistance);

   // Call the IR sensor component interface
   TestUtil::start_timer(1);
   double ret = pTargetServiceObj->GetInfraredData(tcIndexOfSensor, tcNumOfSensor);
   elapsedTime = TestUtil::get_timer(1);

   // Comparing test result
   string strRealValueOfTestResult = DoubleToString(ret);;
   double expectedValueOfTestResult = StringToDouble(testCase->getExpectedTestResult());
   if (ret == expectedValueOfTestResult)
      testCase->setTestResult("PASS");
   else
      testCase->setTestResult("FAIL");
   testCase->setRealTestResultValue(strRealValueOfTestResult);

   // save the processing time of the IR sensor component interface
   testCase->setMethodProcessTime(elapsedTime);
}
```

Fig. 16. Test driver code for testing *GetInfraredData* Interface

Two types of test result view are provided and shown in Fig. 17. The table-based view shows the test result of each test case. And the graphic-based view shows the summary test result, including the percentage of test success and the processing time of the IR sensor component. The example used shows that the proposed framework is working well.

Table View GraphView

Number of Total Cases : 60
Number of PASS : 60
Number of FAIL : 0
Number of DON'T : 0

Min Time(us) : 3.711075
Max Time(us) : 6.379784
Average(us) : 4.569742
Std Deviation(us) : 5.183028

Statement(%) :
Branch(%) :

No.	IndexOfSensor	NumOfSensor	Distance	ER	MR	PF	Exetime
1	7	4	9.5	-1	-1	TRUE	4.605137
2	1	4	5.0	5.0	5.0	TRUE	6.494382
3	6	1	0.5	-1	-1	TRUE	5.875869
4	4	4	0.5	0.5	0.5	TRUE	5.566613
5	6	4	0.0	-1	-1	TRUE	5.566613
6	1	1	0.0	0.0	0.0	TRUE	4.020332
7	1	2	10.0	10.0	10.0	TRUE	5.875869
8	7	1	10.0	-1	-1	TRUE	5.257357
9	6	7	5.0	5.0	5.0	TRUE	5.257357
10	2	6	5.0	5.0	5.0	TRUE	4.948101
11	4	2	5.0	-1	-1	TRUE	4.948101
12	1	6	9.5	9.5	9.5	TRUE	6.494382
13	2	2	0.5	0.5	0.5	TRUE	5.257357
14	6	6	10.0	10.0	10.0	TRUE	5.875869
15	2	1	9.5	-1	-1	TRUE	5.257357
16	2	4	10.0	10.0	10.0	TRUE	5.875869
17	7	2	0.0	-1	-1	TRUE	5.566613
	Test cases			Expected results	Actual results	Pass(true) /Fail(false)	Processing time

(a) Table view of test result

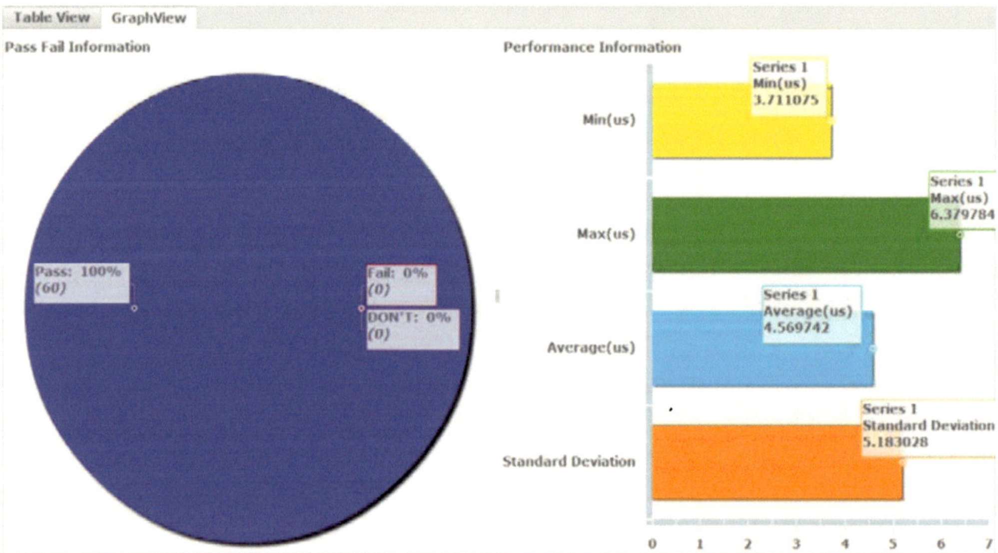

(b) Graphic view of test result

Fig. 17. Test result views of *GetInfraredData* interface

5. Conclusions and future works

This paper proposes and develops the interface testing framework, SITAF, based on simulation and specification-based test for robot software components and develops the automatic test case generation technique for interface testing. SITAF uses three types of the test parameters, which are the input parameter, the test stub parameter, and the simulation dependent parameter and applies specification-based test techniques. SITAF also performs the automatic interface testing to identify errors caused by CUT interactions with an external environment.

SITAF is evaluated via the example of the test of the IR sensor component used in the distance measurement. The example shows that the SITAF generates test cases and performs the automatic interface testing by interactive simulation.

As future works, are considering an automatic regression test by applying software configuration management, and the mixed test environment of a simulation-based environment and a real environment for testing robot software component.

6. References

A. Martin, A. & Emami, M. R. (2006). An Architecture for Robotic Hardware-in-the-Loop Simulation, *Proceedings of the International Conference on Mechatronics and Automation*, pp.2162-2167, June 2006

Bundell, G. A.; Lee, G., Morris, J. & Parker, K. (2000). A Software Component Verification Tool, *Proceedings of the Conf. Software Methods and Tools*, pp. 137-146, 2000

Buy, U.; Ghezzi , C., Orso , A., Pezze, M. & Valsasna, M. (1999). A Framework for Testing Object-Oriented Components, *Proceedings of the 1st International Workshop on Testing Distributed Component-Based Systems*, 1999

Edwards, S. H. (2001). A framework for practical, automated black-box testing of component-based software, *International Journal of Software Testing, Verification and Reliability*, Vol. 11, No. 2, pp. 97-111, June, 2001

Hoffman, D.; Strooper, P. & White, L. (1999). Boundary Values and Automated Component Testing, *Journal of Software Testing, Verification, and Reliability*, Vol. 9, No. 1, 1999, pp. 3–26,

http://www.opros.or.kr/

Hu, X. (2005). Applying Robot-in-the-Loop-Simulation to Mobile Robot Systems, *Proceedings of the 12th International Conference on Advanced Robotics*, pp. 506-513, July 2005

Jawawi, D.N.A.; Mamat, R. & Deris, S. (2007). A Component-Oriented Programming for Embedded Mobile Robot Software, *International Journal of Advanced Robotics Systems*, Vol.4, No.2, 2007, pp.371-380, ISSN 1729-8806

Michel, O. (2004). Webots: Professional Mobile Robot Simulation, *International Journal of Advanced Robotic Systems*, Vol. 1, No. 1, 2004, pp. 39-42, ISSN 1729-8806

Momotko, M. & Zalewska, L. (2004). Component+ Built-in Testing: A Technology for Testing Software Components, *Foundations of Computing and Decision Sciences*, pp. 133-148, 2004

Ntafos, S. (1998). On Random and Partition Testing, *Proceedings of International Symposium on Software Testing and Analysis (ISSTA)*, 1998, pp. 42–48

Ostrand, T.J. & Balcer, M.J. (1998). The Category-Partition Method for Specifying and Generating Functional Tests, *Communications of the ACM*, Vol. 31, No. 6, 1988, pp. 676–686

Williams, A.W. (2000). Determination of test configurations for pair-wise interaction coverage, *Proceedings of the 13th Conf. Testing of Communicating Systems*, pp. 59-74, August 2000

Zamli, K. Z.; Isa, N. A. M., Klaib, M. F. J. & Azizan, S. N. (2007). A Tool for Automated Test Data Generation(and Execution) Based on Combinatorial Approach, *International Journal of Software Engineering and Its Applications*, Vol. 1, No. 1, pp. 19-35, July, 2007

Zhu, H.; Hall, P.& May, J. (1997). Software Unit Testing Coverage and Adequacy, *ACM Computing Surveys*, Vol. 29, No. 4, December 1997, pp. 366–427

24

Advanced Bit Stuffing Mechanism for Reducing CAN Message Response Time

Kiejin Park and Minkoo Kang
Ajou University
Republic of Korea

1. Introduction

As customer requirements for safety and convenience in automobiles increases, so does the quantity of electronics and software installed in them. The amount of signal data from the electronics systems needs to be managed, making the design of communication protocols for in-vehicle networks (IVN) more important (Navet et al., 2005). The IVN protocols can be classified into two paradigms: event-triggered and time-triggered (Obermaisser, 2004). The event-triggered protocols are efficient in terms of network utilization, because the messages transmitted within event-triggered protocols are only transmitted when specific events occur. This differs from time-triggered communication in that the response time of message transmission is not predictable (Fabian & Wolfgang, 2006).

The controller area network (CAN) is a well-known event-triggered protocol originally developed in the mid-1980s for multiplexing communication between electronic control units (ECUs) in automobiles (ISO 11898, 1993). In recent years, CAN has been used in embedded control systems that require high safety and reliability because of its appealing features and low implementation costs (Navet et al., 2005; Johansson et al.; 2005). Appealing features of CAN protocols are that the error detection mechanisms can identify multiple types of error (e.g. bits error, bit stuffing error, cyclic redundancy checksum error, frame error, and acknowledgement error). Moreover, error counters in a CAN controller can be used to represent which states of the controller are associated with specific errors, which include an error-active state, an error-passive state, and a bus-off state (Gaujal & Navet, 2005).

In spite of low implementation costs and wide acceptance of the CAN protocol in automotive control systems and industrial factory automation, limited bandwidth and nondeterministic response time have restricted the wider use of CAN in safety-critical real-time embedded control systems such as x-by-wire applications (Rushby, 2003; Wilwert et al., 2004). To mitigate the effects of these problems, the worst-case response time of a CAN message should be reduced as much as possible. Calculating the worst-case response time of CAN messages has been studied in order to guarantee its schedulability (Tindell et al., 1994, 1995), and this approach has been cited in over 200 subsequent papers. More recently, the schedulability analysis of CAN has been studied as the revised version of the original approach (Davis & Burns, 2007).

To reduce the length of CAN messages, the pre-processing mechanism using bitwise manipulation before bit stuffing has been suggested (Nolte et al., 2002, 2003). According to this mechanism, the worst-case response time can be reduced by minimizing stuffing bits in CAN messages. However, this mechanism cannot be applied to CAN network systems because the problem of message priority inversion has not been addressed. In our previous work, to resolve the problem of message priority inversion, a mechanism with a new bit mask for reducing the length of CAN message as well as preserving message priorities has been proposed (Park et al., 2007). Subsequently, we found that the mechanism has a problem which causes the frame shortening error and proposed the advanced bit stuffing (ABS) mechanism for resolving the problems with the previous approach (Park & Kang, 2009). In this paper, we describe the ABS mechanism in detail and extend the generation procedure of the bit mask of the ABS mechanism for the extended 2.0B frame format.

The outline of this paper is as follows. Section 2 presents a summary of the CAN protocol, and describes the impossibility of the worst-case bit stuffing scenario proposed by Nolte et al. Then calculating response time of CAN messages is presented. In Section 3, the ABS mechanism for reducing CAN message response time using generation of a new mask is described in detail. Also, we describe the examples of problems with priority inversion and frame shortening error. In Section 4, we evaluate the performance of the ABS mechanism with various CAN message sets. Finally, Section 5 concludes the paper.

2. Background

2.1 CAN message frame format

Controller area network (CAN) is the ISO standard for communication in automotive applications. It is designed to operate at network speeds of up to 1 Mbps for message transmission. Each CAN message contains up to 8 bytes of data (Farci et al., 1999). The frame format of a CAN message is classified into two categories that include the standard 2.0A with 11-bit identifier and the extended 2.0B with 29-bit identifier. Furthermore, message transmission over a CAN is controlled by four different types of frame: data frame, remote transmit request (RTR) frame, overload frame, and error frame (Etschberger, 2001). Fig. 1 shows the format of a CAN standard 2.0A data frame.

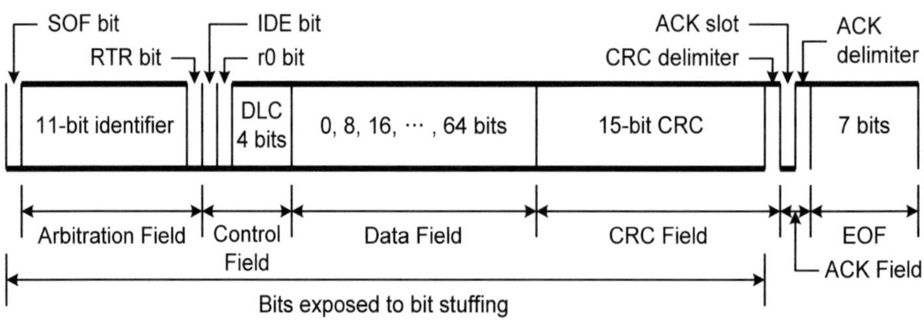

Fig. 1. Standard 2.0A data frame format of a CAN message.

As shown in fig. 1, a data frame consists of start-of-frame (SOF), arbitration field, control field, data field, acknowledgement (ACK) field, and end-of-frame (EOF). An SOF bit marks the beginning of a data frame, and is represented by one dominant bit (value=0). The arbitration field consists of an 11-bit identifier and a dominant delimiter bit. The identifier indicates the priority of the message. A message with identifier '00000000000' has highest priority, and a message with identifier '11111111111' has lowest priority. The last four bits of the control field are called the data length code (DLC). Its value represents the length of data field. Data field contains up to 8 bytes of data to be transmitted. The CRC field consists of a 15-bit CRC code and a recessively transmitted delimiter bit. The ACK field has two delimiter bits. The EOF consists of a sequence of 7 recessive bits (Etschberger, 2001).

In recent years, a luxury car may incorporate as many as 2500 signals exchanged by up to 70 ECUs (Albert, 2004). In the standard 2.0A frame format of a CAN message, the length of the identifier is 11 bits. This means that 2048 different CAN messages are distinguishable in the CAN communication system, so, the 11-bit identifier is insufficient to distinguish all signals. For this reason, the extended 2.0B frame format with 29-bit identifier has been defined. Fig. 2 shows the format of a CAN extended 2.0B data frame (Pfeiffer et al., 2003)

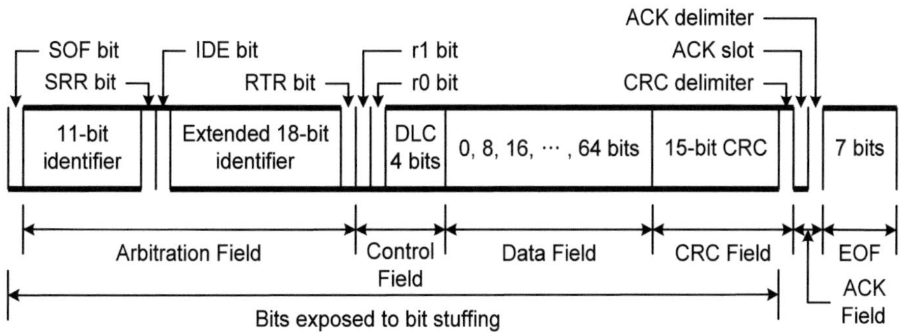

Fig. 2. Extended 2.0B data frame format of a CAN message.

2.2 Worst-case bit stuffing scenario

When a CAN node detects an error in a transmitted message, it transmits an error flag which consists of six bits of the same polarity. The bit stuffing mechanism prevents six consecutive bits from having the same polarity by inserting a bit of opposite polarity after the fifth bit. Moreover, the main purpose of the bit stuffing mechanism is used to synchronize transmitter and receiver when the same values are to be transmitted consecutively (Nolte et al., 2001). Bits exposed to bit stuffing are from an SOF bit to a 15-bit CRC code without a CRC delimiter (see Fig. 1 and Fig. 2). The stuffing bits of the received frame are removed at the receiving node before the message is processed (Wolfhard, 1997).

The worst-case scenario of the bit stuffing has been presented as shown in Fig. 3 (Nolte et al., 2007).

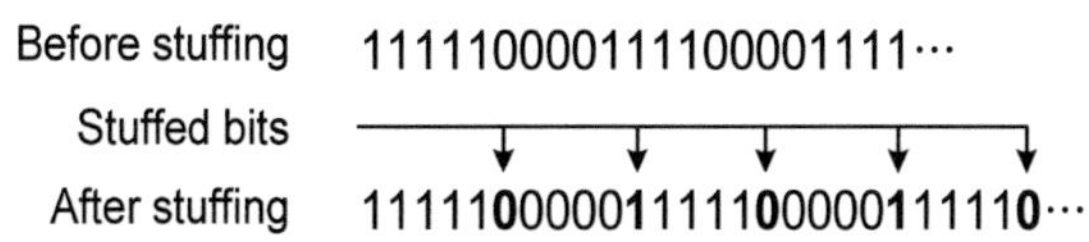

Fig. 3. The worst-case scenario of the bit stuffing.

According to the worst-case scenario, the number of bits of a CAN message is given by:

$$g + 8s + 13 + \left\lfloor \frac{g+8s-1}{4} \right\rfloor \quad (1)$$

where g is 34 for the standard format or 54 for the extended format, s is the number of data bytes of a CAN message (Nolte et al., 2007). However, it is impossible that stuffing bits be inserted in the worst-case scenario. The causes are:

1. Several bit values are fixed by the CAN frame format (e.g., the SOF bit and delimiter bits in the arbitration field and the control field).
2. The DLC in the control field depends on its number of data bytes.
3. The CRC field of a CAN message depends on the bit sequence from an SOF bit to the data field.

Accordingly, the worst-case number of stuffing bits in a standard 2.0A data frame is reduced by 21~40% from previous values.

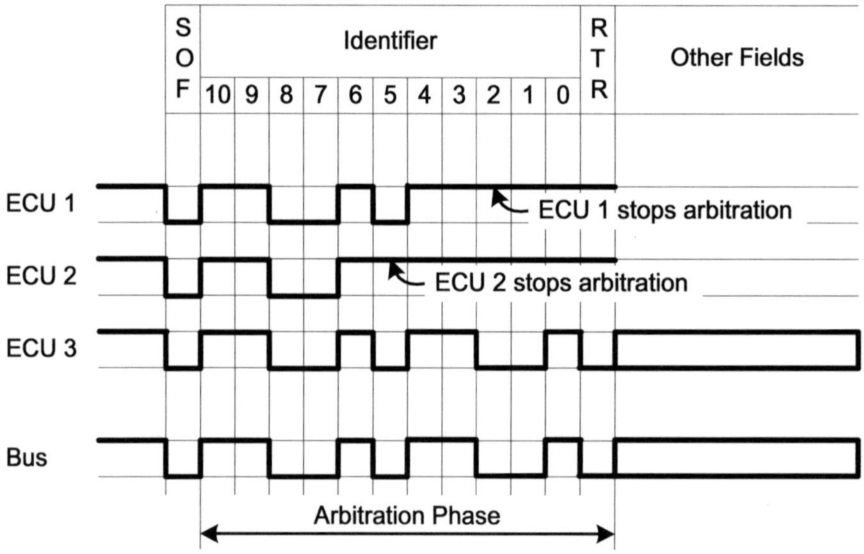

Fig. 4. The CAN arbitration process.

2.3 Bitwise bus arbitration

Each ECU of a CAN network can initiate the transmission of a message as soon as the bus is free. Because it may happen that more than one ECU begins a message transmission at the same time, an arbitration process is necessary. To prevent the ECUs from destroying each other's transmitted data, the message with the highest priority of all simultaneously arbitrating messages is determined in an arbitration phase. As mentioned in Section 2.1, the message having the lowest value message identifier is assigned highest priority (Etschberger, 2001). Fig. 4 shows the CAN arbitration process. Each ECU monitors the signal level on the bus during the arbitation phase. The arbitation phase consists of the transmission of the message identifier and of the RTR bit. If an ECU detects a dominant bus level, although it has switched a recessive level itself, it aborts the transmission process immediately, as in this case a message with higher priority is obviously being transmitted at the same time, and goes into the receive state (Etschberger, 2001).

2.4 Response time model of CAN messages

The worst-case response time of a CAN message m can be calculated as the sum of the queuing delay t_m and the transmission delay C_m as follows:

$$R_m = t_m + C_m \tag{2}$$

The queueing delay t_m is composed of the blockng time B and the interference and is given by:

$$t_m = B + \sum_{\forall j \in hp(m)} \left\lceil \frac{t_m + J_j + \tau_{bit}}{T_j} \right\rceil C_j \tag{3}$$

where the set $hp(m)$ consists of all the messages in the system of higher priority than message m, J_j is the jitter on the queueing of the message j, T_j is the transmission period of the message j, and τ_{bit} is the transmission time for a single bit.

The blocking time B can be calculated by the transmission time of the longest CAN message within the system. The transmission delay C_m can be calculated by multiplying the number of bits of the message m as in (1) and τ_{bit} (Tindell et al., 1995).

3. Advanced Bit Stuffing (ABS) mechanism

As mentioned in Section 1, the mechanism proposed by Nolte et al. cannot be applied to reduce the length of CAN messages because the message transmission priorities can be shuffled as shown in Table 1.

	High priority CAN ID	Low priority CAN ID	Description
Original CAN Message	00001010111...	01111110101...	Low bit value has higher priority
XORing (by Nolte et al.)	010111111101...	00101011111...	Mask: 01010101010
Bit Stuffing Message	0101111110101...	001010111110...	Priority inversion occurs.

Table 1. Counter example of priority inversion problem.

To solve the problem of priority inversion, a mechanism for minimizing the length of CAN messages in bit stuffing, and for preserving message priorities, has been proposed (Park et al., 2007). However, the previous mechanism contains a flaw which causes frame shortening errors. The frame shortening error means that the receiver anticipates a frame of different length than the original (Charzinski, 1994; Tran, 1999). It can occur when the first bit and the last four bits of the control field are changed by the XOR operations with the bit mask. An example of the frame shortening error is shown in Fig. 5.

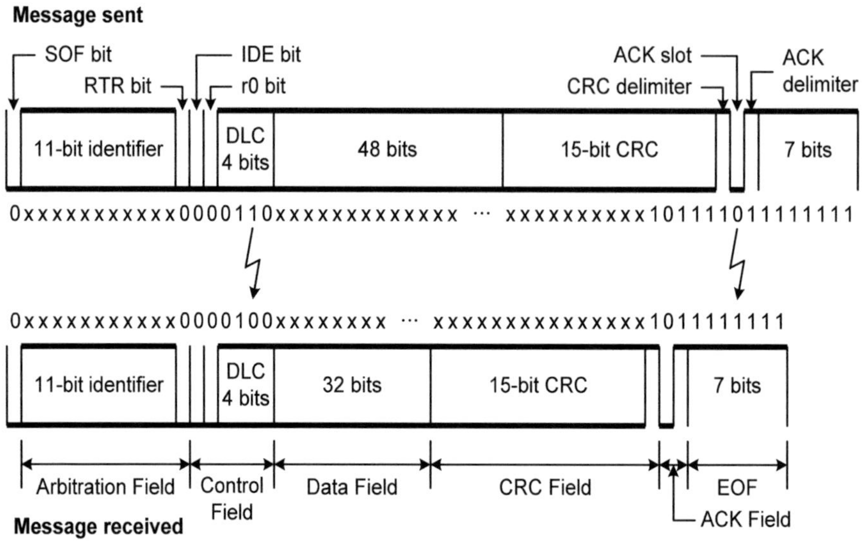

Fig. 5. Example of the frame shortening error.

As shown in Fig. 5, two bits are changed by bit errors. The first error is the bit in the DLC and the second one is the ACK slot. Thus the receiver can expect a message with a smaller than the original message.

In this section, we propose an advanced bit stuffing (ABS) mechanism which adopts XOR operations and prevents priority inversion and frame shortening errors at the same time. In order to develop the ABS mechanism, an assignment scheme for CAN message identifiers and generation rules of a new XOR bit mask are presented.

3.1 Message identifier assignment

To better understand the number of bits used for message identifiers in CAN-based control systems, we assumed that there are two assignment schemes of message identifiers. The first scheme is to assign to messages consecutive identifiers starting from 1. In this scheme, when the number of message identifiers is n, the number of used bits is $\lceil \log_2 n \rceil$. For instance, if the system requires 256 messages, than the number of used bits for message identifiers are 8. On the other hand, the second scheme is based on the grouping of message identifiers in accordance with their level of importance. In this scheme, the number of used bits can be calculated by:

$$n_{used} = \lceil \log_2 m \rceil + \lceil \log_2 n \rceil \tag{4}$$

where m and n represent the number of groups and the maximum number of identifiers in a group, respectively. For example, the system requires 4 message groups and each group consists of up to 32 message identifiers. In this case, only 7 bits are used for message identifiers. 2 bits out of 7 bits are required for representing message groups and 5 bits out of 7 bits are required for representing message identifiers of each group. Because the first scheme is a special case (i.e., $m = 1$) of the second scheme, in this paper, we have applied the second scheme in order to assign message identifiers.

3.2 XOR mask generation for standard 2.0A frame format

In order to generate a new mask for standard 2.0A frame format, we assumed that the bits for group and those for group identifiers are assigned to the most significant bit (MSB) and the least significant bit (LSB) field in the arbitration field, respectively. When the number of data bytes (s), the number of groups (m) and the maximum number of identifiers in a group (n) are determined, the following mask generation procedure is constructed for the standard 2.0A frame format.

1. The length of a mask is the length of the bits exposed to bit stuffing, $8s + 34$. The mask is initially set to "010101…"
2. A value of 0 is assigned to $\lceil \log_2 m \rceil$ bits of the MSB and $\lceil \log_2 n \rceil$ bits of the LSB of the 11-bit identifier in the mask.
3. A value of 0 is assigned to the RTR bit of the arbitration field in the mask.
4. The string "010000" is assigned to 6 bits of the control field in the mask. Then, the mask can be generated as depicted in Fig. 6.

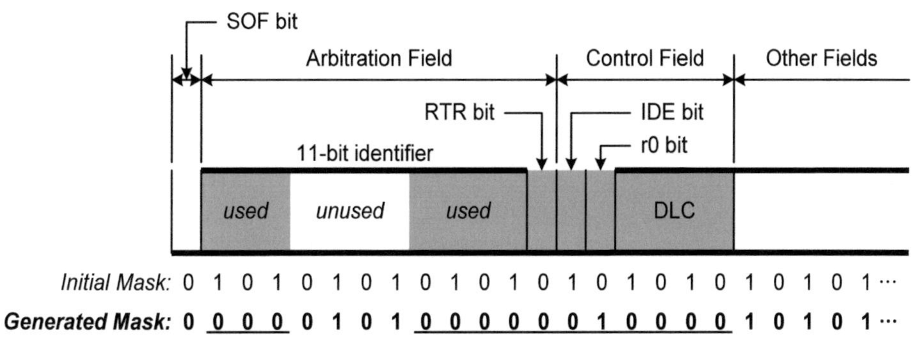

Fig. 6. XOR mask generation for standard 2.0A frame format.

3.3 XOR mask generation for extended 2.0B frame format

We extended the mask generation procedure in Section 3.2 for extended 2.0B frame format. In the same manner, we assumed that the bits for a group and those for group identifiers are assigned to the MSB and the LSB field in the arbitration field of the extended 2.0B frame format, respectively. When the number of data bytes (s), the number of groups (m) and the maximum number of identifiers in a group (n) are determined, the following mask generation procedure is constructed for the extended 2.0B frame format.

1. The length of a mask is the length of the bits exposed to bit stuffing, $8s + 54$. The mask is initially set to "010101…"
2. A value of 0 is assigned to $\lceil \log_2 m \rceil$ bits of the MSB and $\lceil \log_2 n \rceil$ bits of the LSB of the 29-bit identifier in the mask.
3. A value of 00 is assigned to 2 medial bits of the arbitration field, and a value of 0 is assigned to the RTR bit of the arbitration field in the mask.
4. The string "010000" is assigned to 6 bits of the control field in the mask. Then, the mask can be generated as depicted in Fig. 7.

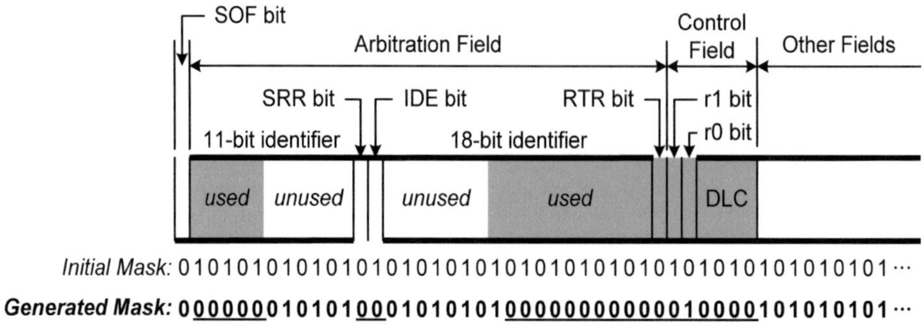

Fig. 7. XOR mask generation for extended 2.0B frame format.

In both standard and extended frame format, r0 bit should be a dominant bit. But the bit of r0 bit location in the generated mask is assigned to a recessive bit in the generation procedure for both standard and extended frame format. This assignment is for separating XOR masked CAN messages from original CAN messages. If an ECU receives a CAN message with dominant bit of r0 bit location, the received message is unmasked, and otherwise (i.e., a CAN message with recessive bit level of r0 bit location), a received message is masked.

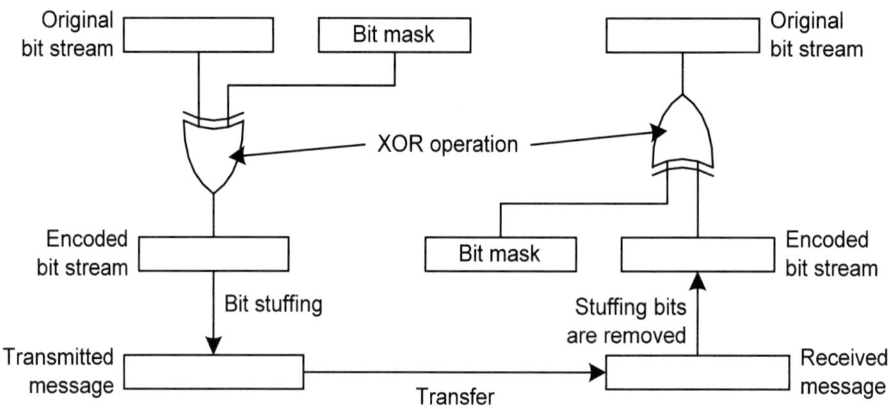

Fig. 8. Implementing the ABS mechanism

3.4 Guidance for implementing the ABS mechanism

In Section 3.2 and Section 3.3, the XOR masks are generated for standard and extended frame format. The ABS mechanism reduces the number of stuffing bits in the CAN message by a bitwise manipulation using the XOR masks (Fig. 8).

4. Performance evaluation

As in our previous work, the example of the case of $\lceil \log_2 m \rceil = 2$ illustrates the efficiency of the proposed mechanism (Park et al., 2007). For the SOF bit, the arbitration field, and the control field, the expected number of stuffing bits has been calculated with a variable number of $\lceil \log_2 n \rceil$. The expected number of stuffing bits in the original messages is shown in Table 2. For comparison, the expected number of stuffing bits in the XOR masked messages is shown in Table 3, where $P(X = i)$ represents the probability that the message has i stuffing bits, and $E(X)$ is the expected number of the stuffing bits. In order to calculate the probability, we assume that the probability of a bit having value 1 or 0 is equal.

$\lceil \log_2 n \rceil$	$P(X = 0)$	$P(X = 1)$	$P(X = 2)$	$P(X = 3)$	$E(X)$
1	0	0.234	0.594	0.172	1.938
2	0	0.305	0.609	0.086	1.781
3	0	0.324	0.633	0.043	1.719
4	0.086	0.402	0.477	0.035	1.461
5	0.157	0.477	0.341	0.025	1.234
6	0.199	0.509	0.268	0.024	1.117
7	0.251	0.561	0.176	0.012	0.949
8	0.260	0.572	0.159	0.009	0.917
9	0.260	0.572	0.159	0.009	0.917

Table 2. The expected number of the stuffing bits in the original messages.

In order to evaluate the performance of the ABS mechanism, the number of stuffing bits of masked message frames is compared with the number of stuffing bits of the original message frames. The original and masked message frames are generated by a simulation program. The flowchart of the simulation program is shown in Fig. 9. Simulation procedures are as follows: 1) Initially, the simulation program obtains input parameters. In this simulation, the number of runs is 1000 and each CAN message has four data bytes; 2) Then, the original message frame is generated and the number of stuffing bits is calculated; 3) Next, the bit mask is generated by using mask generation as shown in Section 3.2 and Section 3.3, and masked message frame generated by XOR operation in both the bit mask and the original message frame; 4) Finally, the number of stuffing bits of the masked message frame is calculated. When a simulation is complete, the average and maximum number of stuffing bits in the original and masked message frames are calculated, respectively.

In this experiment, the number of runs is 1000 and the assumptions that are made in these experiments are as follows: 1) the probability that a bit which does not have a specific value can be set to 0 or 1 is the same, and 2) it is independent of other bit values, and finally, 3) there is no transmission error during experiment.

$\lceil \log_2 n \rceil$	$P(X=0)$	$P(X=1)$	$P(X=2)$	$P(X=3)$	$E(X)$
1	1	0	0	0	0
2	1	0	0	0	0
3	0.875	0.125	0	0	0.125
4	0.875	0.125	0	0	0.125
5	0.875	0.125	0	0	0.125
6	0.844	0.156	0	0	0.156
7	0.754	0.234	0.012	0	0.258
8	0.728	0.254	0.018	0	0.291
9	0.728	0.254	0.018	0	0.291

Table 3. The expected number of the stuffing bits in the XOR masked messages.

Fig. 10 and Fig. 11 show the average and maximum number of stuffing bits in the standard data frame as the number of used bits in the arbitration field changes, in both the original frame and the masked frame. In the same manner, the number of stuffing bits in the extended data frame is shown in Fig. 12 and Fig. 13. Here we can see that the more the number of used bits decreases, the more does the number of stuffing bits of the original message frame increase. However, the number of stuffing bits of the masked message frame showed little variation regardless of the change in the number of used bits.

From the experimental results, it can be found that the expected number of stuffing bits is reduced by the maximum of 58.3% with an average of 32.4%. This may effect the response time of CAN messages because message response time is in proportion to the length of message frame.

The average and maximum number of stuffing bits in the standard data frame with $\lceil \log_2 m \rceil = 2, \lceil \log_2 n \rceil = 4$ as the number of data bytes changes, in both the masked frame and the original frame, are shown in Fig. 14. In the same manner, the number of stuffing bits in the extended data frame is shown in Fig. 15. As shown in Fig. 14 and Fig. 15, the number of stuffing bits of the masked message frame is smaller than the number of stuffing bits of the original message frame.

As mentioned in Section 2.4, the response time of CAN messages is determined by the queuing delay and the transmission delay. If the number of stuffing bits decreases, both the queuing delay and the transmission delay can be reduced. Furthermore, this may have effects on the reduced network utilization of the embedded systems using CAN network protocols.

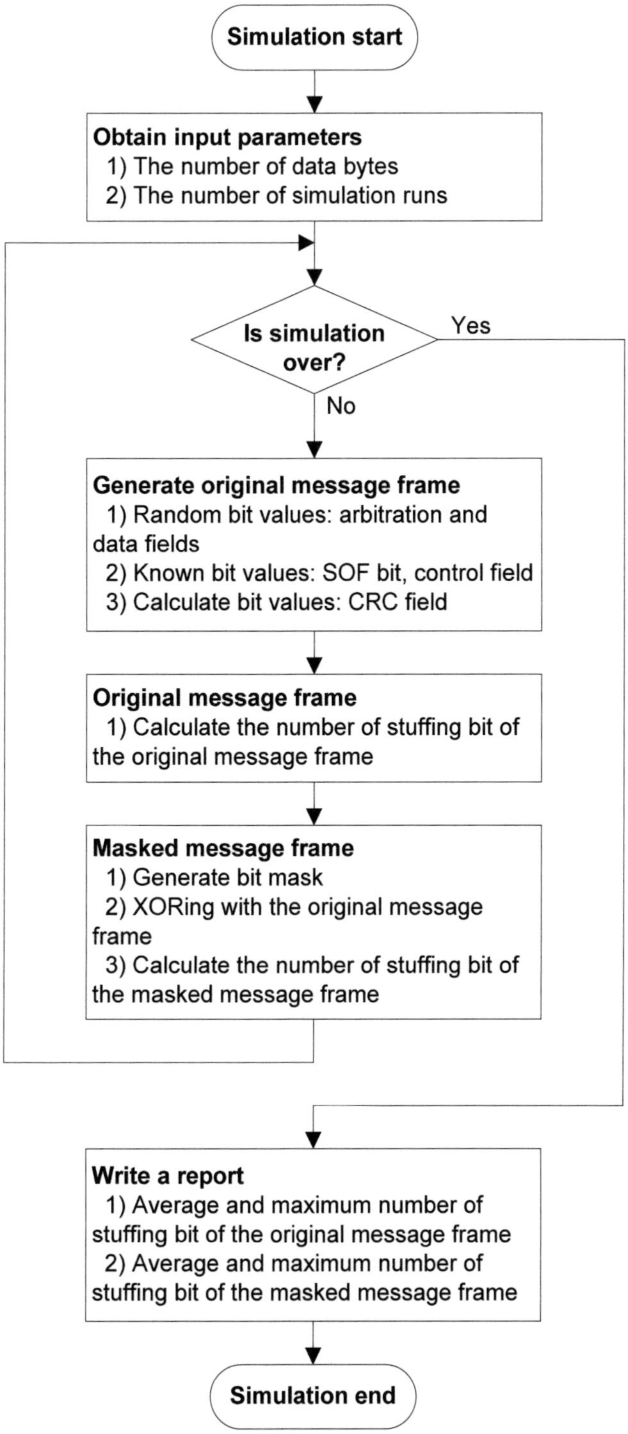

Fig. 9. Flowchart of the simulation program.

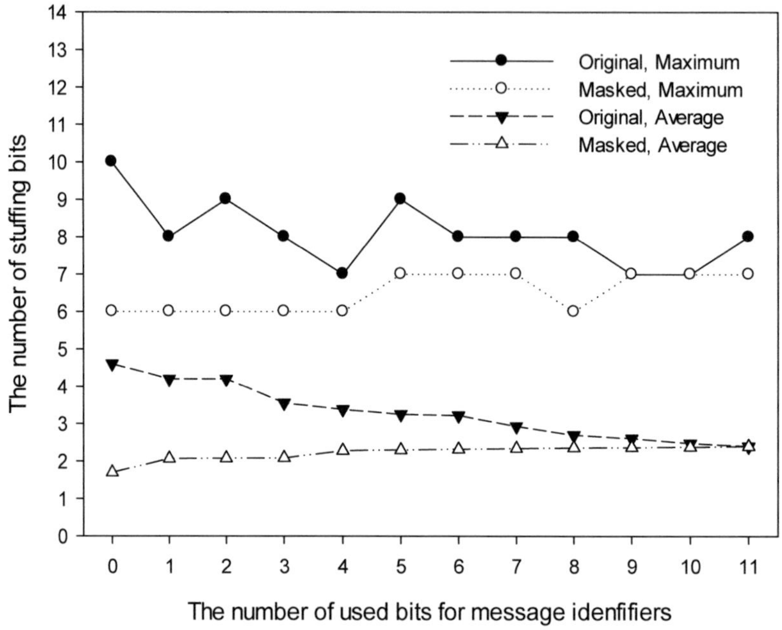

Fig. 10. The number of stuffing bits in the standard frame with 4 data bytes.

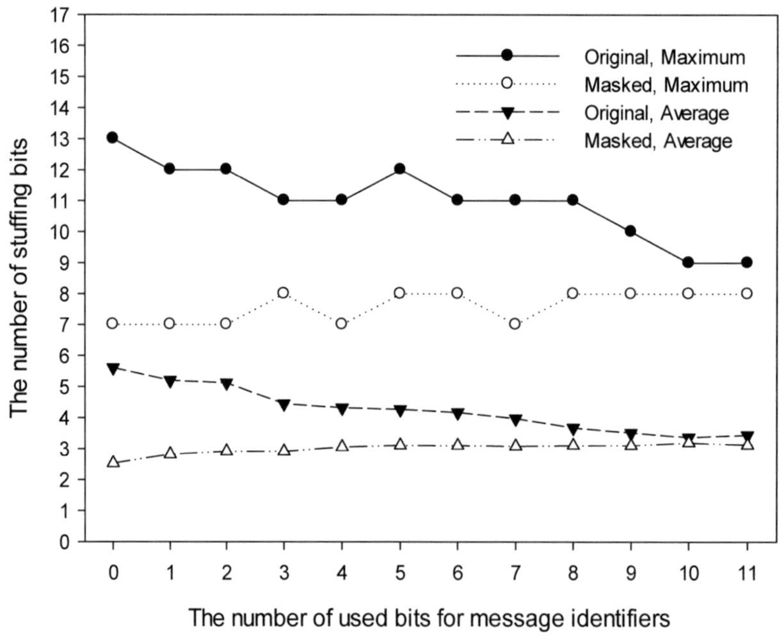

Fig. 11. The number of stuffing bits in the standard frame with 8 data bytes.

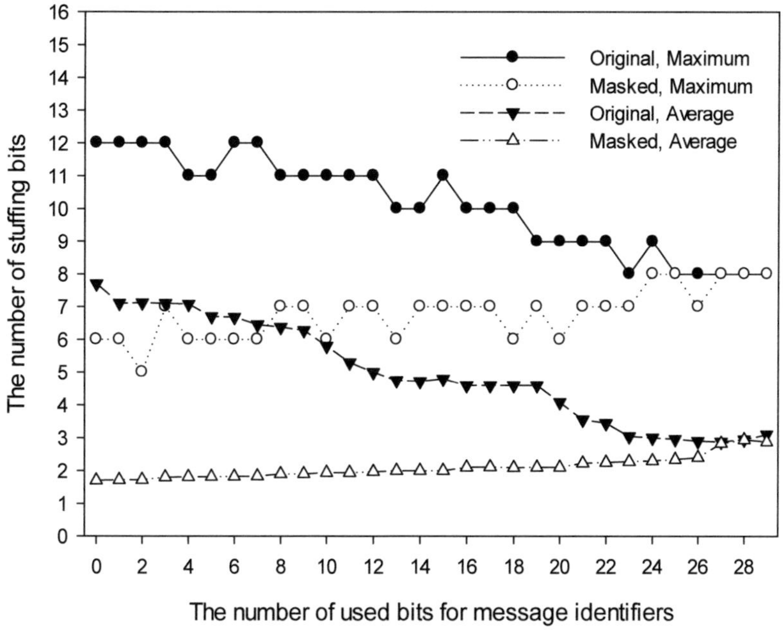

Fig. 12. The number of stuffing bits in the extended frame with 4 data bytes.

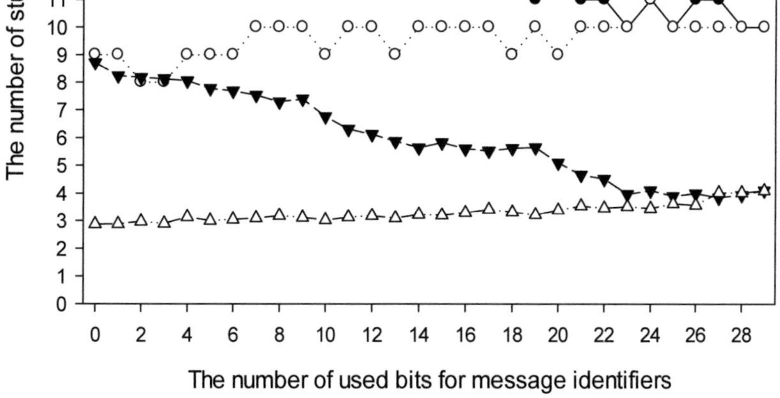

Fig. 13. The number of stuffing bits in the extended frame with 8 data bytes.

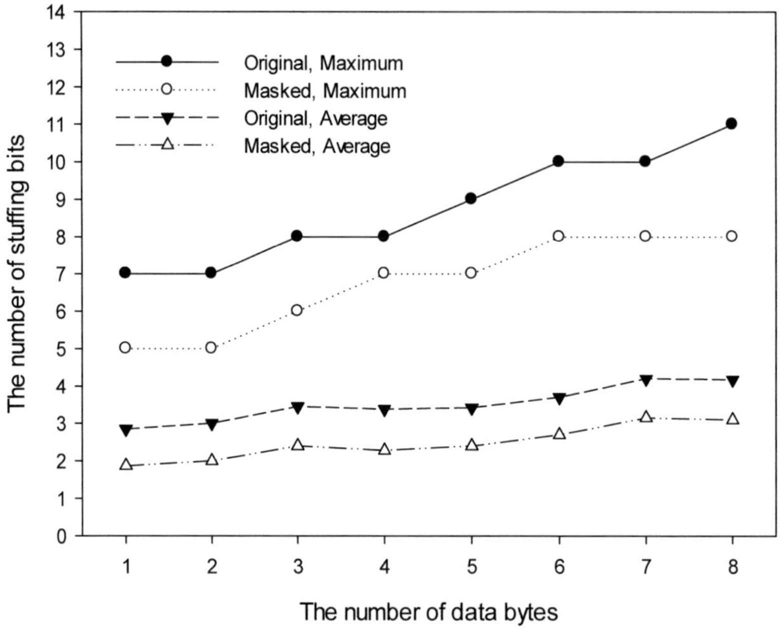

Fig. 14. The number of stuffing bits in the standard frame with $\lceil \log_2 m \rceil = 2, \lceil \log_2 n \rceil = 4$

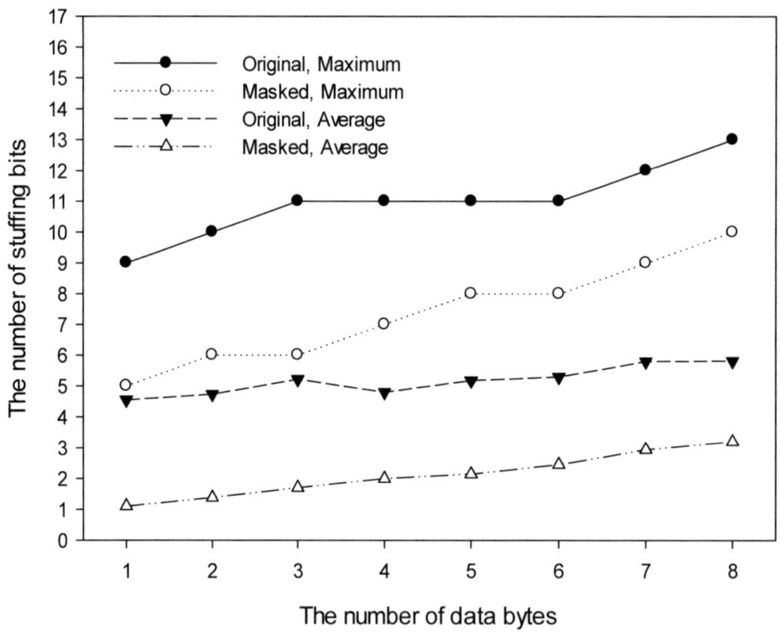

Fig. 15. The number of stuffing bits in the extend frame with $\lceil \log_2 m \rceil = 6, \lceil \log_2 n \rceil = 9$

5. Conclusions

Minimizing the response time of the CAN message is necessary to guarantee real-time performance improvement. In this paper, an effective mechanism called advanced bit stuffing (ABS) mechanism is presented. The ABS mechanism develops an assignment scheme of CAN message identifiers and generation rules of a new XOR bit mask, to prevent the problems with message priority inversion and frame shortening error. From the experimental result, the number of stuffing bits of the masked message frame showed little variation regardless of the change of the number of used bits. It has been also found that it is more effective in embedded systems in which the number of CAN messages is less than that of bits that are used for message priority.

6. References

Albert, A. (2004). Comparison of Event-triggered and Time-triggered Concepts with Regards to Distributed Control Systems, *Embedded World Conference,* Nürnberg, Germany.

Charzinski, J. (1994). Performance of the Error Detection Mechanisms in CAN, *Proceedings of the 1st International CAN Conference,* Mainz, Germany.

Davis, R. & Burns, A. (2007). Controller Area Network (CAN) Schedulability Analysis: Refuted, Revisited and Revised, *Real-Time Systems,* Vol. 35, No. 3, pp. 239-272.

Etschberger, K. (2001). *Controller Area Network (CAN): Basics, Protocols, Chips, and Applications,* IXXAT Automation GmbH, ISBN 978-3000073762, Weingarten, Germany.

Fabian, S. & Wolfgang, S. (2006). Time-Triggered vs. Event-Triggered: A Matter of Configuration, *Proceedings of the GI/ITG Workshop on Non-Functional Properties on Embedded Systems,* pp. 1-6.

Farsi, M.; Ratcliff, K.; & Barbosa, M. (1999). An Overview of Controller Area Network, *Computing & Control Engineering Journal,* Vol. 10, pp. 113-120.

Gaujal, B. & Navet, N. (2005) Fault Confinement Mechanisms on CAN: Analysis and Improvements, *IEEE Transactions on Vehicular Technology,* Vol. 54, pp. 1103-1113.

International Standards Organization. (1993). Road Vehicles - Interchange of Digital Information - Controller Area Network (CAN) for High-Speed Communication. ISO 11898.

Johansson, K.H.; Torngren, M.; & Nielsen, L. (2005). Vehicle Applications of Controller Area Network, *Handbook of Networked and Embedded Control Systems,* ISBN 0-8176-3239-5, pp. 741-766.

Navet, N.; Song, Y.; Simonot-Lion, F.; & Wilwert, C. (2005). Trends in Automotive Communication Systems, *In Proceeding of the IEEE,* Vol. 93, No.6, pp. 1204-1223.

Nolte, T.; Hansson, H.; Norstrom, C.; & Punnekkat, S. (2001). Using bit-stuffing distributions in CAN analysis, *IEEE/IEE Real-Time Embedded Systems Workshop (RTES'01).*

Nolte, T.; Hansson, H.; & Norstrom, C. (2002). Minimizing CAN Response-Time Jitter by Message Manipulation, *Eighth IEEE Real-Time and Embedded Technology and Applications Symposium (RTAS'02),* pp. 197-206.

Nolte, T.; Hansson, H.; & Norstrom, C. (2003). Probabilistic Worst-Case Response Time Analysis for the Controller Area Network, *Ninth IEEE Real-Time and Embedded Technology and Applications Symposium (RTAS'03),* pp. 200-207.

Obermaisser, R. (2004). *Event-Triggered and Time-Triggered Control Paradigms,* Sptinger-Verlag, ISBN 978-1441935694.

Park, K.; Kang, M.; & Shin, D. (2007). Mechanism for Minimizing Stuffing-bits in CAN Messages, *The 33rd Annual Conference on the Industrial Electronics Society (IECON '07),* pp. 735-737.

Park, K. & Kang, M. (2009). Advanced Bit Stuffing Mechanism for Reducing the Length of CAN Messages, *International Conference on Convergence Technologies and Information Convergence (CTIC'09).*

Pfeiffer, O.; Ayre, A.; & Keydel, C. (2003). *Embedded Networking with CAN and CANopen,* RTC Books, ISBN 978-0929392783.

Rushby, J. (2003). A Comparison of Bus Architecture for Safety-Critical Embedded Systems, *NASA/CR, Technical Report,* NASA/CR-2003-212161.

Tindell, K.W. & Burns, A.K. (1994). Guaranteed Message Latencies for Distributed Safety-critical Hard Real-time Networks, *Technical Report YCS 229,* Department of Computer Science, University of York.

Tindell, K.; Burns, A.; & Wellings, A. J. (1995). Calculating Controller Area Network (CAN) Message Response Times, *Control Engineering Practice,* Vol. 3, No. 8, pp. 1163– 1169.

Tran, E. (1999). *Multi-bit Error Vulnerabilities in the Controller Area Network Protocol,* Carnegie Mellon University, Thesis.

Wilwert, C.; Navet, N.; Song, Y.-Q.; & Simonot-Lion, F. (2004). Design of automotive X-by-Wire systems, *The Industrial Communication Technology Handbook,* R. Zurawski, Ed. Boca Raton, FL: CRC.

Wolfhard, L. (1997). *CAN System Engineering: From Theory to Practical Applications,* Springer, ISBN 978-0387949390.

25

A Systematized Approach to Obtain Dependable Controller Specifications for Hybrid Plants

Eurico Seabra and José Machado
Mechanical Engineering Department, CT2M Research Centre, University of Minho
Portugal

1. Introduction

This chapter focuses on the problem that a designer of an automation system controller must solve related with the correct synchronization between different parts of the controller specification when this specification obeys a previously defined structure. If this synchronization is not done according to some rules, and taking some aspects into consideration, some dependability aspects concerning the desired behaviour for the system may not be accomplished. More specifically, this chapter will demonstrate a systematized approach that consists of using the GEMMA (Guide d`Etude des Modes de Marches et d`Arrêts) (Agence Nationale pour le Developpement de la Production Automatisée) [ADEPA], 1992) and the SFC (Sequential Function Chart) (International Electrotechnical Commission [IEC], 2002) formalisms for the structure and specification of all the system behaviour, considering all the stop states and functioning modes of the system. The synchronization of the models, corresponding to the controller functioning modes and the controller stop states, is shown in detail and a systematized approach for this synchronization is presented. For this the advantages and disadvantages of the vertical coordination and horizontal coordination proposed by the GEMMA formalism are discussed and a case study is presented to explain the proposed systematic approach. A complete safe controller specification is developed to control a hybrid plant. Also this chapter presents and discusses a case study that applies a global approach for considering all the automation systems emergency stop requirements. The definition of all the functioning modes and all the stop states of the automation system is also presented according the EN 418 (European Standard [EN], 1992) and EN 60204-1 (EN, 1997) standards. All the aspects related to the emergency stop are focused in a particular way. The proposed approach defines and guarantees the safety aspects of an automation system controller related to the emergency stop. For the controller structure the GEMMA methodology is used; for the controller entire specification the SFC is used and for the controller behaviour simulation the Automation Studio software (FAMIC, 2003) is used.

In order to achieve the goals presented above, the chapter is organized as follows: section 1 presents the challenge addressed to this chapter; in section 2 the main formalisms and methodologies used to define the controller behaviour specification are presented, namely, showing how to deal with complex specifications before the implementation into a physical controller device to help the designer to improve the specifications performance; section 3 discusses different possible approaches for using the coordination of a controller specification when this specification is previously structured by the use of the GEMMA method. This is often applied because the complexity of the specification behaviour demands a separate modular specification, named "task", corresponding to the functioning or failure modes or stop states of the automation system; section 4 presents a case study and shows in detail defining and structuring a controller specification; section 5 presents and illustrates how to coordinate a complex specification applied to a case study with possible extrapolation for similar cases; section 6 is exclusively devoted to the discussion of different possibilities for emergency stop application; further section 7 presents and discusses an example of emergency stop application and, finally, section 8 presents some conclusions.

2. Formalisms used to develop the controller behaviour specification

From the desired behaviour specifications, until the implementation of a controller program for an automation system, the controller designer needs to use some different and complementary methods, formalisms and tools that help him in all the necessary steps. Taking into account aspects related to the systems' dependability, the designer must be able to use together these formalisms and tools in order to achieve the desired behaviour for the system. There are many methods, formalisms and tools for helping the designer during all necessary steps. For the structure of the controller it is possible to use the GEMMA method (ADEPA, 1992), Multi-Agent formalism (Sohier, 1996) can be used: for the specification Petri Nets (Murata, 1989), SFC (IEC, 2002), Statecharts (Harel, 1987), UML (Booch et al., 2000) can be used; for the implementation, the PLCs (Programmable logic controllers) (Moon, 1994), Industrial computers (Koornneef and Meulen, 2002), Microprocessors (Brusamolino et al., 1984) and others can be used.

From the analysis of needs, passing by the conception, realization into the implementation and exploitation of an automation system there are several steps that must be realized (Fig. 1). During each step of the controller design a corresponding step of the development of the plant (physical part of the system: motors, cylinders, sensors etc.) exists. For instance, step 3 corresponds to the specification of the controller and step 3' corresponds to the specification of the plant.

The main objective of this chapter is to show how to deal with complex specifications before the implementation into a physical controller device. Usually there are some methods, formalisms and tools that help the designer to improve the specifications performance, but if the coordination of all the parts of the specification is not well done, some aspects related to the dependability of the system may not be accomplished.

This paper applied a case study and then extrapolated to systems of the same kind, in this it is more detailed and more related to steps 3 and 4 presented in Figure 1, than related to the design of a controller.

Currently, some suitable methods and formalisms for the development and creation of the structure and specification of an automated production system controller exist. Among them there are GEMMA (Guide d'Étude des Modes de Marche et d'Arrêt) and SFC (Sequential Function Chart), both developed in France. GEMMA is well adapted to defining the controller structure and SFC is well adapted to complete controller specification.

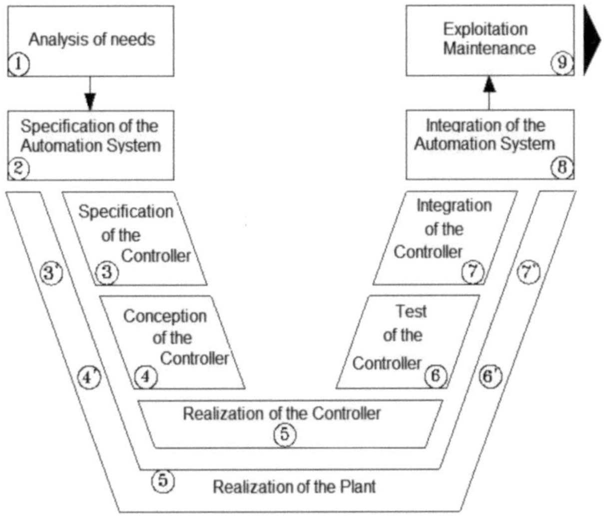

Fig. 1. Steps considered on the design of an automation system

According to SFC rules, the implementation of the automation system requires, in particular, a description relating to cause and effect. To do this, the logical aspect of the desired behaviour of the system will be described. The sequential part of the system, which is accessed via Boolean input and output variables, is the logical aspect of this physical system. The behaviour indicates the way in which the output variables depend on the input. The object of the SFC is to specify the behaviour of the sequential part of the systems. This formalism is characterized, mainly, by its graphic elements, which, associated with an alphanumerical expression of variables, provides a synthetic representation of the behaviour based on a description of the situation of the system.

GEMMA was developed in France by ADEPA (Agence Nationale pour le Developpement de la Production Automatisée) and is a method that on the basis of a very precise vocabulary proposes a simple structured guide for the designer based on a graphical chart that contains all the functioning modes and stop states that a machine or an automated system can assume. It is a tool for helping with system analysis, being used for its supervision, maintenance and evolution definition.

The GEMMA method is based in three basic concepts:

- The operating modes are seen, from the point of view of the command module, as always available. All the systems are composed by a command module and an operative module. In the application of GEMMA, it is assumed that the command module is always on power.
- The production criteria. Two states are considered for the production systems: ON production and OUT of production. Those states are shown on the graphical chart of the method.
- The three groups of functioning modes and stop states of the plant, namely:
  - States "A": Stop states
  - States "D": Failure modes
  - States "F": Running modes

The graphical chart of GEMMA is composed of three parts, each one corresponding to each group of functioning modes and stop states described in Figure 2.

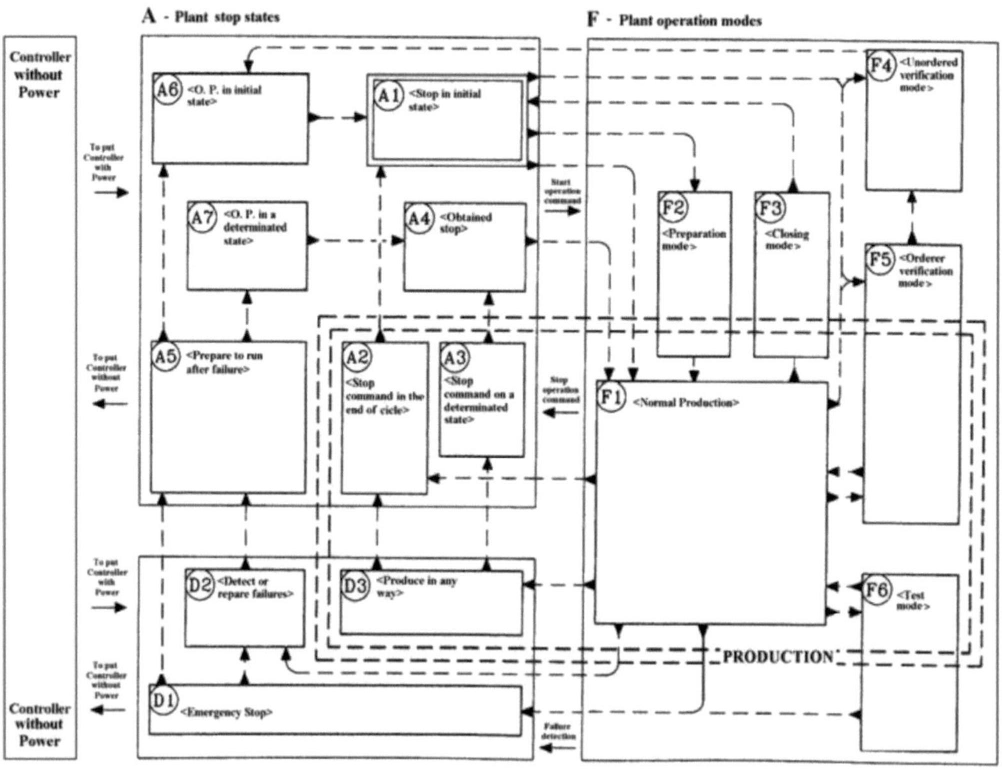

Fig. 2. Functioning modes and stop states considered on the design of an automation system

3. Coordination of a complex specification

Very often modes other than F1 (Normal Production Mode) demand a specification behaviour with complex cycles and this complex specification demands also a specific treatment for each of the functioning or failure modes or stop states of the automation system.

So, it seems useful to separate each modular specification corresponding to each of the functioning or failure modes or stop states of the automation system. Each specification module is named "task": a task is associated with the F1 mode, other tasks to the F2 mode and so on for all the functioning and failure modes and stop states of the automation system.

From a practical point of view, and for implementation of SFC, the division of tasks is particularly adapted to these needs and it is possible to make the correspondence between a mode/state, the respective SFC and the respective task, based on the specification SFC for the corresponding mode (Fig. 3).

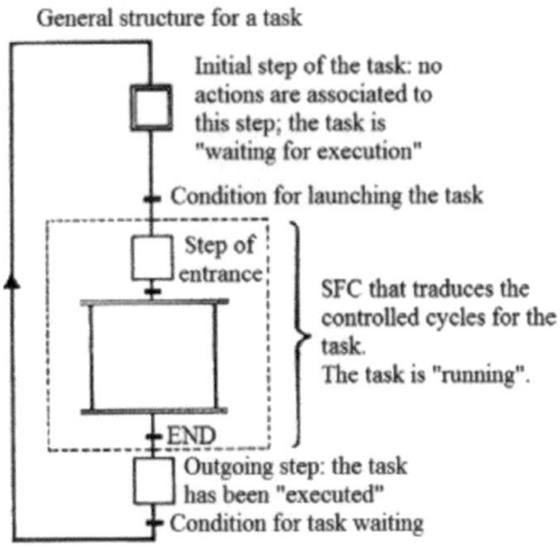

Fig. 3. Structure of a task

3.1 Horizontal coordination

This is a very interesting way of coordinating tasks because any task can be dominant over the others and also each task may launch other tasks (Fig. 4).

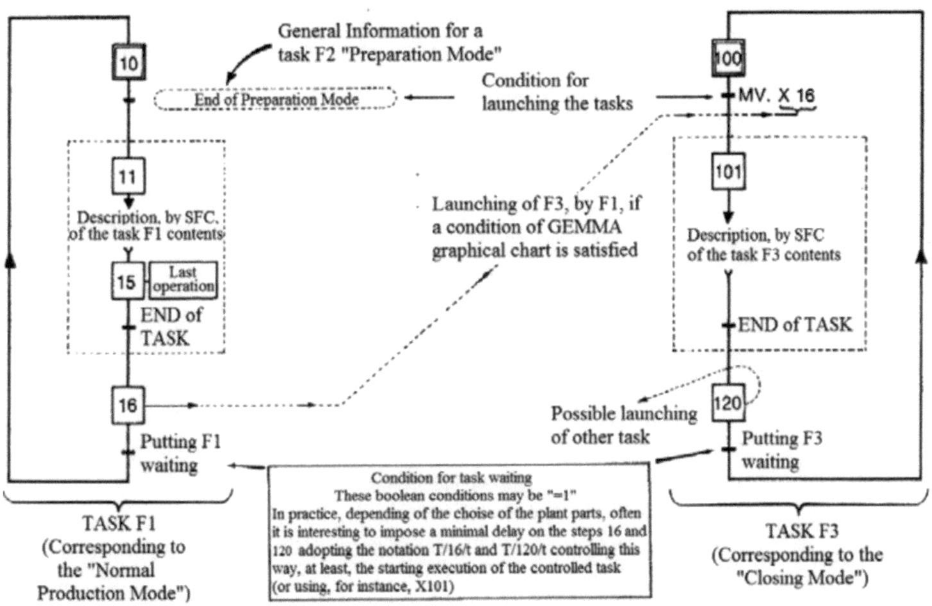

Fig. 4. Horizontal coordination

Let's consider a generic task F1 (normal production mode) and a generic task F3 (closing mode). As F3 appears after task F1, F1 must launch F3. When F1 ends (step 16) the Boolean variable X16 makes possible the evolution of the task F3, from the step 100 to the step 101. At the end of task F3 (step 120) the next task is launched by the Boolean variable X120 and so on with similar behaviour, task by task. The variables X*i* are Boolean variables associated to step *i* defined by (IEC, 2002).

3.2 Vertical coordination

This kind of coordination is hierarchic and there are several levels of abstraction. Each task of an abstraction level may launch any other task at a lower level, but - on the same abstraction level – one task cannot launch other tasks that belong to the same level (Fig. 5).

With this hierarchical approach the designer may have a global overview of the system and also, if he intends so, a very detailed local view of the system.

The synchronization process is illustrated in a very detailed way in Figure 5 and the SFC of higher level coordinates, at the specific order, indicates the evolution of each task. After the end of each task, the higher level SFC evolutes and, on its next steps, it will launch another task – of the inferior abstraction level - and so on.

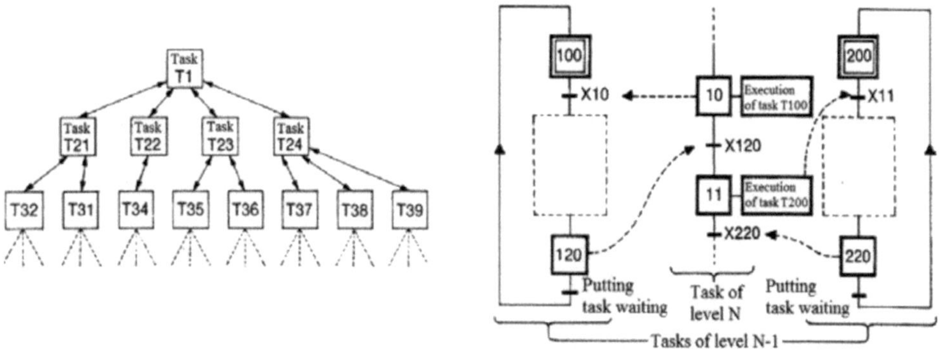

a) b)

Fig. 5. Vertical coordination

This approach is easier to systematize and also better adapted to treat more complex systems because of the use of different abstraction levels.

4. Case study

The case study corresponds to an automatic machine for filling and capping bottles (Fig. 6). This is divided in three modules, transport and feeding, filling and capping. For increasing productivity, a conveyor is used with several alveoli for the bottles, allowing the operation simultaneously at three working stations (modules of the automation system).

The transport and feeding station is composed of a pneumatic cylinder (A) that is responsible for feeding the bottles to the conveyor and another pneumatic cylinder (B) that executes the step/incremental advance of the conveyor.

The filling module is composed of a volumetric dispenser, a pneumatic cylinder (C) that actuates the dispenser and an on/off valve (D) for opening and closing the liquid supply.

The capping station has a pneumatic cylinder (G) to feed the cover, a pneumatic motor (F) to screw on the cover and a pneumatic cylinder (E) to advance the cover. The cylinder (E) moves forward until the existent cover retracts with this cover during the retraction of (G), and it moves forward again with rotation of the motor F to screw on the cover.

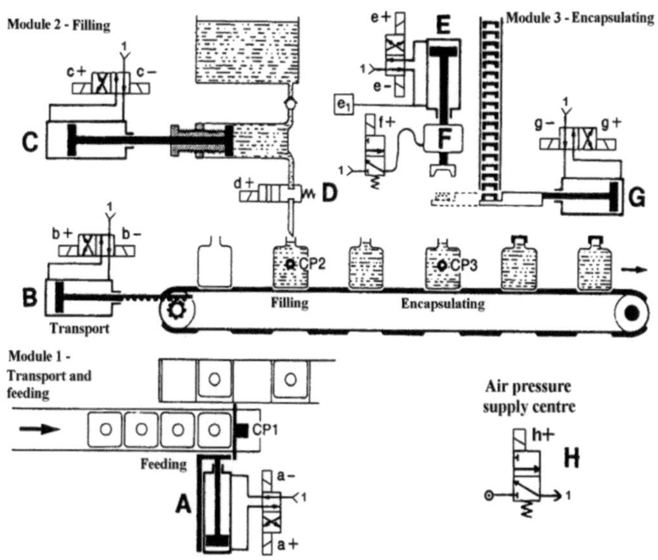

Fig. 6. Case study plant

4.1 Base controller behaviour specification

Figure 7 shows the base SFC of the system controller, corresponding only to the "normal production" mode. The basic sensors involved are: two end-course-sensors for each cylinder (example: cylinder A, sensor a0 and a1, respectively, retracted and forward) and a sensor of pressure e1, which detects the point of contact/stop of the cylinder E in any point of its course.

Valve D and motor F do not have position sensors because they are difficult to implement. On the other hand, in order to obtain the total SFC controller, which includes all the operation modes required for the correct operation of the system, the graphic chart of GEMMA was used because it allows definition of the functioning (operation) modes and stop states of the machine.

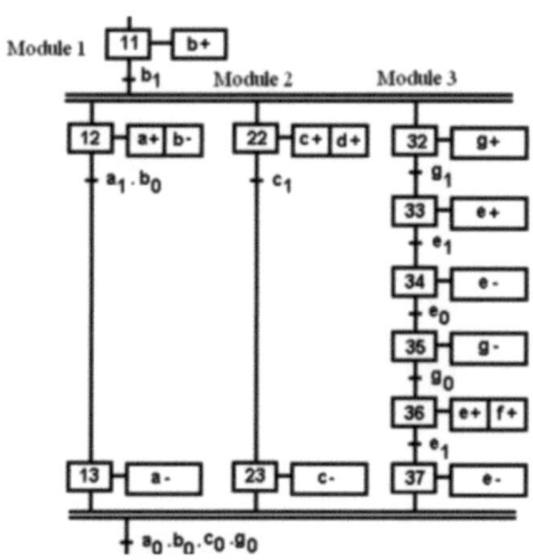

Fig. 7. Base SFC controller specification

4.2 Global controller structure

Figure 8 shows the GEMMA graphic chart developed for the presented case study. The considered tasks are described as follows:

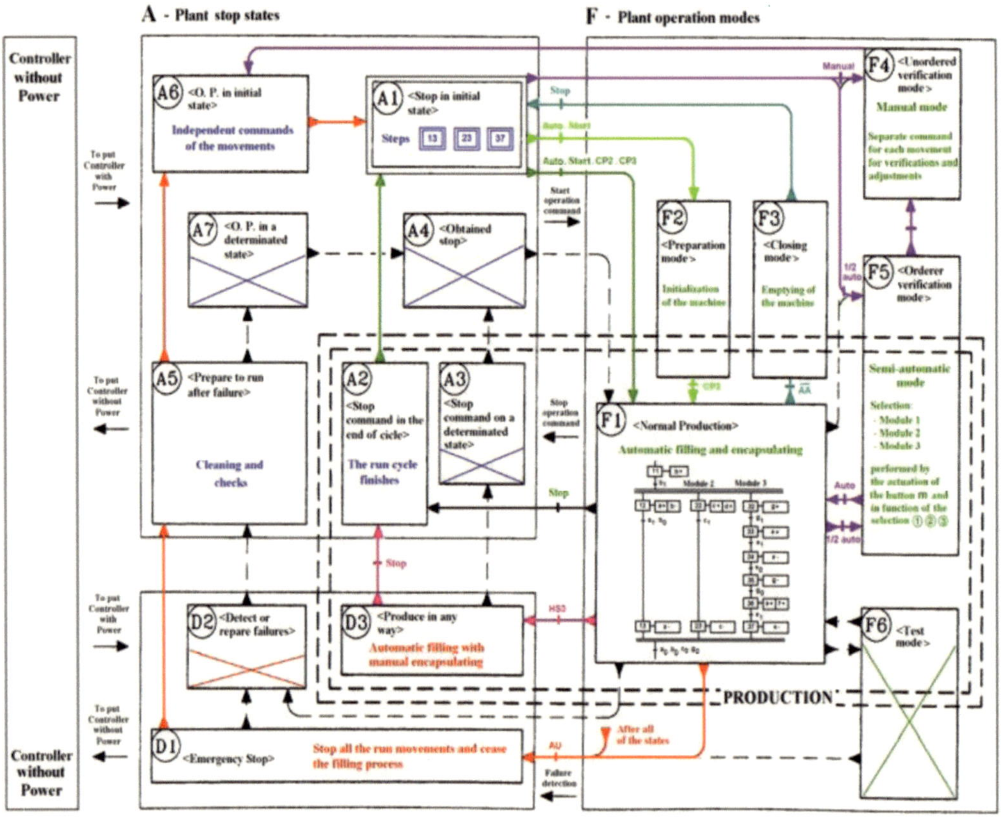

Fig. 8. GEMMA of the plant controller

A1 - The task A1 "stop in the initial state" represents the task of the machine represented in Figure 8.

F1 - Start of task A1, when the start command of the machine occurs, the change for the task F1 "normal production" happens (filling and automatic capping) with the consequent execution of base SFC presented in Figure 7.

A2 - When happens the stop command of the machine happens, the run cycle finishes in agreement with the condition described at task A2 "stop command at the end of cycle".

F2 - When the machine is "empty" (without bottles) it is necessary to feed bottles progressively, the machine being ready to begin the normal production (task F1) when it has bottles in the conveyor positions of the production modules 2 and 3, respectively. This operation is defined by task F2 "preparation mode".

F3 - The "closing mode" of task F3 allows the reverse operation, that is, the progressive stop of the machine with the exit of all of the bottles (emptying of the machine).

D3 - When the capping module is out of service it can be decided to produce in another way, that is, to perform the bottle filling in an automatic way and posterior manual capping, this is the main purpose of task D3 "production in another way".

D1 - In the case of a situation emergency, task D1 "emergency stop" is executed. This stops all the run actions and closes the filling valve to stop the liquid supply.

A5 - After the emergency stop (task D1), cleaning and verification are necessary: this is the purpose of task A5 "prepare to run after failure".

A6 -After the procedures of cleaning and verification finish it becomes necessary to perform the return to the initial task of the machine, as described at the task A6 "O.P. (operative plant) in the initial state".

F4 - For example, for volume regulation of the bottle liquid dispenser and adjustment of the bottle feeder, a separate command for each movement is required, according to task F4 "unordered verification mode".

F5 - For detailed operation checks, a semiautomatic command (only one cycle) it is necessary to check the functioning of each module: task F5 "ordered verification mode".

To make this possible GEMMA evolution becomes necessary, creating transition conditions for the run and stop operation modes, as described previously.

These transition conditions will be accomplished using GEMMA, as presented, to proceed:

- To allow the progressive feeding demanded in the preparation way (F2) and the progressive discharge required in the closing way (F3) it will be necessary to consider sensors that detect the bottles' presence under each one of the modules 1, 2, 3, respectively, CP1, CP2, CP3 (see Fig. 6);
- Also, it will be necessary a command panel that supplies the transition conditions given by an operator (Fig. 9).

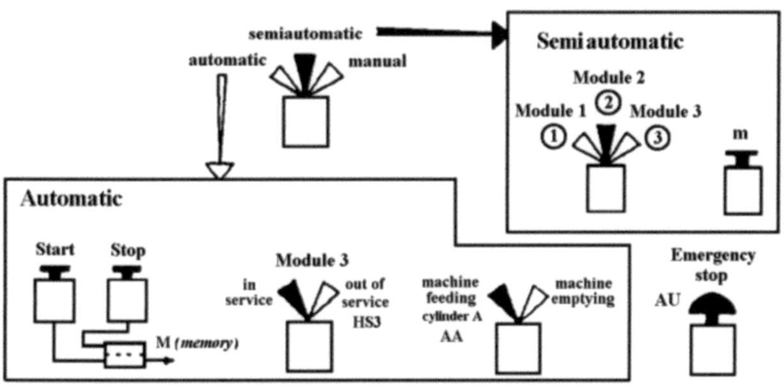

Fig. 9. Command panel of the system controller

In the command panel, there is a main switch that allows selecting the "automatic", "semiautomatic" and "manual" operation modes.

"Automatic" option corresponds to:

- Two buttons, "start" and "stop", whose actions are memorized in memory M;
- A switch HS3 to put module 3 "in service" or "out of service";
- A switch AA to control the bottles' feeding permission (cylinder A), to allow the emptying of the machine.

These switches/buttons and sensors CP1, CP2 and CP3 are the transition conditions of the tasks A1, F1, F2, F3, A2 and D3, as shown in Figure 8.

The "semiautomatic" option corresponds to task F5 "ordered verification mode" which allows the actuation of button (m) to check one cycle operation of each module selected by the "semiautomatic" switch ①,②, or ③.

The "manual" option corresponds to tasks F4, A5 and A6, which require a separate command from each movement using a direct command on the directional valves.

Finally, the AU button (emergency stop) allows passing to task D1 which starts from all of the tasks.

5. Coordination of the case study's complex specification

The implementation of total controller specification, based on GEMMA presented in Figure 8, can be realized using the following two alternative methods, when one SFC for each task is developed (Multiple SFC):

- Horizontal coordination;
- Vertical coordination.

As shown in section 3, there are several aspects/benefits for each described implementation (vertical coordination and horizontal coordination). However, it seems to be more systematic for vertical coordination because two levels of abstraction can be defined and when the system is really complex, this aspect seems to be very helpful.

Figure 10 shows the schema of the adopted approach for the case study (vertical coordination).

According Figure 10, GEMMA implementation is performed based on the following main stages:

1. Elaboration of a high level SFC that directly translates the base GEMMA of the system behaviour;
2. Elaboration of multiple low level SFCs corresponding to each functioning mode and/or stop state;
3. Synchronization of the SFCs using the vertical coordination methodology.

5.1 High level SFC

This is the first stage of the vertical coordination implementation of total controller specification. Figure 11 shows the high level SFC that directly corresponds to the base GEMMA of the case study plant controller presented previously in the Figure 8.

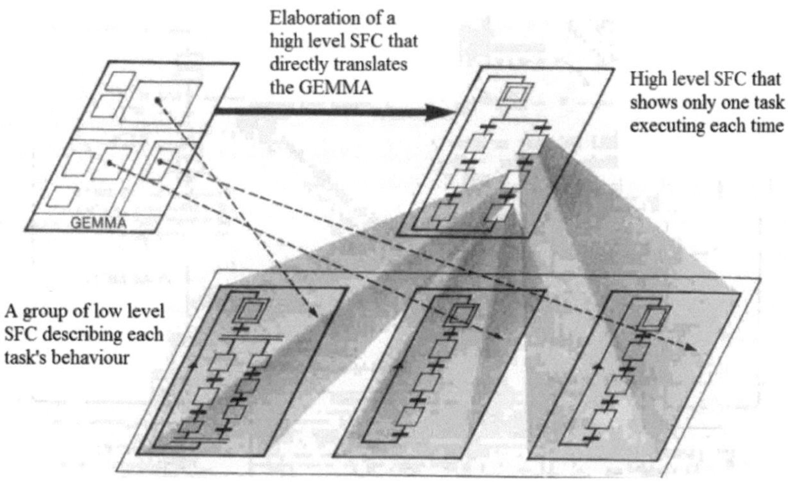

Fig. 10. GEMMA implementation with vertical coordination of multiple SFC

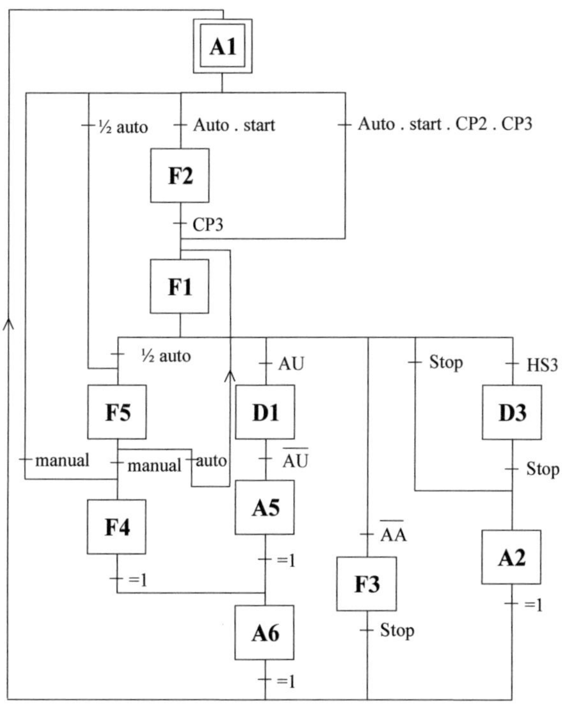

Fig. 11. High level SFC

5.2 Low level SFCs

The development of multiple low level SFCs specifications, corresponding to each one of the functioning modes and/or stop states considered for the case study plant controller, is the second stage of the vertical coordination implementation of total controller specification.

In this chapter, the SFCs' specifications corresponding to each one of the functioning modes and/or stop states will be shown. As mentioned before, when using the GEMMA approach, each SFC corresponding to each functioning mode and/or stop state is treated as a task. In this way, Figure 12 shows the SFC specification for the tasks F1 "normal production" and F2 "preparation mode". The SFC of the task F3 "closing mode" is not shown because it is similar to that presented for task F2. Additionally, Figure 13 shows the SFC specification for tasks F5 "ordered verification mode" and D3 "production in another way".

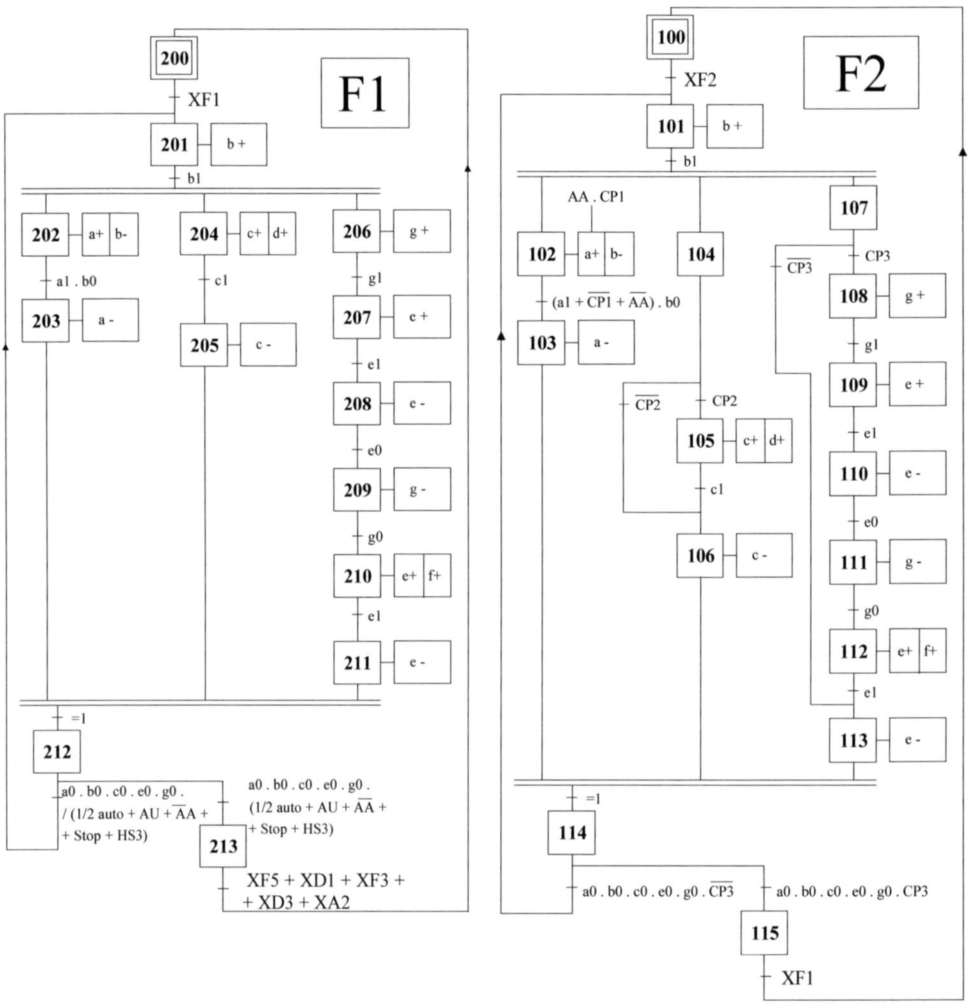

a) b)

Fig. 12. a) Low level SFC for the normal production task, and b) low level SFC for the preparation task

It is of importance to note that the "emergency stop" related to GEMMA task D1 is not treated in this section due to its complexity. The "emergency stop" controller behaviour specification will be presented in detail in sections 6 and 7 of this chapter. In particular, section 7 presents and discusses the implementation of GEMMA task D1 of the same case study (Fig. 8), with the aim of applying a global approach considering all automation system emergency stop requirements.

The definition of all functioning modes and all stop states of the automation system were performed according European standards EN 418 (EN, 1992) and EN 60204-1 (EN, 1997).

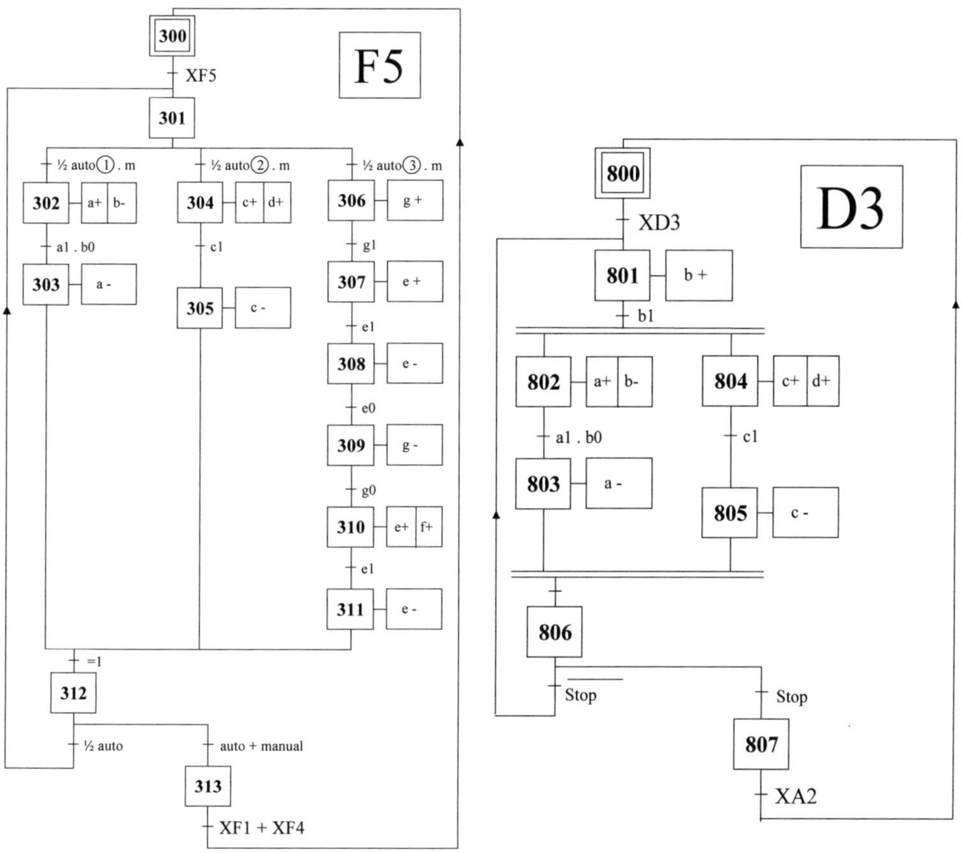

a) b)

Fig. 13. a) Low level SFC for the ordered verification task, and b) low level SFC for the production in another way task.

5.3 SFCs synchronization

The last stage of the vertical coordination implementation of total controller specification is related to the synchronization of the low level SFC specifications.

To achieve this as described in section 2, the high level SFC presented in Figure 14 was completed with the SFC step activity/action (Xi - i step number) that correspond to the low level SFC execution stop. Figure 14 shows the complete high level SFC obtained for vertical coordination implementation (the SFC step activities added are represented in red).

All the controller specifications, presented in the previous figure, were simulated on Automation Studio software. The obtained results led to the conclusion that all the requirements defined on the Emergency Stop Standards were met.

Further, the specification was translated to Ladder Diagrams according to the SFC algebraic formalization and implemented on a Programmable Logic Controller (PLC) adopted as the controller physical device. This part of the developed work is not detailed in this chapter.

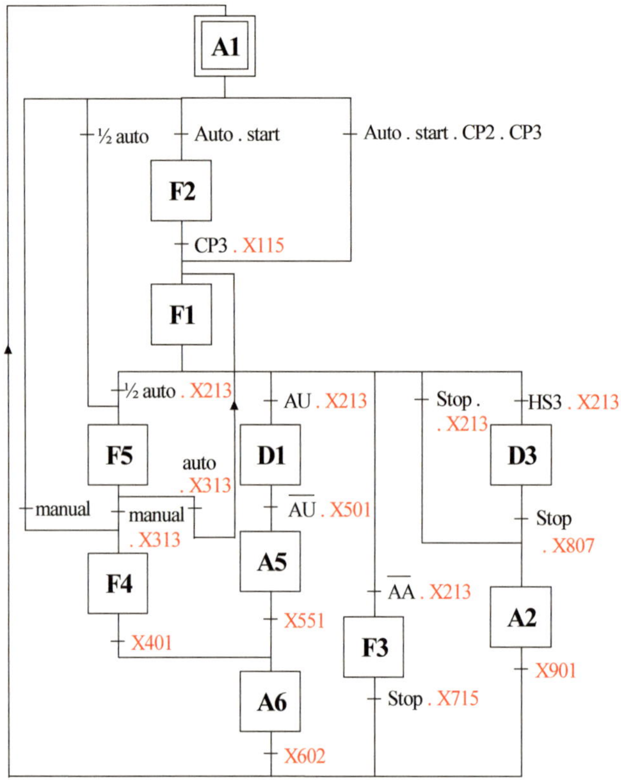

Fig. 14. High level SFC completed with low lever SFC step activities (in red).

All the SFC controller specifications, presented in the previous Figure, were simulated on Automation Studio software (FAMIC, 2003). The obtained results led to the conclusion that all of the automation system requirements were met.

Further, the specification was translated to Ladder Diagrams according to the SFC algebraic formalization (IEC, 2002) and implemented on a Programmable Logic Controller (PLC) adopted as the controller physical device. This part of the developed work is not detailed in this publication.

6. Emergency stop controller behaviour specification

The Emergency Stop is one of the most important aspects related to the safety of people, goods and equipments that interact with automated systems.

In order to obtain safe controllers, it must obey some rules (EN, 1992), (EN, 1997):

- A fault in the software of the control system must not lead to hazardous situations;
- Reasonably foreseeable human error during operation must not lead to hazardous situations;
- The machinery must not start unexpectedly;
- The parameters of the machinery must not change in an uncontrolled way, where such change may lead to hazardous situations;
- The machinery must not be prevented from stopping if the stop command has already been given;
- No moving part of the machinery or piece held by the machinery must fall or be ejected;
- Automatic or manual stopping of the moving parts, whatever they may be, must be unimpeded;
- The safety related parts of the control system must apply in a coherent way to the whole of an assembly of machinery and/or partly completed machinery;

The above mentioned rules can be seen as very important and they must be accomplished by the system behaviour and must be guaranteed by the controller program. However, the ways that designers use to achieve these goals can change. For instance, it can depend on the complexity of the system: if the system is more complex, then the implementation of the emergency stop requirements can be harder. Some indications can be done to the designers, but the final decision depends always of his/her scientific and technical background. This means that different solutions – for application of the emergency stop requirements - can lead to the same practical results. Also, some indications can be done - according to the different ways of guaranteeing the emergency stop requirements – such as if it is necessary, or not, a specific SFC for modelling the behaviour of the system after the emergency stop actuation. Sometimes a specific SFC for modelling the behaviour of the system is necessary after the emergency stop actuation and sometimes it is not.

From this last point of view, the types of emergency stops are divided in two main groups:

- Without emergency sequence - the actuation of the emergency button stops the system/automatism through the inhibition of the outputs and/or for stop the evolution of SFC.
- With emergency sequence - the actuation of the emergency button starts a particular predefined procedure.

As guarantee that the developed controller will always react according the expected behaviour, it is only necessary to model the controller and the plant discretely. Indeed, our system has a hybrid plant, but the behaviour properties that we intend to guarantee for our system are only related to discrete behaviour.

6.1 Without emergency sequence

The emergency without emergency sequence can be performed in three alternative modes:

- Outputs inhibition;
- Evolution stop;
- Outputs inhibition and evolution stop.

In the case of outputs inhibition the actuation of emergency button does not stop by itself the evolution of the SFC controller, but it inhibits the outputs associated with their steps, as shown in Figure 15. The ON outputs (state 1) are turned OFF (state 0), as well as the evolution of SFC usually being stopped by the non-fulfilment of the logical conditions associated with SFC transitions.

This can be obtained through the insertion of inhibition functions in the interface with the machine plant. In this case, after the occurrence of an emergency stop, the actuator's command should be particularly well studied in agreement with the type of expected response.

For instance, for the cylinders directional valves:

- One stable state valve (single control with spring return), if a cylinder return for a given position is demanded.
- Two stable state valve (double control), if a stop at the end of the cylinder movement is demanded.
- Valve with three positions (double control and spring return), if a cylinder stop in the actual position is demanded.

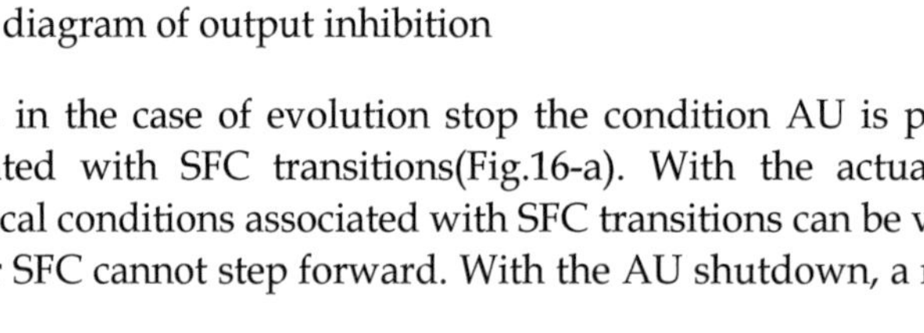

Fig. 15. Functional diagram of output inhibition

In the other hand, in the case of evolution stop the condition AU is present in all logical conditions associated with SFC transitions(Fig.16-a). With the actuation of emergency button AU, no logical conditions associated with SFC transitions can be validated and in this way, the controller SFC cannot step forward. With the AU shutdown, a new cycle evolution is allowed.

It is of importance to note that in this situation the outputs associated to the active steps stay validated. This way, the started actions can be maintained, if dangerous situations are not to occur.

Finally, also it is possible to use the described types of emergency stop simultaneously, without emergency sequence, outputs inhibition and evolution stop (Fig. 16-b). This situation is used more in practice, if a specific emergency sequence is not necessary. Seen that has the advantage of allowing, after the emergency button shutdown, the pursuit of the evolution of the system starting from the same position at which it was stopped.

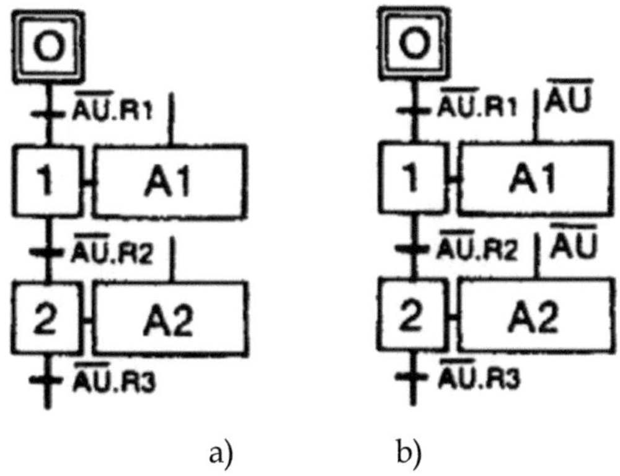

Fig. 16. a) Evolution stop, and b) evolution stop and outputs inhibition

6.2 With emergency sequence

This type of emergency implies the introduction of an emergency sequence. Through the activation of the emergency button AU, an emergency sequence can be added to the normal run SFC (Fig. 17).

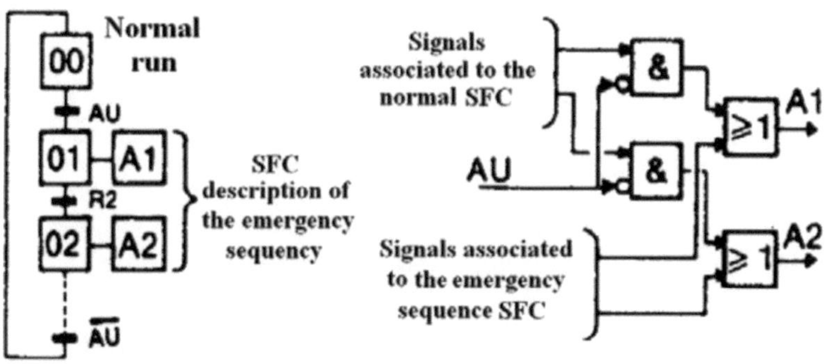

Fig. 17. Introduction of an emergency sequence

7. Emergency stop adopted solution: case study

The implementation of total controller specification, based on GEMMA presented in Figure 8, can be realized using the following two main alternative methods:

- Multiple SFC – developing one SFC for each task (implementation of horizontal or vertical coordination);

- Single SFC – developing one SFC for all tasks.

Although this last alternative is possible from a theoretical point of view, for most if not all automation systems practice shows that if the application of emergency stop requirements is done on a complex system, the first solution (multiple SFC) is better.

However, the specification of a specific behaviour for the emergency stop requirements (with emergency sequence) and its linking with other specified behaviours for the system (it doesn't matter if by single or multiple SFCs) is similar. For this reason, this section considers one single SFC for specification of all the desired behaviours for the system. Although the Single SFC method was used for the implementation of the "emergency stop" to allow a better global visualization and understanding of the implementation of the total controller specification that includes the "emergency stop" (GEMMA task D1), it must be highlighted that step 100 of the SFC presented in Figure 5 corresponds to an emergency stop sequence.

The emergency stop adopted for the case study presented was obtained according the EN 418 and EN 60204-1 standards.

According to the behaviour of the case study, the emergency stop with emergency sequence was selected. The considered requirements that need to be accomplished by the emergency sequence are:

- Stop all of the movements;
- Stop the filling operation.

To obtain these procedures the selection of the type of the appropriate directional valves to accomplish, simultaneously, the requirements of the emergency stop and of the plant behaviour was crucial.

The directional valve specifications used were the type of control (single solenoid control with spring return or double solenoid control) and number of ways/ports.

The first security requirement relates to the stop of the movements, obtained by stopping the air compressed supply to the directional valves of the cylinders A, B, C, E, G and of motor F. For that, as shown in Figure 6, the air supply will be centralized and controlled through a directional valve 3/2 way normally closed with spring return (H).

The second security requirement relates to the stopping of the filling operation; this was performed through the turn OFF of the filling directional valve 2/2 way normally closed with spring return (D).

Figure 18 shows the total controller SFC specification based on GEMMA implementation with the single SFC method.

All the SFC controller specifications, presented in Figure 18, were simulated on Automation Studio software. The obtained results led to the conclusion that all the requirements defined in the Emergency Stop Standards were met.

Further, the specification was translated to Ladder Diagrams according to the SFC algebraic formalization and implemented on a Programmable Logic Controller (PLC) adopted as the controller physical device. This part of the developed work is not detailed on this publication.

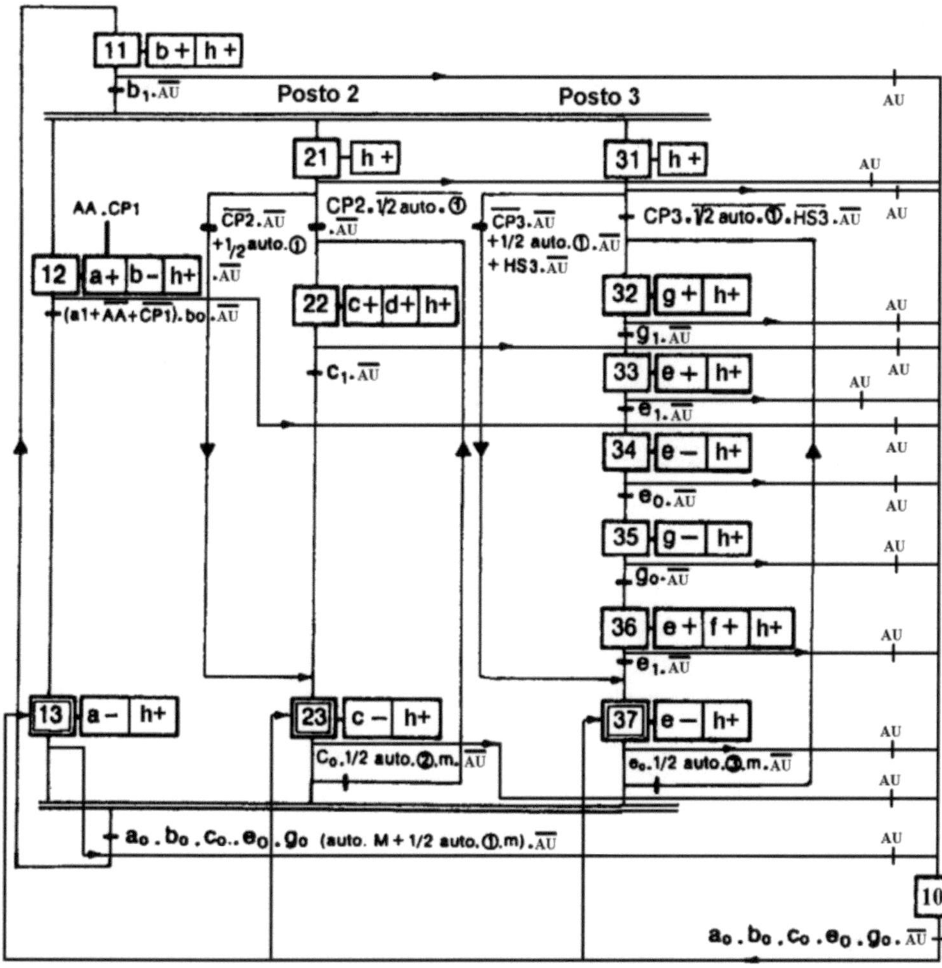

Fig. 18. Total SFC controller specification with emergence sequence

8. Conclusions

A systematic way was presented using the adopted techniques for the implementation of complex specifications of automation systems.

First the use of GEMMA and SFC for the structure and specification of all the system's behaviour was explained, considering its stop states and functioning modes.

Further, the vertical coordination implementation of a complex total controller specification, based on the GEMMA graphical chart, was also presented and discussed.

Also, the adopted techniques for the emergency stop behaviour specification of automation systems were presented in a systematic way.

The ways to translate the GEMMA graphical chart to low level specification were also presented and discussed.

The standards (EN418, EN60204-1) related to the stop emergency specifications were considered and all the requirements were accomplished.

The obtained results, by simulation with Automation Studio software, showed that the adopted approach is adequate.

9. References

ADEPA (1992). GEMMA - Guide d`Étude des Modes de Marches et d`Arrêts, Ed.2, In: *Agence Nationale pour le Developpement de la Production Automatisée*, France

Booch, G.; Jacobson, I. & Rumbaugh, J. (2000). OMG Unified Modeling Language Specification., In: *Object Management Group edition*, Object Management Group

Brusamolino, M.; Reina, L. & Spalla, M.F. (1984). An example of microprocessor's application in minicomputer systems: a copy volume design and implementation, In: *Microprocessing and Microprogramming*. Vol.13, Issue 5, pp. 331- 339

EN 418 (1992). Safety of machinery. Emergency stop equipment, functional aspects. Principles for design, In: *European Standard*

EN 60204-1 (1997). Safety of Machinery - Electrical Equipment of Machines - Part 1: General Requirements-IEC 60204-1, In: *European Standard*

FAMIC (2003). Automation Studio – User´s Guide, *Famic Technologies Inc*, Canada.

Harel, D. (1987). Statecharts: a visual formalism for complex systems, In: *Science of Computer Programming*, Vol.8, pp 231-274, North Holland, Netherlands

IEC (2002). IEC 60848 - Specification language GRAFCET for sequential function chart, Ed.2, In: *International Electrotechnical Commission*

Koornneef, F. & Meulen, M.V.D. (2002). Safety, reliability and security of industrial computer systems, In: *Safety Science*, Vol.40, Issue 9, pp. 715-717

Moon I. (1994). Modeling Programmable Logic Controllers for Logic Verification, In: *IEEE Control Systems Magazine*, pp. 53-59

Murata, T. (1989). Petri Nets: Properties, Analysis and Applications, *Proceedings of the IEEE*, Vol. 77, No. 4, pp. 541-580, 1989

Sohier, C. (1996). Pilotages des Cellules Adaptatives de Production: Apport des Systemes Multi-Agents, *PhD Thesis*, École Normale Supérieure de Cachan, Paris, France

26

Answering Causal Questions and Developing Tool Support

Sodel Vazquez-Reyes and Perla Velasco-Elizondo
Autonomous University of Zacatecas, Centre for Mathematical Research
México

1. Introduction

People explore the world by asking questions about what is seen and felt. Thus, Question Answering is an attractive research area as a distinctive combination from a variety of disciplines, including artificial intelligence, information retrieval, information extraction, natural language processing and psychology. Psychological approaches focus more on theoretical aspects, whereas artificial intelligence, information retrieval, information extraction and natural language processing approaches investigate how practical Question Answering systems can be engineered.

A simple query to a search engine will return hundreds of thousands of documents. This raises the need for a new approach to allow more direct access to information. Ideally, a user could ask any question (open domain) and instead of presenting a list of documents, Question Answering technology could present a simple answer to the user.

In contrast to Information Retrieval systems, Question Answering systems involve the extraction of answers to a question rather than the retrieval of relevant documents.

At present, there are textual Question Answering systems working with factual questions. These systems can answer questions of the type: *what, who, where, how many* and *when*. However, systems working with complex questions, such as *how* and *why,* are still under research, tackling the lack of representations and algorithms for their modelling (Burger et al., 2001; Maybury, 2003; Moldovan et al., 2003).

One kind of complex question is *"why"* (causal). One reason why causal questions have not been successfully treated is due to their answers requiring more elaboration (explanations) instead of short answers.

The idea for our research *"detecting answers to causal questions through automatic text processing"*, was born, and the hypothesis is that, whilst *"why"* questions appear to require more advanced methods of natural language processing and information extraction because their answers involve opinions, judgments, interpretations or justifications, they can be approached by methods that are intermediate between Information Retrieval and full Natural Language Understanding. We advocate the use of methods based on information retrieval (bag-of-words approaches) and on limited syntactic and/or lexical semantic analysis as a first step towards tackling the problem of automatically detecting answers for causal questions.

For this research, a *"why"* questions, is defined simply as interrogative sentences in which the interrogative adverb *why* occurs in the initial position. For example: *Why is cyclosporine dangerous?*

The topic of a *"why"* question is the proposition that is questioned and it is presupposed to be true according to the document collection; otherwise, the *"why"* question cannot be answered. The proposition of the *"why"* question shown previously is: *Cyclosporine is dangerous.*

The topic of the *"why"* question that is questioned should be the event that needs an explanation according to the questioner. Identifying the question's topic and matching it to an item (event, state, or action) in the text is a prerequisite for finding the answer. The response to the *"why"* question stated previously is:

Why is cyclosporine dangerous?
It can harm internal organs and even cause death.

Inappropriate answers to *"why"* questions are mainly due to misunderstanding the questions themselves (Galambos & Black, 1985). For example, the answer to the following question varies depending on how the question is understood.

Why did Sodel dine at the Mexican Restaurant?

If we understand that this question concerns Sodel's motivation for eating, we could reply that he ate there "because he was hungry". If we understand that the question relates to why Sodel chose that particular restaurant, we could answer that "He had heard that it is a good restaurant and he wanted to try it". If we understand that the question is about Sodel going to a restaurant instead of eating at home, we could answer that "Sodel's wife is out of town and he can't cook". Neither having the same topic nor being expressed by the same sentence constitutes a criterion of identity for *"why"* questions. In other words, the same sentence can express different *"why"* questions. There are contextual factors and background knowledge for the description of its interpretation. These answers to *"why"* questions may be subjectively true and each answer have boundary conditions for itself. The accuracy of the answer is in the mind of the perceiver.

The ultimate goal in textual Question Answering systems should be to answer any type of question; consequently, the research direction should move towards that goal, and a focus on causal questions serves this agenda well.

2. Definition of problem

Many members of the information retrieval and natural language processing community believe that Question Answering is an application in which sophisticated linguistic techniques will truly shine, owing to it being directly related to the depth of the natural language processing resources (Moldovan, et al., 2003).

However, Katz and Lin (2003) have shown that the key to the effective application of natural language processing technology is to employ it selectively only when helpful, without abandoning the simpler techniques. They proved that syntactic relations enable a Question Answering system successfully to handle two linguistic phenomena: semantic symmetry and ambiguous modification. That is, the incorporation of syntactic information in Question Answering has a positive impact.

Voorhees shows that the research in Question Answering systems has made substantial advances (Voorhees, 2001, 2002, 2003, 2004), answering factual questions with a high degree of accuracy. Several forms of definition questions are processed appropriately, and list questions retrieve sequences of answers with good recall from large text collections, although it is necessary to answer hard and complex questions too, such as procedural, comparative, evaluative and causal questions.

On the other hand, (Moldovan, et al., 2003) showed that the main problem with Question Answering systems is the lack of representations and algorithms for modelling complex questions in order to derive as much information as possible, and for performing a well-guided search through thousands of text documents.

Some of the complex questions types seem to need much semantic, world knowledge and reasoning to be handled properly, e.g. for automatically resolving ambiguities or finding out which measure or granularity a user would prefer. This is beyond the scope of the current research and even those in the near future. However, in this research, we advocate the use of methods based on information retrieval (bag-of-words approaches) and on limited syntactic and/or lexical semantic analysis as a first step towards tackling the problem of causal questions.

Our goal was not to implement a functional (fully-fledged) textual Question Answering system but to investigate how methods based on information retrieval (bag-of-words approaches) and on limited syntactic and/or lexical semantic analysis can contribute to the real-world application of causal text and causal questions. This enables us to focus on the key matter of how the answer is contained in the document collection.

3. Related work

In the literature on Question Answering, the system developed in the Southern Methodist University and Language Computer Corporation (Harabagiu et al., 2000) has been considered to have the most sophisticated linguistic techniques due to the depth of its natural language processing resources. This system classifies questions by expected answer type, but also includes successive feedback loops that attempt to make progressively larger modifications to the original questions until they find an answer that can be justified as abductive proof —semantic transformations of questions and answers are translated into a logical form for being analysed by a theorem prover.

The system of the Southern Methodist University and Language Computer Corporation first parses the question and recognises the entities contained in it to create a question semantic form. The semantic form of the question is used to determine the expected answer type by finding the phrase that is most closely connected to other concepts in the question. The system then retrieves paragraphs from the corpus, using boolean queries and terms drawn from the original question, related concepts from WordNet, and an indication of the expected answer type.

Paragraph retrieval is repeated using different term combinations until the query returns a number of paragraphs in a pre-determined range. The retrieved paragraphs are parsed into their semantic forms, and a unification procedure is run between the question semantic form and each paragraph semantic form. If the unification fails for all paragraphs, a new set of paragraphs is retrieved using synonyms and morphological derivations of the previous query.

When the unification procedure succeeds, the semantic forms are translated into logical form, and a logical proof in the form of an abductive backchaining from the answer to the question is attempted. If the proof succeeds, the answer from the proof is returned as the answer string. Otherwise, terms that are semantically related to important question concepts are drawn from WordNet and a new set of paragraphs is retrieved.

While research in Question Answering mainly focussed on responding to factual questions, definition questions and list questions using stochastic processes, a more recent trend in Question Answering aims at responding to other types of question that are of great importance in everyday life or in professional environments such as procedural, causal, comparative or evaluative questions. These have not yet been studied in depth; they require different types of methodologies and formalisms, particularly at the level of the linguistic models, knowledge representation and reasoning procedures.

However, the ideal system does not exist yet although approaches to support that goal have been created. We will demonstrate a small number of approaches to Question Answering working with complex questions or advanced methods. The explanations are based on their general ideas.

In order to answer *"why"* questions, the aim of an ideal system should be to address a form of Question Answering that does not focus on finding facts, but rather on finding the identification and organisation of opinions, to support information analysis of the following types: *(a)* given a particular topic, find a range of opinions being expressed about it; *(b)* once opinions have been found, cluster them and their sources in different ways, and *(c)* track opinions over time.

Verberne et al. (2007) demonstrate an approach to answering *"why"* questions, based on the idea that the topic of the *"why"* question and its answer are siblings in the rhetorical structure of the document, determined according to Rhetorical Structure Theory (RST) (Mann & Thompson, 1988), connected by a rhetorical relation that is relevant for *"why"* questions —"discourse-based answer extraction". They implemented an algorithm that: *(a)* indexes all text spans not from the source document but from a manually analysed representation of it into RST relations that participate in a potentially RST relation relevant; *(b)* matches the input question to each of the text spans in the index; and *(c)* retrieves the sibling for each of the found spans as the answer. The result is a list of potential answers, ranked using a probability model that is largely based on lexical overlap. For the purpose of testing their implementation, they created a test collection consisting of seven texts from the RST Treebank and 372 *"why"* questions elicited from native speakers who had read the source documents. From this collection, they obtained a recall of 53%, with a mean reciprocal rank of 0.662. On the basis of the manual analysis of the question-answer pairs, they argued that the maximum recall that can be obtained for this data set, from the use of RST relations as proposed, is 58.0%. They declare that, although there are no reference data for the performance of automatic Question Answering working with *"why"* questions, they considered a recall of 53% (and a maximum recall of 58%) to be mediocre at best.

Waldinger et al. (2004), Benamara and Saint-Dizier (2004), and McGuinness and Pinheiro da Silva (2004) delve into knowledge-based Question Answering and support inferential processes for verifying candidate answers and providing justifications. That is, systems that target the problem of Question Answering over multiple resources have typically taken the approach of first translating an input question into an intermediate logical representation, and, in the realm of this intermediate representation, matching parts of the question to the content supplied by various resources.

Light et al. (2004), provide an empirical analysis of a corpus of questions that enables the authors to identify examples of reuse scenarios, in which future questions could be answered better by using information previously available to the system (e.g., in the form of previously submitted questions or answers already returned to the users). The authors acknowledge that some of the proposed categories of reuse are very difficult to implement in working system modules.

Schlaefer (2007) has used ontologies for extracting terms from questions and corpus sentences and for enriching the terms with semantically similar concepts. In order to improve the accuracy of Question Answering systems, semantic resources have been used. Semantic parsing techniques are applied to transform questions into semantic structures and to find phrases in the document collection that match these structures.

Vicedo and Ferrandez (2000) have demonstrated that their evaluation improvements when pronominal references are solved for IR and Question Answering tasks. That is, they are solving pronominal anaphora.

Mitkov (2004) describes that coreference resolution has proven to be helpful in Question Answering, by establishing coreferences links between entities or events in the query and those in the documents. The sentences in the searched documents are ranked according to the coreference relationships.

Castagnola (2002) shows that for the purpose of improving the performance of Question Answering, he resolves pronoun references via the use of syntactic analysis and high precision heuristic rules.

Galitsky (2003) introduces the reasoning mechanism as the background of the suggested approach to Question Answering, particularly, scenario-based reasoning about mental attitudes. Default logic is used for correction of the semantic representations. He describes the process of representing the meaning of an input query in the constructed formal language.

Setzer et al. (2005) address the role that temporal closure plays in deriving complete and consistent temporal annotations of a text. Firstly, they discuss the approaches to temporal annotation that have been adopted in the literature, and then further motivate the need for a closed temporal representation of a document. No deep inferencing, they argue, can be performed over the events or times associated with a text without creating the hidden relations that are inherent in it. They then address the problem of comparing the diverse temporal annotations of the same text. This is far more difficult than comparing, for example, two annotations of part-of-speech tagging or named entity extent tagging, due to the derived annotations that are generated by closure, making any comparison of the temporal relations in a document a difficult task. They demonstrate that two articles cannot be compared without examining their full temporal content, which involves applying temporal closure over the entire document, relative to the events and temporal expressions in the text. Once this has been achieved, however, an inter-annotator scoring can be performed for the two annotations.

Nyberg et al. (2004) aim to capture the requirements of advanced Question Answering and its impact on system design and the requirements imposed on the system (e.g., time-sensitive searches and the detection of obscure relations). The challenges that face the push towards the development of Question Answering systems of increased complexity are especially the challenges of practicability and scalability. Indeed, such issues become important for any system that would actually attempt to perform planning in a broad domain. Similarly, it may be very challenging to find common linguistic representations to use across highly modular systems for encoding internal information, as the information sources themselves can vary widely, from unstructured text at one end of the spectrum to full-blown knowledge bases at the other end.

Planning structures explicate how a person does certain things, and how he or she normally tries to achieve some goal. Plans cannot be built from the story itself but have to be taken from some world knowledge module. Studying instructional texts seems to be very useful for answering procedural questions —"how". Aouladomar (Aouladomar, 2005a, 2005b; Aouladomar & Saint-Dizier, 2005a, 2005b) incorporated concepts from linguistics, education, and psychology to characterise procedural questions and content to produce an extensive grammar of the ways in which a procedural text may be organised, a framework that appears to show much promise. Although her work is on French texts, the procedural features she identified included general ones, e.g. the distinct morphology of verbs in procedures. However, her research did not directly address the task of classifying texts as either procedural or non-procedural.

Aouladomar mentioned that questions beginning with how should not be neglected, since, according to recent usage data from a highly-trafficked web search engine, queries starting with how alone is the most popular category of queries beginning with question words.

The approaches to Question Answering mentioned above are likely to become relatively more language-dependent, as they require larger and more complex resources of various kinds.

3.1 Challenges of Questions Answering

The ultimate goal in textual Question Answering systems is to answer any type of question. If the information needs are very simple ones (e.g. factoid, definition or list), then the answer can be simple word(s), phrase(s) or sentence(s). If the information needs are more complex, then the answers may come from a deep documentary analysis, or from multiple documents. Where candidates answer from different corpora, these could be merged or possibly summarised.

Moreover, if we recognise that users can obtain valuable information through inference and construction from new material, combined with what they know already, the scope for Question Answering is far wider. This can be far more than simply deriving the kind of exact answer that is required due to the rich knowledge or complex inference it requires.

Alternatively, candidate answers from different languages could be translated into the user's native language. For example, if we retrieve answers from Spanish, French and Italian and translate them into English, we could compare the nature of the answers drawn from different geographic and cultural contexts.

The research direction is moving gradually towards these goals and it is our hope that the Question Answering research groups can collaborate in order to achieve these goals.

4. Approach

Most causal questions are of the form "*Why Q?*", where *Q* is an observation or fact to answer (which we have identified as an effect). If a *"why"* question is an effect, then we are searching for its explanations (which we have identified as causes). So, we have called the cause and its effect a causal relation.

The *"why"* question (effect) has an infinite number of different answers. Each answer contains an explanation of a cause for the question. In particular, causes explain their effects. For this reason, a cause tells us why its effect occurs.

The natural complexity of a question depends on how the question is understood, and the accuracy of the answer is in the mind of the perceivers, depending mainly on their knowledge level (contextual factors and background knowledge) for the description of its interpretation. Some users prefer a more accurate explanation, while others look for explanations with a broader perspective and better explanatory resources. Any answer that appeals to a cause is taken to be highly relevant and, therefore, to provide an explanation of the effect – a *"why"* question.

In order to get answers to a *"why"* question, we should try to detect causal relations. Although textual Question Answering systems are evolving towards providing exact answers only, for *"why"* questions the answers should be surrounded by some context, with the purpose of supporting the answer.

4.1 Methodology

We have used the lexico-syntactic classification for *"why"* questions proposed by Verberne et al. (2007). The categories used are existential "there" questions, process questions, questions with a declarative layer, action questions and have questions. The result of question analysis task is not used in the answer candidate extraction task. However, it gives a category to each *"why"* question.

The answer candidate extraction task provides an approach to tackling a subset of causal questions. We used the following procedure for detecting possible answers to *"why"* questions:

Identify the topic of the question.
In the list of sentences of source document, identify the clause(s) that express(es) the same proposition as the question topic.

Select the best three clauses as answers.
Detect cause-effect information expressed in the answers selected.

Step 1. The topic of the question (which we have identified as an effect) is the observation or fact that is questioned. In other words, it is the premise of question.

Step 2 and 3. We suggest that decomposition of the complex task of recognizing which source text expresses the same proposition as the question topic would make a step towards better understanding the process for answering causal questions. This should involve making use of set of measures (see 4.1.1), and using each one as a weighting factor within the whole evaluation for ranking of possible answers. The sum of factors is the final value. To be precise, each measure is applied to the words belonging to the question-text pair. The best three answers are selected.

Step 4. We used a rule-based approach to identify and extract cause-effect information expressed in the answers selected.

4.1.1 Matching formulae

The answer extraction process relies on the computation of four measures:

1. *Simple matching*. The stop words are not removed; for this reason, non stop words are weighted with *1.9* and stop words with *0.1*. The final weight is calculated as the sum of all values and normalized dividing it by the length of the question and text (total number of words). In which, *Q* is a question and *T* is a text with possible answer, see (1).

$$dist_{jcm} = IC(N1) + IC(N2) - 2 * IC(LCS) \quad (1)$$

However, if simple matching is not possible, we are working with stems. All the occurrences in the question's stems set that also appear in the text's stems set will increase the accumulated weight in a factor of one unit. The stems are weighted with *1.9* for non stop words and *0.1* for stop words.

Longest consecutive subsequence. This process measures the surface structure overlap between the text with possible answer and the question (only consecutive words). In order to compute this overlap we extract the longest consecutive subsequence (LoCoSu) between the question and the text with possible answer, *LoCoSu(Q, T)*, see (2).

$$structure\_Overlap = |LoCoSu(Q,T)| \quad (2)$$

For example, if we have *Q*= {"AA", "BB", "CC", "FF" } and *T*= {"AA", "BB", "DD", "FF"} then *LoCoSu* = {"AA", "BB"} = 2.

In order to calculate *LoCoSu*, we have used a third party implementation, the longest common substring tool (Dao, 2005).

This feature indicates the presence of the same word with *1*, or otherwise *zero*. We are removing stop words. We are using stems if simple matching is not possible.

One should note that this measure assigns the same relevance to all consecutive subsequences with the same length. Furthermore, the longer the subsequence is, the more relevant it will be considered.

We have used a threshold of 2, that is, *LoCoSu (Q, T)* bigger or equal to 2.

Sorensen's similarity coefficient. This only considers non stop words. The value is increased by one per word in the intersection or union. Sorensen's similarity coefficient is a distance measure, see (3).

$$Sorensen = 1 - D \tag{3}$$

Where *D* is Dice's coefficient, which is always in [0,1] range, see (4).

$$D = \frac{2|Q \cap T|}{|Q| + |T|} \tag{4}$$

Where *Q* is a set of words of question and *T* is a set of words of sentence, possible answer. Note that if simple matching is not possible, we use their stems for evaluation.

WordNet-based Lexical Semantic Relatedness. The measure uses WordNet (Miller, 1995) as its central resource. Here, we are in fact considering similarities between concepts (or word senses) rather than words, since a word may have more than one sense. Measures of similarity are based on information in is-a hierarchy. WordNet only contains is-a hierarchies for verbs and nouns, so similarities can only be found where both words are in one of these categories. WordNet includes adjectives and adverbs but these are not organised into is-a hierarchies, so similarity measures cannot be applied.

Concepts can, however, be related in many ways apart from being similar to each other. These include part-of relationships, as well as opposites and so on. Measures of relatedness make use of this additional, non-hierarchal information in WordNet, including the gloss of the synset. As such, they can be applied to a wider range of concept pairs including words that are from different parts of speech.

If we want to compute lexical semantic relatedness between pairs of lexical items using WordNet, we can find that several measures have been reported in the literature. According to the evaluation of Budanitsky and Hirst (2006), the measure proposed by Jiang and Conrath (1997) is the most effective. The same measure is found the best in word sense disambiguation (Patwardhan, Banerjee, & Pedersen, 2003). We confirm that the Jiang and Conrath measure was the best for the task of patter_induction for information extraction (Stevenson & Greenwood, 2005).

The Jiang and Conrath metric (jcm) uses the information content (IC) of the least common subsumer (LCS) of the two concepts. The idea is that the amount of information two concepts share will indicate the degree of similarity of the concepts, and the amount of information the two concepts share is indicated by the *IC* of their *LCS*. Thus, they take the sum of the *IC* of the individual concepts and subtract from that the *IC* of their *LCS*, see (5).

$$dist_{jcm} = IC(N1) + IC(N2) - 2 * IC(LCS) \tag{5}$$

Where *N1* is the number of nodes on the path from the *LCS* to concept *1* and *N2* is the number of nodes on the path from the *LCS* to concept 2.

Since this is a distance measure, concepts that are more similar have a lower score than the less similar ones. The result with the smallest distance is taken to disambiguate the senses between two words. In order to maintain consistency among the measures, they convert this measure to semantic similarity by taking its inverse, see (6).

$$sim_{jcm} = \frac{1}{IC(N1) + IC(N2) - 2 * IC(LCS)} \tag{6}$$

In order to calculate *jcm* similarity, we have used a third party implementation, the WordNet Relatedness tool (Pedersen, Patwardhan, & Michelizzi, 2004).

We use three types of words for calculating this feature in order to discover lexical semantic relatedness:

a. *jcm* similarity between question-word against text-word.
b. *jcm* similarity between question-word against synonyms of text-word. The idea is to show that if synonyms are different words with identical or at least similar meanings, then we can use them to calculate *jcm* similarity in order to provide extra resources for disambiguation process.
c. *jcm* similarity between question-word and the synonyms of text-word antonyms. We think that a word pair where the individual words are opposite in meaning could help in the disambiguation process, identifying cause-effect (text-question) described with opposite words. In other words, an additional exploration of potential associative relations for text-question pairs.

4.2 Implementation

The system architecture is depicted in Fig. 1. As can be seen, it has a base client-server architecture (Shaw & Garlan, 1996) hosting several components that support different duties. There are three main actors in the environment surrounding the system: (*i*) the User —which is the person who issues the questions to be answered by the system, (*ii*) the Administrator —which is the person whose main duty is to update the system's databases and (*iii*) CAFETIERE —which is an external tool used to support the query/document processing work. All the communication among the main architectural parts of the system, is carried out in a synchronous request-response mode, i.e. a "source part" submits a request and waits until the response is returned from the "target part". All the components of the architecture are written in Java.

As depicted in Fig. 1, the client-side of the architecture hosts two user interface (UI) components: the User UI and Administration UI. These UI components enable the communication of the User and Administrator with the system.

A User issues a question to the system via the User UI component. This question is passed up to the server as a user request. In the server side, all user requests are processed by the Query Processing component. Internally, this component has a Pipe-and-Filter like architecture (Shaw & Garlan, 1996), which allows splitting the query-processing job into a series of well-defined low-coupled sequential steps. Three filter components constitute the Query Processing component: Question Analysis, Answer Candidate Extraction and Cause-Effect Detection. The Question Analysis filter enables the classification of the issued question with regard to a category that corresponds to a syntactic pattern. The Answer Candidate Extraction filter automatically maps the question onto the sentences of document, mainly by measuring lexical overlapping and lexical semantic relatedness between the question topic and the sentences of document to detect possible answers for the question evaluated. Finally the Cause-Effect Detection filter uses a rule-based approach in order to identify the cause and effect information expressed in the selected answers. Both the Question Analysis and Cause-Effect Detection filters interact with the CAFETIER tool (Black et al., 2003). Specifically, they use CAFETIER's lexico-syntactic analysis pipeline.

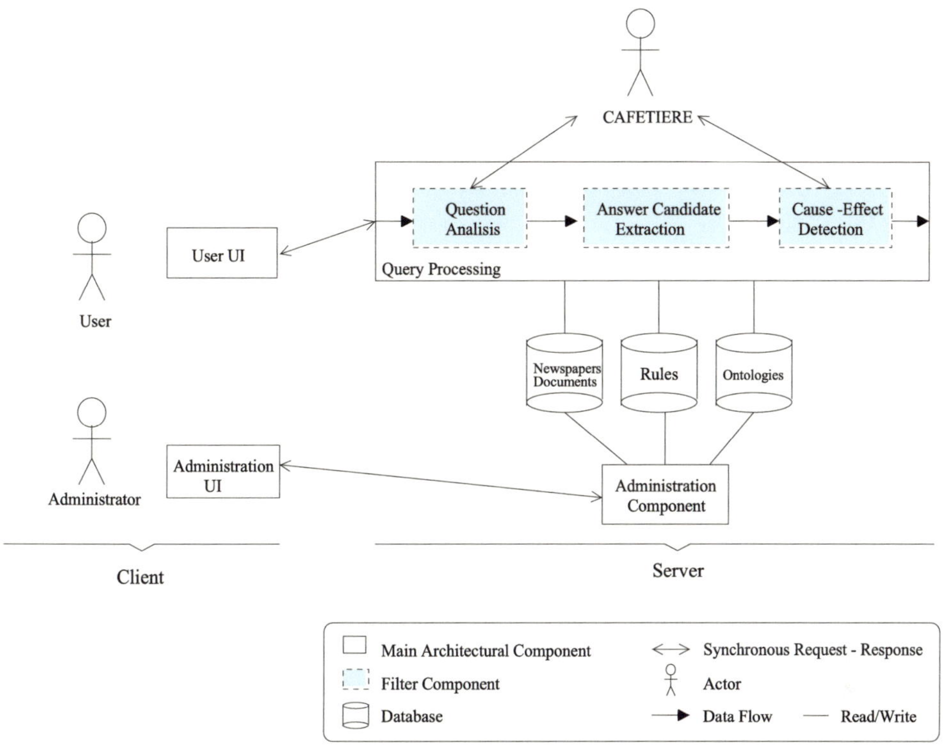

Fig. 1. The architecture of the system.

The Administration UI component is the means by which the Administrator maintains the system's databases. As shown in Fig. 1, there are three databases: (*i*) Newspaper Documents —which contains a list of candidate text passages (possible answers) that, in all likelihood, match the original question, (*ii*) Rules —which contains a set of lexico-syntactic and basic semantic rules for the English language that are used to produce phrasal and conceptual annotations as well as representations of elements of interest, events and relations and (*iii*) Ontologies —which are lists of known names of places, people, organizations, artifacts, etc.; that help to assign conceptual classes to single and/or multi word phrases as additional information for the information extraction analysis.

In the following sections we will focus on describing the elements of the architecture supporting Query Processing.

4.2.1 Question Analysis

As mentioned before, we use CAFETIERE's lexico-syntactic analysis pipeline for the classification of *"why"* questions. This pipeline is constituted by: sentence splitter, tokenizer, orthography tagger, stemmer, POS tagger, gazetteer lookup (single and multi-word names and terms), and rule-based analyzer (context sensitive rule-based analysis).

The Question Analysis filter that implements the required logic to interact with CAFETIERE in the following way:

1. *Modifying the resources for an already existing analysis engine.* For example, we have done the following:

- added words to the lexicon used by the part-of-speech tagger.
- added rules to the file used by the part-of-speech tagger.
- added patterns to the file used by the part-of-speech tagger.

The previous three points have been executed for improving the result of part-of-speech tagger to our research.

2. *Creating a new instance of an existing analysis engine type, with its own set of resources.* This does not involve changing any of the code of the analysis engine, only declaring what specific resource instance(s) it will use. We have for example:

- created one lookup analysis engine for using our six gazetteers: list of cue words, modal verbs, auxiliaries, process verbs, declarative verbs and agentive nouns. The objective was to support the phrase-level analysis of our research.
- created three different rule-based analysis engines, each with its own set of rules for performing higher level analysis up to the clause and sentence level. They have been divided into three levels: tags, phrase level and clause level.

3. *Creating a new aggregate analysis engines to run modules in different sequences, or to run different permutations of modules that are used in the question analysis.* Basically, the analysis engines integrating the aggregate are: sentence splitter, tokenizer, orthography tagger, stemmer, POS tagger, gazetteer lookup, rule-based analyzer and concept collector.

This question analysis process relies on CAFETIERE tool in order to assign a question category that corresponds to the types of entities, which constitute the category —each question category corresponds to a syntactic pattern. The question focus is its premise.

The process is a rule-based approach, which uses hand-crafted rules that look for lexical and syntactic clues in the question. We have sets of rules for tags, phrases, and clauses (which include the question types). The rules use six gazetteers as knowledge source: cue words, modal verbs and auxiliaries, process verbs, declarative verbs and agentive nouns. We have 143 rules, which are constituted by 32 rules for tags, 76 rules for phrases and 35 rules for clauses.

The output generated for this filter, which is the annotated and original question, are passed up to the answer candidate extraction filter.

4.2.2 Answer candidate extraction

We implemented an answer candidate extraction from text to identify the three best answers to the question. In order to do that, this component implements the logic to perform the computation of four matching formulae: simple matching, longest consecutive subsequence, Sorensen's similarity coefficient and WordNet-based lexical semantic relatedness. This process matches the question against the sentences in the text to identify the clauses that express the same proposition as the question.

Within the entire set of measures, each one of them is considered as a feature with the same weight. For selecting the best three possible answers for each question, we have combined the four metrics, see (7).

$$matching\_formulae = \frac{f_{sm} + f_{lcs} + f_{ssc} + f_{lsr}}{4} \tag{7}$$

The extraction of possible answers is accomplished by selecting the top three possible answers for each question. The answers are saved into a file, which is passed up to the Cause-Effect Detection filter to do the corresponding processing.

4.2.3 Cause-Effect Detection

The detection of cause-effect information expressed in the identified best three possible answers is done by the Cause-Effect Detection filter. As depicted in Fig. 1, this filter contains the required logic to use the CAFETIERE tool to support this job.

We have investigated how cause-effect information could be extracted from newspaper text using rule-based approach without full parsing of sentences. A set of rules that usually indicate the presence of a causal relationship was constructed and used for the extraction of cause-effect information.

No inferencing from common sense knowledge or domain knowledge was used. Knowledge-based inferencing of causal relationships requires a detailed knowledge of the domain, and newspaper text covers a very wide range of topics. Only linguistic clues were used to identify causal relationships. For example, we are using explicit linguistic indications of cause and effect, such as *because, however, due to, so, therefore, but, as a result of this,* and so on, instead of inferencing from common sense knowledge or domain knowledge.

We identified the following two ways of explicitly expressing cause-effect:

1. Using causal links to link two phrases, clauses or sentences.
2. Using causative verbs.

The detection of cause-effect information (rule-based approach) expressed in the identified answers was implemented in the same way that our question analysis (see Section 4.2.1), *(a)* modifying the resources for an already existing analysis engine; *(b)* creating a new instance of an existing analysis engine type, with its own set of resources; and *(c)* creating a new aggregate analysis engine.

Rules were created to identify sentences containing *causal links and causative verbs*. The cause-effect information was then extracted. The implementation can identify causal relations in newspaper text when it focuses on the causal relations that are explicitly indicated in the text using linguistic means.

A complete example is the question, "Why did researches compare changes in oxygen concentration?" which belongs to the document wsj_0683. It is an action question. The best three possible answers detected by our answer candidate extraction task are:

1. Researchers at Ohio State University and Lanzhou Institute of Glaciology and Geocryology in China have analyzed samples of glacial ice in Tibet and say temperatures there have been significantly higher on average over the past half-century than in any similar period in the past 10,000 years.
2. According to greenhouse theories, increased carbon dioxide emissions, largely caused by burning of fossil fuels, will cause the Earth to warm up because carbon dioxide prevents heat from escaping into space.
3. To compare temperatures over the past 10,000 years, researchers analyzed the changes in concentrations of two forms of oxygen.

The correct answer is the third one. The verb phrase *to compare temperatures over the past 10,000 years* is the cause, and the clause *researchers analyzed the changes in concentrations of two forms of oxygen* is the effect.

5. Evaluation

We have shown previously that a *"why"* question has more than one answer since there does not exist only one correct way of explaining things; therefore, it is quite difficult to determine whether a string of text provides the correct answer. Human assessors have legitimate differences of opinions in determining whether a response actually answers a question. If human assessors have different opinions, then eventual end-users of the Question Answering technology have different opinions as well because some users prefer a more accurate explanation, while others look for explanations having a broader perspective and better explanatory resources.

We can highlight that the time and effort required to manually evaluate a Question Answering application is considerable, owing to the need for human judgment. This issue is compounded by the fact that there is no such thing as a canonical answer form. Assessors' decision on the correctness of an answer makes resulting scores comparative, not absolute.

For our evaluation, we used the collection of Verberne et al. (2007), It has a relative preponderance of questions (typically expressed in the past tense) about specific actions of their motivations or relations to them, because their source documents are Wall Street Journal articles (news) and they describe a series of events that are specific to the topic, place and time of the text. The collection consists of seven texts from the RST Treebank of 350-550 words each and 372 *"why"* questions.

Throughout the evaluation of the answers detected, we adopt a manual approach whereby an assessor determines if a response is suitable for a question, with two possible outcomes. The response is either correct (i.e. the answer string must contain exactly the information required by the question), or incorrect (i.e. it is a wrong answer or no answer at all). To evaluate a question, an assessor was required to judge each answer string in that question's answer pool (set of three answers).

The kind of evaluation executed for our research is a post-hoc evaluation on analyzed data —question-answer pairs. It is divided in three sections: question classification, which is a lexico-syntactic classification where each question category corresponds to a pattern, answer candidate extraction, which maps the question onto the correct source text in order to detect possible answers for the question evaluated, and the detection of cause-effect information expressed in the answers selected.

5.1 Question classification

We classified the 372 *"why"* questions of the collection using the question analysis filter (rule-based approach). The lexico-syntactic classification is constituted of 5 categories: *existential "there" questions, process questions, questions with a declarative layer, action questions and have questions.*

The following questions are examples that were classified with the existential "there" category:

- Why is there resistance to the Classroom Channel?
- Why is there a reference to the musical "The Music Man"?
- Why is there controversy in this Dallas suburb?

Examples of questions in the process question category are:

- Why did Cincinnati Public Schools reject the subscription offer?
- Why did the US Coast Guard close part of the Houston Ship Channel?
- Why did the petroleum plant explode?

The following questions are examples of declarative layer category:

- Why does Whittle think he can reach subscription goals within one year?
- Why did the report say that advertisers were showing interest?
- Why does Mr Hogan think the company will be successful?

Examples of questions in the action question category are:

- Why do researches conclude that the earth is warming?
- Why did Dr. Starzl advise against buying Fujisawa stock?
- Why were town officials embarrassed?

The following questions are examples that were classified with the have question category:

- Why did firefighters have difficulties getting the fire under control?
- Why does Whittle have reason for concern?
- Why did the research team have no financial stake in the drug?

For some categories, the question analysis filter only needs fairly simple cues for choosing a category. For example, the presence of the word *there* with the syntactic category *EX* leads to an the category existential "there" question.

For deciding on questions with a declarative layer, action questions and process questions, complementary lexical-syntactic information is needed. In order to decide whether the question contains a declarative layer, the filter checks whether the main verb is in declarative verbs list, and whether it has a subordinate clause. The distinction between action and process questions is made by looking up the main verb in a list of process verbs. This list contains the 529 verbs from Levin verb index (1993). If the main verb is not determined to be process, declarative or have, it is assigned to the action verb category.

Questions with a declarative layer need further analysis because they are ambiguous. For example the question "Why did they say that migration occurs?" can be interpreted in two ways: "Why did they say it?" or "Why does migration occur?". Our answer candidate extraction filter should try to find out which of these two questions is supposed to be answered. In other words, the filter should decide which of the clauses contains the question focus. For this reason, questions with a declarative layer are most difficult to answer.

Table 1 shows the results of our question classification. We observe that the five categories of classification (existential "there" questions, process questions, questions with a declarative layer, action questions and have questions) had a performance of 54.56%. In other words, for the 372 questions of collection, 203 questions were classified correctly. We want to highlight that question classification uses only questions, without answers.

Category	Question per Category	Questions Classified Correctly	Performance
Existential "there" questions	11	5	45.45%
Process questions	145	102	70.34%
Questions with a declarative layer	93	17	18.27%
Action questions	102	65	63.72%
Have questions	21	14	66.66%
TOTAL	372	203	54.56%

Table 1. Lexico-syntactic classification for the 372 *"why"* questions of collection.

We have assigned correctly the existential "there" category to 45.45% of the questions; 70.34% were labelled as process questions; 18.27% of the questions had a declarative layer; the category of action questions was assigned to 63.72% of the questions because if the main verb of the questions is not a process, declarative or a have verb, then we are assumed that its type is action. And 66.66% were labelled, as have questions.

We observe that question with a declarative layer are most difficult to identify because of clausal object, that is, a subordinate clause must be detected after declarative verb.

The general rules (only a lexico-syntactic analysis) for categories of classification are:

- *Existential "there"* category:

WRB + VP + EX + NP + [ADVP | PP | NP] + ?

- *Process question* category:

WRB + VP + + NP + Process-VERB + [ADVP | PP | NP] + ?

- *Declarative layer* category:

WRB + VP + NP + Declarative-VERB + S + ?

- *Action question* category:

WRB + VP + NP + Action-VERB + [NP] + [ADVP | PP | NP] + ?

- *Have question* category:

WRB + VP + NP + Have-VERB + NP + [ADVP | PP | NP] + ?

If the lexico-syntactic pattern of the question did not correspond to any category, then it was assigned to the default category. However, we detected that questions can also be classified in the default category due to *(a)* errors of part-of-speech tagger; and *(b)* errors of rule-based analyzer. The default category contained 45.43% of the questions. A few examples of questions assigned to the default category are:

- Why is the sago expensive?
- Why is the Sago a pricey lawn decoration?
- Why is rowdy behavior unlikely at the Grand Kempinski?

5.2 Answer candidate extraction

In order to evaluate the answer candidate extraction filter, most previous Question Answering work has been evaluated using traditional metrics as recall, and mean reciprocal rank (Voorhees, 2003, 2004). We follow the standard definition of recall, see (8).

$$recall = \frac{c}{t} \tag{8}$$

Where c is the number of correct annotations produced, and t is the total number of annotations that should have been produced.

Using this formula, the recall obtained by our lexical overlapping and lexical semantic relatedness approach is 36.02%. In our test corpus, t=372, the number of *"why"* questions in the collection, and c=134, the number that were answered correctly by those techniques.

We hypothesized some of the *"why"* questions could have been unanswered because the collection's questions were created by native speakers who might have been tempted to formulate *"why"* questions that did not address the type of argumentation that one would expect of questions posed by persons who needed a practical answer to a natural *"why"* questions.

For this reason, working from the premise that "our lexical overlapping and lexical semantic relatedness approach can only answer *'why'* questions with explicit and ambiguous causation because it uses basic external knowledge for disambiguation", we recalculated recall for the 218 *"why"* questions with explicit and ambiguous causation. The rate of our recall increased to 61.46% of the former. The rate of recall thus increases considerably and we reach similar results as the RST method (Verberne, et al., 2007), which relies on texts where all causal relations have been pre-analysed.

The second metric used was mean reciprocal rank (MRR), see (9). The original evaluation metric used in the Question Answering tracks TREC 8 and 9 (Voorhees, 2000) was mean reciprocal rank (MRR), which provides a method for scoring systems which return multiple competing answers per question. We used MRR because our implementation returns a list with the best 3 possible answers that have been found to each question.

$$MRR = \frac{\sum_{i=1}^{|Q|} \frac{1}{r_i}}{|Q|} \tag{9}$$

Where Q is the question collection and r_i the rank of the first correct answer to question i or 0 if no correct answer is returned.

We showed that our answer candidate extraction filter found answers for 134 *"why"* questions on undifferentiated texts. The distribution for 134 *"why"* questions is 48 correct answers located in the first position, 29 correct answers located in the second position and 57 correct answers located in the third position. Consequently, the MRR is 0.219

Working from the previously mentioned premise that "our lexical overlapping and lexical semantic relatedness approach can only answer *'why'* questions with explicit and ambiguous causation because it uses basic external knowledge for disambiguation", then for the 218 *"why"* questions with explicit and ambiguous causation, the rate of our MRR increases to 0.373 of the former.

5.3 Cause-effect information

The cause-effect information in the answers selected is mainly expressed by causal links. The reason for this could be the fact that the events discussed in newspaper texts use connectives between two adjacent clauses. We detected the following causal links: *before, after, where, due to, because of, but, about, because, for,* and *from*. Three examples are presented:

> Why are the Mayor and two members of the Council worried?
>
> Mayor Lynn Spruill and two members of the council said they were worried ***about*** setting a precedent that would permit pool halls along Addison's main street.
>
> Why would the interior regions of Asia heat up first?
>
> Some climate models project that interior regions of Asia would be among the first to heat up in a global warming ***because*** they are far from oceans, which moderate temperature changes.
>
> Why could the number of people known injured increase?
>
> Nearby Pasadena, Texas, police reported that 104 people had been taken to area hospitals, ***but*** a spokeswoman said that toll could rise.

The 89.01 % of question-answer pairs of the collection contain causal links.

In order to evaluate cause-effect information, we manually identified 50 question-answer pairs in the collection. After that, we used the answer candidate extraction filter for evaluating the cause-effect information detected, using the same 50 question-answer pairs manually identified. Table 2 shows we detect 34% of cause-effect information.

	Total questions-answer pairs	% of questions-answer pairs
Question-Answer pairs analysed manually	50	100%
Questions for which we identified a text (possible answer)	39	78%
Questions for which the identified text is a correct answer	17	34%

Table 2. Outcome of cause-effect evaluation.

Questions, which contain modals, constitute 7.52% of the total of collection. The function of modals is important in defining the semantic class of question. We cannot solve this issue because our answer candidate extraction filter works with lexico-syntactic level. For example:

1. Why did Nando's not use actors to represent chefs in funny situations?
2. Why can Nando's not use actors to represent chefs in funny situations?

Answer to Question *1* is a motivation, and answer to question *2* is a cause.

6. Conclusion

We introduced an approach which draws on linguistic structures, enabling the classification of *"why"* questions and the retrieval of answers for *"why"* questions from a newspaper collection. The steps to summarize our approach are:

1. Assign one category to each *"why"* question, using lexico-syntactic analysis. Each question corresponds to a syntactic pattern (rule-based approach).

   The lexico-syntactic classification is constituted of 5 categories: *existential "there" questions, process questions, questions with a declarative layer, action questions* and *have questions.*

2. Detecting three possible answers to each *"why"* questions:

   2.1. Identify the topic of the question (effect).
   2.2. In the list of sentences of source document, identify the clause(s) that express(es) the same proposition as the question topic (making use of a set of measures).
   2.3. Select the best three clauses as answers.
   2.4. Detect cause-effect information expressed in the answers selected (rule-based approach).

The output for each question is a question category and three possible answers.

6.1 Contributions

We have hypothesised that these methods will also work for *"why"* questions, and have attempted to discover to what extent methods based on information retrieval ('bag of words' approaches), and on limited syntactic and/or lexical semantic analysis can find answers to *"why"* questions. So, our research contributes new knowledge to the area of automatic text processing by the following:

- We have developed an analysis component for feature extraction and classification from questions (a rule-based approach as a first step towards tackling the problem of question analysis of *"why"* questions using an Information Extraction Analyzer.
- An original answer candidate extraction filter has been developed that uses an approach that combines lexical overlapping and lexical semantic relatedness (lexico-syntactic approach) to rank possible answers to causal questions. On undifferentiated texts, we obtained an overall recall of 36.02% with a mean reciprocal rank of 0.219, indicating that simple matching is adequate for answering over one-third of *"why"* questions. We analyzed those question-answer pairs where the answer was explicit, ambiguous and implicit, and found that if we can separate the latter category, the rate of recall increases considerably. When texts that contain explicit or ambiguous indications of causal relations are distinguished from those in which the causal relation is implicit, recall can be calculated as 61.46% of the former, with a mean reciprocal rank of 0.373, which is comparable to results reported for texts where all causal relations have been pre-analyzed. This plausible result shows the viability of our research for automatically answering causal questions with explicit and sometimes ambiguous causation.
- We have found that people have conflicting opinions as to what constitutes an acceptable response to a *"why"* question. Our analysis suggests that there should be a proportion of text in which the reasoning or explanation that constitutes an answer to the *"why"* question is present, or capable of being extracted from the source text. Consequently, the complexity of a *"why"* question depends on the knowledge level of users. While some users prefer a more accurate explanation, others look for explanations with a broader perspective and better explanatory resources. Any answer that appeals to a cause is taken to be highly relevant and, therefore, to provide an explanation of the effect – a *"why"* question. In order to provide a context with which to support the answer, the paragraph from which the answer was extracted should be returned as the answer.

We conclude that this research offers a greater understanding of *"why"*. It provides an approach to tackling a subset of *"why"* questions (with explicit and sometimes ambiguous causation) which combines lexical overlapping and lexical semantic relatedness. It further considers the detection of cause-effect information that is explicitly indicated in the text using causal links and causative verbs. For these reasons, this research contributes to a better understanding of automatic text processing for detecting answers to *"why"* questions and to the development of future applications for answering causal questions.

6.2 Further work

To improvements the answer candidate extraction, we could experiment with ambiguous and implicit causation since our lexico-syntactic approach has not been successful for these types of causation. In order to generate correct answers, we would need to go beyond the co-occurrence of terms and lexical semantic relatedness due to the mismatch between the expressions used in the question and the expressions used in the source text.

We should contribute to the implementation and evaluation of fundamental techniques representing knowledge and reasoning.

When we use a causal relation to describe the interaction between two sentences, it would be more interesting and informative if we could present an answer that offers chain of explanations connecting the two events, that is, the entire answer to the *"why"* question should be a chain of explanation of its causal relation.

When considering the relevance of answers to causal questions, we should involve carrying out inferences (we view inference as the process of making implicit information explicit) to arrive at the required answer. A relevant answer requires the provision of an appropriate explanation, according to the questioner. The explanation should increase the questioner's existing knowledge rather than duplicate it. A filter for explanation generation that takes into account the descriptions already presupposed by the question could be implemented by using descriptions to generate explanations, which are themselves answers to the *"why"* question. The algorithm could begin to make use of the information explicitly encoded by the lexical and syntactic analysis.

Three questions need to be considered in order to advance our reasoning about the descriptions embedded in *"why"* question, (1) Where can we find the descriptions presupposed by the question?, (2) How can we recognise the descriptions presupposed by the question?, and (3) How can we represent and compute the descriptions presupposed by the question?

In order to understand this process in more detail, an analysis of epistemology (the theory of knowledge) would be necessary. This is the branch of philosophy concerned whit the nature, origin, and scope of knowledge. It addresses the question "how do you know what you know?"

7. References

Aouladomar, F. (2005a). A Preliminary Analysis of the Discursive and Rhetorical Structure of Procedural Texts, *Proceedings of the Symposium on the Exploration and Modelling of Meaning,* pp. 21-24, Biarritz, France, November 14-15, 2005.

Aouladomar, F. (2005b). Towards Answering Procedural Questions, *Proceedings of the Workshop on Knowledge and Reasoning for Answering Questions,* pp. 21-31, Edinburgh, Scotland, July 30, 2005.

Aouladomar, F., & Saint-Dizier, P. (2005a). An Exploration of the Diversity of Natural Argumentation in Instructional Texts, *Proceedings of the 5th International Workshop on Computational Models of Natural Argument,* pp. 12-15, Edinburgh, Scotland, July 30, 2005.

Aouladomar, F., & Saint-Dizier, P. (2005b). Towards Generating Procedural Texts: an exploration of their rhetorical and argumentative structure, *Proceedings of the 10th European Workshop on Natural Language Generation,* pp. 156-161, Aberdeen, Scotland, August 8-10, 2005.

Benamara, F., & Saint-Dizier, P. (2004). Advanced Relaxation for Cooperative Question Answering. In: *New Directions in Question Answering*, M. Maybury, (Ed.), 263-274, ISBN 0-262-63304-3, Cambridge, Massachusetts USA.

Black, W., McNaught, J., Vasilakopoulos, A., Zervanou, K., Theodoulidis, B., & Rinaldi, F. (2003). *CAFETIERE: Conceptual Annotations for Facts, Events, Terms, Individual Entities, and Relations*, Parmenides Technical Report No. TR-U4.3.1, In: The University of Manchester.

Budanitsky, A., & Hirst, G. (2006). Evaluating WordNet-based Measures of Lexical Semantic Relatedness. *Computational Linguistics*, Vol. 32, No. 1, (March 2006), pp. 13-47, ISSN 0891-2017.

Burger, J., Cardie, C., Chaudhri, V., Gaizauskas, R., Harabagiu, S., Israel, D., et al. (2001). *Issues, Tasks and Program Structures to Roadmap Research in Question & Answering*, Technical Report, In: NIST, Available from http://www-nlpir.nist.gov/projects/duc/roadmapping.html.

Castagnola, L. (2002). *Anaphora Resolution for Question Answering*, Master Degree Thesis, Department of Electrical Engineering and Computer Science, Massachusetts Institute of Technology, USA, June, 2002.

Dao, T. (2005). The Longest Common Substring with Maximal Consecutive, In: *The Code Project*, Retrieved on 30/09/2011, from http://www.codeproject.com/KB/recipes/lcs.aspx?display=Print

Galambos, J. A., & Black, J. B. (1985). Using Knowledge of Activities to Understand and Answer Questions. In: *The psychology of questions*, A. C. Graesser & J. B. Black (Eds.), 148-169, ISBN 0898594448, Lawrence Erlbaum Associates.

Galitsky, B. (2003). *Natural Language Question Answering system*, Advanced Knowledge International Pty Ltd, ISBN 0868039799, London, UK.

Harabagiu, S., Moldovan, D., Pasca, M., Mihalcea, R., Surdeanu, M., Bunescu, R., Girju, R., Rus, V. and Morarescu, P. (2000). FALCON: Boosting Knowledge for Answer Engines. *Proceedigns of the Ninth Text REtrieval Conference (TREC-9)*, pp. 479-488, online publication, 06/10/11, http://trec.nist.gov/pubs/trec9/t9_proceedings.html, Gaithersburg, Maryland, USA, November 13-16, 2000.

Jiang, J., & Conrath, D. (1997). Semantic similarity based on corpus statistics and lexical taxonomy. *Proceedings of the International Conference on Research in Computational Linguistics*, pp. 19-33, Taipei, Taiwan, August, 1997.

Katz, B., & Lin, J. (2003). Selectively Using Relations to Improve Precision in Question Answering, *Proceedings of EACL-2003 Workshop on Natural Language Processing for Question Answering*, pp. 43-50, Budapest, Hungary, April 12-17, 2003.

Levin, B. (1993). *English Verb Classes and Alternations: a preliminary investigation*, University of Chicago Press, ISBN 0226475336, USA.

Light, M., Ittycheriah, A., Latto, A., & McCracken, N. (2004). Reuse in Question Answering: A Preliminary Study. In: *New Directions in Question Answering*, M. Maybury, (Ed.), 162-182, ISBN 0-262-63304-3, Cambridge, Massachusetts USA.

Mann, W. C., & Thompson, S. A. (1988). Rhetorical Structure Theory: Toward a functional theory of text organization. *Text*, Vol. 8, No. 3, (December 1998), pp. 243-281.

Maybury, M. T. (2003). Toward a Question Answering Roadmap. In: *New Directions in Question Answering*, M. Maybury, (Ed.), 3-14, ISBN 0-262-63304-3, Cambridge, Massachusetts USA.

McGuinness, D., & Pinheiro da Silva, P. (2004). Trusting Answers from Web applications. In: *New Directions in Question Answering*, M. Maybury, (Ed.), 275-286, ISBN 0-262-63304-3, Cambridge, Massachusetts USA.

Miller, G. A. (1995). WordNet: A Lexical Database for English. *Communication of the ACM*, Vol. 38, No. 11, (November 1995), pp. 39-41, ISSN 0001-0782.

Mitkov, R. (2004). Anaphora Resolution. In: *The Oxford Handbook of Computational Linguistics*, R. Mitkov (Ed.), 265-283, Oxford University Press, ISBN 0198238827, USA.

Moldovan, D., Pasca, M., Harabagiu, S., & Surdeanu, M. (2003). Performance Issues and Error Analysis in an Open-Domain Question Answering System. *ACM Transaction on Information Systems*, Vol.21, No.2, (April 2003), pp. 133-154, ISSN 1046-8188.

Nyberg, E., Burger, J., Mardis, S., & Ferrucci, D. (2004). Software Architectures for Advanced QA, In: *New Directions in Question Answering*, M. Maybury, (Ed.), 19-30, ISBN 0-262-63304-3, Cambridge, Massachusetts USA.

Patwardhan, S., Banerjee, S., & Pedersen, T. (2003). Using Measures of Semantic Relatedness for Word Sense Disambiguation, *Proceedings of the Fourth International Conference on Intelligence Text Processing and Computational Linguistics*, pp. 241-257, Mexico City, Mexico, February 16-22, 2003.

Pedersen, T., Patwardhan, S., & Michelizzi, J. (2004). WordNet::Similarity - Measuring the Relatedness of Concepts, *Proceedings of the 9th National Conference on Artificial Intelligence, (Intelligent Systems Demonstration)*, pp. 38-41, Stroudsburg, PA, USA, May 2-7, 2004.

Schlaefer, N. (2007). *A Semantic Approach to Question Answering*, VDM Verlag Dr. Mueller e.K., ISBN 3836450739.

Setzer, A., Gaizauskas, R., & Hepple, M. (2005). The Role of Inference in the Temporal Annotation and Analysis of Text, *Journal Language Resources and Evaluation*, Vol. 39, No. 2-3, (May 2005), pp. 243-265, ISSN 1574020X.

Shaw, M., & Garlan, D. (1996). *Software Architecture: Perspectives on an Emerging Discipline*, Prentice Hall, ISBN 0131829572, USA.

Stevenson, M., & Greenwood, M. (2005). *A semantic approach to IE pattern induction, Proceedings of the 43rd Annual Meeting of the Association for Computational Linguistics*, pp. 379-386, Stroudsburg, Pensilvania, USA, June 25-30, 2005.

Verberne, S., Boves, L., Oostdijk, N., & Coppen, P. (2007). Discourse-based answering of why-questions, *Journal Traitement Automatique des Langues. Special issue on Computational Approaches to Discourse and Document Processing*, Vol. 47, No. 2, (Octuber 2007), pp. 21-41.

Vicedo, J., & Fernandes, A. (2000). Applying Anaphora Resolution to Question Answering and Information Retrieval Systems, *Proceedigns of Web-Age Information Management*, pp. 344-355, ISBN 3-540-67627-9, Shanghai, China, June 21-23, 2000.

Voorhees, E. M. (2000). Overview of the TREC-9 Question Answering Track, *Proceedings of the Ninth Text REtrieval Conference*. pp. 1-13, online publication, 06/10/11, http://trec.nist.gov/pubs/trec9/t9_proceedings.html, Gaithersburg, Maryland, USA, November 13-16, 2000.

Voorhees, E. M. (2001). Overview of the TREC 2001 Question Answering Track, *Proceedings of the Tenth Text REtrieval Conference*. pp. 1-15, online publication, 05/10/11, http://trec.nist.gov/pubs/trec10/t10_proceedings.html, Gaithersburg, Maryland, USA, November 19-22, 2001.

Voorhees, E. M. (2002). Overview of the TREC 2002 Question Answering Track, *Proceedings of the Eleventh Text REtrieval Conference*. pp. 1-15, online publication, 04/10/11, http://trec.nist.gov/pubs/trec11/t11_proceedings.html, Gaithersburg, Maryland, USA, November 18-21, 2002.

Voorhees, E. M. (2003). Overview of the TREC 2003 Question Answering Track, *Proceedings of the Twelfth Text REtrieval Conference*. pp. 1-13, online publication, 03/10/11, http://trec.nist.gov/pubs/trec12/t12_proceedings.html, Gaithersburg, Maryland, USA, November 17-20, 2003.

Voorhees, E. M. (2004). Overview of the TREC 2004 Question Answering Track, *Proceedings of the Thirteenth Text REtrieval Conference*, pp. 1-12, Special Publication 500-261, Gaithersburg, Maryland, USA, November 16-19, 2004.

Waldinger, R., Appelt, D., Dungan, J., Fry, J., Hobbs, J., Israel, D., et al. (2004). Deductive Question Answering from Multiple Resources, In: *New Directions in Question Answering*, M. Maybury, (Ed.), 253-262, ISBN 0-262-63304-3, Cambridge, Massachusetts USA.

27

An Intelligent System for Improving Energy Efficiency in Building Using Ontology and Building Automation Systems

Hendro Wicaksono, Kiril Aleksandrov and Sven Rogalski
FZI Research Centre for Information Technology
Germany

1. Introduction

Energy efficiency is a keyword that can be found nowadays in all domains in which energy demand exists. The steadily rising energy demand, the consequent energy scarcity and rising prices of energy resources are forcing companies and people to redefine their activities in a more energy efficient way. Besides industry and transportation, the building sector is the most important energy consumer, as a study for European countries showed that up to 40% of the total energy consumption in the EU can be ascribed to this segment. The residential use represents around 3% of the total energy consumption in the building sector (Balaras et al., 2007).

There are many technical possibilities to improve energy usage efficiency. The European Union has issued the Directive 2002/91/EC concerning the overall energy efficiency of buildings. The directive aims to improve energy efficiency by taking into account outdoor climatic and local conditions, as well as indoor climate requirements and cost-effectiveness (Cox & Fischer Boel, 2002). The most widespread energy-saving measures in buildings are conventional and are based on improved thermal insulation, energy-efficient air conditioners and appliances as well as the usage of energy saving light bulbs and smart meters. Yet there is more potential for energy saving in buildings than the abovementioned measures. Based on the directive, a new DIN V 18599 "Energetische Bewertung von Gebäuden" describing an integral calculation method for energy requirements for heating, acclimatisation and lighting has been published. Furthermore, in the future, energy savings in buildings can be increased by utilizing a building automation system. This kind of method is at least as important as the conventional thermal insulation of walls or insulating glazing for improving energy efficiency in buildings (Spelsberg, 2006a, 2006b)

A look at the operation of residential and commercial buildings has also shown that there is a wide range of technical possibilities to reduce energy consumption, as demanded by the European Union through the European Directive 2002/91/EC on the energy efficiency in buildings. An important aspect that can improve the energy efficiency in buildings is the use of building automation systems. However, building automation systems are usually not considered for energy conservation, as they are mostly used for comfort and safety. This often causes great problems due to an inefficient use of these systems and unawareness of energy consumption. It is therefore crucial that the existing system solutions are adapted to focus on energy conservation.

This book chapter describes our research approach in developing an intelligent system to improve energy efficiency in buildings, which takes into account the different technical infrastructures of buildings, the occurring events in the buildings and the relevant environmental factors with the help of building automation systems and ontology. By using different building automation technologies, data in a variety of detail and quality could be collected. Thus, it allows the user to have an overall view of the energy consumption inside the entire building. Ontology offers the generic model for different building automation technologies and allows logical inference as well.

Modern building automation solutions for the private or commercial sector offer a great potential for the comprehensive improvement of energy efficiency. Building automation systems that are composed of computer aided networked electronic devices are mostly used for improvement of the individual quality of life, and safety. Building automation technology offers an easy way to control and monitor the building conditions. However, the existence of building automation systems does not imply an optimal usage of energy. They are predominantly applied for comfort and security reasons and do not lead to an improvement in energy consuming behaviour.

Low cost and low energy consuming building automation technologies have already been developed in recent years. Technologies such as digitalSTROM[1] and WPC[2] offer energy measurement and sensors by using small chips that consume less than 10 mW (Watteco, 2009). These chips communicate through power line communication (PLC). Therefore modifications of walls and other building structures as well as extra wiring in the building are not necessary for installation. By using these devices, extra energy consumption can be avoided. Thus, an improvement of energy efficiency can be achieved.

In this paper we proposed a framework of intelligent system for energy management in buildings by utilizing building automation systems. The framework uses a knowledge-based approach to providing an intelligent analysis based on the relations between energy consumption, activities and events in the building, building related information and surrounding factors, such as temperature and weather conditions.

2. State of the art and related works

The importance of the use of building automation systems in assisting energy saving in buildings, especially households, has been proven in recent years. In 2009 the Electric Power Research Institute USA conducted research about the impact of energy consumption feedback information on energy conservation levels in households. They categorized the feedback mechanism based on information availability into standard billing, enhanced billing, real-time feedback, real-time plus feedback, etc. (Neenan et al., 2009). The research showed that real-time plus feedback leads to the biggest improvement of energy conservation compared to the other feedback mechanisms, despite its higher cost of implementation. Real-time plus feedback allows users to monitor their energy consumption

[1] http://www.digitalstrom.org
[2] http://www.watteco.com

and/or control appliances in their home in real-time by utilizing building automation system (BAS) and home area network (HAN).

In Europe, there are several projects that aim to improve energy efficiency in buildings by using building automation technologies. Most of the projects do not have the functionality to intelligently synchronize the energy consumption with other energy consumption impacting parameters, such as temperature, occupancy or the state of the building. Some of the projecs developed knowledge based approachs to enable the intelligent analysis of the collected data. In the project IntUBE, a knowledge base containing semantic building objects, their properties and their relationships, is developed in their energy management integration platform (Madrazo et al., 2009). The knowledge base to provide an intelligent system is not only developed in energy management system for buildings. The project AmI-MoSeS develops an (ambient) intelligent monitoring system for energy consumption, dedicated to manufacturing SMEs, to provide comprehensive information about the energy efficiency and a knowledge-based decision support system for the optimisation of energy efficiency (ATB Institut für angewandte Systemtechnik Bremen GmbH et al., 2008). In these projects the knowledge bases are mostly generated manually. This prevents the system from being able to react to and solve problems accordingly by a given specific situation, since all the situation-reaction causal relations are given by a human, based only on his knowledge and experience. This might be too general.

Currently there are some research projects that investigate and develop ontology based approaches to the building automation domain. Ontology is used as the generic application model facilitating an integration of heterogeneous building automation networks. The ontology provides a single access point for configuration and maintenance tasks, as well as automatic calculation of gateway configuration data (Reinisch et al., 2008). Ontology is also used to build an expert system to control home automation devices by incorporating SWRL rules that regulate the system behaviour (Valiente-Rocha & Lozano-Tello, 2010). In requirement engineering of building automation systems, ontology provides a semantic representation of the requirements, containing possible and specific requirements and their causal dependencies. By means of the ontology, consistent requirements for an individual building automation system can be achieved (Runde et al., 2009). Ontology that allows a representation of a sophisticated model and an intelligent reasoning is used to support complex interoperation, generalization and validation tasks in the building automation environment (Bonino et al., 2008).

Ontologies represent knowledge in a particular domain. They are commonly created manually by experts in the corresponding domain. For example biology ontologies are created by biologists. However, a manual ontology creation is considered a very time-consuming task. To overcome this problem, researchers have developed methods to generate ontology components in a semi-automatic way. Researchers from the University of Leipzig, Germany have developed a semi-automatic OWL generation from XML files by converting XSD items to ontology's classes and attributes. To perform this they developed four XSLT instances to transform XML files to OWL without any semantic and structural intervention during the transformation (Bohring & Auer, 2005). Dragan Gasevic et al. have also developed an XSLT based approach to automatically generate OWL ontology from UML. The resulting ontology still needs to be refined using ontology development tools. However, the approach helps software engineers who want to participate in ontology development without having a deep knowledge of ontology development tools. Ontology can be generated semi-automatically from a set of free text documents. It was proven by research conducted by the University of Karlsruhe, Germany in 2001 (Maedche & Staab, 2001). They developed a tool called TextToOnto[3] for semi-automatic ontology construction encompassing ontology import, extraction, pruning, refinement, and evaluation. The framework helps ontology engineers to construct ontologies.

3. Overview of the approach

In this paper we propose an IT framework for intelligent energy management in a building using building automation technologies. The framework considers different aspects of the building that influence the energy consumption in the building and are to be considered in the analysis of the energy consumption. The framework allows the user to have a holistic and integrated view of the energy consumption in their apartment, office, as well as in entire buildings. The building automation system offers the possibility to perform an energy consumption evaluation in different levels of detail and quality, for instance, energy consumption per appliance, per group of appliances, per zone in the building, or per user event. These evaluations are carried out by considering the relation between building elements, energy consuming appliances, surrounding circumstances such as temperature and weather conditions, as well as user events.

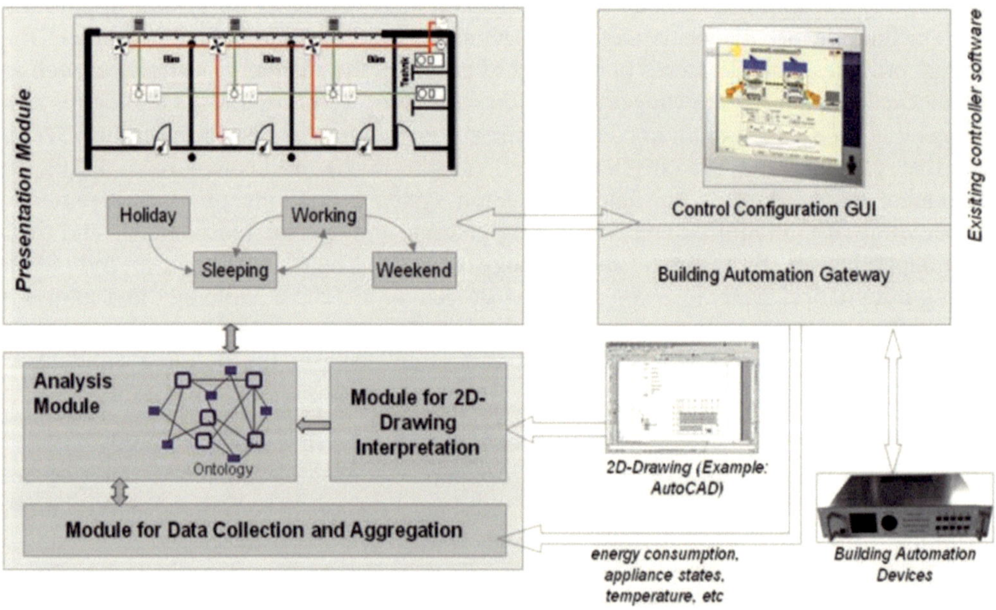

Fig. 1. Functional architecture of SERUM-iB

The energy consumption can be evaluated based on activities or events that occurred in the building, for example breakfast, cooking, sleeping, work day and holiday. This kind of

[3] http://texttoonto.sourceforge.net

evaluation offers the occupants the possibility to get a view of their activity-specific energy consumption. Therefore, they can consider, whether their activities are energy-efficient enough and how can they reorganize their activities to improve the energy efficiency. The framework offers the functions for energy monitoring, energy data analysis and manual and automatic controlling of energy consuming appliances using building automation systems. Fig. 1 shows the designed functional architecture of the framework whose main modules are explained briefly in the following sub sections. We named the developed framework "SERUM-iB (**S**mart **E**nergy and **R**esource **U**sage and **M**anagement **i**n **B**uilding) Framework". The work described in this paper is part of our research project KEHL[4].

3.1 2D-drawing interpretation module

CAD software applications that find application in construction engineering are commonly used to draw building layouts in two dimensional graphics. But unfortunately, the two dimensional drawings only contain geometric information of the objects, for instance lines, points, curves, circles, symbols, etc. They are not able to describe the semantics of the building components contained in the drawings themselves. People could still understand the semantics of the drawing, if they had the knowledge in interpreting the semantics of the symbols, shapes and other geometric information, for example symbols of doors, double lines representing walls, rectangles constructed by four double lines representing rooms, etc. We have developed a method to semi-automatically interpret the semantics of 2D-drawings to support the life-cycle management of building automation systems (Krahtov et al., 2009).

In this module we use configurable JavaScript based rules to semi-automatically interpret 2D-CAD-drawings. These rules define the relation between CAD-symbols and the semantics of building components and energy consuming appliances. The results of the 2D-drawing interpretation are stored as ontology individuals. This is explained further in section 5.

In existing energy analysis systems, users have to manually configure the system in order to perform an energy consumption analysis for each room or appliances. Another configuration mechanism is through automatic discovery of appliances. But it has the drawback that users have to manually relate the appliance to the building environment. The mechanism is dependent on a specific building automation technology as well. The interpretation of 2D-drawings helps users to configure the software framework, so they can have a model that represents their building layout with minimal effort.

3.2 Sensor data acquisition module

This module collects data from different building automation systems. Building automation technologies have different bus systems to exchange data. Most of them provide gateways to IP networks. This module contains an interface to communicate with different building automation logic control units or gateways via Simple Object Access Protocol (SOAP), which uses Extensible Markup Language (XML) for its message format and relies on Hypertext Transfer Protocol (HTTP) for message negotiation and transmission. The collected data, for instance energy consumption and the relevant sensor data such as that from the contact sensor, occupancy sensor, and weather station are converted into a uniform

[4] http://kehl.actihome.de

format and unit, for example, the energy consumptions are converted into kWh, the values from occupancy sensors are converted to true (room is occupied) and false (room is empty). These data are then aggregated and stored in a relational database.

In order to allow the visual representation of energy consumption data, we perform the necessary data pre-processing such as removing erroneous values, data transformation and data selection. The data are prepared to enable an energy consumption analysis of different criteria based on relation between rooms, appliances, time, and user events as mentioned before. This module provides data in such a form as to enable the execution of a data mining algorithm for finding the energy usage pattern.

3.3 Module for energy analysis

The energy consumption and other related data which are collected and stored in the SQL database as mass data are analysed in this module. Through this module, energy consumption can be related to device levels, room and time, which in addition, can be combined with the relation to user events and surrounding conditions. In this module we consider two methods of energy analysis, i.e. data-driven analysis and knowledge-driven analysis. Data-driven analysis means the analysis is conducted directly on the collected data by performing SQL-query, simple calculation, or visualization, for instance, energy consumption per time unit and each appliance. Knowledge-driven analysis is not conducted directly on the data, but by utilizing ontology that represents the knowledge.

The module consists of three sub-modules. They are a learning engine, a knowledge base consisting of ontology and rules and an energy analyser. The learning engine is responsible for extracting the knowledge hidden in the database. The learning engine utilizes data mining algorithms to get the association rules representing causal relations between the data. These rules are used, for example, for detecting energy usage anomalies and energy wasting. Furthermore, the rules are used to detect the activities that occur in the building based on data from different sensors, for instance, occupancy sensors, windows and door contact sensors, pressure sensors, etc. The data mining algorithms are also used for appliance classification based on their energy consumption. They are also used to detect the operation states of appliances. These functions will be explained further in section 6.

The knowledge base is the central component that provides the intelligent mechanism of the system. The knowledge base containing OWL ontology and SWRL rules is initially created by an expert. The knowledge base is then enriched with knowledge components (rules and instances), which are resulted from the 2D-Drawing Interpretation module and data mining algorithm.

The analysis module evaluates energy consumption data that were collected and aggregated by the data acquisition module, which is then presented in the presentation module. Through this module, energy consumptions can be related to device levels, room and time. In addition, they can be combined with the relation to user events and surrounding conditions.

In this module we consider two methods of energy analysis, i.e. data-driven analysis and knowledge-driven analysis. Data-driven analysis means the analysis is conducted directly on the collected data by performing SQL-query, simple calculation, or visualization, for instance energy consumption per time unit and each appliance. Knowledge-driven analysis is not conducted directly on the data, but by utilizing ontology that represents the knowledge. The ontology is created manually by an expert and enriched with ontological elements generated from the data through the ontology generation process. This is explained further in section 5 and 6.

3.4 Presentation module

This module gives the users a visual overview of their energy consumption in the building related to the building environment and their behaviour. Within the module, the energy consumption and other relevant information are presented intuitively. The energy consumptions in a building and in single rooms or for appliances are visualized via a terminal or smart client placed in the building. At the same time the configuration and control of the building automation systems can be performed within the developed GUI.

4. Ontology for knowledge representation

In the SERUM-iB framework we use ontology to build the knowledge base. It contains concepts representing rooms, energy consuming appliances, building automation devices, environment parameters, events and their relations. It is represented in OWL (Web Ontology Language), a W3C[5] specified knowledge representation language (Smith et al., 2004). In our work, we use the knowledge-driven analysis to find energy wasting conditions and to recognize energy usage anomalies in buildings. We also use it for identifying the operational state of an appliance, to identify appliance classes based on their energy consumption and to identify the user event or consumption behaviour based on the sensor data.

Within the SERUM-iB framework, we develop the three following methods to generate the ontology components:

1. manual ontology generation by an expert,
2. semi-automatic through interpretation of 2D-drawings, and
3. semi-automatic by data mining algorithms.

The ontological classes as well as their attributes and relations representing building automation devices, for instance sensor and energy meter, building environments such as door, window, room, are created manually by experts. The ontology containing these hand-crafted elements builds the knowledge called *building domain knowledge* in the knowledge base. It contains only the ontological classes or Tbox components that terminologically describe the knowledge. It provides a common conceptual vocabulary of building domains. It does not contain any ontological individuals or Abox components. The building domain knowledge represents the meta model of buildings, therefore it owns the validity for all buildings. It does not contain any building instance specific information.

As occupants, we usually do not realize that our current energy usage leads to energy wasting. For instance, the light on the balcony is still on, even though the sun has already risen. In our work, we try to resolve this problem by modelling in the ontology the general

[5] http://www.w3.org

situations in the building that lead to energy wasting. Thus, a computer supported reaction can be performed.

Energy wasting situations are modelled by experts in the base ontology. These situations are described in a rule representing language called SWRL (Semantic Web Rule Language). SWRL is a language combined by OWL (Web Ontology Language) and RuleML (Rule Markup Language) that enables the integration of rules in OWL ontology (Horrock et al, 2004).

Here we provide examples of modelling energy waste situation with SWRL. We assume that a smart meter attached to an appliance can give information on whether the appliance is currently in use. Rule (1) models the situation that turning the heating appliance on and opening the window located in the same room leads to energy wasting.

$$\begin{aligned}&\text{Window}(?w) \wedge \text{isAttachedTo}(?cs,?w) \wedge \text{ContactSensor}(?cs) \wedge \text{hasState}(?cs,\text{"true"}) \wedge \text{Heater}(?h) \wedge \\ &\text{hasState}(?h,\text{"true"}) \wedge \text{E}\textit{nergy}\text{Meter}(?em) \wedge \text{hasState}(?em,\text{"true"}) \wedge \text{isAttachedTo}(?em,?h) \wedge \\ &\text{Room}(?r) \wedge \text{isIn}(?w,?r) \wedge \text{isIn}(?h,?r) \rightarrow \text{EnergyWaste}(?h)\end{aligned} \quad (1)$$

The ontology representing the building domain knowledge is enriched with ontological individuals or Abox components from building specific information. The individuals are created by configuring the framework based on building specific information. They are also created automatically by interpreting the 2D construction drawing of the building layout depicting building structures, such as room, floor, door, windows and contained energy consuming appliances, as well as building automation devices, for instance temperature sensors and occupancy sensors. The semantic information of building environments such as rooms, lights, and appliances is obtained from geometric data through the interpretation of 2D-drawings. The concept of 2D-drawing interpretation is further presented in section 5.

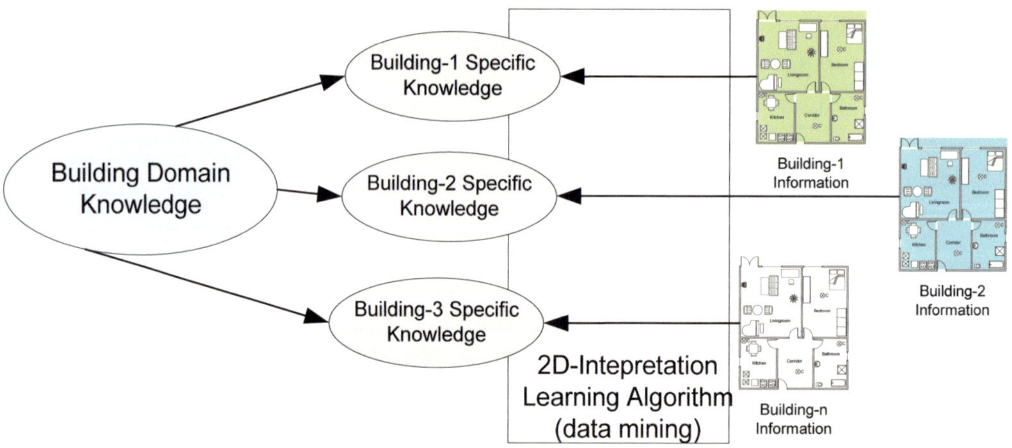

Fig. 2. Creating building specific knowledge from building domain knowledge

The building domain knowledge is added with building specific SWRL rules. These SWRL rules are created semi-automatically from the data mining algorithm. The data mining algorithm generates the association and classification rules as the hidden knowledge that is extracted from the collection of sensor data. The rules are converted to SWRL rules and integrated into the building domain knowledge. Both processes create the new version of knowledge called building specific knowledge. Fig. 2 depicts the concept of creating building specific knowledge from building domain knowledge using 2D-drawing interpretation and the data mining algorithm.

5. Knowledge acquisition through interpretation of 2D construction drawings

For its correct usage, this approach requires geometric and semantic information about building layout and appliances as an initial knowledge base. Geometric information could include the exact shape and measurements of rooms in a building, their topological relations to each other (and/or to the energy appliances) as well as the position of the appliances and their dimensions. Semantic information is identification keys, names, consumption values, etc. Manually inputting this type of data, especially for large buildings or facilities, could be a time-consuming, error-prone and inefficient task. For these reasons we adopted an approach for rule-based, semi-automatic interpretation of 2D construction drawings to create the initial knowledge base.

In order for the intelligent system to improve the energy efficiency in buildings, the following domain elements are considered important:

1. Zones that describe various parts of a building like rooms, corridors, etc.
2. Zones can also be nested (i.e. apartment, floor, building), as they can comprise other zones. They have geometric information about their layout and their size as well as semantic information (names, ids etc.). Zones can be topologically connected to energy consumption devices, furniture, etc.
3. Appliances such as any electrical device or energy consumption device like heaters.
4. Sensors and smart meters.

A layout of a building is commonly created as a two dimensional drawing or sketch using CAD software applications, such as AutoCAD[6]. Further AutoCAD-based software tools are used to plan and model many domains of a building, such as ventilation, heating, access controls and photovoltaic (Krahtov et al., 2009). The number of elements in a sketch and its complexity may vary (Donath, 2008). Some examples of variations are the degree of the details drawn (information density Fig. 3 a, b) or their representation (representation depth Fig. 3 c, d).

The diversity of the 2D-drawings creates problems in the energy management approach described in this chapter. There are currently two open issues linked to knowledge acquisition from 2D building plans. The first issue is their overall high complexity and the lack of explicit semantics of drawn objects, due to the fact that 2D building plans are mainly used by professional designers and engineers who have the background knowledge to interpret them. The second issue is the difference in the structure of the various CAD-drawings and in the representation styles used in it (Kwon et al., 2009). This high variability obstructs any attempts at an automatic interpretation of an arbitrary building plan.

To handle high complexity we need an abstraction from the unimportant elements of a building plan that would allow us to concentrate on the energy consumption-related objects

[6] http://usa.autodesk.com/autocad

of a building. For the second problem we need a semi-automatic solution that allows us to use our background knowledge about the structure and representation style of a 2D building drawing.

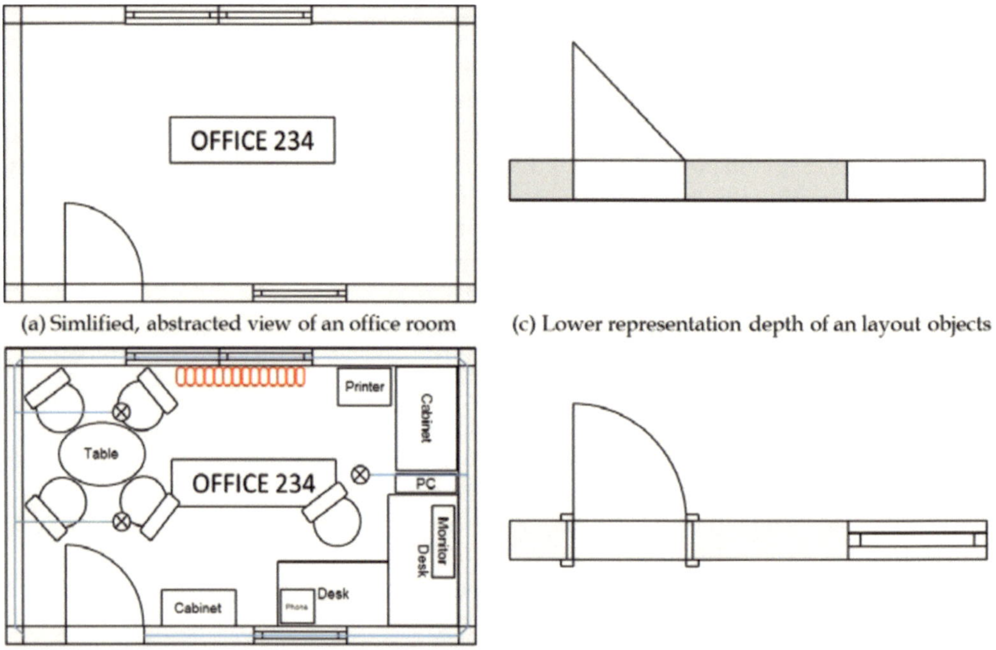

(a) Simlified, abstracted view of an office room

(b) More detailed view of an office room

(c) Lower representation depth of an layout objects

(d) Higher representation depth of the layout objects

Fig. 3. Variety in the degree of information density and level of representation depth in 2D-CAD-drawings

We developed an approach that deals with the described issues and allows us to semi-automatically interpret 2D drawings with an abstraction level defined by the user. The workflow of the approach consists of the following steps:

1. First step is the identification of relevant entities from a 2D drawing. Those can be building elements as well as building automation devices.
2. The second step is to define a diagram for the relevant elements from step one.
3. The third step is the definition of a mapping between the visual representations in the 2D drawing and the elements of the diagram via pattern based recognition rules.
4. The fourth step is the execution of the defined recognition rules.
5. The last step is the transforming of the recognized elements to knowledge base individuals.

We implemented the approach, creating the software framework presented on Fig. 4. The framework consists of several parts (i.e. AutoCAD export function, diagram editor, interpretation rule engine, transformation utility). These framework parts aim to solve the described issues, at the same time ensuring the modularity of the approach and its extensibility.

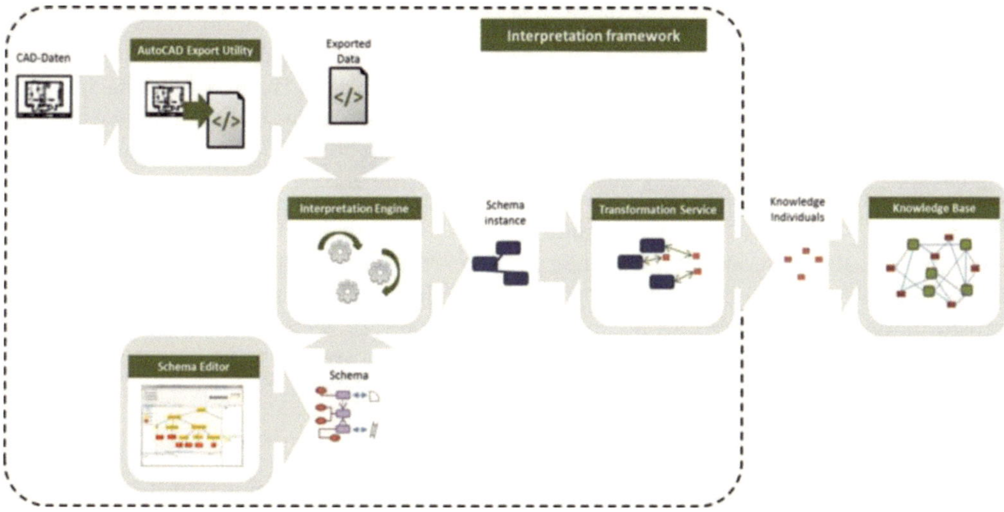

Fig. 4. Simplified architecture of the 2D drawing interpretation framework

The AutoCAD export function is the first component of the developed framework. This component was needed to prepare the 2D drawings for the interpretation process. As part of it, an XML-based transfer format was developed that can be used directly in the interpretation process. This is a key feature of the interpretation because it allows the development of a generic recognition engine that works with the neutral format and that can be extended by additionally exporting modules for various CAD applications. For verification an exemplary conversion function was implemented that exports a native AutoCAD-file (.dwg format) into the defined intermediate data format. The function was developed using ObjectARX®, a programming interface provided by Autodesk. The export function allows a variable degree of detail for the exported geometric primitives.

The second part of the developed framework is the diagram editor. It is developed as an Eclipse RCP[7] program and provides an intuitive configuration process for the source domain of the interpretation framework. Via the diagram editor, the user can define all the important building parts and elements from the given 2D drawing, their semantic properties and their topological relations. They all build up a diagram also called a domain model. By the definition the user can configure the abstraction level for the interpretation process, ignoring all irrelevant drawing details. He can also define various semantic properties, which are not drawn on or not directly derivable from the 2D drawing (i.e. names and ids of the appliances, etc.). The most important part of the diagram that is defined through the editor is the interpretation rule base. These rules are defined as JavaScript routines, which can read and manipulate elements from the neutral transfer format, created by the AutoCAD export function. They describe patterns in the XML format that represent elements of the 2D construction drawing. They are based on the layers, attribute information and relations between AutoCAD primitives that build up a construction element. By defining rules and attaching them to the elements of the domain model created by the diagram editor, the user creates a mapping between the drawing elements and the elements of the domain.

[7] http://www.eclipse.org/home/categories/rcp.php

The rules are used in the third major component of the interpretation framework – the interpretation rule engine. This is a Java based application that can execute the defined interpretation rules over the extracted intermediate XML data, recognizing the building elements and building automation appliances. For the execution of the rules we use the Rhino[8] framework, which is an open-source implementation of JavaScript for Java and enables the end user to use scripts inside of Java applications.

The last part of the interpretation framework is the transformation service. It is used subsequently after the end of the interpretation process, to transform the recognized elements into ontology individuals, creating the initial state of the knowledge base. The transformation service provides a configurable mapping of recognized building elements, building automation or other electronic elements to zones, appliances or sensors and smart meters.

The following models participate in the interpretation process:

1. The domain model for information about the environment, defined by the knowledge experts familiar with the application domain and the specifics of the given 2D drawing. The domain model is the diagram for the recognition process and is defined through the diagram editor.
2. The logical model is the instance of the domain model, created via the recognition process. It represents an abstraction of the 2D drawing.
3. The knowledge model is generated from the logical model through the transformation service.

The described approach for interpretation of two-dimensional building drawings successfully handles the issues that were presented earlier in this book chapter, providing the following benefits in context of the knowledge acquisition for the creation of an initial knowledge data base:

1. No manual input of geometric and topological information should be done. The building-related information as well as the appliance positions and their topological relations to the environment or to each other are provided by the interpretation process.
2. User defined level of abstraction is guaranteed. All unnecessary elements of a two dimensional drawing can be ignored.
3. The approach provides flexibility in both dimensions of types and attributes. That means that new energy appliances (new building automation or household devices) that haven't been considered yet can easily be defined and integrated into an existing energy efficiency system.

6. Knowledge generation through data mining algorithm

In our work, data mining algorithms are used to identify energy consumption patterns and their dependencies. Data mining is defined as the entire method-based computer application process with the purpose to extract hidden knowledge from data (Kantardzic, 2003). In the SERUM-iB framework we used different data mining procedures to generate knowledge for recognition of energy usage anomalies, determining classification of appliances based on

[8] http://www.mozilla.org/rhino/

their energy consumption, and identifying operation states of an appliance based on energy usage. We illustrate these procedures in the following sub sections.

6.1 Recognition of energy usage anomalies

In this work, we relate the energy consumption with the occupant's behavioural pattern in the building. We aim to recognize energy consumptions that do not occur normally in the building. To perform this task, we have to know how energies are consumed normally regarding occupant events and surroundings. For example, normally when an occupant is sleeping and the outside temperature is comfortable, let us say greater than 20 degrees Celsius, total energy consumption in the building is low, for instance lower than 10 kWh. If in the same pre condition total energy consumption in building is more than 10 kWh, then it is considered a usage anomaly.

It is difficult for users if they always have to log their activities. In our work we use simple sensors to automatically recognize user activities. Simple sensors can provide important hints about user activity. For instance, an occupancy sensor in a kitchen can strongly give a clue whether somebody is currently cooking. Of course it should be combined with information of appliance states in the kitchen. The method for automatic recognition of user activity is explained in section 6.2.

The rules representing normal energy consumptions are obtained through a data mining classification rules algorithm. The algorithm is based on a divide-and-conquer approach. The created rule (2) shows the probability of 67% of how often it could happen if the event is sleeping while outside temperature is greater than 20 degree with total energy consumption is lower than 10. This value is called confidence. The rules described in (2) represent a condition that normally occurs. The rule is transformed to (3), in order to represent an anomaly condition, by negating the consequent part of the rule.

$$Event = "Sleeping" \wedge OutsideTemperature \geq 20 \rightarrow TotalEnergyConsumption < 10 (conf: 0.67) \quad (2)$$

$$Event = "Sleeping" \wedge OutsideTemperature \geq 20 \rightarrow TotalEnergyConsumption \geq 10 (conf: 0.67) \quad (3)$$

The rules created by the data mining algorithm are stored in ontology as SWRL. SWRL rule (4) represents the transformed rule (3), which is stored in ontology.

$$\begin{aligned} &Event(?e) \wedge hasName(?e, "Sleeping") \wedge OutsideThermometer(?ot) \wedge hasValue(?ot, ?otv) \wedge \\ &swrlb: greaterThanOrEqual(?otv, 20) \wedge SmartMeter(?sm) \wedge hasValue(?sm, ?smv) \wedge \\ &swrlb: greaterThanOrEqual(?smv, 10) \rightarrow UsageAnomaly(?e) \end{aligned} \quad (4)$$

The rules resulting from the data mining algorithm do not always have 100% confidence. Therefore we represent the rules as SWRL in order to enable verification by using a rule editor provided in the Presentation Module or using a SWRL editor such as Protégé. The editor enables users to add, modify and delete the resulted rules. Fig. 5 illustrates the whole process to enrich the knowledge base with SWRL rules coming from data mining to recognize the energy usage anomalies.

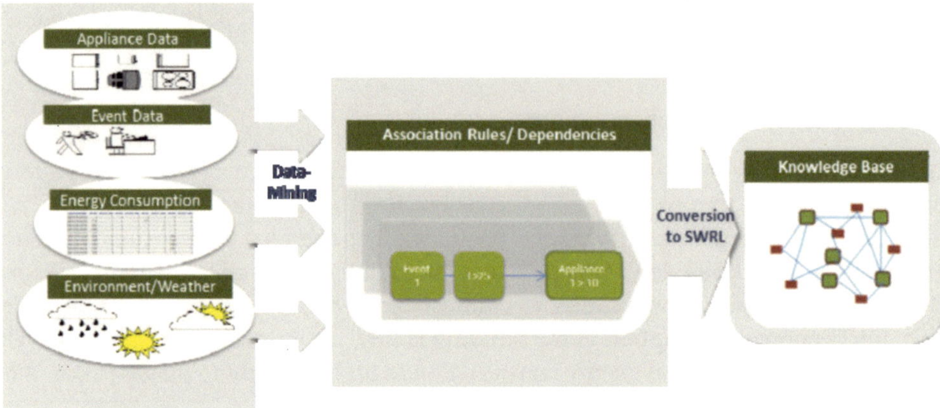

Fig. 5. Data mining process for recognizing an energy usage anomaly

6.2 Recognition of user activities in the building

In our work, we use occupancy sensors, window contact sensors, door contact sensors, as well as energy meter to collect the data that will be used to recognize user activities in a building. An energy meter attached to an appliance can give a clue as to the appliance's state. It means we can get the information, for example whether the appliance is on or off.

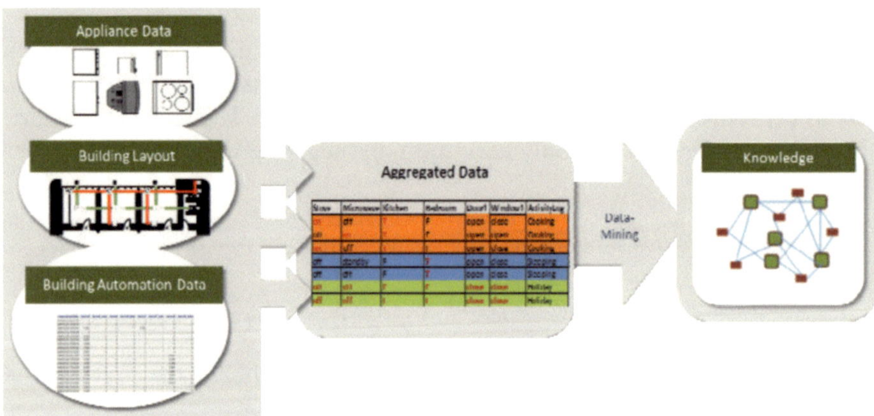

Fig. 6. Data mining process for recognizing user activities in a building

The data from these different sensors are collected and stored in database. Similar to the process for recognizing energy usage anomalies, the data mining algorithm is executed on the data. The algorithm generates rules that it uses to recognize the user activities. Fig. 6 depicts the whole process to generate these rules and store them in a knowledge base. The simplified rule (5) is an example of a rule used to recognize the user activity "Holiday" based on the data depicted in Fig. 5.

$$Stove(?s) \wedge hasState(?s,"off") \wedge Microwave(?m) \wedge hasState(?m,"off") \wedge ... \wedge UserActivity(?u) \wedge hasName(?u,"Holiday") \rightarrow isTakingPlace(?u,true) \quad (5)$$

6.3 Determining classification of appliances

Classification is the process of finding a set of models or functions that describe and distinguish data classes or concepts for the purpose of being able to use the model to predict the class of objects whose attribute is unknown (Han & Kamber, 2001). The model is based on the analysis of a set of training data, i.e. data objects whose attribute is known.

In our work, appliances are classified based on their energy consumption by using classification algorithms, for instance Naive Bayes. Such a grouping determines which devices consume more energy or if a household appliance works properly due to its energy demand. The grouping even allows the occupants to understand which of their activities need more energy. The resulted classes are transformed into ontological individuals of class `Energy_Cosumption_Class`. A relation `is_member_of` is added. It relates instances of appliance classes created by the interpretation of 2D-drawings module and corresponding instances of class `Energy_Consumption_Class`.

6.4 Generating knowledge to identify operation states of appliances

In many cases an appliance consumes energy differently depending on its operational states. We are often not able to determine in which state our appliances consume more energy. Therefore we cannot react respectively, for example, to reduce or even to avoid using appliances operating modes that consume more energy.

We use cluster analysis algorithms to recognize operation states of appliances, for instance, a washing machine has the operation states standby, water heating, washing and dry-spinning. A clustering algorithm analyses and puts data objects in certain clusters so that objects have high similarity to the others within a cluster but dissimilarity in comparison to objects in other clusters. In our case, we measure the similarity of energy consumption of an appliance. Fig. 7 shows the approach to generate the knowledge to identify operation states of appliances using data mining algorithm.

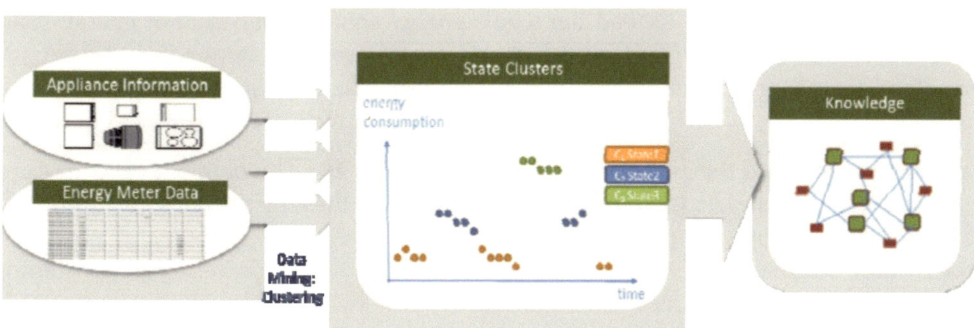

Fig. 7. Data mining process to identify operation states of appliances

7. Intelligent analysis through ontology processing

In the knowledge base represented in ontology, all conditions of energy wasting and anomalies are represented as SWRL. Periodically the data acquisition module requests real-time data from a building automation gateway. These data contain states of all the installed building automation devices. SWRL rules are used to decide whether these incoming data correspond to an energy waste or usage anomaly condition. We develop a rule engine based on `SWRLJessBridge` to support the execution of SWRL rules combined with Protégé API that provides functionality in managing OWL ontology.

First the attribute values of relevant ontological instance are set to values corresponding to incoming data. For example, if contact sensor attached to window gives a state "Open" then the attribute `hasState` of the corresponding ontology instance of concept `ContactSensor` is set to "Open". After that the rule engine executes the SWRL rules and automatically assigns individuals to the ontology classes defined in the rule's consequent. For example for rule (1) the instance of concept `Heater` is additionally assigned to `EnergyWaste` class and for rule (4) the instance of class `UserActivity` to class `UsageAnomaly`.

SPARQL (SPARQL Protocol and RDF Query Language) is used to evaluate whether energy wasting conditions or energy usage anomalies occurs. Which appliances cause the energy wastage can be retrieved as well. It is performed by querying all the individuals of `EnergyWaste` or `UsageAnomaly` class. If individuals of these classes are found, a predefined automatic notification to the user such as sending an SMS, email or visually on the GUI can be performed. With this mechanism, the user can react more quickly in order to avoid more energy wasting. SPARQL is also used to get appliances belonging to a certain energy consumption class. The SPARQL processing is implemented in analysis module.

8. Conclusion

In this paper we have presented a system of comprehensive intelligent energy analysis in buildings. In the developed system, we combined classical data-driven energy analysis with novel knowledge-driven energy analysis that is supported by ontology. The analysis is performed on information collected from building automation devices. The ontology supported analysis approach provides intelligent assistance to improve energy efficiency in households, by intensively considering individual user behaviour and current states in the building. Users do not have to read the whole energy consumption data or energy usage profile curves in order to understand their energy usage pattern. The system will understand the energy usage pattern, and notify user when energy inefficient conditions occur.

We have also presented the methods to semi-automatically generate ontological components i.e. individuals and rules, in building automation domains through the interpretation of 2D-drawing and data mining approaches. The approaches can reduce the ontology generation process, since the manual ontology components creation process is very time-consuming. However, the base ontology representing terminological knowledge in building automation is created manually.

The approach described in this paper can be adopted in more complex systems such as production systems by integrating the base ontology with other ontologies in production system domains. Additional relative ontological components representing knowledge in production systems, for example product, production tasks, machine related information should be created.

9. Acknowledgment

This work has been partially funded by the German government (BMBF) through the research project KEHL within the program KMU-Innovativ. We would like to thank Enrico Kusnady, Ariel Guz, Mariusz Kosecki, and Youssef Aitlaydi for their contribution in implementing the prototype of this work.

10. References

ATB Institut für angewandte Systemtechnik Bremen GmbH; Fundación LABEIN-Tecnalia; RIFOX - Hans Richter GmbH Spezialarmaturen; Vicinay Cadenas Sociedad Anonima; MB Air Systems Limited; Reimesch Kommunikationssysteme GmbH; Disseny I Subministres Tecnics, S.A.; Solvera Lynx d.d. & Valtion Teknillinen Tutkimuskeskus (2008). Ami-MoSES Concept, In: AmI-MoSES Deliverable 7.4, 08.09.2011, Available from http://www.ami-moses.eu/fileadmin/templates/amimoses/files/AmI-MoSES_D7.4_Concept_Public_v.1.0.pdf

Balaras, C.A.; Gaglia, A.G.; Georgopoulou, E.; Mirasgedis, S.; Sarafidis, Y. & Lalas, D.P. (2007). European Residential Buildings and Empirical Assessment of the Hellenic Building Stock, Energy Consumption, Emissions and Potential Energy Savings. *Journal Building and Environment,* Vol. 42 No. 3, pp. 1298-1314

Bohring, H. & Auer, S. (2005). Mapping XML to OWL Ontologies, In: *Leipziger Informatik-Tage 2005,* pp. 147-156

Bonino, D.; Castellina, E. & Corno, F. (2008). DOG: an Ontology-Powered OSGi Domotic Gateway, In: *Proceeding 20th IEEE International Conference on Tools with Artificial Intelligence,* Vol. 1, pp. 157-160, ISBN 978-0-7695-3440-4, IEEE Computer Society Washington, DC, USA

Cox, P. & Fischer Boel, M. (2002). DIRECTIVE 2002/91/EC OF THE EUROPEAN PARLIAMENT AND OF THE COUNCIL of 16 December 2002 on the energy performance of buildings, In: Official Journal of the European Communities, 13.10.2011, Available from http://eur-lex.europa.eu/LexUriServ/LexUriServ.do?uri=OJ:L:2003:001:0065:0065:EN:PDF

Donath, D. (2008) *Bauaufnahme und Planung im Bestand,* pp. 35-36, Vieweg+Teubner Verlag, Wiesbaden

Gasevic D.; Djuric, D.; Devedzic, V. & Damjanovic, V. (2004). From UML to Ready-To-Use OWL Ontologie, In: *Proceedings of the IEEE International Conference Intelligent Systems*

Han, J. & Kamber, M. (2001). *Data Mining Concepts and Techniques,* Morgan Kaufmann Publishers, ISBN 978-155-8609-01-3, San Francisco, USA

Horrock, I.; Patel-Schneider, P.; Boley, H.; Tabet, S.; Grosof, B. & Dean, M. (2004). SWRL: A Semantic Web Rule Language Combining OWL and RuleML, In: *W3C Member Submission,* May 2004

Kantardzic, M. (2003). *Data Mining: Concepts, Models, Methods, and Algorithms,* John Wiley & Sons, ISBN 978-047-1228-52-3, New Jersey, USA

Krahtov, K.; Rogalski, S.; Wacker, D.; Gutu, D. & Ovtcharova, J. (2009). A Generic Framework for Life-Cycle-Management of heterogenic Building Automation

Systems, In: *Proceedings to Flexible Automation and Intteligent Manufacturing, 19th International Conferemce (FAIM 2009)*, July 6th-8th, 2009

Kwon, D.Y.; Gross, M.D. & Do, E.Y.L. (2009), ArchiDNA: An interactive system for creating 2D and 3D conceptual drawings in architectural design, *Computer-Aided Design*, Vol. 41, No. 3, pp. 159-172

Madrazo, L.; Sicilia, A.; Cantos, S.; Böhms, M.; Chervier, B.; Decorme, R.; Dawood, N.; Karhela, T. & Plokker W. (2009). IntUBE Integration Platform Specification, In: *IntUBE Executive Summary of Deliverable 6.2*, 08.09.2011, Available from http://www.intube.eu/wp-content/uploads/COI/2010/executive_summary_D6.2.pdf

Maedche, A. & Staab, S. (2001). Ontology Learning for the Semantic Web, In: *IEEE Intelligent System*, vol. 16, pp. 72-79

Neenan, B.; Robinson, B. & Boisvert R. N. (2009). Residential Electricity Use Feedback: A Research Synthesis and Economic Framework. EPRI, Palo Alto, CA. 1016844

Reinisch, C.; Granzer, W.; Praus, F. & Kastner, W. (2008). Integration of Heterogeneous Building Automation Systems using Ontologies, In: *Proceedings of 34th Annual Conference of the IEEE Industrial Electronics Society (IECON '08)*, pp. 2736-2741

Runde, S.; Dibowski, H.; Fay, A. & Kabitzsch K. (2009). A Semantic Requirement Ontology for the Engineering of Building Automation Systems by means of OWL, In: *Proceedings of the 14th IEEE international conference on Emerging technologies & factory automation*, pp. 413-420, ISBN 978-1-4244-2727-7, IEEE Press Piscataway, NJ, USA

Smith, M.; Welly, C. & McGuinness D. (2004). OWL Web Ontology Language Guide, In: *W3C Recommendation*, February 2004

Spelsberg, J. (2006). Chancen mit dem Gebäudeenergiepass. Energie sparen durch Gebäudeautomation. *Technik am Bau*, Vol. 37, pp. 74-77

Spelsberg, J. (2006). Künstiliche Intelligenz im Büro senkt Energiekosten. *UmweltMagazin*, Oct-Nov 2006, pp. 62-63

Valiente-Rocha, P.A. & Lozano-Tello, A. (2010). Ontology-Based Expert System for Home Automation Controlling, In: Proceeding IEA/AIE'10 Proceedings of the 23rd international conference on Industrial engineering and other applications of applied intelligent systems, Volume Part I, pp. 661-670, ISBN 3-642-13021-6 978-3-642-13021-2, Springer-Verlag Berlin, Heidelberg